COMPUTER INTEGRATED MANUFACTURING

COMPUTER INTEGRATED MANUFACTURING

Proceedings of the Sixth CIM-Europe Annual Conference
15–17 May 1990   Lisbon, Portugal
CEC DG XIII: Telecommunications, Information Industries and Innovation

Edited by

L. Faria and W. Van Puymbroeck

Springer-Verlag
London Berlin Heidelberg New York
Paris Tokyo Hong Kong

Sixth CIM-Europe Annual Conference

An international event organised by:

CIM-Europe, Commission of the European Communities
200 Rue de la Loi, 1049 Brussels, Belgium

Conference Committee

Anderson J.S., ICI, Cleveland, United Kingdom
Bastos J., CEC
Beyer A., Mannesmann Kienzle GmbH, Ratingen, Germany
Duarte-Ramos H., Universidade Nova de Lisboa, Portugal
Faria L., Instituto Superior Técnico, Lisbon, Portugal
Hirsch B., BIBA, Bremen, Germany
Morgado R.J.R., EFACEC, Porto, Portugal
Morin D., SYSECA, Saint-Cloud, France
Motta P.G., TECNATION, Torino, Italy
Pinto Basto E., CTT, Aveiro, Portugal
Puente E.A., Universidad Politecnica de Madrid, Spain
Rebuffel V., LETI, Grenoble, France
Skjolstrup C.-E., Odense Shipyard, Odense, Denmark
Stanford-Smith, B., CEC
Temudo de Castro J., Laboratorio Nacional de Engenharia e Tecnologia Industrial, Lisbon, Portugal
Van Puymbroeck W., CEC

Published by: Springer-Verlag, Berlin Heidelberg New York London Paris Tokyo

L. Faria
Mechanical Engineering Department
Instituto Superior Técnico
Technical University of Lisbon
Av. Rovisco Pais
P-1096 Lisboa Codex, Portugal

W. Van Puymbroeck
Commission of the European Communities
Rue de la Loi 200
B-1049 Brussels, Belgium

British Library Cataloguing in Publication Data
CIM Europe Conference (6th: 1990: Lisbon, Portugal)
Computer integrated manufacturing.
1. Manufacture. Applications of computer systems
I. Title II. Oliveira Faria, Luciano de III. Puymbroeck,
W. Van (Willy), 1955– 670.427
ISBN-13:978-3-540-19616-7

ISBN-13:978-3-540-19616-7   e-ISBN-13:978-1-4471-1786-5
DOI: 10.1007/978-1-4471-1786-5

Publication No. EUR 12785 of the Commission of the European Communities, Scientific and Technical Communication Unit, Directorate-General Telecommunications, Information Industries and Innovation, Luxembourg

Preface

The impact of CIM (Computer Integrated Manufacturing) on the competitiveness of industry is nowadays well acknowledged. Significant increases in productivity, reduction of production costs and the ability to modify operations quickly are amongst the gains made when applying CIM technologies. The integration of automation islands and the application of information technology throughout manufacturing and engineering environments constitute key tasks for European industry.

ESPRIT (European Strategic Programme for Research and Development in Information Technology) is a pre-competitive industry-oriented collaborative research and development programme in information technology. The programme is managed and co-funded by the European Community and is organised in close liaison with industry, national administrations and the research Community. ESPRIT has the following three objectives:

- To provide the European information technology industry with the basic technologies to meet the competitive requirements of the 1990s;
- To promote European industrial cooperation in information technology;
- To pave the way for standards.

The CIM part of the ESPRIT programme addresses the application of information technology in industrial environments.

CIM-Europe is an information and awareness activity of ESPRIT. Its aim is to consolidate and enhance the effects of ESPRIT CIM by disseminating information on progress and achievements in the programme. It stimulates interaction between project teams in CIM and other areas, encouraging the development and the application of CIM techniques to the benefit of European industry. CIM-Europe's main activities are meetings (Study Groups, Workshops and its Annual Conference) and publications (Notices and Proceedings).

CIM-Europe Conferences have taken place in Toulouse, France (1985), Bremen, Germany (1986), Knutsford, England (1987), Madrid, Spain (1988), and Athens, Greece (1989). This, the sixth CIM-Europe Conference, was held in Lisbon, Portugal on 15–17 May 1990.

Further information on CIM-Europe's activities can be obtained from: The Commission of the European Communities, CIM-Europe Secretariat, 200 Rue de la Loi (office: Breydel 9/54), B-1049 Brussels, Belgium.

Acknowledgements

Many people contributed to the success of this conference, but the Commission of the European Communities wishes particularly to thank the Programme Committee for managing the programme and the administration, and the Portuguese Ministry of Industry and Energy who were the hosts.

Contents

Enabling Technologies – Part I

Methods and Tools for Integrating CIM Elements – Part I

Enabling Technologies – Part II

Methods and Tools for Integrating CIM Elements – Part II

Human-organisational and Managerial Aspects of CIM

CIM Technology in Various Industrial Sectors

Speakers' Organisations and Addresses

R. Bernhardt
IPK, Pascalstrasse 8–9
D-1000 Berlin 10, West Germany
Tel: +49 30 39 00 61 56
Fax: +49 30 391 10 37

A.J. Billington
Sunderland Polytechnic
School of Electrical Engineering and
Applied Physics
Edinburgh Building, Chester Road
Sunderland, Tyne & Wear SR1 3SD, UK
Tel: +44 91 515 28 21
Fax: +44 91 515 24 23

H. Bolk
Intervisie
Schipholweg 88
NL-2316 XD Leiden, The Netherlands
Tel: +31 71 22 04 41
Fax: +31 71 22 72 35

V.A. Branco
INESC-Norte
Largo de Mompilher, 22
P-4000 Porto, Portugal
Tel: +351 2 32 10 06
Fax: +351 2 31 86 92

D. Brankley
SAC Taylor Hitec Ltd
77 Lyons Lane
Chorley, Lancs
PR6 0PB, UK
Tel: +44 2572 65 825
Fax: +44 2572 64 712

A. Campbell
ICI Engineering
PO Box No 6
Billingham, Cleveland TS23 1LD, UK
Tel: +44 642 55 36 01
Fax: +44 642 55 90 14

L.O. Faria
Mechanical Engineering Department
Instituto Superior Técnico
Technical University of Lisbon
Av. Rovisco Pais
P-1096 Lisboa Codex, Portugal
Tel: +351 1 80 20 45 ext: 14 64
Fax: +351 1 89 92 42

P.A. Fehrenbach
GEC – Marconi Research Centre
West Hanningfield Road
Great Baddow, Chelmsford, Essex
CM2 8HN, UK
Tel: +44 245 73 331
Fax: +44 245 75 244

J. Frick
TESA as
PO Box 851, Krossen
N-4301 Sandnes, Norway
Tel: +47 4 66 43 90
Fax: +47 4 62 38 66

R. Guiffrey
Syntax Factory Automation SpA
Via V. Vela, 27,
I-10128 Torino, Italy
Tel: +39 11 56 11 888
Fax: +39 11 56 60 679

A. Hars
Institut für Wirtschaftsinformatik
Im Stadtwald
Gebäude 14.1
D-6600 Saarbrücken, West Germany
Tel: +49 681 302 36 44
Fax: +49 681 302 36 96

M.S.J. Hashmi
Dublin City University
School of Mechanical Engineering
Dublin 9, Ireland
Tel: +353 1 37 00 77
Fax: +353 1 36 08 30

U. Hertel
Mannesmann Kienzle Software
Breitscheider Weg 117
D-4030 Ratingen (Lintorf), West Germany
Tel: +49 2102 307 268
Fax: +49 2102 307 107

E. Hirsch
ENSPS
University of Strasbourg
7 rue de l'Université
F-67000 Strasbourg, France
Tel: +33 88 35 51 50
Fax: +33 88 60 75 50

H.F. Jacobi
IPA FhG
Postfach 800 469
Nobelstrasse 12
D-7000 Stuttgart 80, West Germany
Tel: +49 711 68 68 02
Fax: +49 711 68 68 399

P. Kidd
Cheshire Henbury Research & Consultancy
Tamworth House
PO Box 103, Macclesfield SK11 8UW, UK
Tel: +44 625 619 313
Fax: +44 625 616 991

H. Kreppel
Siemens AG
Dept. AUT E 514
PO Box 3220
D-8520 Erlangen, West Germany
Tel: +49 9131 7 23 353
Fax: +49 9131 7 33 193

D. Kumpel
Instituto de Automatica Industrial C.S.I.C.
Km 23 de N3, La Poveda
Arganda del Rey
E-28500 Madrid, Spain
Tel: +34 1 87 11 900
Fax: +34 1 87 17 050

M.A. Lahoud
British Maritime Technology Ltd.
Orlando House
1 Waldegrave Road, Teddington, Middlesex
TW11 8LZ, UK
Tel: +44 1 94 35 544
Fax: +44 1 94 35 347

V.B. Mazzola
LAAS du CNRS
7 Avenue du Colonel-Roche
F-31077 Toulouse Cédex, France
Tel: +33 61 33 62 44
Fax: +33 61 33 64 11

G. Mazzocchi
Mandelli SpA
Via Caorsona 35
I-29100 Piacenza, Italy
Tel: +39 523 79 33
Fax: +39 523 67 775

A. Meier
Krupp Atlas Datensysteme
Niederlassung Bremen
Am Wall 142/143
D-2800 Bremen 1, West Germany
Tel: +49 421 14 214
Fax: +49 421 14 073

L. Nazaré
INESC-Norte
Largo Mompilher, 22
Apartado 4433
P-4007 Porto Codex, Portugal
Tel: +351 2 32 10 06
Fax: +351 2 31 86 92

C. Parker
Loughborough University of Technology
HUSAT Research Institute
The Elms, Elms Grove, Loughborough,
Leicestershire LE11 1RG, UK
Tel: +44 509 61 10 88
Fax: +44 509 23 46 51

A. Pezzinga
FIAR SpA
Via Montefeltro 8
I-20156 Milano, Italy
Tel: +39 2 35 79 03 14
Fax: +39 2 33 40 09 81

V.B. Prabhu
Newcastle upon Tyne Polytechnic
Ellison Building, Ellison Place, Newcastle
upon Tyne NE1 8ST, UK
Tel: +44 91 232 6002
Fax: +44 91 235 8017

M. Isabel Ribeiro
CAPS/LRPS-IST
Av. Rovisco Pais, 1
P-1096 Lisboa Codex, Portugal
Tel: +351 1 35 24 303
Fax: +351 1 89 92 42

M.J. Schachter-Radig
NTE NeuTech
Neue Technologien Entwicklungsgesell-
schaft mbH
Dachauer Strasse 44
D-8000 München 2, West Germany
Tel: +49 89 59 43 18
Fax: +49 89 59 46 95

P. Schoonjans
WTCM-CRIF Machinebouw
Campus Arenberg
Celestijnenlaan 300 C
B-3030 Heverlee (Leuven), Belgium
Tel: +32 16 28 66 11
Fax: +32 16 23 76 78

J. Sentieiro
CAPS/LRPS-IST
CAPS - Complexo I - IST
Av. Rovisco Pais, 1
P-1096 Lisboa Codex, Portugal
Tel: +351 1 35 24 303
Fax: +351 1 89 92 42

I.P. Tatsiopoulos
National Technical University of Athens
Dept of Mechanical Engineering
Sector of Industrial Management & O.R.
28is Octovriou 42
GR-10682 Athens, Greece
Tel: +30 1 36 11 955
Fax: +30 1 36 26 792

V. Tschammer
GMD Fokus Berlin
Hardenbergplatz 2
D-1000 Berlin 2, West Germany
Tel: +49 30 25 499 226
Fax: +49 30 25 499 202

S. Ulfsby
Noratom A/S
Ringsveien 3
PO Box 185
N-1321 Stabekk, Norway
Tel: +47 2 12 55 60
Fax: +47 2 12 54 01

H. Ulrich
Institute for Operations Research
ETH Zürich
ETH-Zentrum
CH-8092 Zürich, Switzerland
Tel: +41 1 25 64 026
Fax: +41 1 26 23 973

F. Vernadat
INRIA-Lorraine
Campus Scientifique
BP 239
F-54506 Vandoeuvre Cédex, France
Tel: +33 83 91 23 73
Fax: +33 83 27 83 19

R. Vio
FIAT SpA
Sistemi e Telecomunicazioni
Corso Marconi 10
I-10125 Torino, Italy
Tel: +39 11 6565 422
Fax: +39 11 6565 708

I. With
Computer Resources International A/S
Space Division
Bregnerodvey 144
DK-3460 Birkerod, Denmark
Tel: +45 45 82 21 00
Fax: +45 45 82 17 66

ARCHITECTURES AND COMMUNICATIONS

Migration to OSI in Factory Communication

Hubert Kreppel
Siemens AG, Erlangen, FRG

1. INTRODUCTION

During the past few years the importance of industrial communication has steadily increased. The reasons for this are:
o higher decentralized intelligence has been made available
o computerized integrated production systems are getting more and more important
o local networks have been installed.

In the wake of this development a complex structural change has made it impossible for the user to rely on systems from a single manufacturer only. The predominance of a manufacturer even in a single area, for example commercial data processing, is unsatisfactory, because of the inherent disadvantages of commercial dependancy on one supplier. Although the field of integrated production taken as a whole those restrictions are no less undesirable.

Standardization has brought open, manufacturer-independent communication in office and factories within reach. Compared with the situation only two or three years ago, considerable progress has been made, thanks not least to the success of MAP (Manufacturing Automation Protocol) in the U.S. and CNMA (Communications Network for Manufacturing Applications) in Europe.

It is often forgotten, however, that the complexity of the subject requires certain conditions to be fullfilled in order to progress from first concepts to broad application. The important conditions are:
o Maturity of the technology
o Acceptance by vendors and users
o Test tools
o Clarification of system responsibility
o Experience
o Profitability

Only when a wide range of vendor products are available and these are being used in the majority of new systems we can talk about a breakthrough of open communications. To reach this objective a number of activities have developed in Europe which can be seen as an indicator for the future acceptance and the growing degree of maturity of open communications.

3

2. CNMA - A SHORT OVERVIEW

For cost effective Computer Integrated Manufacturing facilities to be implemented it is essential that equipment from many different vendors is able to be integrated both simply and cheaply. Several international initiatives are underway to develop standards aimed at achieving this - the most publicised examples are the General Motors MAP (Manufacturing Automation Protocol) and the Boeing TOP (Technical and Office Protocol) projects based in the USA. CNMA is the complementary European initiative, supported by the Commission of the European Communities under their ESPRIT programme (European Strategic Programme of Research and Development in Information Technology - see figure 1).

CNMA aims to specify, implement, validate and promote communication standards for manufacturing which are emerging within the framework of the International Standards Organisation (ISO) model for Open Systems Interconnection (OSI). This activity is intended to make it easier for vendors of computer, communication and control equipment to quickly develop and market equipment which is truly compatible, thereby meeting the user requirement of interoperability as quickly as possible.

The CNMA project, which commenced work in January 1986, comprises currently a consortium of sixteen major European users, vendors and system organisations to provide a very effective combination of skills, interests and resources. The companies are: Aeritalia, Aerospatiale, Alcatel/TITN, British Aerospace (Prime Contractor), Bull, ComConsult, Fraunhofer Institute, GEC, Magneti Marelli, Nixdorf, Olivetti, Renault, Robotiker, Siemens and the universities of Porto and Stuttgart.

The CNMA programme is complementary to the work of the MAP programme. Many of the objectives of the two groups are identical, but whilst the MAP programme develops a manufacturing communications profile which has to accept vendor pressure and adopt a specifications' freeze from time to time, the CNMA consortium is not so constrained and continues to track international standards as they mature.

2.1 Objectives and Results

The primary objectives of the CNMA project are
o Specification
o Implementation
o Validation
o Demonstration
of Open Systems communications protocol profiles for factory automation.

It is worthwhile to examine the issues that surround each of these primary objectives:

o Specification
The project strategy is to identify those communications services required to satisfy CIM user needs. The suitable protocols and services are selected from current International Standards Organisation papers. The normal area of operation for CNMA is those protocols and services which are still immature.

An example has been MMS. CNMA partners worked with ISO early draft proposals through DIS and into the International Standard it has now become. The same situation occurs in the current phase of the project with the emerging and immature standards of Network Management.

At each stage, the project results are fed back into the standardization process at national and international level by the members of the CNMA team. Through these personal liaisons CNMA strongly influences e.g. the NC Companion Standard or the network management specifications of ISO, NIST and Network Management Forum.

o Implementation
Proof of the specification work comes only from implementation. Implementation results are fed back into the CNMA specification work and into the Standards bodies. Until beginning of 1991 several products based on CNMA implementations will be officially released by the relevant vendors.

o Validation/Pilot Projects
Validation of results is achieved in two very different ways. Firstly, by the use of conformance test tools an implementation can be validated against the specification. As a side effect the tools developed in CNMA and the spin-off project TT-CNMA are now being marketed worldwide by SPAG-CCT. Licences have been sold to USA and Japan. Secondly, the implementations are proven in real production environments thus proving that the user requirement has been accomplished.

Installations in real production environment in 1988:
BMW (FRG), British Aerospace (UK) and Aeritalia (I).

Planned for autumn 1990:
Renault (F), Aerospatiale (F), Magneti Marelli (I).

o Demonstration
In order to promote the Open System Interconnect principles and to give positive proof of successful implementations, CNMA believes that significant public demonstrations of the work are vital. By such demonstrations it is possible to promote the development of internationally agreed protocols for communications and the value of conformance test tools and to disseminate the project results adequately.

Major demonstrators: Hannover Fair 1987, Enterprise Networking Event 1988, University of Stuttgart/AMB Fair 1990.

Most of the achievements were only possible because of a well balanced composition of the consortium: Vendors (of controllers and computers), users (from automotive and aircraft industry), system houses and research institutes provide complementary skills and know how.

3. SERVICES AND PROTOCOLS

Automation technology for shop-floor and technical office applications consists of a variety of functionally staged components which are interconnected via a communications network, to form the basis for Computer Integrated Manufacturing (CIM). Such a network is made-up of various controllers (PLCs, robot controls, numerical controls, drive controls, etc.), cell control computers (minicomputer or PC), PCs and workstations for programming and document handling, mainframes for order handling, accounting and corporate planning. All these components are highly heterogeneous in terms of hardware, system software (i.e., operating systems) and application software.

The *ISO Reference Model for Open Systems Interconnection (OSI)* [1] has been defined to form a framework for the development of communication protocol standards. Layer 1 to 4 cover reliable date transmission with error detection and correction, and layers 5, 6 and 7 govern the application-oriented dialog between users. CNMA has been designed according to those principles and references the latest versions of international standards and drafts which are available.

Probably the most significant goal of CNMA is to define *protocol profiles* through close cooperation between vendors and users. To do so, a consensus must be reached on all parameters and options that are needed for unambiguous, open systems communication.

The CNMA *Implementation Guides* [2] contain exact protocol profiles for each project phase. Using these profiles, vendors develop system interfaces for their equipment (programmable controllers, NCs, minicomputers), in order to evaluate implementations, to gather experience, to uncover loopholes in the norms, and to find ways to close them. Finally prototype CNMA systems based on those guides are being integrated, interconnected and commissioned as pilot installations.

Options for implementation are provided at the top and bottom of the profile. These are the aspects which are of concern and apparent to users: What services do I get? What networks can I use?

3.1 Networks

CNMA currently uses Local Area Networks (LANs) as the means to link devices. Three types of LAN are supported:
1) Carrier Sense Multiple Access with Collision Detection (CSMA/CD), operating at 10 Mbps with 500 m cable segments.
2) Token-bus, operating at 10 Mbps modulated onto a *broad-band* cable system.
3) Token-bus operating at 5 Mbps modulated onto a *carrier-band* system.

Providing a range of networks in the profile, allows devices to be specified to use the network which most suits the user's implementation requirements. Requirement considerations will include: the topology which has to be constructed; whether services have to share the same cable; the geographic area to be spanned; the installed base; cost and preferences, environment, etc.

In any application, CNMA allows any of its selected LAN's to be combined

6

to form a single logical network. This means that a device on one LAN can communicate with a device on any other LAN: Individual LAN's are combined using devices known as bridges or routers.

In addition to that access to remote sites is necessary via Wide Area Networks. CNMA therefore caters also for X.25 links.

Routers have two or more ports, one on each subnet, and protocols up to layer 3. The layer 3 protocol is responsible for receiving data from one subnet and transmitting it to another subset if it is required there. Figure 2 illustrates a possible network structure.

Bridges have two ports and protocols up to layer 2.

3.2 Standard applications

At layer 7 of the OSI reference model CNMA offers a choice of different application protocols: FTAM, MMS , Network Management and Directory Service. These protocols are detailed in the following chapters.

3.2.1 MMS

The *Manufacturing Message Specification (MMS)* supports communications to and from programmable devices such as Numerical Controllers (NCs), Programmable Logic Controllers (PLCs) and Robot Controllers (RCs) in a CIM and Process Control environment.

The basic concept of MMS is the so-called Virtual Manufacturing Device (VMD). A VMD represents the standardized view of the structure and behaviour of real manufacturing devices. It defines objects (e.g. variables, programs) and operations (e.g. read, download).
Examples of typical operations are:
- up/download of programs and data to/from a controller
- read/write of variables in remote nodes of the network
- status reports from the controllers
- event reports (e.g. when a threshold is reached which has been defined before)
- semaphores
- journals.

A separate set of standards, called *companion standards*, are needed to supplement MMS and 'tune' it to specific devices. The companion standards define: the meaning to be attached to data, how MMS messages should be used to achieve the real-world functions and relevant subsets of MMS to be supported by particular types of device. Companion standards are currently being defined for: robot controllers, numerical controllers, process controllers, programmable logic controllers, cell controllers etc.

3.2.2 FTAM

FTAM, the application service for file transfer, access and management supports the transmission of various types of files within a heterogeneous environment. FTAM is mainly targeted for systems with peripheral filestores. Therefore, within CNMA, FTAM is to be used for computer to computer communications at factory backbone level (i.e., among cell controllers, workstations, PC's) rather than for communications between cell-controllers and controlling devices.

Since well established and widely agreed functional standards/profiles (ENV 41 204 and NIST which is referenced in the MAP/TOP 3.0 specification) are already available for the functionality needed for this phase of CNMA, the IG rather references those profiles instead of defining a new profile.

3.2.3 Network Management

The CNMA Network Management is based on the OSI Management Framework. It is responsible for collecting information related to network usage by the network devices, ensuring the correct operation of the network, providing notification of errors, and providing reports. The NM information can be used to carry out such functions as network maintenance, performance monitoring and configuration planning.

The Network Manager Application and its interface to the Network Administrator provide the mechanism for the network administrator, a human, to read or alter data, control the network and access reports.

The CNMA Network Manager Application typically has a number of functions associated with it:

o *Configuration Management*:
  - to identify and describe managed network topology, network components and their relationship,
  - to set/examine characteristics at installation time, to facilitate network extension and network tuning,
o *Performance Management*:
  - to monitor in realtime the network communication quality of service (workload, throughput, reliability),
  - to adjust the network performance,
o *Fault Management*:
  - to identify and locate a faulty component,
  - to improve diagnosis and determine corrective actions through use of knowledge base techniques.

MAP 3.0 network management is functionally a subset of CNMA. The network managers developed in CNMA are capable of driving both MAP 3.0 and CNMA 4.0 management agents.

3.2.4 Directory Service

The Directory is a facility which supports the storage and interrogation of information about named objects (things or people), in order to provide services such as network "White pages" (to extract information from an explicity identified object) and "Yellow Pages" (to obtain a list of subordinate information of an identified object). In an Industrial LAN the Directory Service provides access to a user friendly data-base of *information about objects in the network, their addresses and attributes.*

3.3 Profiles/Implementation Guides

Besides these base standards which contain a variety of subsets and options, another set of specification, called functional standards or profiles are needed to reduce the functionality of the base standards to a useful and implementable subset, and to ensure maximum interoperability of different implementations. A number of different groups have elaborated profiles. Multivendor projects like MAP, TOP and CNMA contributed significantly to this profile definitions work through specification and early prototype implementation.

During the definition phase where CNMA generated detailed implementation specifications working from standards documents, there were areas of confusion and misinterpretation and also the need for stability. Without this detail in the CNMA Implementation Guide (see figure 2), a lot of extra work would have been needed during conformance testing and interworking testing to resolve problems.

Note: The CNMA Implementation Guides are Public Domain documents and can be obtained from the Commission (ESPRIT CIM).

3.4 Standardized Application Interfaces

In order to allow easy porting of application software written for host computers, MAP and CNMA specifiy standardized interfaces to MMS and FTAM.

Changes of the underlaying protocol can be hidden to the application, and the so called "High Level" interface frees the application programmer from being familiar with the underlaying protocol. The interface is a library of procedures, currently defined for "C"-language bindings, but with future support of other programming languages. The CNMA application interfaces are derived from MAP/TOP3.0 and adjusted to the CNMA subset.

4. STATUS OF STANDARDIZATION

Progress and acceptance of open communication architectures have suffered from the lack of mature and stable protocol standards in the past. Tremendous progress has been made in recent years in the standardization area, since the necessity for manufacturer-independent communication has been widely recognized. This progress was helped significantly by the MAP/TOP projects in the U.S. and CNMA in Europe. In addition to the progression of the base standards - primarily developed by ISO, IEEE, IEC, and CCITT - vendor and user groups have been established to agree on achievable subsets of the base standards and to develop Functional Standards or profiles.

Figure 3 gives an overview of the status of standardization of those application services which are the most important ones for CIM. See figure 4 for a complete list of base standards that are referenced in the CNMA specification [2].

5. MIGRATION TO OSI

An obvious pre-condition for the provision of OSI products is the existence of stable standards. At ISO the status of individual standards is defined by the stages Working Draft (WD)/Draft Proposal (DP)/Draft International Standard (DIS)/International Standard (IS). Our experience so far shows that up to DP the changes of standards may lead to complete reimplementations and changes of the architecture. After DP, greater changes to the communications software take place which may well affect the interface. Stability for the communications software is only really assured after IS.

Figure 3 shows history and forecasts for the further development of important standards for factory automation.

First products can be expected approximately one year after publication of the standards. If we require a complete profile on the basis of IS including Companion Standards and NMT, broad application may not become reality until the mid-90s.

5.1 What Does a User Expect?

The basic condition to create successfull OSI products is of course openess, i.e. the ability to interwork with other vendor's products. But this is of course not sufficient. The user expects more:

o   The functionality of OSI products must be comparable to that already provided in existing proprietary solutions.

o   OSI products must be stable enough to be used in the factory.

o   The performance must meet the requirements of the production process and

o   OSI products must be competitive in terms of costs.

In a limited range a user will probably tolerate deviations from these rules, since it is clear, that the first generations of OSI products cannot meet all of those requirements.

Since a user normally invests large amounts of money in the development of application software, he is highly interested in the protection of his investments. That means he requires upwards compatible application interfaces.

In very rare cases a user can build up a new facility without the need to integrate already existing installations. So they must be integrated into the new environment.

5.2 What Are the Problems to Provide OSI Products ?

Vendors are faced with several problems when they consider to develop and market OSI products:

o   OSI is complex. E.g. more than 10.000 pages of standards documents were relevant for the compilation of the CNMA Implementation Guide.

o   Large investments are necessary for the development of the new family of products and the maintenance of two product lines in parallel (the existing family and the new OSI family) for some years.

o   OSI products are medium-term more expensive than proprietary solutions (more memory necessary for the OSI stack; smaller quantities can be sold in the short term).

o   A smooth migration path and the integration of existing product families are necessary which need additional development efforts.

5.3 The Solution: a Stepped Approach

Only a stepped approach solves the above mentioned problems and meets the user's requirements.

Step 1:   Early prototype implementations of OSI
- provides the unique opportunity to influence the standardization in order to get standards that meet the real requirements
- to study the impact on communication hardware and software and the endsystems (e.g. the real integration of a new communication technology and especially of new application protocols in controllers for factory automation is a nontrivial task)
- to evaluate new technologies in pilot projects in order to get early feedback from real environments.

Step 2:   Early provision of application interfaces on all endsystems which can be kept stable when migrating from the proprietary architecture to OSI (see figure 5).
This means
- development of OSI interfaces on top of the existing proprietary architecture
- cost-effective solution (existing hardware and major parts of the system software can still be used)
- early access to OSI functionality for the user and
- protection of the user's investment in application software.

Step 3: Development a gateway between the proprietary architecture and OSI.

This provides

- the option to integrate OSI products from the same or different vendor into the existing proprietary network and is also
- a means to maintain interworking between existing product families and future OSI products.

Step 4: The final step is the provision of more and more OSI products (maintaining the proprietary product line in parallel for some years).

Users with large installed LAN's with vendor-specific architectures will only switch to CNMA/MAP/OSI, if a convincing migration strategy is offered to them. The path described above appears to be a key to success.

6. CONCLUSION

o Open Communication is a basic pre-requisite on the long way to CIM.
o Projects like CNMA including vendors and users were (and are still) necessary to
- identify and specify user needs
- define a vendor-independant Implementation Guide which is based on the state-of-the-art in OSI communication and which provides a consistent basis for implementations
- develop independant test tools in order to validate implementations against the agreed specification
- develop communication software to connect the controllers and computers of the respective vendors to the OSI networks
- use these implementations in real production facilities in order to
  - prove that the user requirements have been met
  - provide the users with the opportunity to get experiences with OSI communication technology in their own factories
- promote the results through open demonstrators
- feed the results into standardization in order to come to an internationally agreed and stable set of norms for the OSI communication in CIM.
o There are still areas which need further work within the next few years (especially MMS Companion Standards and Network Management).
o But in contrary to the situation in the second half of the 1980's (where technology and market were premature), it is now evident that Open Communication in the factory is leaving the status of prototypes and pilot implementations. More and more products appear on the marketplace and users (especially in the automative industry) are planning OSI installations in 1991/92.
o The key issue for a successfull implementation of OSI is that vendors offer clear migration strategies for their existing product families.

REFERENCES

[1] OSI-Basic Reference Modell
     International Standard ISO 7498

[2] CNMA Implementation Guide V4.0
     September 1989

[3] Manufacturing Automation Protocol
     Specification Version 3.0
     August, 1, 1988

SIEMENS

CNMA Partners

Users:
- Aeritalia
- Aerospatiale
- British Aerospace
- Fiat/Magneti Marelli
- Renault

Vendors:
- Bull
- CGE/TITN
- ComConsult
- GEC
- Nixdorf
- Olivetti
- Robotiker
- Siemens

Institutes:
- Fraunhofer-Institute IITB
- University of Stuttgart
- University of Porto

Highlights Phase 4:

- MMS and Companion Standards
- Network Management
- FTAM
- Gateways to vendor specific networks
- Application Interfaces

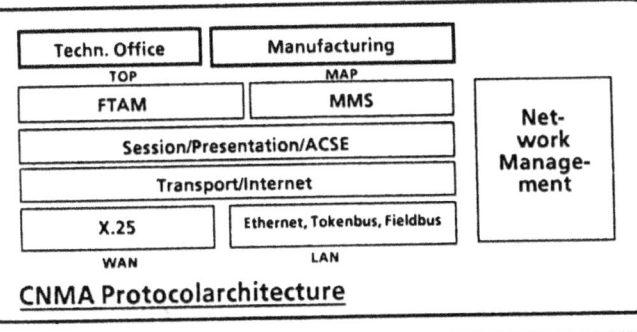

CNMA Protocolarchitecture

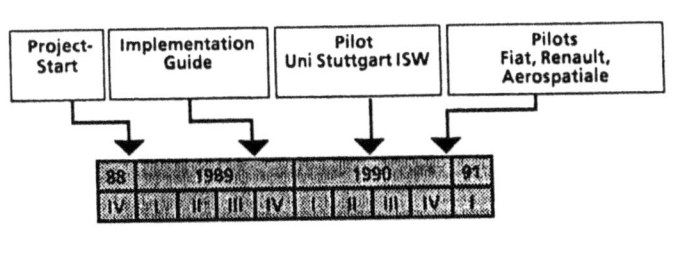

CNMA

Multi-Vendor-Project CNMA - Phase 4
(Communications Network for Manufacturing Applications)

8/89
18.8.89

Figure 1

Vendors

o Background Knowledge

o Experiences of existing
   communication architectures/products

o Profile work of ISO, EWOS, NIST-WS

Users

o Pilot Functionality Specifications

o Requirement Study
   (Renault on NMT)

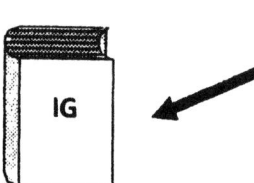

IG

meets the requirements of the pilots as a minimum
<u>but</u> is beyond the pilots a solid basis for general-purpose-products

---

AUT E 514　　　　　**Design Approach (Figure 2)**　　　　　19.03.90
　　　　　　　　　　　　　　　　　　　　　　　　　　　　　Kreppel

13　　　　　　　　　　　　　　　　　　　　　　　　　　　**SIEMENS**

Layer							
			1) oriented on ISO/IEC WD/DP Companion Standards for NC, PLC				
7. Application			MMS Companion Standards 1)	File Transfer, Access and Management (FTAM)	Networkmanagement		Direct. Serv. (ISO 9594)
					CM WD N3295 / PM WD N3313 / FM WD N3312		
			Manufacturing Message Specification (MMS) (ISO IS 9506)	(ISO 8571)	CMIS/CMIP (ISO DIS9595/9596)		
					ROSE (ISO DIS 9072)		
	Association Control Service Element (ACSE): (ISO 8649/8650)						
6. Presentation	Presentation (ISO 8822/8823) Kernel; Abstract Syntax Notation One (ASN.1): (ISO 8824/8825)						
5. Session	Session (ISO 8326/8327) Kernel, Full Duplex						
4. Transport	Transport (ISO 8072/8073) Class 4						
3. Network	PLP (CCITT X.25)	Connectionless Internet (ISO DIS 8348/8473)					
		ES/IS (ISO DIS 9542)					
2. Data Link	HDLC LAP B (CCITT X.25)	LLC 1 (ISO 8802/2)					
1. Physical	X.21/X.21 bis	CSMA/CD 10 MBit/s (ISO 8802/3)		Token Bus (ISO DIS 8802/4) Broadband 10 MBit/s or Carrierband 5 MBit/s			
	MAP, TOP, CNMA	TOP, CNMA		MAP, CNMA			

CNMA　　　　　**CNMA Phase 4: Protocolarchitecture**　　　　　**8/89**
　　　　Figure 3　　　　　　　　　　　28.8.89

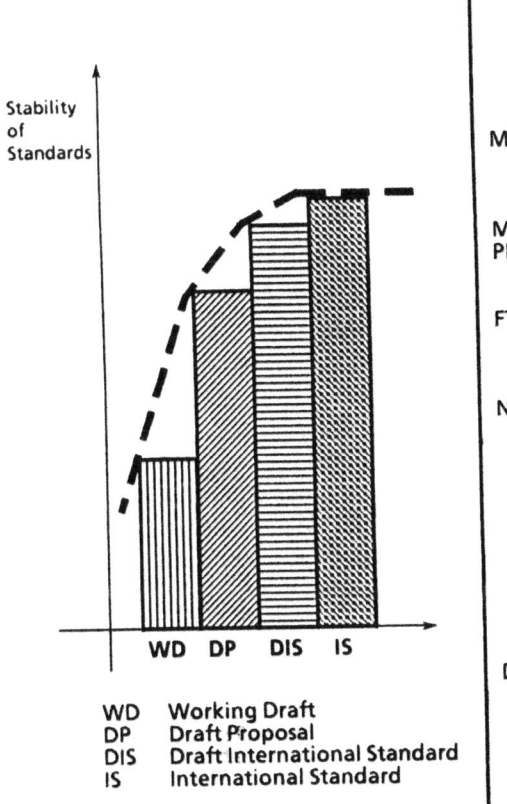

WD, DP, DIS, IS bars — Stability of Standards

WD — Working Draft
DP — Draft Proposal
DIS — Draft International Standard
IS — International Standard

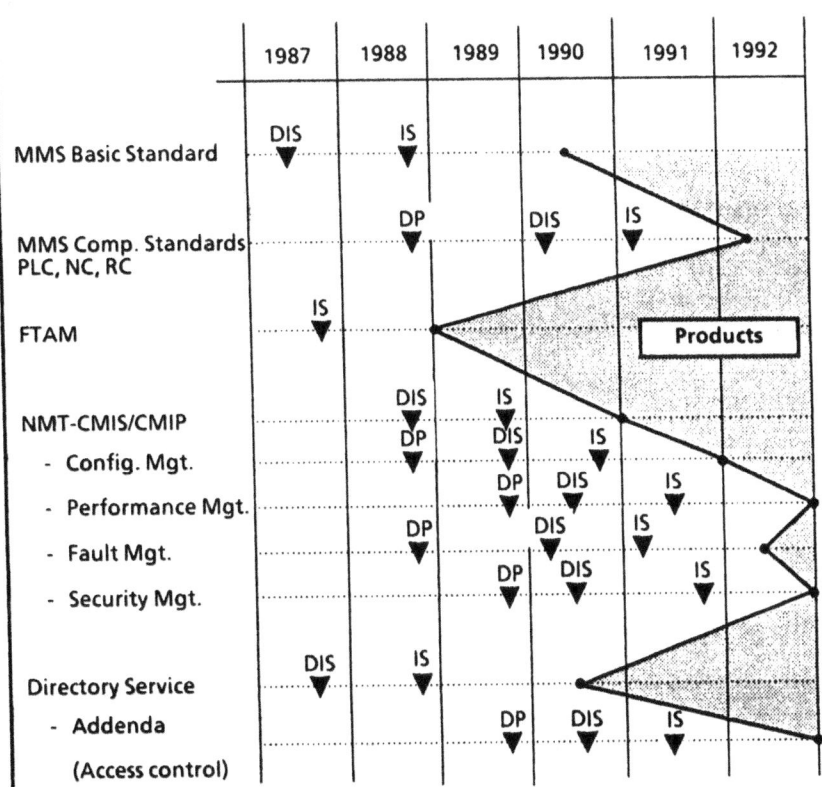

	1987	1988	1989	1990	1991	1992
MMS Basic Standard	DIS	IS				
MMS Comp. Standards PLC, NC, RC		DP	DIS		IS	
FTAM	IS					
NMT-CMIS/CMIP						
- Config. Mgt.		DIS DP	IS DIS	IS		
- Performance Mgt.			DP	DIS	IS	
- Fault Mgt.			DP	DIS	IS	
- Security Mgt.			DP	DIS	IS	
Directory Service	DIS	IS				
- Addenda (Access control)			DP	DIS	IS	

Products

CNMA

Status of International Standardization
Figure 4

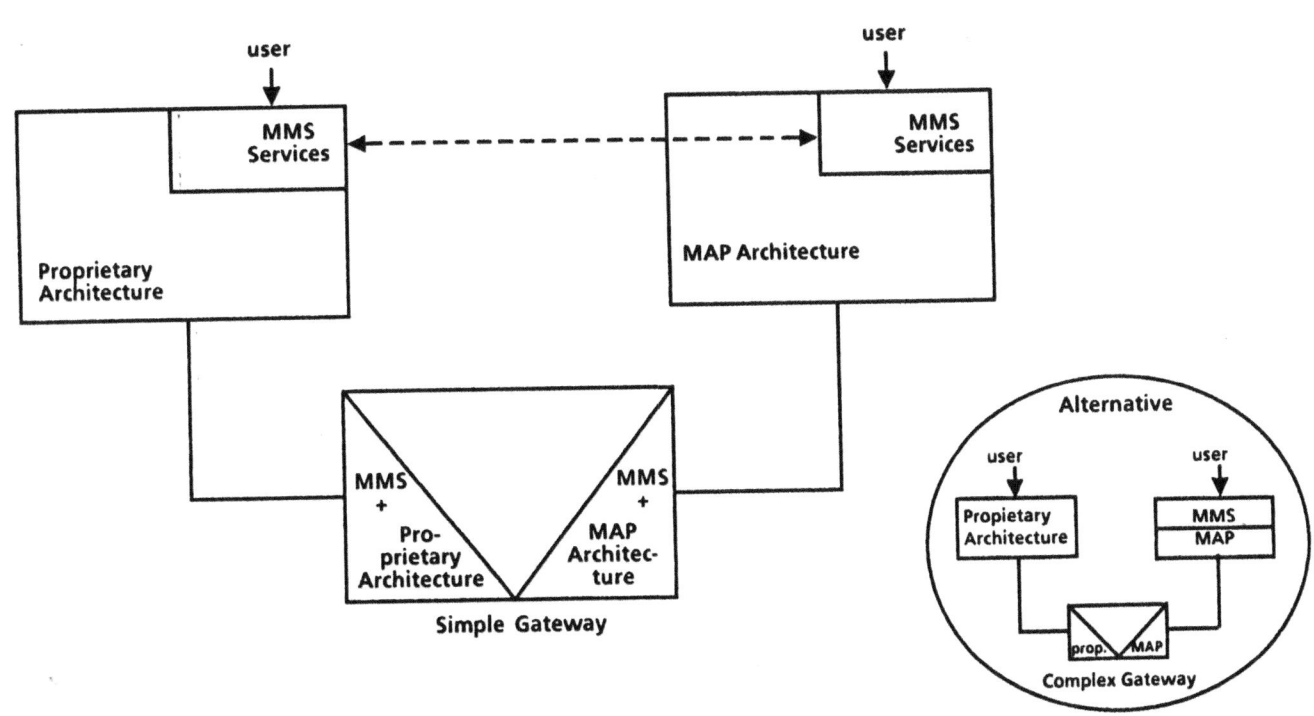

CIM/ILAN

Migration path from proprietary architectures to ISO-Architecture
Figure 5

MODELLING AND ANALYSIS
OF ENTERPRISE INFORMATION SYSTEMS WITH CIM-OSA

F. Vernadat
INRIA-Lorraine, France
and AMICE Consortium

INTRODUCTION

Enterprise modelling and analysis methods and tools to support system design and to prepare system implementation according to system requirements are definitively required for the implementation of CIM systems. Also required to achieve full system integration is an integrating infra-structure, i.e. a software layer implemented on-top of heterogeneous operating systems, which can provide a common shared platform on which diversified system components (i.e. information technology components, manufacturing technology components and human operators) can be interfaced and through which they can communicate.

The AMICE Consortium, which groups 21 major European companies in a common ESPRIT effort, is developing CIM-OSA, an Open Systems Architecture for CIM, to address these requirements. CIM-OSA is made of a Reference Architecture for modelling the Particular Architecture of a given enterprise (or part of it) and of an Integrating Infra-Structure (IIS), which is a set of basic services used to achieve systems integration and communication built on-top of OSI-based communications facilities. CIM-OSA advantages and basic principles (Beeckman, 1989), the CIM-OSA modelling framework (Jorysz and Vernadat, 1990a, 1990b) and the CIM-OSA Integrating Infra-Structure (Klittich, 1990) have already been discussed in previous papers.

The Modelling Framework developed in CIM-OSA is based on three orthogonal modelling principles (Figure 1):
- the instantiation process based on the recognition of
  * Generic Building Blocks or basic constructs
  * Partial Models
  * Particular Models
- the derivation process consisting of
  * a Requirements Definition Modelling Level
  * a Design Specification Modelling Level
  * an Implementation Description Modelling Level
- the generation process involving four modelling views:
  * the Function View
  * the Information View
  * the Resource View
  * the Organisation View

Particular Models are models of a particular enterprise. They can be built from previously defined, incompletely instantiated, Partial Models stored in the CIM-OSA Reference Architecture and developed for well-defined industrial sectors. Partial and Particular Models are specified in terms of basic Building Blocks, also called modelling constructs.

At the Requirements Definition Modelling Level a user-specified model of the enterprise is built which defines WHAT has to be done in terms of business requirements. At the Design Specification Modelling Level consistent and non ambiguous models are developed for the four Modelling Views.

They represent possible solutions to the enterprise problems and the types of components required. Design criteria, system requirements and simulation are used to determine the "best" solution. At the Implementation Description Modelling Level an <u>executable model</u> is produced which indicates HOW things will be performed on implemented components to fulfil system requirements.

The CIM-OSA Function View is a modelling standpoint which allows the specification, design, analysis and implementation description of the structure, behaviour and functionality of the CIM enterprise functions.

The CIM-OSA Information View is another modelling standpoint which allows the specification, design, analysis and implementation description of the information aspects of the CIM enterprise.

The CIM-OSA Resource View and Organisation View are respectively concerned with physical components and individual responsibilities.

In this paper, we discuss how to model the information system of an enterprise according to CIM-OSA principles. However, since the analysis of the Function View of a given enterprise is a prerequisite to the analysis of its Information View, we first present concepts of the CIM-OSA Function View.

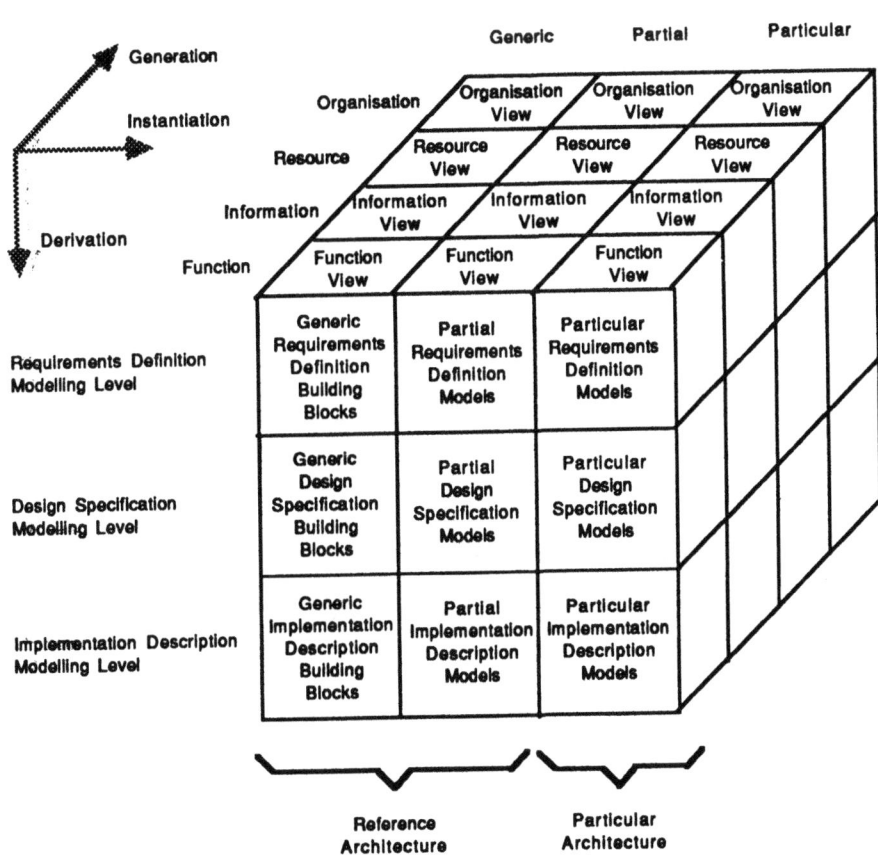

Figure 1: CIM-OSA Modelling Framework (known as the CIM-OSA Cube)

CIM-OSA FUNCTION VIEW

The purpose of the CIM-OSA Function View is to provide tools and methods to support the development of that part of the enterprise model describing system functional structure, functionality and behaviour. It concerns modelling and analysis of enterprise functions. It is based on the functional decomposition principle and largely extends previous techniques such as SADT (Ross, 1977), IDEF0 (Bravoco and Yadav 1985a, 1985b) and the like.

Basic Concepts

In CIM-OSA, any enterprise can be decomposed into a number of *Domains*, i.e. non-overlapping subsets of the enterprise realising functions of the enterprise in terms of processes (e.g. product engineering, manufacturing, production planning and control, etc.). A Domain must never be confused with an enterprise department, which means that CIM-OSA banishes the traditional Taylorism approach to enterprise decomposition which has led in the past to the creation of many so-called islands of automation.

A CIM-OSA Domain is the part of the enterprise which will be the focus of a CIM-OSA analysis (it defines the universe of discourse of the analysis). It is made of a set of *Domain Processes*, each one contributing to the realisation of some enterprise objectives under a given set of enterprise constraints, respectively known as *Domain Objectives* and *Domain Constraints*. The scope of each Domain is clearly identified by its set of *Domain Relationships* (defining the Domain Boundary) described in terms of *Object Classes* received from or sent to other Domains. Object Classes are families of enterprise objects created, processed or used by Domains. A Domain Relationship is always defined between two Domains, which are said to be adjacent.

Domain Processes are high-level constructs used to represent the major tasks to be performed in a Domain. They are composed of *Business Processes* and *Enterprise Activities*, which respectively describe the Domain behaviour (i.e. the dynamic part of the model) and the Domain functionality (i.e. the static part of the model). Domain Processes and Business Processes are triggered by *Enterprise Events* which represent external happenings (arrival of a customer order), human orders (decision to start a task) or timed actions (a process is started each day at 5:00 pm) occurring in the enterprise.

Domain Processes, Business Processes and Enterprise Activities (which appear in the Particular Model of the enterprise, i.e. the right-hand slice of the CIM-OSA cube) and their types (i.e. Partial Models) are described in terms of a unified modelling construct called *Enterprise Function* (which belongs to the CIM-OSA Building Blocks, i.e. the left-hand slice of the cube). The Enterprise Function construct (Figure 2) is used to describe each enterprise process, task, subtask, and so-on to a level of decomposition satisfactory to model and control the CIM system operations. Thus, this modelling construct can keep track of the enterprise functional decomposition (structure part) as well as the enterprise *Objectives and Constraints* decomposition (which drives the functional decomposition process). It also allows to record *Declarative Rules* of the task (i.e. combinations of objectives and constraints linked by conditions which might influence the task execution), *Procedural Rules* (which describe the behaviour of the task, i.e. how low-level tasks are used to perform that task), *Events* (which trigger the execution of the task), *Required Capabilities* (which define a set of technical limitations on the operational, functional and performance capabilities of the task), and *Inputs and Outputs* (namely function, control and resource inputs and outputs).

Enterprise Activities (Figure 3) describe the functionality of basic tasks which can be performed in various enterprises (such as move, make, verify and control) and tailored to specific business requirements (such as procurements, metal cutting or shipping and receiving activities). Their inputs, outputs and resources are well identified as views of Enterprise Objects (see Information View). They operate according to control inputs, and report about their status as control outputs, in order to transform function inputs into function outputs using resources. They are further described in the enterprise implementation model in terms of *Functional Operations*, i.e. elementary sub-tasks which can be executed via the Integrating Infra-Structure by the enterprise components, called *Functional Entities*. Functional Entities are active elements which can perform a defined set of

Functional Operations. CIM-OSA recognises data storage, application, machine, human and communication Functional Entities and Operations.

Business Processes (Figure 4) are used to describe the way Enterprise Activities are grouped into processes and how activities and processes are procedurally chained to form larger processes (such as design processes, production planning processes, manufacturing processes, etc.) to realise sub-objectives of larger enterprise objectives of a complete Domain Process.

The distinction between the concepts of Business Process and Enterprise Activity is considered as very important in CIM-OSA since it is assumed that what makes enterprises different from one another is the way they use Enterprise Activities to form Business Processes (this represents their know-how). Enterprise Activities (such as operation scheduling, FEM analysis, assembly activities, robot welding, etc.) are usually performed the same way by two competing enterprises and are subject to standardisation for a well-identified industrial sector while Business Processes are not necessarily standard.

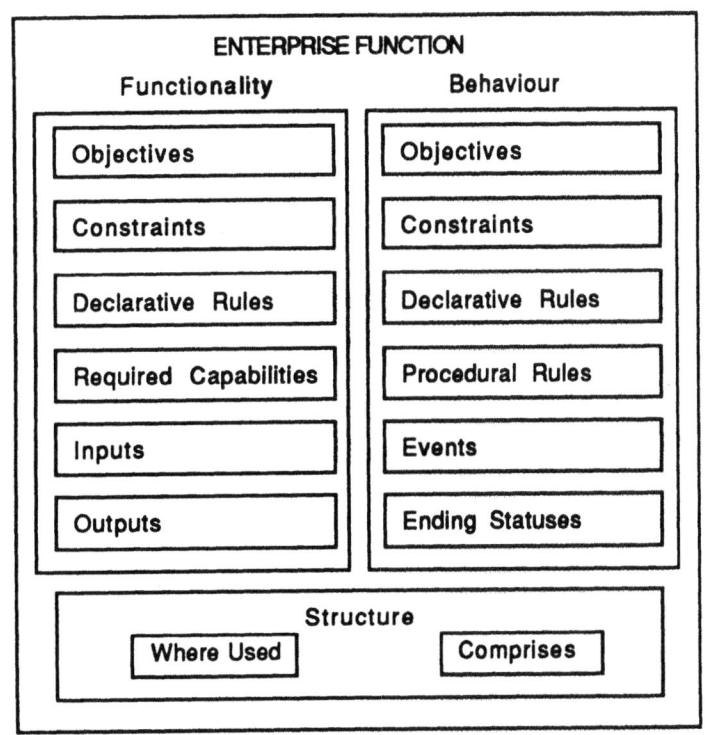

Figure 2: Enterprise Function Concept

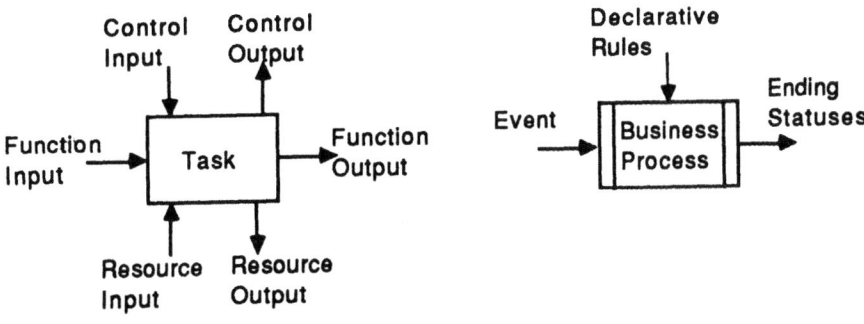

Figure 3: Enterprise Activity Diagram   Figure 4: Business Process Diagram

19

Example: Manufacturing Workshop Activity Control

Manufacturing workshop activity control can be considered as a Domain as defined in the COSIMA Project (Trentin, 1990), another ESPRIT Project. This includes functions such as:

- **Order Scheduling** to schedule production activities on the shop floor on the basis of planned orders generated by an MRP system and to comply with due dates, priorities, availability of resources, etc.
- **Order Dispatching** to send real time instructions to moving and producing machines according to the detailed schedule produced by the scheduler.
- **Producer Activity Control** which controls specific types of production equipment of the workshop such as CNC machines, machine centres, robots and manual operations through standard protocols. It also sends status and performance data to Activity Monitoring.
- **Mover Activity Control** which controls workshop transport devices such as carousels, handling robots, automated-guided vehicles (AGVs) and manual handling operations. It also sends status and performance data to Activity Monitoring.
- **Activity Monitoring** is the feedback function. It collects real time data on equipment utilisation, materials, stock status and quality management and reports back to the order scheduler and order dispatcher or to the workshop controller.

As an example, let us assume that the CIM-OSA Domain is a FMS producing turbine blades with complex sculptured surface for gaz turbine:

Domain: Workshop Activity Control

Domain Objectives:
- to produce turbine blades made of aluminum (max. weight 1.0 Kg, max. length 1.0 m)
- to keep low work-in-process inventories
- to meet customer due dates
- to minimise lead times

Domain Constraints:
- to maintain inventory level <= $ 300 000
- to work with no more than two shifts
- cost limit (budget <= $ 700 000)

Domain Processes:
- Activity Planning
- Activity Control
- Activity Monitoring

Object Classes:
- (1) Planned Orders
- (2) Parts
- (3) Workpieces
- (4) Machines
- (5) Toolsets
- (6) Batches
- (7) Operators
- (8) Process Plans
- (9) Materials
- (10) Tools
- (11) Order Status
- (12) Performance Reports
- (13) Time Reports
- (14) Machine Instructions

Domain Relationships:
  R1, R2, R3, R4.

They are described by the diagram of Figure 5 showing the object class exchanges between adjacent Domains (Domains are depicted by squared boxes with their name inside).

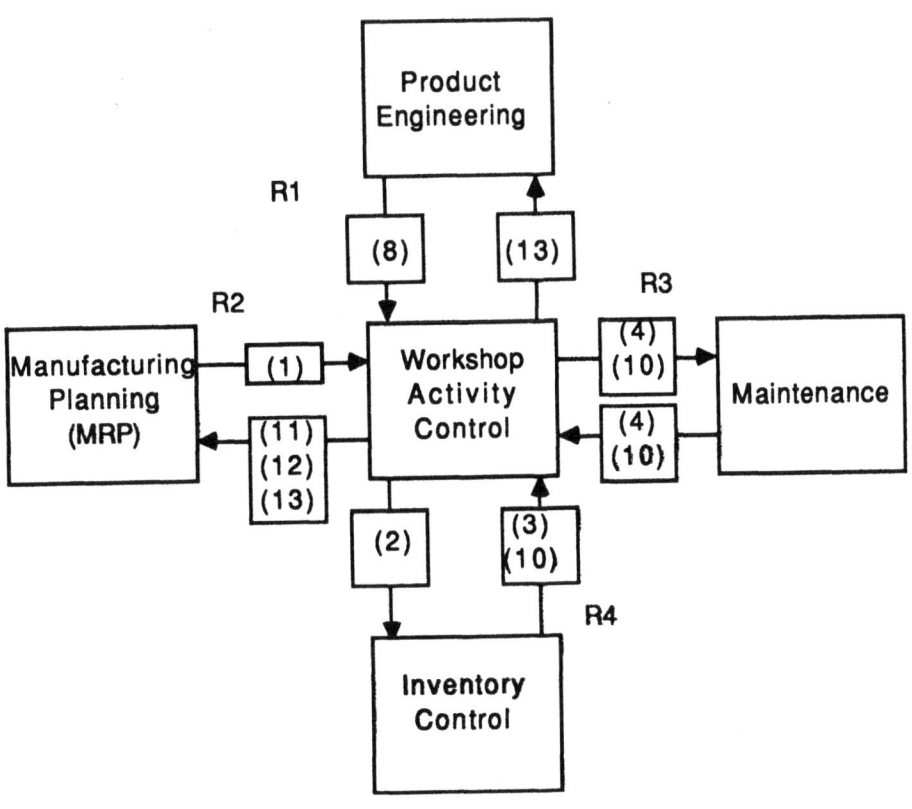

Figure 5: Domain Relationships

For this example, Domain Processes can be further decomposed into Business Processes and Enterprise Activities as follows:

DP1: Activity Planning
   BP11: Order Scheduling
     EA111: Plan Capacity
     EA112: Allocate Operations
     EA113: Schedule Operations
   BP12: Order Dispatching
     EA121: Analyse Schedule
     EA122: Dispatch Producer
       Operations
     EA123: Dispatch Mover
       Operations

DP2: Activity Monitoring
   EA201: Collect Data
   EA202: Report to Dispatcher
   EA203: Report to Scheduler
   EA204: Report to User

DP3: Activity Control
   EA301: Control Producer
      Operations
   EA302: Control Mover
      Operations

All these Domain components are described in more details in CIM-OSA by means of templates. As an example, the Domain Process template is given for the Activity Planning process.

DOMAIN PROCESS
   Identifier: DP1
   Name: Activity Planning

   A. Functional Description
   Objectives: O11: to prepare the detailed schedule for daily
                      workshop operations
                 O12: to produce instructions for mover and producer
                      components
   Constraints: C11: Use scheduling program ABC
                 C12: No sub-contracted operations allowed

	C13: Produce with two working shifts
Declarative Rules:	D11: scheduling for second shift must reschedule unfinished operations resulting from first shift
Tasks:	- schedule detailed orders
	- dispatch detailed orders
Required Capabilities:	
	RC11: must be able to schedule up to 200 operations on 60 machines in less than 15 minutes

Inputs
Function Input:	Planned Orders, Process Plans, Standard Times
Control Input:	Scheduling policy
Resource Input:	Scheduling program ABC

Outputs
Function Output:	Detailed Schedule, Mover Instructions, Producer Instructions
Control Output:	Activity Planning status
Resource Output:	Nil

B. Behaviour Description
| Objectives: | O13: schedule and dispatch detailed production orders |
| Constraints | C14: Order scheduling precedes order dispatching |
| Declarative Rules: Nil |

Procedural Rules:

No.	Wait For	Ending Status	Trigger
1.	START Scheduling Request		Order Scheduling
2.	Order Scheduling	done	Order Dispatching
3.	Order Dispatching	abandon	Order Scheduling
		done	FINISH

Events: EV1: Scheduling Request

C. Structure Description
Where Used:	D1: Workshop Activity Control Domain
Comprises:	BP11: Order Scheduling
	BP12: Order Dispatching

CIM-OSA makes use of six types of Procedural Rules to control the behaviour of enterprise processes. They include:
- Forced rule: control is passed to next task irrespective of the ending status value of the finishing task
- Go/NoGo rule: is a IF THEN conditional statement
- Conditional rule: control is passed to a subsequent task selected from a set of possible tasks according to the value of the ending status of the finishing task
- Spawning rule: allows for parallel execution of several tasks
- Rendezvous rule: control is passed to next task when all preceding tasks are finished
- Loop rule: allows for iterative execution of some task(s)
    The flow of control (Procedural Rules) or the flow of information and materials (inputs and outputs) of Enterprise Functions can be illustrated using symbols of Figures 3 and 4. Figures 6 a) and b) illustrate the behaviour (i.e. the set of Procedural Rules) of Domain Process DP1 (Activity Planning) and Business Process BP12 (Order Dispatching). Figure 7 illustrates the flow of information for BP12.

<u>Nota:</u> Domains must not be regarded as "islands of automation" since (1) Domain Relationships (i.e. Domain interactions) are clearly established and specified, (2) Domains must contain entire Business Processes, and (3) CIM-OSA provides the necessary integrating infra-structure to support information exchange between the various enterprise Domains.

CIM-OSA INFORMATION VIEW

The purpose of the CIM-OSA Information View is to provide tools and methods to support the development of the information model of the Domains analysed. It makes use of three modelling paradigms, one for each modelling level. At the Requirements Definition Modelling Level, a semantic object-oriented modelling approach is used. At the Design Specification Modelling Level, an extended entity-relationship approach is used which is based on the M* methodology (Vernadat et al., 1989). At the Implementation Description Modelling Level conventional data modelling techniques are used. The global modelling framework is compliant with the three-schema approach proposed by ANSI (ANSI/X3/SPARC, 1976), which advocates for the use of a global conceptual schema implemented in terms of an internal schema and presented to system users via external schemata.

Basic Concepts

At the Requirements Definition Modelling Level, enterprise requirements are decribed in terms of *Enterprise Objects* and *Object Views*. In fact, what users use and manipulate in their day-to-day operations are Object Views rather than true Enterprise Objects, i.e. a description of a particular aspect of an Enterprise Object. Furthermore, we assume that inputs and outputs of any kind of Enterprise Functions are Object Views only. Therefore, analysis of the enterprise information system must start with functional analysis to identify all enterprise object views and then to derive the structure of enterprise objects. Both Enterprise Objects and Object Views are defined in terms of their *Information Elements*, i.e. any items of information which, for the purpose they are being used, are indivisible and which are characerised by a type (simple or complex data type). Any kind of *Integrity Rules* can be defined on values of Information Elements to describe existence, conformity or validity constraints. Enterprise Objects are connected to one another by means of *Object Relationships*, i.e. user-defined links, or *Object Abstraction Mechanisms*, i.e. natural semantic links. Four abstraction mechanisms are being used in CIM-OSA (Peckam and Maryanski, 1986):
- Generalisation (or ISA link)
- Aggregation (or PARTOF link)
- Classification (or INSTANCEOF link)
- Association (or MEMBEROF link)

Graphically the model is a semantic network in which nodes are squared boxes which represent Enterprise Objects and oriented arcs are Object Relationships. Object Abstraction Mechanisms are arcs labelled with G for generalisation, Ag for aggregation, C for Classification and As for association.

At the Design Specification Modelling Level, a Conceptual Schema must be defined as a consistent and non ambigous data structure representing static and dynamic properties of data and information. The static part is described in terms of an entity-relationship-attribute (ERA) model as defined in the methodology M* (Vernadat et al., 1989). This formalism is based on the concept of entities, relationships along with their cardinalities, attributes, internal and external identifiers, and two abstraction hierarchies which are special cases of the ISA link (Figure 8). The dynamic part is described by (1) *Database Transactions* which are sets of operations to be executed on the

23

database and considered as a whole, and (2) by *Integrity Constraints* which are formal expressions of Integrity Rules in the ERA formalism. Furthermore, External Schemata must be derived from the Conceptual Schema to describe the Object Views in the ERA formalism or to specify detailed user views of the data. CIM-OSA provides translation rules to convert the object-oriented model into an ERA model. The ERA model can be fully formally described and used for simulation purposes.

At the Implementation Description Modelling Level, an Internal Schema of the information system must be described in an executable form. This is achieved by a two-stage process. First, a *Logical Data Model* is produced. This is a direct translation of the structure of the Conceptual Schema and its External Schemata in ERA form into classical data model formalisms (relational, hierarchical, network) (See Date, 1986). Next, a *Physical Data Model* is produced as the final form of the Internal Schema. It consists of an optimised data structure of the information system with index definitions, user access authorisations, partition definitions and integrity constraints specification, all specified in the data definition languages of the implementation data storage systems (such as relational database management systems).

Example

As mentioned earlier, function inputs and outputs of Enterprise Functions are Object Views. For a given Domain, they have been identified during the Function View analysis. They need to be specified in the Information View and their underlying Enterprise Objects must be described. As an example, we provide the description template for the Enterprise Object "Process Plan" and for "Opeline", a sub-object of the Process Plan object.

ENTERPRISE OBJECT
Identifier: EO-15
Name: Process Plan
Description: Describes the sequence of
                 operations to manufacture
                 the part
Abstraction Relationships:
  Isa: Nil
  Partof: Part Description
  Memberof: Nil
Properties:
  Partcode
  Designer
  CreationDate
  Version
  Operations: Setof Opeline

ENTERPRISE OBJECT
Identifier: EO-16
Name: Opeline
Description: Describes one
                 line of a Process
                 Plan
Abstraction Relationships:
  Isa: Nil
  Partof: Process Plan
  Memberof: Nil
Properties:
  SequenceNumber
  OpeCode
  OpeDesignation
  MachineType
  SetupTime
  RunTime
  Labour

All object properties are either Information Elements or Enterprise Objects or a set of (Setof) Information Elements or Enterprise Objects. For example the Information Element "OpeDesignation" can be described by:

INFORMATION ELEMENT
Name: OpeDesignation
Short Description: Abbreviated description of a manufacturing operation
Data Type: Character string [30]
Related Objects: Operation, Opeline
Composed of: Operation name, operation instructions
Synonym: OpeDescr

Object Views are incomplete object descriptions and are also defined in terms of Information Elements. Object Views and Enterprise Objects are then transformed into an entity-relationship-attribute model. Figure 9 gives an example of such a model for workshop control. Simple Enterprise Objects (i.e. objects only made of Information Elements) are usually directly converted into entities and Object Relationships into entity relationships. Complex objects need to be converted into several entities and their links must be analysed carefully, resulting in the creation of extra relationships.

This model can then be translated into a relational model using CIM-OSA rules for schema derivation to produce the Logical Data Model of the Internal Schema. A short example of such a model follows for the schema of Figure 9.

Relational Logical Data Model:
Part (partid, type, status, location, material, process_plan_id, NC_program, inspection_pgm)
Plan (process plan id, partid, alternate_plan_id, particularities, designer)
Operation (operationid, Opecode, type, designation)
Plan_Ope (plan id, SeqNumber, Opeid)   Mach_Ope (machineid, opeid)
ProducerOpe (Opeid, Machineid, Toolid, cut_type, setup_time, run_time, labour, rate_code)
MoverOpe (Opeid, Machineid, movetype, from, to, quantity)
Tool (toolid, tool_code, type, condition, location, tool_life)
Tooling (tool code, tool_material, max_speed, min_speed, max_feed, min_feed, max_depth_of_cut, min_depth_of_cut, average_tool_life, tool_geometry)
Machine (machineid, type, condition, status, work_hours)
Standard_Time (Opecode, partid, machineid, std_setup_time, std_run_time, std_labour)
Fixture (fixtureid, type, designation, condition, location)
Fixture_Part (fixtureid, partid)
Part_Fixture (partid, fixtureid, name, mounting_instructions)
Lot (lotid, partid, quantity, priority, status, due_date, start_date, finish_date)
Schedule (cellid, lotid, start_date, finish_date, priority)

CONCLUSION

CIM-OSA is a modelling framework and an integrating infra-structure for CIM environments. In this paper we have introduced the Function View and the Information View of CIM-OSA using a manufacturing example. It is believed in the AMICE Project that the CIM-OSA framework largely enhances previous modelling approaches though the Resource and Organisation Views are still being engineered. The concepts being provided by the modelling framework need to be understood by the underlying CIM-OSA Integrating Infra-Structure (IIS) so that the model can be executed. This issue is currently receiving special attention in the project and demonstration prototypes are under development. CIM-OSA is currently beingconsidered by various standardisation bodies (national and international). Also, several ESPRIT Projects are considering the use of CIM-OSA for modelling purposes.

Acknowledgements
The author is grateful to K. Kosanke, AMICE Project Manager, and to his AMICE colleagues, namely R. Gaches, K. Farman, H.R. Jorysz, G. Müller, P. Russell, P. Viollet and M. Zelm, who contributed to this work.

References

ANSI/X3/SPARC, 1975, ANSI/X3/SPARC Study Group on Data Base Management Systems, ANSI 7514TS01.

R.R. Bravaco and S.B. Yadav, 1985a, Requirements Definition Architecture - An Overview, *Computers in Industry*, Vol. 6, No. 1, pp. 237-251.

R.R. Bravaco and S.B. Yadav, 1985b, A Methodology to Model the Functional Structure of an Organization, *Computers in Industry*, Vol. 6, No. 1, pp. 345-361.

D. Beeckman, 1989, CIM-OSA: Computer-Integrated Manufacturing Open System Architecture, *Int. Journal of CIM*. Vol. 2, No. 2, pp. 94-105.

C.J. Date, *An Introduction to Database Systems, 4th Edition,* Addison-Wesley, Reading, MA.

H.R. Jorysz and F.B. Vernadat, 1990, CIM-OSA Part I: Total Enterprise Modelling and Function View, *International Journal of CIM*. Vol. 3.

H.R. Jorysz and F.B. Vernadat, 1990, CIM-OSA Part II: Information View, *International Journal of CIM*. Vol. 3.

M. Klittich, 1990, CIM-OSA Part III: Integrating Infrastructure, *International Journal of CIM*. Vol. 3.

J. Peckam and F Maryanski, 1988, Semantic Data Models, *ACM Computing Surveys*, Vol. 20, No. 3, pp.153-189.

E.T. Ross, 1977, Structured Analysis for Requirements Definition, *IEEE Transactions on Software Engineering*, SE-3, No. 1, pp. 6-15.

R. Trentin, 1989, Production Activity Control (PAC): Pilot Implementation and Project Evaluation. ESPRIT 89 Conference Proceedings, pp. 626-635.

F. Vernadat, A. Di Leva, P. Giolito, 1989, Organization and Information System Design of Manufacturing Environments: The New M* Approach, *Computer-Integrated Manufacturing Systems*, Vol. 2, No. 2, pp. 69-81.

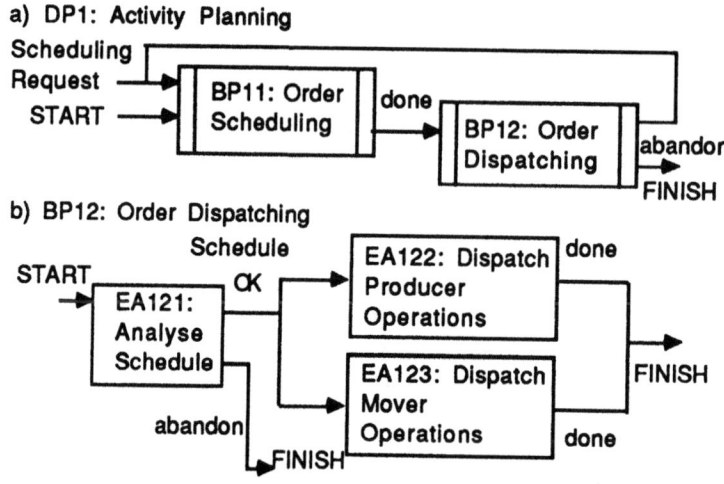

Figure 6: Flow of Control in Activity Planning

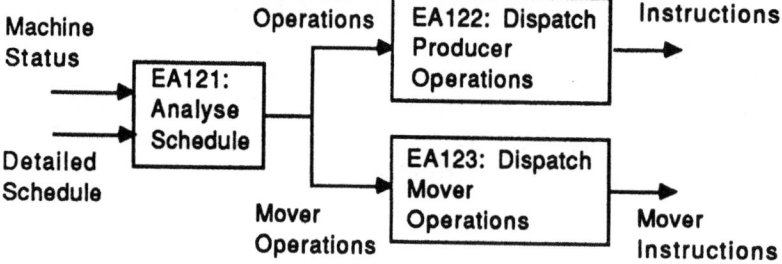

Figure 7: Flow of Information in Order Dispatching

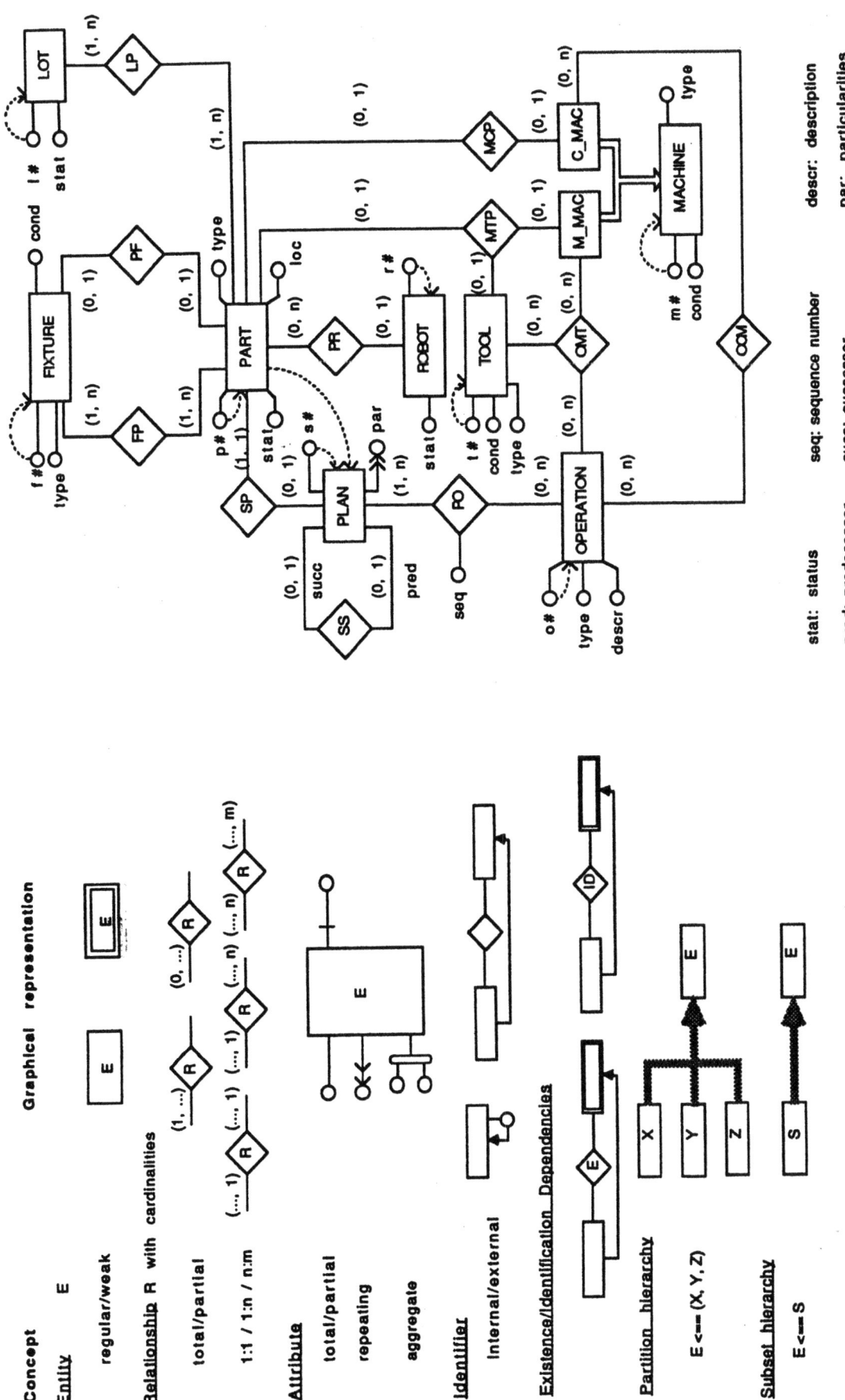

Figure 9: Example of Data Schema

Figure 8: Entity-Relationship-Attribute Model

27

MULTI-MEDIA COLLABORATIVE WORKING IN THE AUTOMOTIVE INDUSTRY - THE ROLE FOR BROADBAND COMMUNICATION SYSTEMS

Sue Joyner, Caroline Parker, Susan Powrie and Carys Siemieniuch
HUSAT Research Institute, Loughborough University

INTRODUCTION

Exponential growth in the generation of new product and process technology coupled with rapidly shortening product life-cycles means successful companies must learn how to respond rapidly and flexibly to changing market demands. In 1987 the US National Research Council reported that international competitors are successfully reducing the lead time from initial product conception to high volume manufacture by conducting design in parallel with manufacturing decision-making. With traditional manufacturing 95% of life-cycle costs are committed during the design phase (Berliner and Brimson, 1988). Good communication with production engineers, specialists and suppliers during this time will maximise 'product producibility' by timely consideration over choice of material, identification of manufacturing capabilities/constraints, whether to make or buy, availability of vendor supplies etc. With competitors spending time and money to reduce production costs, knowledge of process capabilities and limitations during the design phase will be vital. As an example of the challenge facing certain industries, Japanese automobile manufacturers have used this 'simultaneous engineering' approach to lower product development times; by the mid 1990s they aim to design and produce a new model within a year as opposed to the present four. To remain competitive European industry must take advantage of effective, cheap communications coupled with powerful distributed computer systems and advanced infrastructures for information exchange (CEC, 1989). It is firmly believed technological developments in computing and telecommunications, with the resultant integration of data transmission, telephone links and video, are vital to the maintenance of a competitive edge within European industries.

It is not just technical factors, however, that will determine the effectiveness of such technology. It is the manner in which this technology is implemented into companies that will be the ultimate determinant of its success or otherwise. Unless technology is developed to meet specific user and organisational requirements it will fail to produce the desired result.

The RACE programme

RACE (Research and Development on Advanced Communications in Europe) is a major collaborative programme focused on IBC and is aimed at a single European market for commercially viable IBC services by 1995.

It encompasses research and market-oriented development, reflecting all facets of IBC including usability, cost, equipment manufacture, and compatibility.

The CAR project

The acronym CAR stands for 'CAD/CAM in the Automotive industry in RACE' and focuses on potential opportunities for the use of IBC systems in CAD/CAM within the European automotive industry. It is classified as an Application Pilot and covers Work Plan Task T656 which deals with Industrial Design and Manufacturing application pilot schemes.

The aim of the CAR project is to improve the competitiveness of the European automotive industry by accelerating design-to-production time-scales through the concept of 'simultaneous engineering'. Although it uses sophisticated CAD/CAM technology, the industry is characterised by a highly distributed mode of operation, crossing international boundaries and complex company infrastructures. IBC networking offers great potential benefits as a means of overcoming the problems associated with distributed CAD/CAM environments, the transfer of large volumes of data, and the need for frequent dialogue between designers, engineers, production personnel, specialists and suppliers involved at all stages of the 'design to manufacture' of a given component. New opportunities for simultaneous engineering have great potential for reducing delays but, unless carefully implemented, can endanger the integrity of the design, leading to expensive and time-consuming remedial work late in the design cycle. At present there is also little technological support for close working with suppliers and sub-contractors. If available, fast communication links with suppliers would allow their inclusion early in the design cycle to offer advice on, for example, feasibility, costings and design issues; access to design data would also allow sub-contractors to plan their work earlier (Winkler, 1987). This would facilitate the close links required between customers and suppliers, particularly in the increasing number of just-in-time environments. CAR encompasses the use of both public and local networks and, given the potential benefits, there is no doubt the automotive industry will take full advantage of IBC systems provided they meet their needs.

There are a total of ten collaborators in the CAR project. Three of these are end user companies of which two are large automotive manufacturers and the third an internationally recognised design house. The remainder are providers of telecommunications and video technology, universities or companies within the computer industry.

User centred design

To provide structure and a set of clearly defined sub-goals, the CAR project is divided into nine self-contained but interlinked work packages leading towards the development of an application model and prototype.

One of the main principles underlying the four year project is that of user centred design, ensuring user requirements are incorporated into the system design at the earliest stage. Consequently the first year (Work

Package 1) was specifically targeted at identifying user requirements. This was seen as the driving force within the project, allowing the requirements to be iteratively translated into the 'Initial Functional Specification' during Work Package 2. The primary aim of Work Package 1 was therefore to establish a set of organisational and end user requirements; it was also considered important to develop a preliminary set of evaluation criteria for the application pilot. Although this work was completed at the end of 1989, the task of refining and expanding both the user requirements and the evaluation criteria will continue throughout the life of the project.

This paper will concentrate on the identification of user requirements, describing the methodology used and the resultant findings of Work Package 1.

METHODOLOGY

To identify organisational and end user requirements it was necessary to develop a methodology that could be used within the CAR environment. The methodology had to strike a balance between the technocentric and humanistic approaches to design, thereby acknowledging both technical and organisational requirements in the final specification. Socio-technical theory seeks to combine these and has been established as a useful starting point. In its standard form however it can fail to take full account of the technological basis of the organisation and may not fully facilitate the integration of user requirements into a technical specification. The final approach agreed upon was based on established techniques (Gardner and McKenzie, 1988) and the main elements of the methodology shown in Figure 1 are presented below.

Site Identification and Stakeholder Analysis

Stakeholder analysis identifies those people with a legitimate 'stake' in a new system so that their needs can be identified and catered for. Individuals with a stake in the CAR project include primary end users of IBC such as designers, process engineers, material specialists, production engineers; managers of end users; experts in areas such as computing, communications, strategic planning; and possible secondary users who will make use of the data generated.

Sites for data collection were chosen by user companies based on criteria such as the need for frequent interactive communication between sites, a requirement for large amounts of information to be transferred between them, plus willingness to contemplate change and to co-operate with the aims of the CAR project.

Figure 1. The methodology adopted.

THE APPROACH

1. **SITE IDENTIFICATION**
   - Analysis of user groups

2. **STAKEHOLDER ANALYSIS**
   - Identification of users
   - Identification of experts

3. **KNOWLEDGE AND DATA ACQUISITION**
   - Interview methods

4. **ANALYSIS AND SYNTHESIS**
   - Delphi analysis
   - Flowcharting methods
   - Stakeholder reviews

5. **SPECIFICATION GENERATION**

6. **ITERATE**

A series of interviews on eight sites across the three user companies gained progressively more focused and detailed data on each company and their needs. For the first set of interviews a number of 'stakeholders' ranging from senior management to draughtsmen were identified. Topics covered included:

- General company organisation structure.
- Organisation of their design and manufacturing planning.
- Roles and location of different actors in design and manufacturing planning.
- Mechanisms for project control and management.
- Relationships with outside organisations ie suppliers, sub-contractors, clients.
- Role of applications software in design and manufacture.
- Existing means of communication (person-to-person, paper based and system based).
- The degree of individual autonomy and levels of decision making.

These provided a good overview of how design and manufacture was organised in each of the user companies, each company's structure, activities and communication patterns, their current preoccupations and problems, and areas where an IBC system could prove beneficial. Considering the apparent diversity of user company practice, a high degree of commonality was found once terminological differences had been stripped away.

Knowledge and Data Acquisition

After the first round of interviews, a useful insight had been gained into the organisation of design and manufacture in each of the user companies, their individual structure, activities and communication patterns, their current preoccupations and problems, and areas where an IBC system might be of use. Individual company reports were initiated to complement the generic material.

Major stages of the design process after the initial decision to go ahead with a project were identified (Figure 2) and interview data organised into these headings.

Figure 2. Major stages in iterative design in the CAR context.

```
Establish project team
Fix project plan including milestones, budget constraints etc.
Conceptual design/alternative solutions
Test and evaluate
Select and detail design
Test and evaluate
Review and release design
Define manufacturing processes
Select suppliers
Manufacture
```

This exercise allowed some analysis of the data and the identification of major areas of discrepancy and commonality across the 3 user groups. Considering the apparent diversity of user company practice, a high degree of commonality was found once terminological differences had been stripped away.

Analysis and Synthesis

To stimulate informed comment by users, scenarios were developed to illustrate the use of IBC systems at different stages in the design to manufacture cycle. Each detailed a sequence of events in a cooperative design problem, for example, describing the interactions that might take place when a serious fault is discovered in a single bought-out component shortly before final production. It is important to note these scenarios were very much set in the future and pre-supposed high-speed communication links. At each stage the actors concerned, their locations, the nature of data transferred, the type of interactive working, and the means of communication were specified.

Having identified where new data was required, user companies provided appropriate personnel for a second round of more focused interviews directed at eliciting information about specific working practices, use of certain information, and use of communication links. The scenarios were used as the basis of these interviews and proved a successful technique, stimulating discussion on aspects not obvious from previous analysis. The data obtained was collated and user requirements extracted.

Over the two rounds of interviews a total of 53 people from the user companies were interviewed, either singly or in groups, the interviews lasting from one to three hours; the breakdown can be seen in Figure 3.

Figure 3. Type of user interviewed.

Senior management	6
Project management	4
Communications experts	4
Computer systems experts	3
CAD/CAM system development and support	10
CAD/CAM end-users in design and manufacturing planning	17
Specialists from QA, finance, security, legislation, plant, engineering, structural engineering and purchasing	9
Total number interviewed	53

It is possible potential end-users were under-represented in the sample as system support staff and management do not always accurately reflect their view - there is occasionally a suggestion of 'I was a user once, therefore I know' or 'J' étais, donc je sais!'. Users were also frequently interviewed in the presence of more senior members of staff. This may have constrained some discussion of current communications problems and hence limited the range of potential applications for IBC systems considered.

Workshop/Delphi analysis

To confirm the validity and accuracy of the user requirements as an accurate and realistic reflection of potential needs for an IBC system, nominated experts and end users within the consortium were asked to comment on the initial set of user requirements. The majority of feedback was achieved during a three-day workshop attended by representatives, system experts and end-users from user companies plus experts and representatives from other collaborators.

RESULTS

Preliminary findings

Analysis of the data collected made it possible to summarise key problems and bottlenecks in the design process. These could be organised into 3 main groups, the first of which concerns loss of productivity caused by speed limitations of existing communication systems. This covers:

* Time taken for the tape transfer of CAD data between individuals and/or groups.
* Time required to transfer large files or to perform complicated calculations.
* Delays in transferring data because of administrative procedures.
* Delays in transferring information which tends to result in sequential rather than simultaneous working.
* Delays in feedback of information for decisions, design approval, etc.
* Time wasted in travelling long distances for meetings etc.

The second group of problems relates to the users inability to access appropriate information when required and includes:

* Lack of real-time multi-user access to common databases.
* Suppliers being unable to provide detailed specifications at an early stage of design due to a lack of detailed information, or confirmed orders.
* Access to supplier information generally.

The third group concerns incompatibility and covers all situations where work is delayed or halted by difficulties in passing information between differing hardware/software systems.

Potential benefits of IBC systems

Highlighting the problem areas currently causing concern made it possible to focus on the ways by which IBC systems could directly or indirectly alleviate them and thereby improve performance. As can be seen from the section above, an IBC system could offer considerable improvements in certain key areas. It could, for example, offer more rapid transfer of large

CAD/CAM files and large files produced by 'number crunching' applications; enable the provision of on-line suppliers' catalogues and parts databases; and facilitate multi-media conferencing between users, clients and/or suppliers on remote sites. A list of main facilities identified is shown in Figure 4.

On a more global scale, however, it is perhaps the IBC system's ability to facilitate company-wide and inter-company interactive working that is the key benefit. This is seen as central to the introduction of the 'simultaneous engineering' concept, enabling, for example, production information to be incorporated into the design at an early stage; thereby improving efficiency, cutting costs and reducing the length of the design cycle. As previously stated the realisation of this concept is felt to be vital to the maintenance of the European automotive industries competitive edge.

Figure 4. Predicted benefits of IBC system in the automotive industry

Immediate multi-user access to: • drawings • solid and surface models • interactive 'whiteboard' • video • related design data etc.	Multi-media conferencing between: • users, clients and suppliers • local and remote sites • individuals and groups
On-line suppliers catalogues & parts databases with: • high-speed sophisticated search facilities • full graphics representations.	More rapid transfer of: • Large CAD/CAM files • Large files from 'number crunching' applications.
Remote working for design sub-contractors.	Dynamic simulations.

IBC system functionality

In addition to providing existing asynchronous computer functionality such as CAD, electronic mail etc, the CAR project has identified the following synchronous multi-user facilities which must be supported by the IBC system in order for the 'simultaneous engineering' concept to be supported:

- Multi-media messaging tool.
- Multi-user interface to formal design facilities such as CAD systems, solid and surface modellers, drawings etc.
- Interactive informal sketching tool.

- Database mechanisms to facilitate retrieval, data integrity, version management etc.
- Comprehensive change control and electronic signing off systems.
- Real time multi-media conferencing supported by logging and minuting tools.
- Multi-media facilities for incorporating data not held on-line eg scanning.
- On-line suppliers catalogue/parts database requiring high-speed sophisticated search facilities and full graphics representations.
- Flexible and non-intrusive, but effective, security mechanisms.

A detailed list of over 700 user requirements plus specific usability issues covering a wide range of topics from control mechanisms to interface design was agreed by project collaborators and circulated to system designers within the project; a small sub-set of user requirements identified for an interactive sketchpad are shown below:

- Users require the facility to scan hardcopy or video images into the sketchpad.
- There should be an overlay facility, transparent to the user, for writing or drawing on any image imported into the sketchpad.
- Each active user should have their own distinct, identifiable cursor visible to other users.
- The sketchpad should include a full range of sketching facilities eg the ability to add standard entities, freehand drawing, text.

In addition, issues which must be addressed within the project include specification of hardware requirements and software-related issues such as the design of interfaces, database search facilities, and acceptable delays in system response time and speed of file transfer. An extensive literature search has identified work already completed in this area and human factors recommendations of relevance to the project. The leading edge nature of the technology, however, means little previous work has been done in certain areas and these will be the subject of both technical and human factors investigations.

One example of an area to be investigated is the use of video for face to face communication. This has been shown to add to the effectiveness of long distance communication, particularly in relation to decision making (Dutton et al, 1982) and should therefore make a significant contribution to interactive working. However, such studies have concentrated on group conferencing and initial interviews with CAR user companies indicate they do not feel video will be an important workstation component. The provision of such facilities would be an additional financial burden on companies keen to reduce costs and increase competitiveness; the CAR project therefore needs to investigate the particular role of video in a one-to-one situation, both as a communication tool and as a means of transmitting other information eg filming components of interest.

Organisational implications

Some very significant implications for organisations and end users emerged during the first phase of the project. Most importantly, to fully exploit the enhanced speed of IBC, organisations will need more lateral communication links as users may need to contact others previously only reached via an intermediary such as a supervisor. In addition, users at all levels may need quicker and more direct access to individuals at suppliers, sub-contractors and other outside organisations. This in itself has implications in that certain roles will have to change with jobs redesigned to take account of different modes of working, for example, more responsibility and autonomy for individual designers and process engineers. Project management's role may also change to involve more monitoring of data changes held in the system and perhaps less direct human contact with individuals working on projects for which they are responsible. It must therefore be possible for them to keep abreast of all communication and actions agreed on.

Reducing the scope of the user requirements

The initial set of user requirements were generated without reference to cost or current software limitations. This approach was adopted because these constraints are subject to rapid change over time; something considered impossible at the start of the project might become readily available by its conclusion. However this does mean that some of the facilities suggested are unlikely to be implemented during the life of the project. In recognition of this and of perceived resource limitations it was felt that the user requirements should be reduced to more manageable proportions by identifying a 'minimum set'. This set would both address the immediate needs of the user companies and fulfil the remit set by RACE. These were agreed by all partners and can be encapsulated in three main statements:

- The application pilot should demonstrate communication between a number of workstations across Europe.

- The application pilot should demonstrate both synchronous inter-working (ie real time) and asynchronous interworking. Both types of interaction should occur between 'like' with 'like' workstations, 'like' with 'unlike' workstations and 'like/unlike' with a 'low cost' workstation. A suppliers catalogue of standard components should also be included.

- The application pilot should demonstrate communication of each (and any combination) of text, audio, video (both 'full motion' and 'snapshot'), CAD/CAM files, graphics pictures, scanned images.

Facilities covered by these statements will form the main focus of the project although other identified requirements will not be ignored. The 'minimum set' will act as a base on which to concentrate technical work;

additional requirements are seen as equally important and will be incorporated if resources permit.

Prototyping phase

The design and build of prototypes is seen as the next phase of the project. The output of Work Package 1 in the form of the full and minimum set of user requirements feeds into this, both directly via the detailing of requirements and indirectly via the initial functional specification. Results of an on-going assessment of the existing technological base and of the available emerging technology also form an input to this work. Evaluation of the prototypes will be used to verify and expand the initial user requirement specification. The results will be fed into a generic model of user needs and will contribute to the choice and development of the application pilot.

CONCLUSIONS

CAR has taken an innovative approach to the specification and design of new technology aimed at utilising the proposed IBC links in Europe. By concentrating on real problems faced by user companies and looking for ways that the new technology might solve them CAR has avoided the 'solution looking for a problem' pitfall faced by many other ventures. In addition the continued involvement of the user companies throughout the project in developing the pilot specification ensures that it is perceived by them as having direct and continued relevance to their needs.

Analysis of the data to date has revealed a range of issues common to all user companies which the pilot application should attempt to address. These include cost, security issues, change control, requirement for true interactive working, potential for multi-media conferencing on a one to one or one to many basis, client/supplier communications problems, quality assurance of design, the range of potential users, the disparate locations of individuals involved in design and manufacture of products, and the need for improved communications. Additionally a host of potential implementation problems have emerged, amongst them software and hardware issues, and organisational, task and usability considerations.

There are many arguments as to the relative merits of the 'top down' versus the 'bottom up' approach in projects of this kind. CAR for the reasons stated above has opted for a 'top down' approach in the initial stages and is now embarking on the practical 'bottom up' tasks of design and build. It is hoped that the combination of these approaches will lead to a final deliverable that is both innovative and of immediate use to the industrial collaborators involved.

ACKNOWLEDGEMENT

The authors would like to acknowledge the contributions of our collaborators in the CAR consortium.

REFERENCES

Berliner, C. & Brimson, J.A., 1988, **Cost management for today's advanced manufacturing.** (Boston: Harvard Business School Press).

Commission of the European Communities, 1989, **Race Workplan.** Brussels OTR200.

Dutton, W.H., Fulk, J. and Steinfield, C., 1982, Utilisation of videoconferencing, Telecommunications Policy, September, pp. 164-178.

Gardner, A. and McKenzie, J., 1988, **Human Factors Guidelines for the design of computer-based systems, Parts 1-6, Issues 1.** Ministry of Defence (PE) and the Department of Trade and Industry (distributed by HUSAT Research Institute).

US NRC Task Force on Management of Technology, 1987, **Management of Technology: The Hidden Competitive Advantage.** National Academy Press.

Winkler, W.J., 1987, **Gaining competitive advantage through industrial networking.** Proceedings of the 1987 Autofact Conference, Detroit, November 9-12.

Support for Cooperative Work in Distributed CIM-Structures

V. Tschammer, U.W. Brandenburg, J. Hall, D. Strick

GMD Fokus Berlin

INTRODUCTION

Several European countries are developing wide-spread optical fibre infrastructures combined with broadband transmission, switching and access techniques. Further developments are concerned with modular transit systems for the interconnection of local, metropolitan and wide area networks. Broadband-ISDN and multifunctional endsystems will support innovative applications in an Open Service Environment.

Future CIM-applications will profit substantially from these developments. The integration of distributed islands of automation will be facilitated by the significantly increased transfer and reduced error rates of fiber optic networks. New WAN techniques and related standards will integrate telecom and datacom services and will allow the interconnection of widely dispersed, heterogeneous components. Innovative endsystems will allow the presentation and processing of multimedia data, including raster images and audiovisual data.

Standardisation activities have extended their concern from pure communication-oriented issues (OSI) to distributed processing in an open, inter-organisational environment [ISO/ODP] [CCITT/DAF]. The reference models and architectures discussed there all include a so-called "component infrastructure" or "supporting environment" [ECMA/SE-ODP] which provides a set of useful services for communication management and application service access. These services support the integration of existing components into a unified framework and cooperative environment despite component autonomy and heterogeneity.

In parallel CIM-OSA has identified the need for an "Integrated Infrastructure" providing services for communication via OSI and other networks, information management, resource management, etc. [ESPRIT/CIM].

This paper describes an instance of such a supporting infrastructure which is being developed within the framework of BERKOM, the broadband-ISDN project of the German PTT. The services provided are specifically designed to support cooperative work in different types of applications on top of open, interconnected networks and within interorganisational structures. They enable clients to use the services of heterogeneous, autonomous components and support domain-oriented management. Domains are introduced to represent the organisational structures of enterprises and the various types of cooperative ventures between enterprises. A concept for handling qualities of communication as well as of application services forms the basis for the timeliness and accuracy of interactions.

In this way, the developments within this project contribute to the discussion and projection of a component infrastructure as far as heterogeneity, autonomy, and cooperative work within an interorganisational environment are concerned.

THE OPEN SERVICES ENVIRONMENT

Future applications on top of large interconnected networks will involve large communities of components and users. They will be characterised by federations of autonomous entities which cooperate according to a common goal or plan. Such federations usually arise from a loose coupling of existing entities rather than being designed from scratch. The components have developed separately and independently within their local environment and thus have specific characteristics and tasks. Various local factors, such as environmental conditions, load patterns, operating hours, access regulations, etc., may influence the behaviour of components and prevent them from adhering to global concepts. Therefore, the components in an open service environment remain mainly autonomous despite their involvement in cooperative work.

Autonomy

Autonomy has its origins in the organisational and operational environment. Organisations have the right to design, install, operate, administer and develop their systems according to their own internal needs. Design and implementation autonomy results in heterogeneity at all system levels. Examples in the CIM-environment are the local islands of automation which were developed in specific areas for dedicated purposes and which are usually manufacturer specific. The semantic heterogeneity of these entities is usually indispensable for their owners and therefore only the external representation, i.e. the interface, can be submitted to global concepts, as for example standardisation.

Organisations may furthermore have different methods of system management and administration. This creates different domains characterised by different access rights, accounting methods, maintenance procedures, etc. Examples in the CIM environment are components that belong to different departments or companies and therefore have different access regulations and procedures, costs, operating hours, etc.

For each component the organisational conditions may change independently. Examples from the CIM-environment may be equipment replaced as a result of changes in technology or user requirements, or access rights changed due to the end of collaborative projects.

The common aspect of all types of organisational autonomy is the fact that the rules and laws governing the autonomous entities are under complete control of local organisations, i.e. they are more or less negotiable if cooperation is envisaged.

Behavioral autonomy, in contrast, depends on the local execution environment, the load conditions, and the influence of internal faults and external noise or intrusions. This type of autonomy is not negotiable between cooperating entities. At most, worst case assumptions can be made and supervisors installed which control the validity of these assumptions.

The effects of operational and behavioral autonomy are that remote requests for communications or services may or may not be granted and executed. Examples from the CIM-environment are entities that do not accept messages or service requests because local access rights or naming conventions are violated. In this case, the reasons for the failure of global actions are incomplete arrangements between cooperating organisations. Other examples are requests that are lost or ignored due to deadlocks, overloading or timing failures. In this case, the reasons for the failure of a global action are behavioral and cannot be traced back to incomplete regulations about global cooperation.

Cooperation

Cooperation means operating together for a common objective or towards a common result and this is contradictory to autonomy. While autonomous entities follow their own local laws and behave independently, cooperative entities must adhere to global rules and behave in conformity with their partners.

There are different ways of coping with autonomous behaviour and promoting cooperation in a distributed environment. Which of these is the best depends on the type of autonomy that exists and the goals envisaged.

Common agreement on the restriction of autonomy and a set of rules that governs common behaviour and enforces global concepts is the most obvious way. It is possible where organisations are willing to join a cooperation and use their local power to force their entities to adhere to common concepts.

Special design is necessary where the local power of organisations is not sufficient to enforce global concepts, as for example in several cases of design autonomy, e.g. when heterogeneity is inevitable, and in all cases of behavioral autonomy. An example is the introduction of a third party into the communication between two autonomous entities which is able to store a sent message until the receiver is willing to accept it.

Within most CIM-applications we may assume that local islands are autonomous by nature. However, the organisations which they belong to are more or less willing to cooperate. The degree of cooperation is more intensive within specific departments and less intensive between departments or enterprises. Depending on the degree of willingness to cooperate between organisations, cooperative ventures between entities can be installed and rules and agreements enforced. Security concepts must be developed to protect the cooperation between the entities. Special design and self-optimisation methods must be included, because certain forms of heterogeneity and behavioral autonomy as well as malicious behaviour are inevitable.

Within a cooperative environment the following concepts are of particular relevance.

Roles and Services

Co-operation within a system requires the planning and co-ordination of the components' activities so that they can achieve a common result or fulfill a common task. Each component is assigned a specific part within the overall task, which implies dedicated rights and duties and requires particular interactions with other components. Approaches to the modelling of co-operating open systems often use the term "role" to designate a component's part in the overall activity. Roles also allow the establishment of direct relationships between system components and human users, the latter can be the initial source of activity within the system.

Combined with roles are dedicated functions which a component has to perform. From the designer's point of view, it is obvious that these functions are not self-contained but have specific interrelationships with the functions of other components in the course of the overall activity. From the point of view of a component - from which the global context is hidden - the interesting question concerns the assistance that can be expected from others or the actions that can be requested. The term "service" is usually used as an abstract notation of a component's view of a function which another component has made available for external use.

Both abstractions, roles and services, allow the structuring of complex systems and activities by providing basic building blocks for which only the interface to the outside world must be schematised while the internals need no regularisation and may be constructed according to local requirements.

According to the scope and the context of related global agreements, application- and system-specific services can be distinguished. *Application-specific* agreements regulate the co-operation of components within particular application scenarios. Such agreements have local as well as temporal limits and global regulations are needed only in so far as they must inhibit undesirable interdependencies with entities outside the particular scenarios. *System-specific* agreements, in contrast, have a global and long term significance as they intend providing a global platform of system services on which distributed applications can be built. They must be introduced via global design or global standards. The component infrastructure is an instance of a set of such system services.

Negotiations

The most elemental interactions in co-operating open systems are of the client-server type. Due to local autonomy, however, servers need not necessarily service a request at a given rate with the required quality. Instead, they may offer a different quality or rate, reject the call or just be silent. A requesting entity then must ensure beforehand that a particular server is willing to accept and serve the call as required.

Usually, these negotiations will need the support of particular activities as in open systems it is generally not guaranteed that entities always have enough information and trust others sufficiently to initiate and maintain direct relationships and connections. Such supporting activities may be fulfilled by particular system specific roles like those of a "reference" for specific information, a "broker" acting as an agent or intermediary in negotiations, or a "supervisor" monitoring or controlling quality arrangements.

Service Qualities

A further means of supporting the open services environment is a global strategy for defining, valuing and handling the quality of service offers which will be formulated at various locations within open systems and which occasionally may differ only in some detail. In this case, service qualities enable service providers and users to express requirements and offers unambiguously, to balance offers against requests, and to include performance and dependability characteristics in agreements, guarantees, and accounting procedures.

The problem is to define a limited but sufficient set of quality parameters and to reach a global agreement on the syntax and semantics. A further problem is to agree on rules which regulate the processing of quality information and which define the local actions and mutual interactions necessary to establish service qualities and to react to changes in quality.

Many of those agreements will be specific to particular application scenarios. They will be negotiated between the application and communication entities concerned, and these entities will also be responsible for the maintaining of the service qualities which have been negotiated and guaranteed. In order to support negotiations and to certify and supervise quality related agreements, however, system specific services may be provided which could be requested to simplify the general procedures of a quality-oriented service environment, such as the monitoring and management of quality-related information.

Domains

The aspects of free access, heterogeneity, and freedom from restrictions lead to problems of complexity and scale in open systems. Therefore, some means of structuring very large open systems is required which reflect the different organisational boundaries, operational domains, and the co-operation between individuals, groups, and organisations. These means must support the definition of domains, their establishment and demarcation, and the definition of various static and dynamic relationships between domains. This enables system components to be grouped according to common tasks, common agreements, mutual trust, or various other aspects.

Domain-oriented structuring simplifies many procedures within co-operating open systems, including the administration of system-specific services, the design of applications and the management of global information. We may assume that within domains all regulations and agreements have a more local character because of the availability of more detailed information and the greater conformity and trust between components. Consequently, domains provide means for defining "closed areas" within the open environment, where many aspects of scale and complexity and some problems of autonomy, heterogeneity and decentralisation can be relaxed.

CLIENT SERVER INTERACTION IN AN OPEN SERVICE ENVIRONMENT

The Extended Client Server Paradigm

The interactions between service users and service providers and the roles of supporting components can be characterised by the following *extended client server paradigm*. It describes the relationship between a client, i.e. the service-requesting entity, and a server, i.e. the service-providing entity, in an open service environment. The paradigm can be used as the basic building block for more complex relationships which are structured according to the principle of service delegation.

The central notion of the paradigm is that of a service. Each service defines a certain functionality. This forms semantically the basis for a common understanding between a client and a server and defines syntactically a common interface between them. For each interaction between components the paradigm defines an active role, called service user, and a reactive role, called service provider, whereby the latter has been selected from a set of service-offering entities according to certain criteria defined by the service user.

The selection and binding of service providers are system-services which are provided by the component infrastructure. It further supports the definition and administration of domains which forms the basis for the establishment of spheres of interest, in which services are offered, requested, and used.

Services

The paradigm distinguishes between *abstract services* and *concrete services*. An abstract service is the semantic specification of a certain functionality. It describes "what" can be done. By agreements on a set of abstract services the components within a certain sphere of interest reach a common understanding on the functionality available for external use and on the semantics and the syntax of service requests and offers. A concrete service is the implementation of an abstract service. It describes "how" something will be done by a service provider, including run time aspects, access modalities, dynamic service qualities, etc. Concrete services can only be offered and requested if the appropriate abstract service is defined.

Service Users

The term service users denotes those components of cooperating open systems which request services from other components as a result of external or internal stimuli. Typical service users are enterprises, departments, humans and system or application entities. Generally, a request encompasses the service itself, as defined by the abstract service, as well as desired characteristics and qualities which form the basis for the selection of a suitable service among the service offers available. Details of specific server instances are generally not added to the request and the selection of adequate service providers is done by the supporting environment. The service request only defines a sphere of interest in which a service offer is to be selected and accepted.

Service Offers and Service Providers

Service providers are selected from a set of service-offering entities. Such offers represent specific functionalities which are made available on the basis of internal capacities, resources or activities, or by invoking external services. Service offers may be made by enterprises, departments, human users, and system or application entities.

Generally, a service offer denotes the service together with the characteristics and qualities available and the modalities of its use. It further defines a sphere of interest within which negotiations and agreements upon the provision of the service are allowed.

Service-offering entities are autonomous and may decide not to service certain requests according to local rules and conditions. However, they are cooperative with respect to their willingness to restrict their autonomy by negotiations and agreements. These agreements may for example include a mutual understanding of the semantics of the service qualities requested and trust in the identity and good-behaviour of the components involved. Service providers in this sense are specific service-offering entities which have reacted positively to a service request and have committed themselves to providing the requested service according to specific agreements.

Domains and Spheres of Interest

A domain is a set of components which share some common attribute. In particular, it is a set of components to which the same management policy applies [Robinson 1988]. Domains represent organisational (enterprise, site, project), operational (computing center, maintenance area) and technological (manufacturer) areas and structures. Each domain has a domain manager which creates and deletes domains and inserts or removes components.

We distinguish organisational, which are also called hierarchical domains, and federal domains. Organisational domains reflect the association between components and organisations, and are hierarchically structured into domains, subdomains, and components. Each component must belong to at least one organisational domain, called its "home domain". Federal domains represent cooperations between organisational and federal domains and components. They reflect various associations between entities and attributes of components. Intersection and overlapping of federal domains are possible. Components may register in any federal domain and may then request and offer services under the specific conditions and regulations of that domain.

Domains form the basis for the definition of three types of "spheres of interest" which are included in the extended client-server paradigm: definition area, validity area, and search area. The definition area denotes the domains in which an abstract service is defined. The validity area prescribes the domains in which a service offer is valid. The search area delimits the scope where service offers may be selected.

SUPPORT SERVICES FOR CLIENT SERVER INTERACTION

Within a definition area there may be several service providers all offering the same abstract service. Equally, for each service request several service offers may be found within the search area. Service offers and requests are characterised by different qualities and a matching between the qualities of the service request and offers must be made. The component infrastructure supports this selection and matching process with specific services. The instance of such an infrastructure described here provides an information system, including an *info-*, *yellow pages-*, and *domain*-service, together with a *service mediator* and a *service shell*.

Service Shell

The service shell implements the *unified interface* between service users and service providers. It hides the heterogeneity of the concrete services and provides a unique view of communication, storage, management, and run-time aspects. The shell accepts service calls, transfers results, negotiates service access modalities, and supports the communication with other service providers. It also provides operations for the access of status and event information, and the control of internal activities such as process activation and deactivation, queue handling, etc. The interface between service users and service providers is based on the specification of the abstract service, the service offer, and the service request.

The *abstract service specification* contains the name, the attributes, and the qualities of the service. A description of the semantics, i.e. what the service "does", is added, too. The latter informs human users how to use the service. The definition area is also added to the service specification and service names must be unique within that area. If a service is composed of multiple operations which can be invoked separately, a list of operations and parameters is also included in the service specification.

A *service offer* is the implementation of an abstract service by a service provider. Consequently, the service offer defines how the operations are realised and which values are possible for the attributes and qualities defined. The context of the service offer includes the validity area of the offer, the locality of the offering component, and the communication service necessary to reach the component and to invoke the service. In the course of providing a service the service-offering entity may delegate functionality to other components. For that purpose the component becomes in turn a service user, which requests services offered by other components via a unified service request, and supervises the service provision within the agreements negotiated.

The description of the abstract service is the basis for a *service request* characterised by specific values of the attributes, qualities, and parameters. The context of the service request includes the search area, the locality of the service user, and the communication services necessary to reach the requesting component.

The interaction between service user and service provider includes several management aspects, such as request queue management, communication control, access control, process manipulation, and resource management. The service shell encapsulates the service provider in a way that these management aspects are unified and hidden from the outside user as far as possible.

Furthermore, the service shell is the place where negotiations about the service access modalities and service qualities are performed. It makes agreements on service qualities, such as guaranteed response time, supervises further request acceptance and processing, controls internal activities, and manages resources accordingly.

Service Administration and Information System

The service administration and information system encompasses a set of system services for the managing of the information on services, service offers, and domains. The components providing these services obtain the required information via reports from service-offering entities or inquiries from service-requesting entities. The services allow components to register in domains, to make service offers, and to request information on services and service offers, including access modalities and qualities.

The *domain service* provides access to the information on domains. It allows operations on domains and components, including registration and deletion, as well as search and list operations.

The *info service* provides access to the information on abstract services. By means of this service users can check which services are available and how to use them. The operations provided allow registration, modification and deletion of services, search and list operations, and the retrieval of specific information on services, such as descriptions on how to invoke the services, etc.

The *yellow pages service* provides access to the information on service offers. This allows users and components to check which service offers are available and how to request them. The set of operations covers registration, modification and deletion of service offers, searches and lists, and access to specific attributes of the service, such as qualities and localisation of the offering component.

Service Mediator and Binder

The services provided by the service administration and information system are particularly useful for the entities of the component infrastructure which support the interaction between service requesting and service-offering entities. Two of these supporting entities are the service mediator and binder. The *service mediator* implements a selection process among service offers of similar semantics but different qualities. The selection criterion is the grade of conformance between the attributes and qualities of the request and the offer. The selection process is governed by a selection strategy which is either implicitly defined as a characteristic of the definition or search area, or it is explicitly defined by the requesting entity. The selection process can be further supported by specific knowledge assembled within the mediator itself or within the shell of the requesting entity. The following operations are provided:

(i) Analysis and checking of the service request for completeness and correctness. This is done by invoking operations of the info service. A further check analyses whether the request violates matching constraints on attributes and qualities, i.e. whether the values of the parameters are within their limits.

(ii) Selection of appropriate service offers within the defined search area, considering the match between attributes and qualities, and making allowance for the availability of appropriate communication services between the components under consideration.

(iii) Evaluation and ranking of service offers according to the grade of conformance between attributes and qualities requested and offered. Allowance is made for weighting and priorities defined by the selection strategy. This operation provides a list of potential candidates with the optimal choice first.

The *binding process* includes negotiation with service-offering entities aiming at a restriction of their autonomy and at achieving guarantees on service attributes and qualities. The following operations are provided:

(i) Making contact with service-offering entities selected by the mediator and performing negotiations on service modalities and qualities in order to reach agreements. If no agreements can be reached contacts to other entities must be made.

(ii) Selection and establishment of a communication association between service-requesting and -offering entities with respect to the overall qualities required.

IMPLEMENTATION ASPECTS

The component infrastructure described above will be implemented as part of a CIM-scenario which is being developed within a project called BERCIM. The scenario shall demonstrate possible uses of broadband technology in the CIM environment. It will be built on top of local area networks interconnected by the BERKOM testnet. Endsystems, i.e. workstations from different manufacturers will be connected to LANs or directly to the BERKOM testnet, where 2 Mbit/s and 140 Mbit/s channels are available. Operating systems being used in this environment are UNIX and VMS. The LANs are situated in different institutions where they are part of the local infrastructure. Access to the facilities will be allowed according to the organisational structures. These structures are represented by several domains, including organisational and federal domains. One federal domain represents BERCIM as a group of components and users subject to specific regulations and management structures. Other federal domains represent the specific characteristics of the UNIX and the VMS environments. The BERKOM facilities are being made available by a specific Transport System, also being developed within BERCIM. The Transport System includes selected OSI-protocol stacks and new protocols specifically for group communication and fast bulk transfer [BERCIM/TS].

The component infrastructure is being implemented with respect to this environment. In addition it allows for standards to be used and provides the basis for further research:

The service administration and information system is based on the standard X.500 directory services [X.500] so that use of existing standards within a future component infrastructure can be demonstrated.

The mediator service will be treated equally with other system and application services, so that there may be several mediators within a domain, and a meta mediator could be established for the selection of an appropriate mediator service offer. The usefulness of such a recursive structure is a further subject of investigation in the project. Finally, in the long term, the BERCIM supporting environment will be used as a test environment for further developments, primarily aiming at techniques for handling heterogeneity and autonomy and for the use of knowledge-based systems within support services so that components will be able to learn from the outcome of previous decisions and to optimise their future activities.

RELATION TO CIM AND CONCLUSIONS

The paper discusses autonomy and heterogeneity in an open service environment and provides solutions for integrating systems from existing components. Openness and integration are main elements of future CIM-architectures, as described by CIM-OSA (ESPRIT Project 688) and ISA (ESPRIT Project 2267). The Open Systems Architecture or Integrated Systems Architecture, respectively, developed by these projects are generalised models to identify the principal components and information sources, and to integrate them into a unified framework complying with international standards. An essential part of such a framework is an infrastructure which provides a set of useful services which allow the systems to adapt to new devices, applications, and technology. Main elements of such an integrating infrastructure are services which support the execution of enterprise models, provide transparent access to all data and services distributed over different components in the enterprise, and allow for the design of new models and components. Compared to the projects mentioned above it lays more stress on typical solutions for large multiorganisational structures based on future interconnected networks and on the incorporation of innovative communication techniques as represented by large fiber optical networks, broadband-ISDN, and multimedia information.

Although the paper describes a research and development project which results in concepts, prototype implementation, and experience rather than in definite products, the prototype implementation on a network which consists of different types of LANs interconnected by a broadband-ISDN public network, and which integrates components of different vendors and technology (Sun. Appolo, DEC, UNIX, VMS) will provide detailed experience. Furthermore it will integrate resources from three different organisations (university, research institute, firm) with specific access regulations where strict separation of public and private resources is necessary. This will provide further useful results and experience for CIM-projects.

REFERENCES

[BERCIM/TS]   Projekt BERCIM, Bericht zum vierten Meilenstein, Band III, *Das Transportsystem in BERCIM*, FOKUS GMD, Berlin, Dezember 1989

[CCITT/DAF]   Q19/VII CCITT, *Framework for the Support of Distributed Applications (DAF)*, SG VII Plenary Geneva, July 1989

[ECMA/SE-ODP]   *Technical Summary of SE-ODP Configuration Standard*, Contribution to ECMA/TC32-TG2 and ECMA/TC32-TG9, Aug. 1989

[ESPRIT/CIM]   *Open Systems Architectures and Communications*, Workshop Proceedings, WZL Aachen, FRG, June 1989

[ISO/ODP]   ISO/TC97/SC21/WG1 N520-526, *Information processing systems - Open Systems Interconnection (OSI)*, "Open Distributed Processing (ODP)"

[Robinson 1988]   Robinson D.C., Sloman M.S.: *Domain-based access control for distributed computing systems*, Software Engineering Journal, Sept. 1988

[X.500]   CCITT Draft Recommendation X.500, *Open Systems Interconnection (OSI) - The Directory*, "Overview of Concepts, Models and Services"

ROBOTICS

Benefits to CIM of Robot Deburring: *The use of modelling and control*

M.A. Lahoud   BMT Ltd., UK
H. Münch, D. Surdilovic, J. Timm   IPK, FRG
M. Fraile   ENASA, Spain

INTRODUCTION

Consider the motor car: many of its parts are produced by a casting process, which seldom results in a perfectly formed product. The resulting casting imperfections, burrs, need to be removed before the part can be passed to the next stage in the production line. The removal of the burrs is often a dirty and dangerous operation and is usually performed by human operators. The overall production process would benefit in many ways if the deburring operation were automated; for example, a uniform product with an improved surface finish would be produced with resulting smoother mating of parts. The use of robots thus appears a natural way of attempting to solve a subset of the deburring problems.

Although robots have been used for the deburring operation, only highly specialised problems have been solved, and consequently there have been few robotic deburring applications relative to the conventional pick-and-place systems. Furthermore, the problem is complicated by conflicting mechanical stiffness requirements (see later). This suggests the use of sophisticated control procedures in order to achieve the above goals. Intelligent sensors and Artificial Intelligence procedures can also be usefully employed to cope with the probable high shape irregularity of the burrs and variations in the materials. In order to improve the efficiency of current production lines involving the deburring operation, the Robot Deburring Demonstrator was set up within the KB-MUSICA project under the umbrella and philosophies of ESPRIT II in the CIM area.

The application of robot systems to manufacturing, i.e. deburring functions, requires a system capable of performing complicated, but somewhat unspecified tasks. The deburring demonstrator system established in the KB-MUSICA project appears to offer a more integrated and powerful approach to the problems facing current manufacturing systems. The overall system architecture includes knowledge-based functions, a sophisticated sensory system and advanced hybrid force/position robot control. The concept of the common deburring demonstrator with short-, mid- and long-term industrial applications as a concept for integration into CIM will be presented in the first part of the paper. In the second part, a simple model of a robotic deburring system will be outlined and promising control strategies will be discussed. Initial simulation results will be shown and some advantages over the existing methods will be presented.

The Deburring Problem, its Extent and Present Solutions

A burr can be defined as a rough edge left on a work-piece following drilling, cutting or casting et cetera. The reasons for the removal of the burr can be broadly categorised as follows:

- Functional, for example: removal of obstacles to future processing steps.
- Ergonomic, for example: reduction of the possibility of worker injury during handling.
- Aesthetic, for example: production of a "smooth" surface at a cast seam.

At present, most deburring operations are carried out by hand and it appears logical that the robot should be considered for such tasks. ENASA has carried out an intensive information gathering exercise in their factories in order to identify the benefits of an automated deburring cell, the most important being:

- humans not operating in a hazardous environment;
- a consistent deburring quality;
- a reduction in the number of rejects;
- a decrease in the disruption of production cycles and high stock levels;
- the introduction of the integrated production techniques of CIM;
- the introduction of Just-in-Time manufacturing;
- reduced costs.

The range of tasks in a typical deburring cell is large and include among others:

- cutting of the ingates;
- removal of the deadhead;
- removal of the flanges;
- finishing-off of the disconnection points;
- radiusing of edges;
- finishing-off of the cast surface;
- deburring especially at a cast seam.

Consequently the methods and tools used for the removal of the burr are just as diverse. For example, one of the following tools, appropriate for the required task, is used within a typical automotive deburring cell:

- scraper
- hammer
- chisel
- grinding machine
- milling machine
- sand paper
- rubber cone
- sawing

Definition of the Problem to be Solved and Discussed

The previous section has demonstrated the diverse nature of a deburring cell. However, at this stage, it is unrealistic to expect a single robot to achieve all the deburring tasks. Hence, it is necessary to define a reduced-order problem for the robot to solve, in order to show that even an automated solution to a subset of the deburring problems would be beneficial to the implementation of CIM.

The robot system chosen for the investigation is a Manutec R3, whose end-effector holds a motor which drives an end-mill. The deburring task has been chosen to be the removal of a simulated burr along a straight line. The arrangement is shown graphically in Figure 1.

Figure 1   Geometric Model of the Manutec R3 Robot

THE KB-MUSICA ROBOT DEBURRING DEMONSTRATOR

In order to achieve an increase in the application of industrial automation, two basic requirements have to be fulfilled. The first one is the efficient integration of sensor information for use in real-time process control to avoid expensive island solutions for special tasks and to achieve the highest performance possible. The second one is an overall system-oriented approach for intelligent model-based task-execution planning and shop-floor execution to avoid production down-times for programming and reprogramming.

Figure 2 shows the overall system-oriented approach for multisensor-guided robot-based deburring. In principal it consists of a Planning and an Execution Level. On the Planning Level, all planning tasks have to be performed with the use of high planning intelligence, which results in a low execution bandwidth. The prerequisite is a functional decoupling between pre-planning of the nominal behaviour and real-time execution on the shop-floor. The Planning Level includes the basic functions for the generation of the user pre-planned robot actions. Based on this information, the robot system is able to act in a pre-planned manner, that is, vertical feed forward control, displayed on the right hand side of Figure 2.

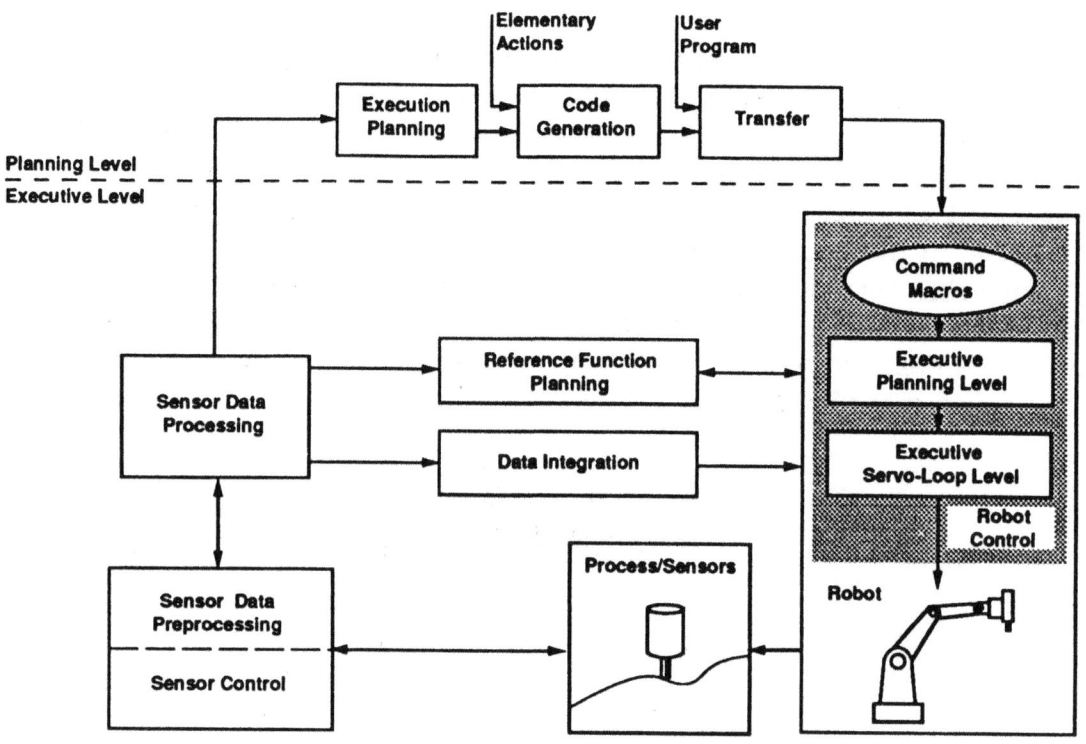

Figure 2   Structure of the System Approach to Deburring

For the automation of deburring processes to be based on robots, two primary problems have to be overcome. They are the process-accompanied adaptation of the system behaviour to varying geometric characteristics based on geometry-detecting sensor system information, and the realisation of task-specific compliance by the application of a force/torque sensor and hybrid force/position control. For the integration of sensor information the feedback of process states to the robot controller and, for special applications (intelligent programming), to the Planning Level, are established. The basic functions to be performed on the feed-back loops are:

- sensor control
- sensor data pre-processing
- process related sensor data processing
- process related reference function planning

The main objective of the Planning Level is a smooth interpolation between the points of the trajectories defined by the planning level (nominal behaviour) or directly obtained from the sensors (reference information). The Planning Level has also to generate "reference trajectories of forces" which the robot has to exert on the work-piece. Based on sensor information at the Planning Level, corresponding control strategies (for example structure switching between free and force-controlled motion, force control strategies, task related strategies for hybrid control etc.) have to be developed.

The Executive Level has to realise trajectories and forces generated by the Planning Level, e.g. generate input signals for the actuators in the robot's joints which will drive the robot along the desired path and produce appropriate forces.

Relationship to Improved CIM

The integration of the Deburring Demonstrator System of the KB-MUSICA Project into CIM-systems is a project objective and has been discussed in broad terms at a previous CIM Europe Conference [Caldeira-Saraiva and Carrapichano, 1989]. The advanced robot control applied to the demonstrator is designed to have interfaces for the adaptation of the control to a host computer system, to other robots working on the same level or to subordinated devices like intelligent measurement systems for adaptation of the robots movement to varying process parameters. The interfaces will enable the robot control to receive information about operations that have to be performed and to send information about the actual process state (i.e. sensor data relating to internal robot parameters) to the host systems. The modules that have to be designed for this will be based on national and international standards like MMS (Manufacturing Message Specification) and related specific standards, especially the robot companion standard to MMS that is in preparation.

THE MODELLING OF THE DEBURRING SYSTEM

The robot and the metal cutting environment form a complex interacting non-linear dynamical system. To investigate the proposed novel control structures discussed in a later section, it is essential that a realistic mathematical model of the robot system exists. This will enable the simulation of the system under a range of conditions similar to those encountered in the industrial scenario.

The Robot

The robot has 6 degrees of freedom and can be described by the following non-linear differential equation in matrix form [Spong, 1989]:

$$\underline{T}_j = [H(\underline{q})]\,\underline{\ddot{q}} + [c(\underline{\dot{q}}\underline{q})]\underline{\dot{q}} + \underline{g}(\underline{q}) + [J]^t\underline{\tau} \tag{1}$$

$\underline{T}_j$ -   vector of joint driving torques
$\underline{q}$ -   vector of joint angles
$[H(\underline{q})]$ -   inertia matrix

$[c(\underline{\dot{q}}\underline{q})]$ -   Coriolis and Centrifugal torque dependent matrix
$\underline{g}(\underline{q})$ -   vector of torques as a result of gravity
$[J]$ -   Jacobian matrix with respect to work-piece reference-frame
$\underline{\tau}$ -   vector of Cartesian reference-frame forces and torques.

The robot's joints are actuated by separately excited dc motors, whose mathematical description adds a further 6 differential equations to the twelve first-order ones of equation (1).

The Environment

Metal cutting is a complex process which involves friction, plastic flow and fracture [Armarego, 1969]. Even though much theoretical research into the physics of the cutting process has been carried out, the empirically derived formulae are still relied upon and it will be this approach that will be considered in this paper [see also Lahoud, 1989].

It is assumed that the burr has the form shown by the three views in Figure 3. The background surface is flat and the mill, with its rotational axis in the z direction, moves in the direction x. It is assumed that the burr is removed so that the surface of the background is level, that is, no trace of the burr remains.

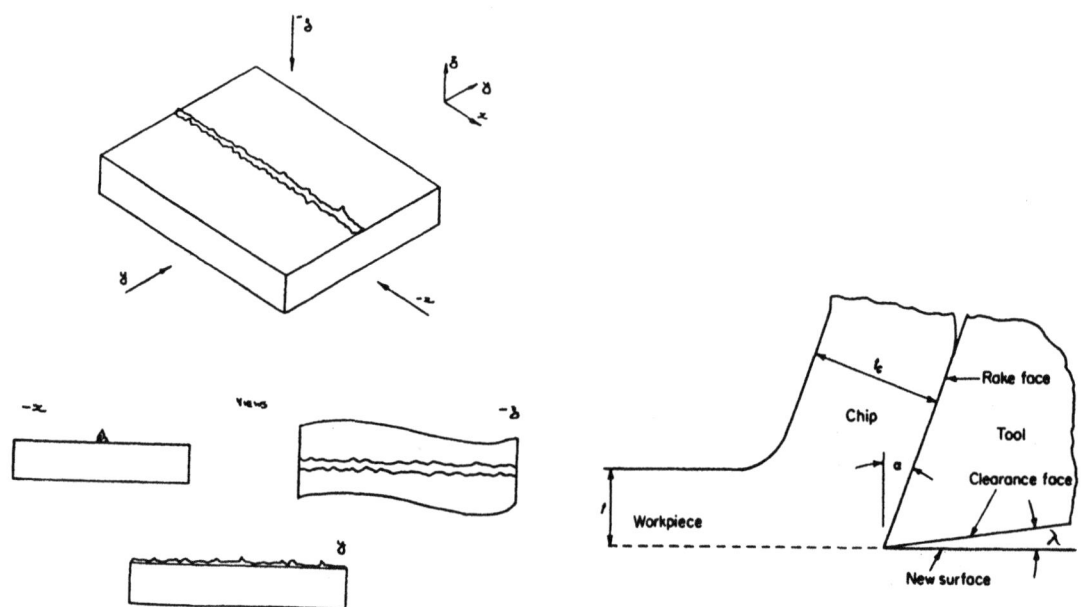

Figure 3   Burr Structure   Figure 4   Orthogonal Cutting Parameters
                                                            [Armarego, 1969]

Figure 4 shows some of the variables in orthogonal cutting, where:

t   -   undeformed chip thickness
t_c -   chip thickness
λ -   clearance angle
α -   rake angle,

and the usual assumptions of the Merchant thin-zone model [Armarego, 1969] are made.

The instantaneous undeformed chip thickness, t, is one of the vital parameters in the calculation of the cutting forces. In order to calculate t, the path of the milling-cutter tooth is required, and it was shown in [Martellotti, 1945] that the path was an arc of a looped trochoid.

The undeformed chip thickness expressions assuming a looped trochoid trajectory are complicated and a simplified version would be useful for initial analysis purposes. Assume a circular tooth path, the feed velocity is 'low' and the angular velocity of the mill is 'high'. These assumptions imply that the undeformed chip thickness is small. Consider Figure 6, from which it follows that (f_t is the feed per tooth):

$$t = f_t \sqrt{\left[1-\left(\frac{(r_m-d)^2}{r_m}\right)\right]} \qquad (2)$$

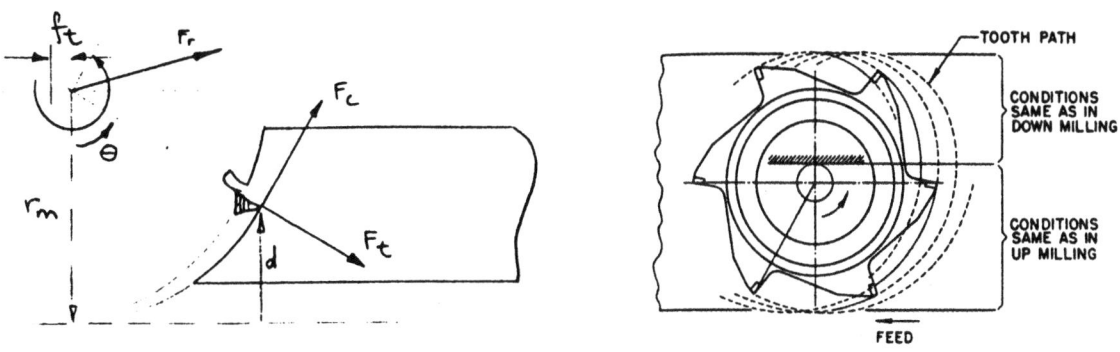

Figure 5   Cutting Geometry and Forces   Figure 6   Tooth paths [Martellotti, 1945]

From [Boothroyd], the power required for machining, P_m, is

$$P_m = p_s Z_w$$

p_s -   specific cutting energy = Ct^ζ in J/m^3
Z_w -   metal removal rate in m^3/s
C, ζ -   empirically derived constants

Hence, the magnitude of the cutting force is (h_b is the height of the burr):

$$|F_c| = t\, h_b\, p_s \qquad (3)$$

The friction of the chip on the mill's tooth creates a force component $|F_t|$ which can be related to the cutting force, $|F_c|$, by an empirical relationship [Shaw, 1989]:

$$|F_t| = \lambda |F_c| \qquad (4)$$

where λ is determined experimentally and depends on the prevailing cutting conditions. The resultant force vector $|F_r|$ on the mill-tool-holder is shown in Figure 6 as well as the forces that the mill exerts on the work-piece.

In the robot deburring system, force controllers (see next section) have been implemented in the x direction and position controllers in the y and z directions. Thus, the components of $|F_r|$ resolved in the y direction will act as a disturbance to the y position controller. Furthermore, this disturbance will vary depending on the position of the mill tooth. Figure 7 shows the variation of the resultant force required of the end-effector for a triangular-shaped burr (burr height varies in a triangular fashion with d), with the z axis of the mill and the centre line of the burr co-incident.

This section has demonstrated the nature of the forces that have to be dealt with by the controllers of the following section and has emphasized the need for adequate environment modelling.

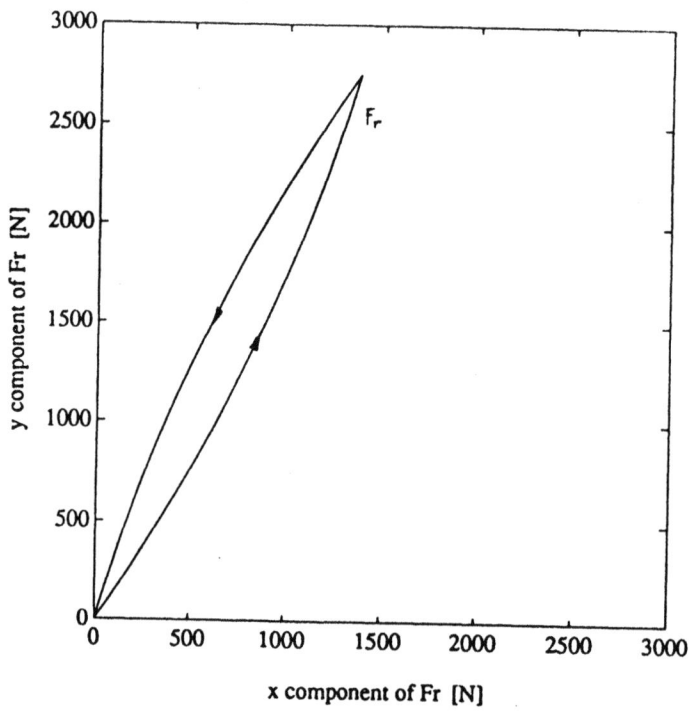

Figure 7   Variation of Force with Tooth Position

THE CONTROL OF THE DEBURRING SYSTEM

The essential control problems arising within robotic deburring are related to the following of very complex edges of the work-piece and to the control of the interaction forces between a tool carried by the robot, and the work-piece. The general control scheme used in the above problems is hybrid force/position control [Raibert, 1981] since both the position of the end-effector and the forces produced by it, have to be controlled.

The hybrid, force position control attracts a great research interest. However, many force control problems are unsolved, caused by only a small number of robots being applied to the machining processes. Some of these problems are related to the general aspects of hybrid control [Kathib, 1987, Whitney, 1987] and others to the specific application. For example, the problem of the opposing requirements of position and force control (position control implies very high stiffness and force control, low stiffness), and the problem of contact between the robot and the environment, belong to the first group. The problems resulting from the progressive wear of the tool, unpredictability of the burr's location and the complex shapes of the machining parts which have to be deburred may be regarded as task specific problems related to deburring by robots.

In the following section the performance of existing control schemes in relation to some of the above mentioned problems will be analysed and a new control approach attempting to overcome their drawbacks will be proposed.

Present Approach

The hybrid force/position control is simple in essence: the control task is split into two orthogonal subspaces in which position and force are to be controlled independently [Raibert, 1981]. In the force controlled direction, the positional degree of freedom is essential for performing the task and the robot's behaviour will be strongly affected by forces acting in this direction. On the contrary, in the position controlled directions, the motion is most critical and must be controlled exactly in order to perform the given task.

For robotic deburring it is obvious to adopt the following strategy: the force produced by the end-effector in the direction of deburring is controlled, and in the other two directions as well as in all the three rotational end-effector's degrees of freedom, position are controlled. Therefore, if the dimensions of the burr vary, the cutting force would change, but due to the force control action, the robot should adapt its speed in order to keep the deburring force constant. This strategy prevents overload or damage to the tool, and an optimised processing time can be achieved by the on-line adjustment of the tool-feed.

Depending on the action provided by the force control loop, two forms of hybrid control algorithms can be distinguished: explicit or torque based, and implicit or position based. In the first approach, the force control loop directly commands a torque signal which corresponds to the controlled force. This signal should be simply added to the output of the position control loop corresponding to the driving torque needed to move the end-effector in the motion controlled directions. In the implicit control schemes the force feedback term is firstly transformed into corresponding position and/or velocity in the force controlled directions, and then realized through a position controller. The transformation is usually based on a simplified relationship between the forces and motion and depends on the nature of the processes.

In order to examine the behaviour of hybrid position/force control with explicit and implicit force control simulations of robotic deburring have been performed. For simulation, the Manutec R3 robot [Münch, 1989] has been used, and all the above presented effects in the deburring process have been included. The roughness of the burrs has been simulated by random functions. Two tasks with demanded forces of 5N and 10N and an average burr depth of 0.1cm and 0.3cm respectively have been simulated. Due to space constraints, only the robustness simulation tests will be presented.

The deburring tasks were simulated for the case of a hybrid controller with explicit force control. The results are shown in Figure 8. It can be seen that if the constant feedback gains are used, satisfactory performance of the system cannot be achieved, that is, the explicit force control is not robust to parameter variations.

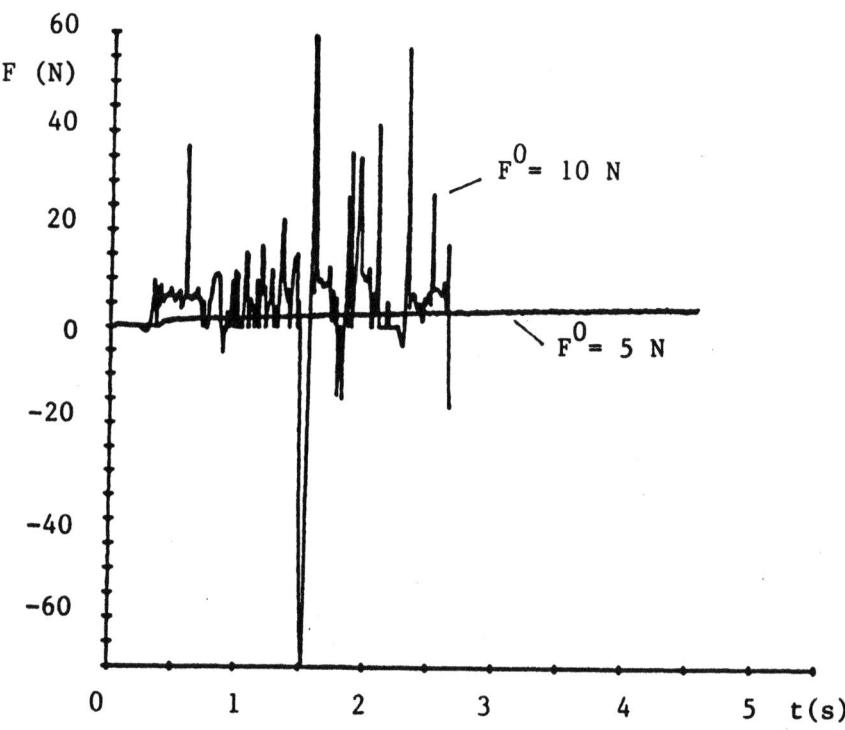

Figure 8   Force Behaviour with Explicit Force Control

The damping force control exhibits better stability properties. The main drawback of such control is the inaccurate determination of the velocities that correspond to desired forces due to inaccurate knowledge and variation of process and environment parameters. Additionally, the position controller behaves as a low-pass filter and therefore the response of the system to the fast variation of forces is slow.

As can be seen from the model of the environment, the process forces are acting in position controlled directions too. If the robot-guided tool is not well centralised with respect to the burr these forces may act as "positive feedback" and the tool may lose contact with the burr.

The New Approach

In order to overcome the above mentioned drawbacks of both explicit and implicit hybrid force/position control we propose to combine both approaches. This scheme involves the on-line identification of the damping of the deburring process by measuring the force and velocity in the deburring direction using a force sensor mounted on the robot's wrist and tachogenerators in the robot's joints. Since damping depends on the burr dimensions, robot's position, etc., the feedback gains have to be adjusted on-line according to the identified values. Explicit force control is added to compensate for fast changes in the process. Therefore, the scheme attempts to use the benefits of both implicit (stability and robustness) and explicit (fast reaction) force control. In order to achieve stable deburring at higher speed, it is proposed to compensate the effects of forces in position controlled directions using force sensor information.

The performance of the new control approach was tested on the same task (Figure 1). The results in Figure 9 show that the controller is robust to the parameter variations even when an error in position of the robot with respect to the burr at the start point is presented. This scheme will be investigated and realized in future work.

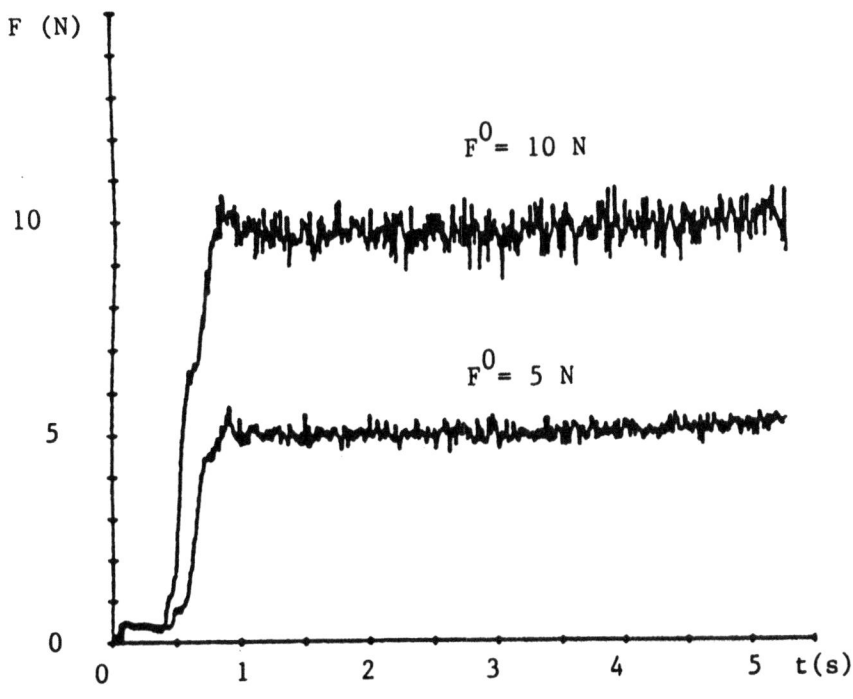

Figure 9   Force Behaviour with New Controller

CONCLUSIONS

Although the pick-and-place configured robot has been well exploited in the automation of factories, the use of robotics in environments in which they must deal with, and control, contact forces, is less widespread. However, many tasks which are not automated on a production line, involve the latter behaviour. This implies that full automation, with its attendant benefits, is not being achieved. Hence, progress in the KB-MUSICA Robot Deburring Demonstrator will produce benefits for automotive CIM environments with applications in many other areas.

REFERENCES

Armarego E. J. A. and Brown R. H., 1969: *"The Machining of Metals"*, Prentice-Hall Inc.

Boothroyd G.: *"Fundamentals of Metal Machining and Machine Tools"*, McGraw-Hill Kogakusha Ltd.

Caldeira-Saraiva F. and Carrapichano M.J., 1989: *"Multivariable control in the Process Industry"*, Proc., 5th CIM Europe Conference, May 17-19, pp. 87-97.

Kathib O., 1987: *"A Unified Approach for Motion and Force Control of Robot Manipulators: The Operational Space Formulation"*, IEEE Journal of Robotics and Automation, Vol. RA-3, No. 1, pp. 43-53.

Lahoud M. A., 1989: *"End-Effector Forces in Robot Deburring Operations using an End-Mill"*, BMT Internal Report, KB-MUSICA Project, DEBUR.220.BMT.01, 15 December.

Martellotti M. E., 1941: *"An Analysis of the Milling Process"*, Transactions of the ASME, November, pp. 671 - 700.

Martellotti M. E., 1945 *"An Analysis of the Milling Process, Part II - Down Milling"*, Transactions of the ASME, May, pp. 233 - 251.

Münch H. and Surdilovic D., 1989: *"Analysis of Control Concepts for Robot Deburring Demonstrator"*, IPK working paper TS130, KB-MUSICA Project, 12 June.

Raibert M. H. and Craig J. J., 1981: *"Hybrid Position/Force Control of Manipulators"*, ASME Journal of Dynamic Systems, Measurement, and Control, Vol. 102, June, pp. 126 - 133.

Shaw M. C., 1989: *"Metal Cutting Principles"*, Oxford University Press.

Spong M. W. and Vidyasagar M., 1989 *"Robot Dynamics and Control"*, John Wiley and Sons.

Whitney E. D., 1987 *"Historical Perspectives and State of the Art in Robot Force Control"*, The International Journal of Robotic Research, No. 1.

ROBOT NAVIGATION STRATEGIES IN A PARTIALLY KNOWN ENVIRONMENT USING A SPACE-TIME LEARNING GRAPH

Daniel M. Kumpel
Francisco Serradilla

Instituto de Automática Industrial. CSIC.SPAIN

1. OVERVIEW

Frecuently, industrial processes are developing in environments which are hostile or harzadous to people. So, the maintenance of the industrial activity in such environments is a difficult and dangerous task.

Autonomous Robotic Systems will be able to perform these tasks, reducing the risk and costs associated with them. Applications of Autonomous Robotic Systems in the industry are numerous and these can be performed with better reliability, accuracy, and efficiency than human personnel.

The ESPRIT II project 2043 MARIE (Mobile Autonomous Robot in an Industrial Environment) aims to develop integrated software/hardware systems to give autonomy to robotic vehicles in two generic situations: automatic manoeuvring and an inspection application. Both situations are very important in an industrial context. These two testbeds will provide a test environment for the research and development results of the projects.

The first test bed is a "docking application". The robotic vehicle will perform complex manoeuvres in a dinamically changing environment autonomously, demonstrating the ability to interpret and reason with information extracted from this environment. In the second test bed, a mobile vehicle with a manipulator mounted on the top and an autonomous diagnostic system must cooperate closely and intelligently to accomplish an inspection of a hydraulic pipe loop, in which defects, typical of chemical and nuclear industry environments, can be generated.

The combination and integration of symbolic and numerical techniques are essential in both test beds. The approach presented in this paper involves only symbolic techniques. The navigation problem is solved in a high level domain, using artificial intelligence techniques.

The problem of obtaining optimum paths for mobile robot navigation in fully-known terrain has been studied extensively by many researchers [Lozano Perez 79] [Brooks 83]. This paper focuses on the problem of navigation in partially-known terrain. This situation is more realistic than the former, and it appears more frequently in practice.

The navigation planner is provided a priori with a model of the factory building. The factory model is comprised of a geometric portion, which includes the factory walls, their size, and a safety zone next to each wall which must not be entered by the robot (expanded wall); and a topological portion, which consists of a graph of doors/rooms which is called the "access graph".

This non-oriented graph defines the connections between rooms.

The nodes on the access graph are mandatory points which must be passed to enter or leave a room. Each door generally has two accesses, belonging to the two rooms communicating with it. The arcs on the graph connect two accesses that are visible from one another when the factory is empty; that is, free of obstacles.

This access graph determines the network of possible paths for the robot to navigate from one room to another in the empty factory. However, the factory building model does not show the contents of rooms or corridors. Furniture, workstations, tool machines, etc., are unknown objects; that is, no a-priori model of them exists. These unknown objects may lie in any position, and their position may change between mobile robot missions. Due to this situation, a method of hierarchical navigation at two levels, global and local is proposed.

The global planner uses the known model of the factory to find a sequence of accesses through which the robot must pass in order to reach the room in which the destination position lies.

The local planner finds an obstacle-free path between accesses (using the sensor system), or, should such a path be impossible, requests a new solution from the global planner. The local planner follows a strategy based on the local information supplied by an ultrasonic sensor system which has been simulated.

The global planner is based on an expanded algorithm A*, and the local planner, on hill climbing with backtracking. Details of both algorithms are in [Kumpel 89] and [Serradilla 89].

With successive missions, the local planner triggers sensorial operations and robot movements which gradually construct a graph which is called a space-time learning graph (STLG). This allows a dynamic model of objects to be established at the same time as navigation efficiency is improved.

The form of the objects will be inferred from successive missions by means of the STLG. This graph not only models the static surroundings in spatial form [Iyengar 85] [Oommen 86], but by describing the arcs of the graph (values of length, reward and penalty) reflects the dynamic variation of objects. It does have the virtue of storing knowledge about the contents of the factory that is useful for better navigation.

In this picture of the world, it is easy to find a possible obstacle-free path, the picture can easily be adapted to observed changes, and its use allows different modes of navigation to be established.

There are four modes of navigation depending on different requirements: Rapid and Reliable navigation using the STL graph, Exploring navigation mode, Relaxed navigation mode, and the combined Global-Local strategy.

2. SPACE TIME LEARNING GRAPH

The mobile robot possesses a limited sensorial capacity. When it enters a room for the first time, it does not have available any initial model of the objects contained in the room. The STLG is a form of representing the knowledge of the surroundings that the robot acquires as it uses its sensors.

The shape of the objects is unknown and their location in the factory can change. The objects can be moved between one mission and another as well as during the execution of a mission, as long as they are moved outside the operation zone. The mobile robot's zone of operation at any given time is not only the zone of its immediate movement, but also that enclosed by the scope of its sensor system.

These changes are brought up to date and incorporated into the STLG. So, the STLG is neither a complete nor a consistent model, but by means of the execution of a number of missions, the exploration and observation allow for the acquisition of new knowledge of the world and for their addition to the STLG. As a result, the STLG is the basis for increasing the navigation's level of efficiency.

With the use of the STLG, it is possible to predict the most promising path, according to some pre-defined criteria, before a new mission. This provides for a new navigational mode that is more reliable and that we call Rapid-Safe Navigation.

In this paper we present two other navigation modes, the purpose of which is to complete the STLG by seeking new knowledge of the world of objects.

Before turning to the navigational modes, we shall describe the kind of knowledge that is stored in the STLG. The STLG is a very useful way of representing the knowledge of the surroundings in which the robot navigates.

The navigation planning module turns to the STLG in order to make decisions. The nodes and arcs of the graph involve spatial knowledge. In [Serradilla 89], the construction of the STLG was explained. The arcs of the graph are qualified by three attributes: length, reward and punishment. The lengths of the arcs allow us to find the shortest paths in the graph.

As the obstacles can change, the nodes incorporated in the spatial-temporal graph in a mission remain in the graph until another mission makes the robot aware of the variation.

Therefore, the paths of the graph must be understood to be possible solutions. The reward and punishment attributes are values that are reflected frequency with which the arc of the graph has been cut by the appearance of an obstacle.

When, during a mission, a mobile robot determines that an arc of the graph continues without intersecting any expanded obstacle, we say that the arc is navigable.

The value of the reward of the arc establishes the times that the arc was navigable, and that of the punishment, the times that it was not.

The forgetting of the vertex of the obstacle that has been moved, is not instantaneous, and for it to be effective, the value of the punishment should exceed a threshold. When the value of a reward exceeds a threshold, one may consider that the arc of the graph limits a relatively immobile obstacle.

In this way, after a number of missions, the STLG allows for the establishment not only of an obstacle map, but of the degree to which navigating next to them is reliable.

The reliability is the difference between the relative frequencies of reward and punishment. This results in a better use of the mobile robot's resources and in a more efficient navigation.

3. NAVIGATION MODES

3.1 Global-Local Navigation Mode

The first navigation mode results from the combination of the local and global strategies. This mode allows for the construction of the

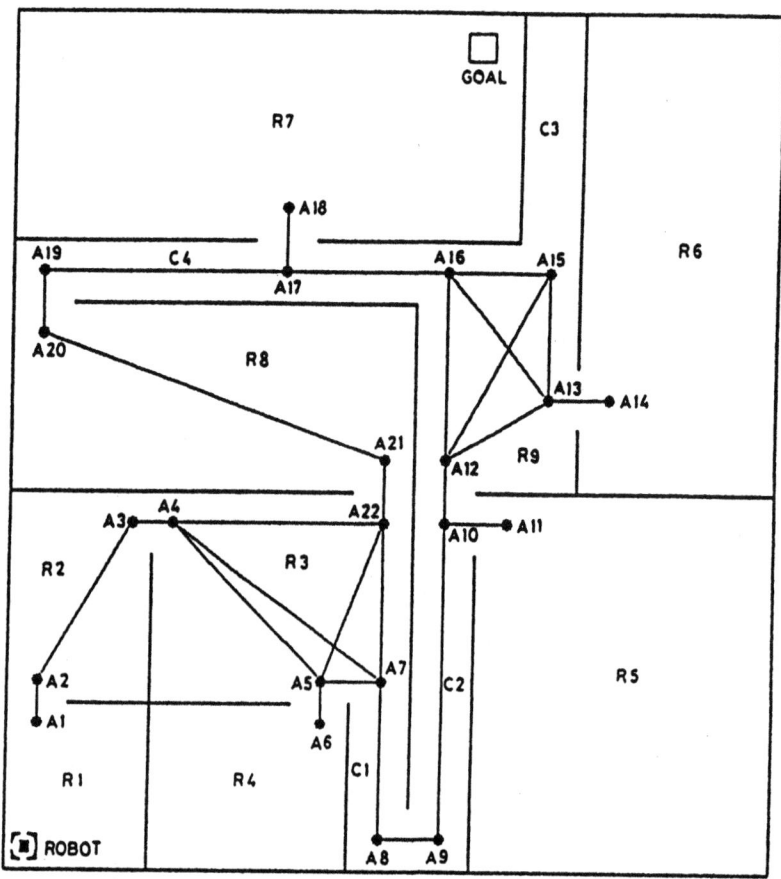

Fig.1 The Access Graph

STLG beginning with the empty graph and, therefore, is fundamental to the robot's first missions, when the contents of the factory are completely unknown.

In order to reach the objective, the global planner only considers the factory building and the access graph (figure 1) [Kumpel 89], and its result is a series of points.

The local planner guides the robot through this sequence, avoiding collisions with the unknown world. Figure 2 shows the simulated surroundings that we have developed. One can observe in the figure the solution path that results from the application of this global-local strategy.

The robot, which is initially in room 7, has to reach the objective, which is in room 5. Figure 3 shows the STLG, which previously was empty, after having completed the mission.

3.2 Rapid and Reliable Navigation Mode

Over a number of missions, the STLG has progressively been completed. When one requires the robot to reach an objective, it is possible to specify the mode in which it should navigate. The rapid and reliable navigation mode uses the STLG, that is, uses the knowledge gathered on previous missions.

Starting with the robot's initial position and the objective, this strategy consists in seeking the STLG node that is closest to

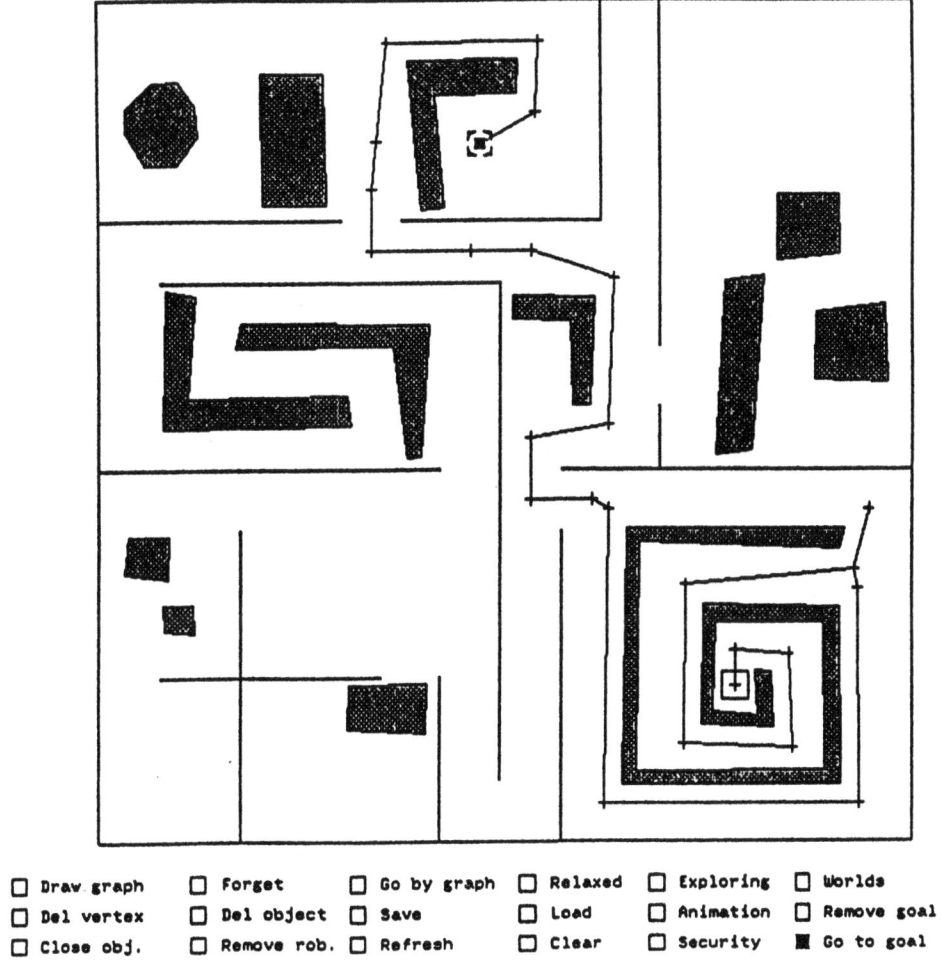

☐ Draw graph	☐ Forget	☐ Go by graph	☐ Relaxed	☐ Exploring	☐ Worlds
☐ Del vertex	☐ Del object	☐ Save	☐ Load	☐ Animation	☐ Remove goal
☐ Close obj.	☐ Remove rob.	☐ Refresh	☐ Clear	☐ Security	■ Go to goal

Fig.2 The global_local mode finds a solution

the objective and the STLG node that is closest to the initial position. Then, it looks for a path in the STLG that connects both nodes.

We use an A* algorithm that finds the optimal path on the graph, assuming that such a path exists. The criterion that we include in the A* algorithm's cost function combines the length, the reward and the punishment of the path. In other words, the A* algorithm selects the shortest path among those that are most rewarded and least punished. The cost function's heuristic component is the Euclidean distance from the current node of the STLG to the objective.

The path selected according to this strategy is the most promising, but it is a tentative path, given that the obstacles might have obstructed a section of the path. This possible solution is developed prior to the robot's performance. The robot follows this path without making use of costly sensorial sweeps, and only detects the presence/absence of obstacles that are before it. This leads to the execution of the most rapid mission, which is also the most secure, since the robot moves among the most static obstacles. The local planner calls the global planner to request a new global solution if an obstacle is blocking a path section.

Figure 4 illustrates this strategy. One notices how the knowledge acquired on a mission (figure 3) are useful for the improvement of the navigation in future missions.

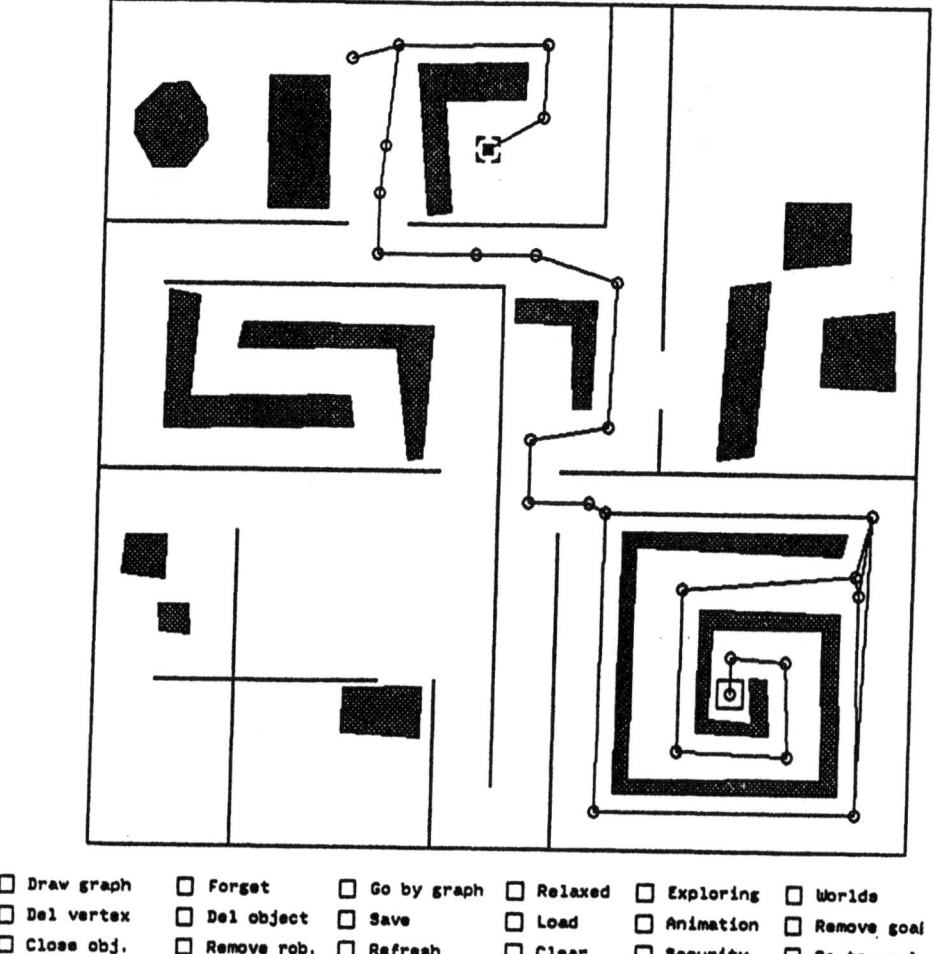

☐ Draw graph	☐ Forget	☐ Go by graph	☐ Relaxed	☐ Exploring	☐ Worlds
☐ Del vertex	☐ Del object	☐ Save	☐ Load	☐ Animation	☐ Remove goal
☐ Close obj.	☐ Remove rob.	☐ Refresh	☐ Clear	☐ Security	☐ Go to goal

Fig.3 STLG after executing fig.2 mission

3.3 Relaxed Navigation Mode

This navigation mode is useful as a means of generating a model of the world of unknown objects a priori. The goal of this strategy is not to guide the vehicle to the final position, but rather to acquire information and incorporate it into the SLTG.

The criterion that is used, is that of guiding the vehicle from its current position to the vertex of the arc with the least reward value. A vertex of an arc with zero reward is one that is on the list of vertices perceived by the sensors, but which is not yet connected to the STLG.

Figures 5, 6, and 7 show an example of the performance of the mobile robot according to this strategy. In figure 5, one sees the robot in room 6 next to two completely unknown objects. The STLG is empty.

The STLG of figure 6 has been obtained after three movements of the robot. It notices that one of the objects has been completely contained in the STLG.

With seven of the robot's movements, one obtains the STLG in figure 7. One can see that paths free of obstacles and a spatial model of the objects have been established. These indicate a suitable navigation path as well as the attributes of the arcs of the STLG.

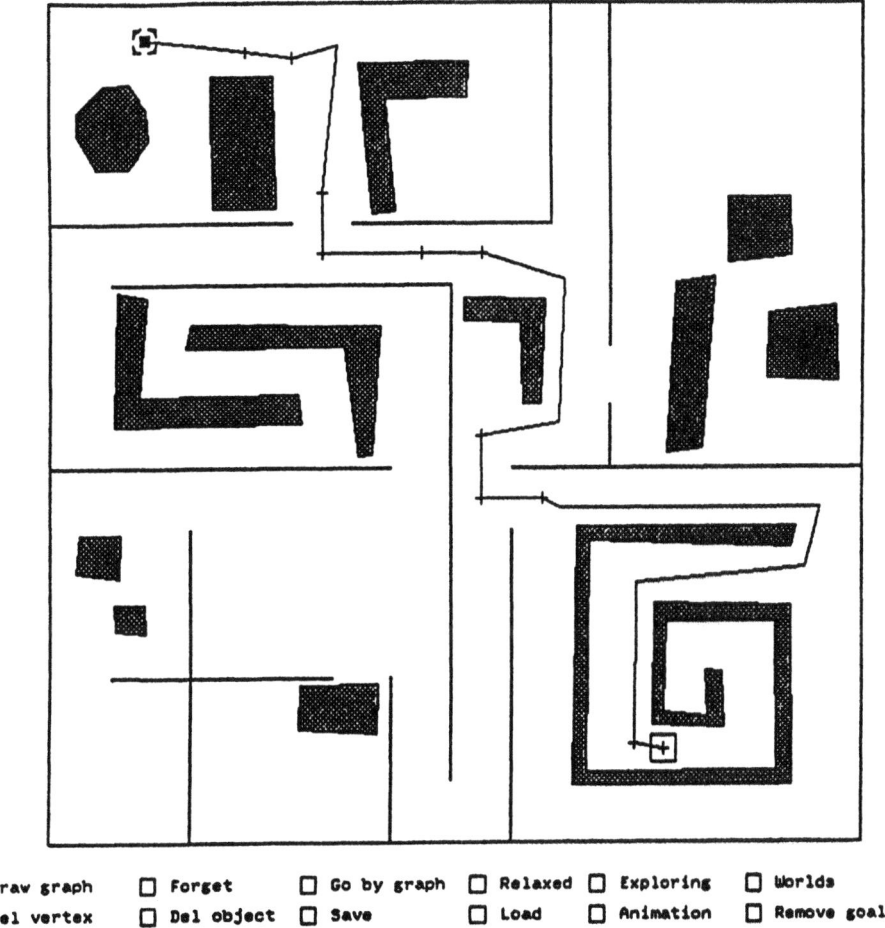

☐ Draw graph	☐ Forget	☐ Go by graph	☐ Relaxed	☐ Exploring	☐ Worlds
☐ Del vertex	☐ Del object	☐ Save	☐ Load	☐ Animation	☐ Remove goal
☐ Close obj.	☐ Remove rob.	☐ Refresh	☐ Clear	☐ Security	☐ Go to goal

Fig.4 Using the STLG is possible to improve
the navigation performance

3.4 Exploring Navigation Mode

This mode is useful as a means of completing an already existing
STLG. The goal of this strategy is not to guide the robot to an
objective position, but rather to explore the less known surrounding
zones.

The STLG allows for the identification of these less known zones.
In contrast to the previous mode, this strategy requires that the
STLG not be empty, and seeks the vertex of least connectivity. In
this case, this will be at least one. The vertex of minimun
connectivity, that may be beyond sensorial reach, is reached by the
vehicle following the local navigation mode. Once this objective is
reached, the navigator selects another vertex.

Figure 8 illustrates this navigation mode. The robot is in room 7
and, because of the execution of a previous mission, there is an
STLG, which is shown. One can see that the object is geometrically
modelled, but only in part. This strategy allows for the completion
of the model.

Figure 9 shows the result of the application of the exploring
mode. The hidden parts of the object on the first mission have been
incorporated into the STLG.

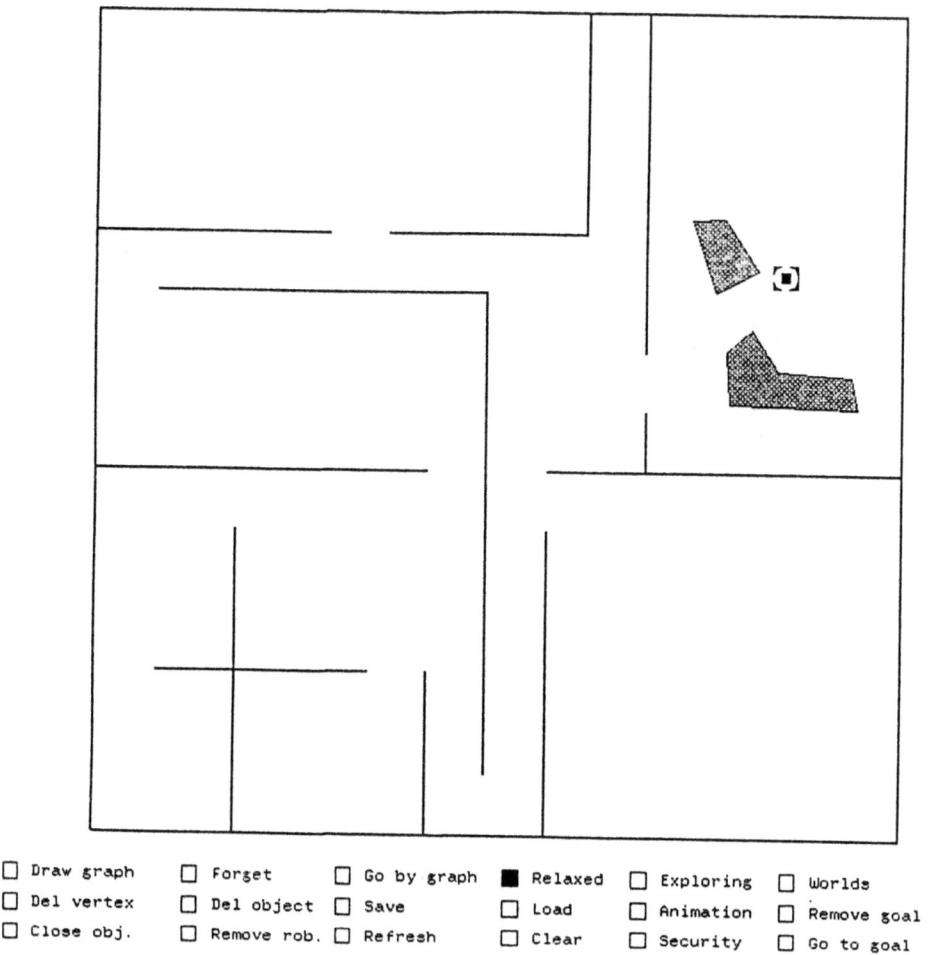

Draw graph	Forget	Go by graph	■ Relaxed	Exploring	Worlds
Del vertex	Del object	Save	Load	Animation	Remove goal
Close obj.	Remove rob.	Refresh	Clear	Security	Go to goal

Fig.5 The relaxed navigation is selected

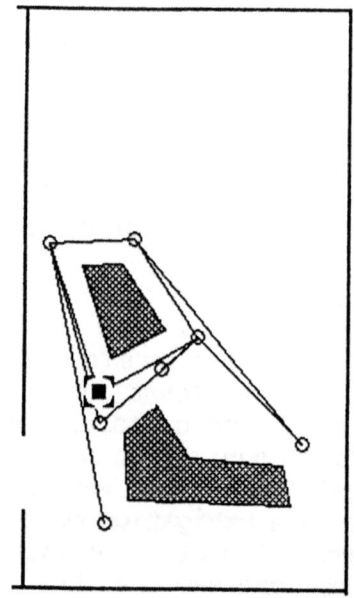

Fig.6 STLG after three robot movements

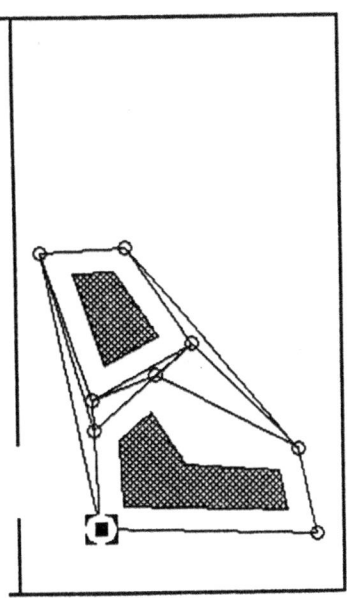

Fig.7 STLG after seven robot movements

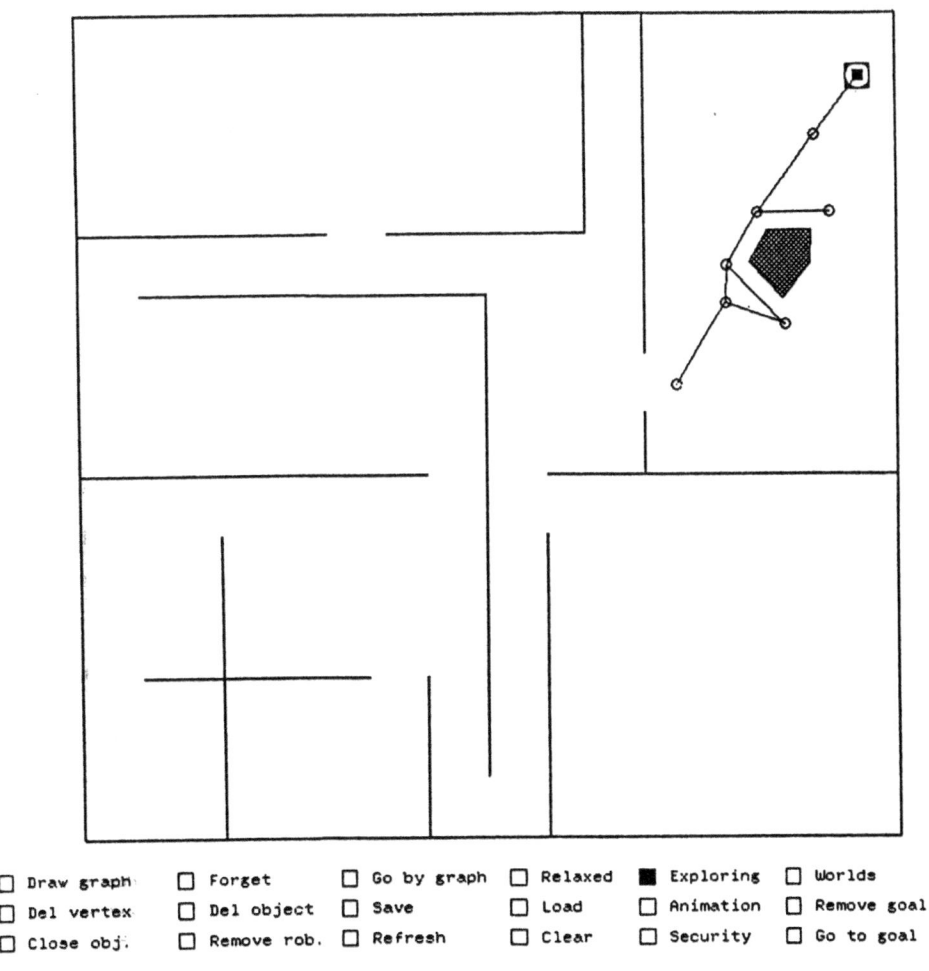

Fig.8 The exploring navigation is selected

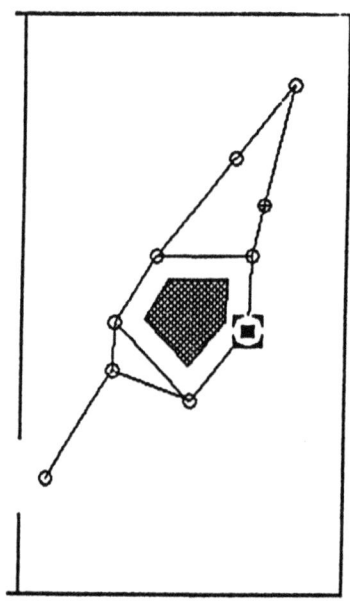

Fig.9 The exploring navigation mode

has completed the model

4. CONCLUSIONS

It has been developed a useful way of representing the knowledge that is gained on an ongoing basis of the world of objects which are, initially, completely unknown.

The knowledge acquired is incorporated dynamically into a structure that is called a Space-Time Learning Graph, and which allows for the increase in the navigation efficiency.

We defined four distinct modes of navigation: the Global-Local mode, the Rapid and Reliable mode, the Relaxed mode and the Exploring mode. The first two are navigation strategies to reach destination position, and the latter two are strategies to better know the world and store the information in the STLG. The global-local mode combines both approaches hierarchically.

The global planner establishes of line a sequence of points that begins from the initial room and arrives at the destination; and the local planner finds, on line, a path that connects the points without having the robot collide with obstacles.

The rapid and reliable strategy allows for the use of the stored information from past missions, so as to navigate more efficiently on each mission.

The relaxed mode is useful as a means of generating an obstacle map, and the exploring mode as a way of increasing the information from the less know surrounding zones.

The next step in this approach would consist in incorporating into the simulation, the physical and kinematic characteristics of the vehicle that is to be used in the context of the ESPRIT 2043 "MARIE" project.

Besides this, it will be necessary to integrate in the same prototype other lower levels of path planning. Other MARIE partners are also working on path planning and trajectory planning and it will be very useful combining the complementary approaches in a short time.

ACKNOWLEDGEMENT

This work is carried out within the context of the ESPRIT II 2043 "Mobile Autonomous Robot in an Industrial Environment" (MARIE)" project.

BIBLIOGRAPHY

[Brooks 83] R. Brooks. Solving the Find-Path Problem by Good Representation of Free Space IEEE Trans. on System, Man and Cybern. Vol. SMC 13 Nº 3, 1983.

[Iyengar 85] S S Iyengar, C.C. Jorgensen, S. V. N. Rao, and C. R. Weisbin. Learned Navigation for a Robot in Unexplored Terrain. A.I. Applications-The Engineering of Knowledge-Based Systems. Second Conference.Miami.1985, Edited by C. Weisbin.

[Kumpel 89] D.M. Kumpel, F. Serradilla
Global Path Planning Avoiding Unknown Obstacles for an Intelligent Autonomous Robot. International Workshop on Sensorial Integration for Industrial Robots: Architectures and Applications. SPAIN. 1989.

[Lozano Pérez 79] T. Lozano Pérez and M.A. Wesley. An Algorithm for Planning Collision-Free Paths Among Polyhedral Obstacles. ACM, Vol 22, Oct 79.

[Oommen 86] B.J. Oommen, S.S. Iyengar, S.V. N. Rao and R.L. Kashyap Robot Navegation in unknown terrains of convex polygonal obstacles using learned visibility graphs Proceedings AAAI-86: Fifth National Conference on A.I., Philadelphia, USA, 1986.

[Serradilla 89] F. Serradilla, D.M. Kumpel
Robot Navigation in a Partially Known Factory Avoiding Unexpected Obstacles. Proceedings of the Intelligent Autonomous System-2 Conference. The Netherlands. 1989.

VISION BASED ONLINE INSPECTION OF MANUFACTURED PARTS: COMPARISON OF CCD AND CAD IMAGES

E. HIRSCH 1) and U. LÜBBERT 2)

1)   University Louis Pasteur
     ENSPS/LSIT
     7, rue de l'Université. Strasbourg. 67000 France

2)   FhG-IITB
     Fraunhofer-Institut für Informations- und Datenverareitung
     Fraunhoferstr. 1. D-7500 Karlsruhe 1
     Federal Republic of Germany

Subject areas: Automated Inspection, Vision, Integrated Systems

This paper presents an automated system as a solution to the 100 % control of manufactured parts in a FMS environment. The technique used is based on comparison between pictures acquired through a vision system and the corresponding data gained from a CAD system. Comparison takes place as well at feature level than at image level. All kind of inspections, ranging from conformity checking up to metrology, can be achieved through use of a user friendly planning system. The paper is based on work of ESPRIT Project P2091 "VIMP".

1 INTRODUCTION

Automated manufacturing of complex and expensive to produce workpieces suffers from the lack of applicable methods to detect faults immediately after critical stages of the manufacturing process. In order to prevent an expensive loss of material and labour, automated inspection must be done for each workpiece after each critical manufacturing step. Manufacturing errors at the machine tool level may originate from break of the tools, fault due to the structure of the manufactured material and malfunctioning of the machine tool. Thus the part which has just been manufactured has to be accepted, rejected, or remanufactured, depending on its state. Today, only a few samples can be inspected with time-consuming methods, which cannot use information about the shape provided in some cases in a CAD/CAM system.

The paper describes a low cost preindustrial prototype to be realized in ESPRIT Project P2091 started on 1/1/1989. This prototype is aimed to replace mechanical sensors with an accuracy of 10 to 100 µm within the overall volume of the part. The system will work online in a Flexible Manufacturing System (FMS) in order to increase productivity and reduce the manufacturing costs, using modern techniques of vision systems. Due to the limited resolution of the imaging sensors used today for digital image processing systems, the sensor is moved in order to scan larger workpieces in their entire extent. CAD-based knowledge is also required and used for efficient performance of such hybrid mechanical and electronical inspection tasks.

P2091 Consortium : ULP/ENSPS, Fhg-IITB (RFA), RPK/UK (RFA), Caption Sarl (F), Speroni (I)

To inspect the geometrical properties, the image of the current field of view is compared with the information stored in the database of the associated CAD system, with or without use of structured light and Moiré Techniques. Comparison takes place after segmentation and registration of the actual image with a synthetic projection of the CAD module. Furthermore, the 3D data coming from the CAD system are used to generate a 2D representation corresponding to the angle of view of the sensor. The output of the inspection stage is used for retrofitting by the manipulator in case of a possible remanufacturing.The whole manufacturing industry is concerned by this new non-contact, CAD connected inspection system.

After description of the inspection system architecture and corresponding hardware modules, the software organisation (and connection to the FMS control computer) will be introduced. The paper is concluded by a summary of the results already achieved. Specific application domains will also be described

2 SYSTEM DESCRIPTION AND ARCHITECTURE

2.1. Objectives of the Inspection System

2.1.1 Present Situation.Today, the major part of the inspection operations are carried out by human operators either visually or through the use of a measuring apparatus, such as an optical or mechanical comparator. In the case of a very high production rate, these controls are often carried out on randomly chosen samples. The necessity to realize these controls with high accuracy led to the development of measuring equipments which allow to measure a part point by point in a programmed fashion. This measuring process requires also the removal of the part from the machine tool to analyse it on a specialized measuring bench. Furthermore, the piece has to be cleaned, positioned in the referential of the measuring machine and reinstalled in the machine tool when remanufacturing is necessary and possible. The measurement is carried out by means of mechanical contacts This process is very slow and can considerably reduce production rates.

2.1.2 Synopsis of the system. Classical inspection methods compare a reference part with the set of parts to be evaluated. The method proposed for implementation in this paper is based on the comparison of a conceptual reference image with the actual images taken from the manufactured parts. The reference image is obtained through a CAD system which stores in its database all the information necessary for the simulation of the part to be inspected. The data-base also contains the input data of the CAM system.

Figure 1 shows the principal steps of the treatment and the functional blocks of the inspection system.

In order to be able to compare the reference image with the real image, it is necessary to:

- transform the 3D model into a 2D representation, taking into account the picture-taking conditions of the camera, in order to generate a synthetic image,
- segment the real image inputed from the camera,
- extract the features (e.g. contours ....) of the part under analysis, or the skeleton of the structured light on the surface of the part,
- compare the synthetic and transformed real images.

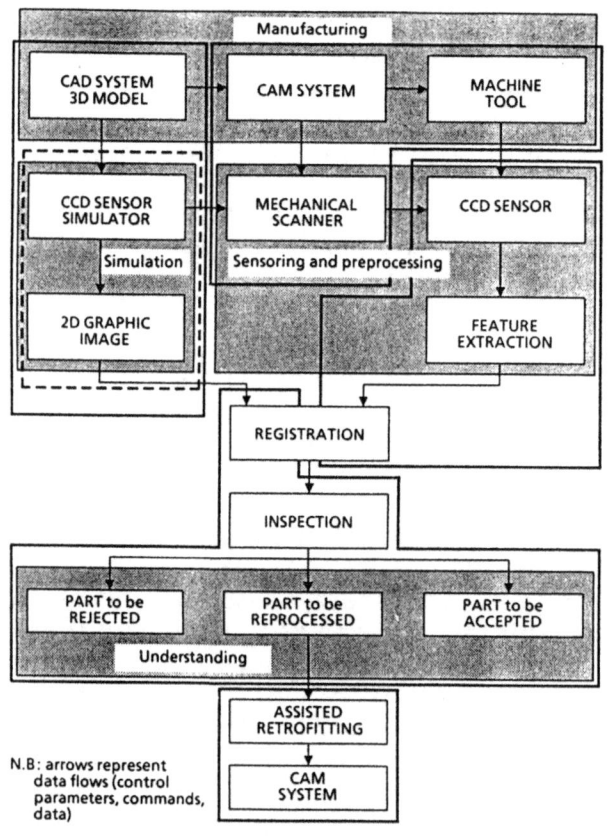

Figure 1   Block diagram of the inspection system

In order to realize dimensional measurements, the comparison involves:
- differences between contours for different object projections,
- differences between the deformation of structured light on the surface of the object and the corresponding deformations due to the structured light simulated on the conceptual reference object.

2.2 Data Processing

Inspection of manufactured parts online during the manufacturing process is for the future FMS integration. Furthermore, input and output information will mostly be visual (CAD/CAM systems), and in some cases spoken (remanufacturing assisted by an operator). The proposed advanced inspection system consists of knowledge processing and engineering at the top level (CAD/CAM and retrofitting after inspection), but the first layers will consist in simulations, preprocessing, feature extraction (e.g. filtering, identification, segmentation) followed by recognition (registration, inspection) and decision taking.

Firstly, the input CCD signal is analysed so as to remove noise, enhance the images and end up with suitably segmented images.

Secondly, suitable combinations (after registration) of these pictures are then chosen as global features to be used in the subsequent stages of processing. The next steps of data processing (some form of matching or comparison) are greatly simplified and undertaken accurately through use of specially designed lightsources (structured light). Furthermore, the scanning of the manufactured part is controlled by the simulation module using CAD data.

The comparison may be a simple distance measure (for example for difference determination between the original and real parts). Alternatively, a more complex processing such as feature descriptor based methods (modified Fourier descriptors for example) may be used to compare the images obtained from the CCD sensor and those synthesized from the CAD database.

It is important that direct measurement techniques are implemented in order to be able to analyse the parts completely. This includes:
-   computation and control of the camera position with reference to the workpiece, making use of multiple images previously segmented and registered,
-   computation and exploitation of the geometrical characteristics or features of the workpiece, using multiple images previously matched with the corresponding CAD image of the part. The output of these measurements leads then to "understanding" and to sustained remanufacturing.

Furthermore, in general, the images are not compared in the representation of grayscale pictures. Also, the synthetic images are gained from the analytic perspective geometric description of the outlines and surfaces of the workpiece adapted to the position of the camera.

As a consequence, in the pixel plane the synthetic image consists of a skeleton of lines which are 1 pixel broad. The pixels on the lines of the skeleton must be calculated. (Also, when structured light such as Moiré is used, synthetic Moiré images are given like this.)

The real image will also assume this format within the vision system. However, if a description plane is used, the synthetic image consists of datasets, describing parts of the outlines and surfaces analytically and additional information about the relationship to other primitives and surfaces. The real image will then also be given like this.

In general, the output from the "comparison" stage is a single decision (good or bad) or a more complex response ("differences" and how to reprocess the part). A further stage of processing can then take place using, for example, a retrofitting technique to the CAD/CAM system or the associated host computer.

The most complex component is the "understanding" part which is some kind of knowledge system (possibly an expert system for remanufacturing) which uses the knowledge about the parts (and the manufacturing process) and the knowledge extracted by the lower level stages. There may be, of course, an interaction and feedback (assisted in a first step by an operator) between the inspection system and the CAM system as shown in figure 1.

For the 2D graphic image generation, the system is based on a conventional CAD/CAM system associated with a fast computer (transputer network). Efforts have been devoted for developing the specific hardware and software modules required by this approach. However, for the acquisition and preprocessing part of the real picture, the processing has to be in real-time and solutions can only be found with highly parallel architectures realized either with specialized modules (wired processors, systolic processors) or with general purpose programmable structures (Transputer networks or array processors). This was also the case for the hardware previously developed by the consortium and used for the described system.

The inspection system described here will be a preindustrial prototype for a real-time inspection system that is designed to be compatible with CAD/CAM systems (through the use of "universal" interfaces, such as defined, for example, in STEP; see, for example, /CADI-88/, /ANDE-89/). Furthermore, the compatibility within the system is ensured through the use of the same kind of computer (Transputer network) for the feature extraction, simulation and inspection or "understanding" stages.

Technical advantages will be obtained by supporting cooperating processing modules at the architectural and software levels through the use of common computer architectures and "universal" interfaces. (This point being suggested by the current trend of this kind of computations.) Furthermore, special attention will be paid to the man-machine interfaces (user friendly, assistance through visual and/or spoken I/O).

3 SOFTWARE ORGANISATION

As a first result (able to change later), the inspection session is divided in two phases : one OFF LINE and the other ON LINE (see figure 2).

1)The aim of the OFF LINE phase, placed before the inspection sequence, outside the FMS line, is to generate data and programs for the ON LINE phase. This is achieved using the so-called planning and simulation systems.
  This planning system creates information for :
  - the simulation (input data)
  - the inspection running (program for the scanner, program for the inspection sequencer, data for the vision system, the processing of images and comparisons)
  The simulation system provides "simulated data", corresponding to the real data to be acquired, stored in a STEP file.

2)During the ON LINE phase, the inspection sequence program itself executes a program for managing the vision system, for synchronising the FMS line and the mechanical scanner and for processing images and comparisons.

Thus, two different software ensembles can be defined for the inspection system:
  - The online inspection software package. It can again be splitted in two parts:
    -The inspection sequencement software,
    -The data processing software packages.
  - The planning and simulation software system offline, aimed to define the inspection procedures for a given part in an interactive way.

3.1 The Inspection Software Package

This software package is build up with two subparts:

- The Inspection Sequence Software.
- The Data Processing Software Package. The algorithmic organisation diagram of this package is indicated and commented in figure 3. This figure is also showing the data structures implied and the hardware used. Figure 4 indicates all the classification/comparison methods which are currently being investigated.

The above software pieces are quite straightforward program packages, relying on well proven methods, on the contrary of the software part to be described in the next section.

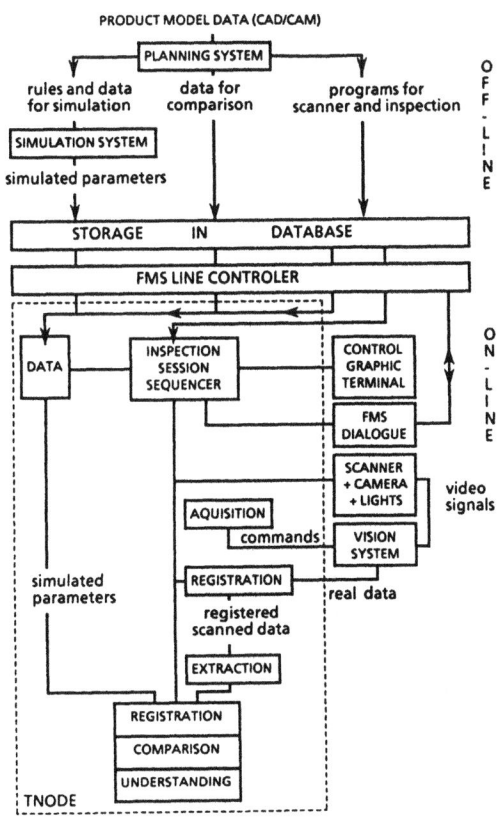

Figure 2   Organization of an inspection session

3.2. The Planning and Simulation Software System

The inspection session planning and simulation of a CCD system on the base of CAD data software package are splitted up into two systems, the

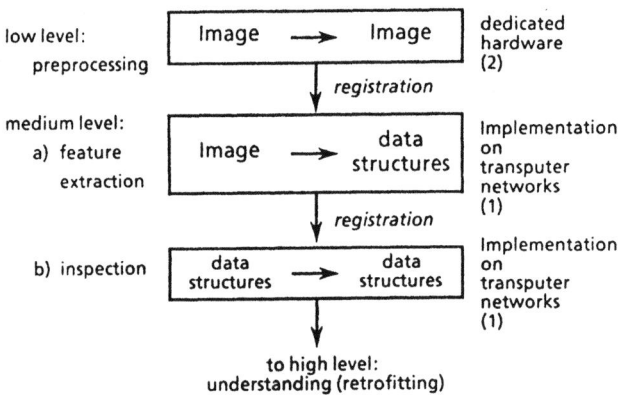

low level:
preprocessing

medium level:
a) feature
   extraction

b) inspection

to high level:
understanding (retrofitting)

(1) currently under study by two partners

(2) under study by two partners in order to be implemented on
Transputer networks

Figure 3 Data structures and flows with corresponding hardware
structures (for segmentation/ registration/ inspection only)

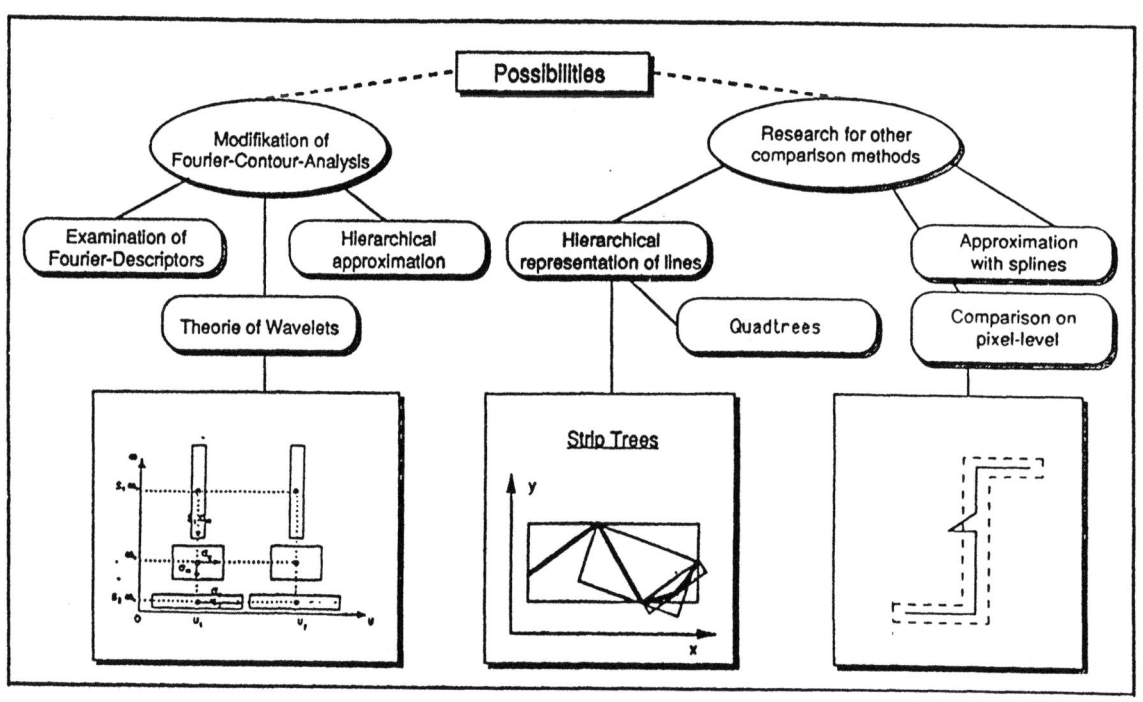

Figure 4   Methods for Classification/Comparison

planning and the simulation system. The role of these two systems in the inspection system architecture and the information flow are shown on Figure 5.

Base of the planning and simulation systems are a STEP-file that contains a 3D geometrical representation of the part to be inspected. The planning system generates on the base of these data, the data for the inspection session by :
- adding tolerance information
- defining areas of interest
- modeling the measurement cell

A measurement strategy must also be implemented in order to generate the camera movement information, the measurement sequence information and the light conditions on the base of the product data and the inspection environment.

The simulation system generates synthetic images of the workpiece on the base of the planning data. Main functions are the simulation of structured light conditions, projection of the 3D part representation in a 2D image representation, see, for example, /SUTH-74/. The result of the simulation system is a STEP file containing a 2D polygonal representation of the part suitable for the comparison.

The planning and simulation systems contain the methods (algorithms) for generating measurement data which are stored in the database.

The workspace is defined once for every inspection environment by using modelling methods similar to methods used and available in CAD systems.

Because of the high accuracy in positioning of the palets in the measurement cell, the calibration of the workpiece in the measurement cell must be done only once. For this reason, one simulation for each type of pieces (offline) is sufficient.

3.2.1 The Planning System. In general such a software system can be divided in three parts: a communication processor, a control system for methods and a database.

The planning and simulation system will use the same data-base. This data-base must be able to store the product model of the piece part which should be measured. That means the 3D geometry received from the CAD system and the technical information like tolerances and surface conditions. In addition to that the data-base must be able to store information about the measurement cell, the camera and camera parameters, the camera movement and the synthetic images.

The planning system contains a database which is able to store tolerancing information. A STEP processor will convert data from the STEP file into the database of the planning system. Figure 6 summarizes how an inspection task is defined.

Figure 7 indicates the architecture of the planning system (and of the simulation system). Planning is concerned with:
- the modelling of the measurement system,
- the inspection planning (choice of the area of interest on the part and how to inspect it),
- the addition of tolerances to standard CAD models.

The planning system generates offline its data, stored in the inspection database and then, in a second time, uses this data online in order to carry out an inspection session.

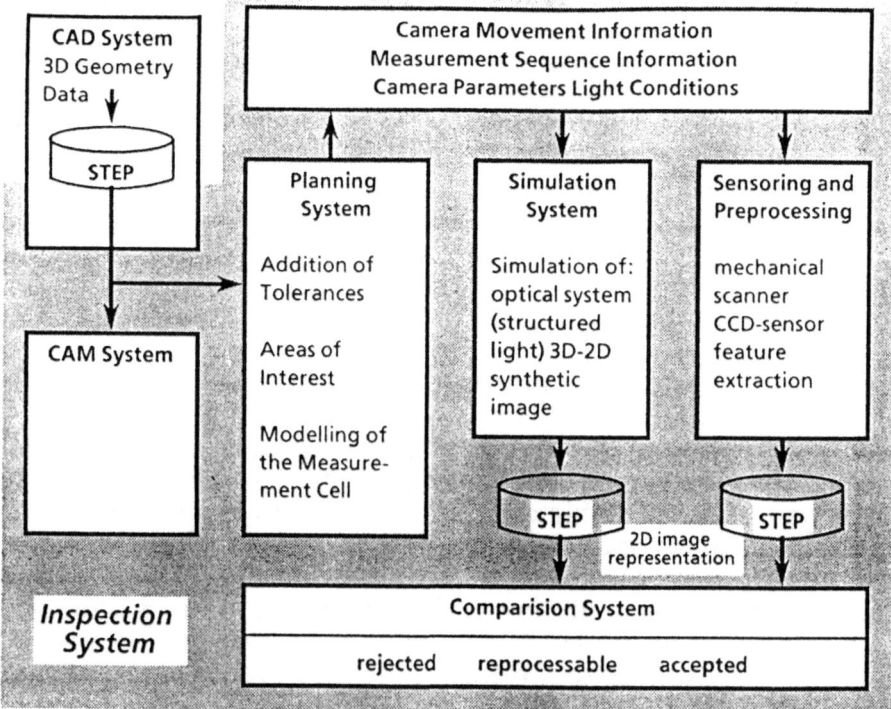

Figure 5 Planning and simulation organization

Description of:

Piece
- Shape, Material, Weight, Temperature
- Surface Characteristics
- Pollution
- Accessibility

Inspection Area
- Area
- Function of the Area
- Special Characteristics of the Area
- Visibility of the Area

Inspection Task
- Characteristics
- Repercussion of Defects
  (Technical Functioning,
  Aesthetic Consequences)
- Problems (Inspection Time,
  Reproducibility)
- Supplementary Tasks
- Marginal Conditions

Figure 6   Inspection task definition

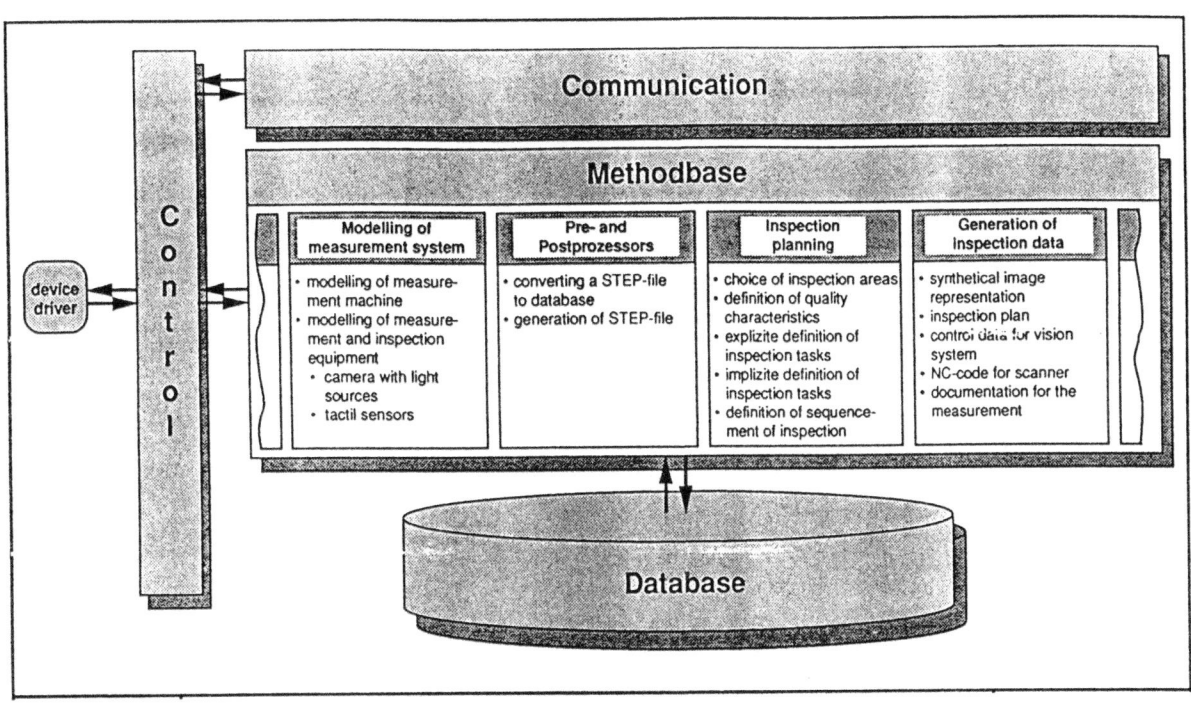

Figure 7   Planning and Simulation Systems Architecture

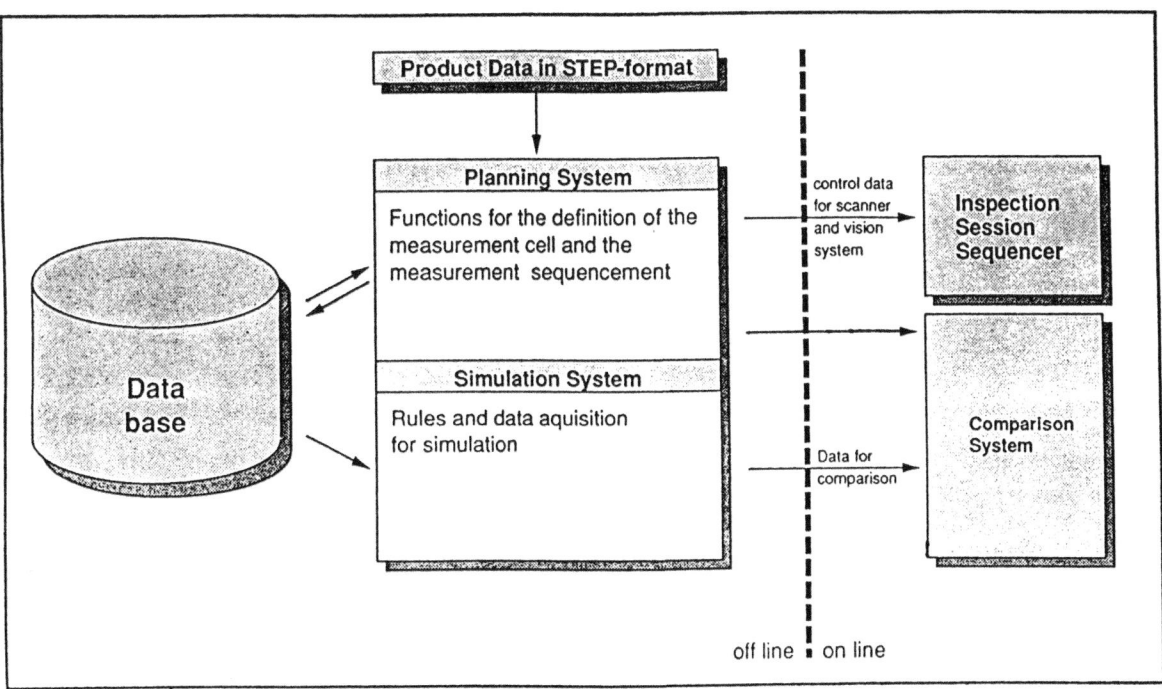

Figure 8   Sequencement of Planning and Simulation Systems Operation

The connection to the simulation system (and the sequence of operation of the two software pieces) is examplified on figure 8. It is noteworthy to mention, that the exchange of data between the different software packages has been standardized through use of STEP files.

3.2.2 The Simulation System. The simulation of a CCD-image requires three different models to represent all the necessary data:
- A geometric model that describes the complete geometry of the object and the measurement cell that are to be simulated. The geometric model is the starting point for constructing the following models which, in turn, are needed for synthesizing an image comparable to the one obtained from the CCD camera,
- a parameterized camera model, that is able to describe the optical system with regard to aperture, focal length, lens, depth of field and distortion,
- a physical model for light-effects, that can describe the patterns projected onto the object surface.

For the representation of these data on the screen or in a STEP-file, two projections are necessary, taking into account the position of camera and workpiece:
- A projection that transforms the object-data corresponding to the camera position,
- a projection that transforms the 3 dimensional object data into a 2 dimensional data.

The research for existing projection algorithms from 3D to 2D data led to a classification of three different projection types:
- object oriented hidden surface algorithms,
- image space based hidden surface algorithms,
- ray tracing algorithms.

Taking into account the different characteristics of the above, the object oriented hidden surface algorithms seem to be suitable because the simulation of structured light is practicable with a tolerable amount of data. Implementation and investigation of existing object oriented Hidden-Surface-Algorithms are to be performed in the future.

3.2.3 Integration of the Inspection System in a FMS. Figure 9 gives the schematic representation of the inspection set-up in a manufacturing cell. Figure 10 then shows what is the foreseen integration in FMS (the general manufacturing organisation makes explicit reference to the analysis model outputed by the COSIMA Esprit project). The manufacturing controler uses a standard MMS library in order, first, to manage the different line components, second, to favour standardization. This part of the work is carried out in cooperation with a FMS manufacturer providing all the facilities needed for the integration of the inspection system in the FMS.

4 ACHIEVED RESULT

4.1 Hardware Components

This section summarizes the hardware modules already defined and currently under realisation.

4.1.1 The Mechanical Scanner. This part of the inspection system is built around a modified version of a standard conventional mechanical

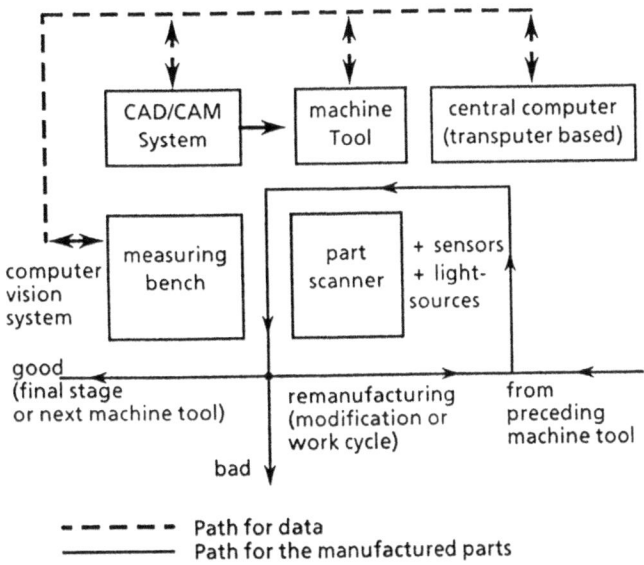

good
(final stage
or next machine tool)

remanufacturing
(modification or
work cycle)

from
preceding
machine tool

bad

- - - - - Path for data
───── Path for the manufactured parts

Figure 9 Inspection set-up in FMS environment

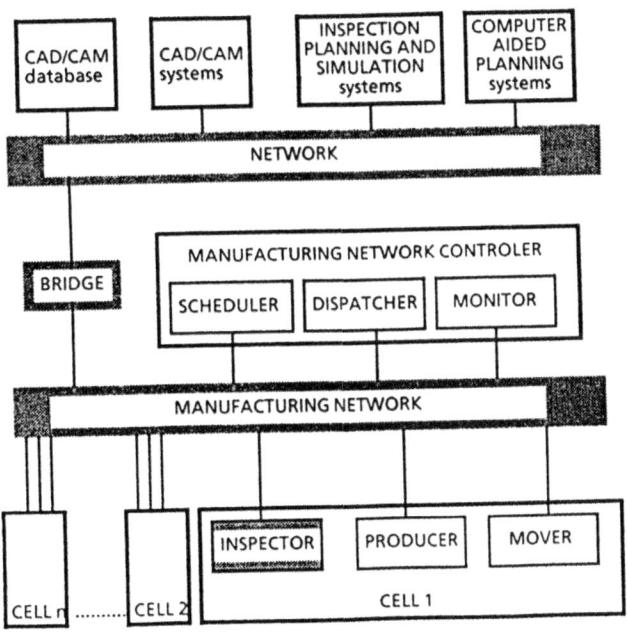

Figure 10   Integration in FMS

inspection robot. Roughly, the probe of this system has been replaced by a 512x512 CCD-camera. The robot also provides facilities to carry the lighting system described later. Due to the use of a proven measuring system, the accuracy of the camera positioning with respect to the piece to analyse, is within a few microns. This system is currently used to define the scanning pathes of a piece in order to be able to analyse the part in its full extent (measurement strategy and definition of the area of interest for specific pieces).

4.1.2 The Vision Systems.

Two systems, already developed and tested, are used:

- One is a modular vision system, built around dedicated hardware modules, having in charge each a specific processing task. This system, described in detail in /PAUL-88/, is able to work at video realtime and its output is fed to the comparison stage (see section 3). This system is aimed for industrialisation, once the processing chains are defined and tested.
- The second is a fully programmable image processing system of general purpose. Its role is to facilitate the development of the processing chains and to validate them. It can be seen as a development tool, used interactively by the user, in order to design its inspection tasks. This system is fully described elsewhere (/PIER-88/, /PERU-88/) and again its output is fed to the comparison stage.

4.1.3 The Sensoric System.

Lighting conditions play a crucial role for the quality of the acquired data. Strong use is thus made of special lightsources and illumination techniques. Among standard methods such as structured light and shadowing techniques, special methods are used:

- Use of diffuse light using a ring of light emitting diodes
- Use of a laser fan
- Use of Moiré techniques. For this latter technique, a Moiré pattern is acquired, whereas the corresponding synthetic pattern is generated using CAD data. The two images are then compared and the shape differences of the piece with its model can thus be estimated (see, for instance, /TEAN-88/, /GASV-89/,/MORS-89/).

4.1.4 The Central Computer System.

The design of the system architecture, the study of input and output techniques in order to achieve good performances (real time) and to avoid bottlenecks for the whole system have been carried out.

As a result, a transputer based system (modified TNODE) has been chosen. This system assumes:
- the synchronisation of scanner movements,
- the acquisition of real images from the vision system,
- the medium and high level processing of real images,
- a part of the simulation process (low level),
- the comparison of real and simulated parameters.

4.2 Software.

The software described under heading 3 is currently being implemented on the different hardware pieces of the inspection system.

4.3 Inspection Tasks (Application domains)

In order to validate the inspection system, the inspection tasks have firstly been classified and secondly representative target applications have been defined.

4.3.1 Classification of Inpection Tasks. Four application tasks have been defined, covering all the inspection tasks one has usually to face in industry:

- Conformity checking: the part is analysed globally in order to verify the conformity of the piece. No accuracy problems are to be solved in this verification.
- Range inspection: A specific part of the piece is analysed and measurements are carried out with an accuracy of 0.1 mm. The attention is only focused on specific features of the piece.
- "Scanning inspection": The part is analysed in its full extent with an accuracy of the order of 10 microns. This inspection task makes full use of the inspection system.
- High resolution inspection: with a precision of the order of 1 µm, specific features of a piece are measured.

All these inspection tasks are oriented toward metrology. In order to test and validate these tasks, a model part has been defined.

4.3.2 Free Form Inspection. An alternative task is to verify the shape of free form pieces such as, for example, turbine blades. In contrast to the tasks defined in the previous section, the position of surface points must be measured which are not situated at particular regions of the object like edges and corners. In this case, the Moiré technique is extensively used. After extraction of the Moiré pattern skeleton, comparison with the corresponding conceptual pattern allows to estimate the shape of the piece. The tangential plane to a given point on the piece can be determined with an accuracy of 0.02°. Carrying this out for a given set of points allows then to define the shape and the deviations with respect to a master piece. This calls for a careful calibration of the system. This part of the work is carried out in collaboration with a turbine blade manufacturer.

5 CONCLUSION

In this paper, a non contact inspection system, based on real time computer vision, has been extensively described. Inspection is carried out through comparison of conceptual images and actual object pictures. The system is oriented toward metrology, measurements having to be carried out with an accuracy of 10 to 100 microns. This necessitates a careful calibration of the inspection system as well as the definition of the references. The method is based on the fact that several images of an object, after combination, lead to a satisfactory object representation. The inspection is online and controls 100 % of the parts. This is aimed to decrease inspection time and increase production rate.

The hardware and software components have been described and some specific application domains have been defined. The project, however, does not aim to developp a restricted system, as the whole manufacturing industry is concerned by this non contact, CAD connected and FMS integrated inspection system.

ACKNOWLEDGMENT

The authors wish to express that they are only acting as a mouthpiece for the P2091 Consortium. They wish also to address their thanks to the people working on this project and for the kindful permission to use the illustrations presented in this paper.

REFERENCES

/SUTH-74/   Sutherland T.E., Sproul R.F., Schumacker R.A., ACM Computing Surveys, Vol 6, N. 1, 1974.

/CADI-88/   Schlechtendahl E.G. (Ed.), " Specification of CAD*I neutral file for CAD geometry, version 3.3 ", Research report, Vol. 1, Springer Verlag, 1988.

/PAUL-88/   Paul D., Hättich W., Nill W., Tatari S., Winkler G., IEEE Transactions - PAMI, Vol. 10, N. 3, 1988, pp 399-407.

/PERU-88/   Perucca G., Giorcelli S., De Couasnon T., Hirsch E., Mangold H., "Advanced algorithms and architectures for speech and image processing ", in "putting Technology to use ", CEC/DGXIII Ed., North-Holland, 1988, pp 543-561.

/PIER-88/   Pierre F., Herve Y., Eugene F., Draman C., Wendling S.,Proc. of 2th PIXIM Conference, Paris, 1988.

/TEAN-88/   Technical Annex of ESPRIT Project P2091 "VIMP", 1988.

/ANDE-89/   Anderl R., Schmitt M., " State of the art of interfaces for the exchange of product model data in industrial applications ", CEN/CENELEC/AMT/WG STEP 8, december 1989.

/GASV-89/   Gasvik K.J., Hovde T., Vadseh T., Proc. 5th Cim-Europe Conference, Halatsis C. and Torres J. Eds., IFS Publications, springer Verlag, 1989, pp 301-308.

/MORS-89/   Morshedizadeh M.R., Wykes C.M., J. Phys. E: Sci. Instrum., Vol. 22, 1989, pp 88-92.

THE IMPACT AND INDUSTRIAL EXPERIENCE
OF CIM FOR SMEs

STRATEGIC INITIATIVES FOR INTRODUCING CIM TECHNOLOGIES IN IRISH SME's

M.S.J. Hashmi
School of Mechanical and Manufacturing Engineering
Dublin City University

and

J. Cuddy
National AMT Programme
EOLAS, DUBLIN

INTRODUCTION

Enhancing competitiveness means reducing the production cost as well as the selling price of a commodity. In view of the wider marketing opportunity for all, this means the cost of producing a product and its selling price should be lower than all other competitors. This is, of course, assuming that the goodwill and quality value of the product is comparable. In the global scenario of highly competitive market place, maintaining and enhancing competitiveness necessitate many companies to make increased use of new developments in manufacturing technologies and concepts.

The key to the long term success of any industry, whether manufacturing or service, is the ability to retain and possibly increase its market share. Only competitiveness in terms of price and quality of the goods or services will ensure the retention of the current and future market share. Application of new technology not only leads to increased productivity and lower unit cost but also improves the product quality and reliability. Thus, the question is not whether companies should employ new technological developments in manufacturing engineering but in what form and when if they are not using them already.

Manufacturing engineering includes functional areas such as product development and design, production planning and control, materials handling and inventory, tooling and process selection, sales and marketing, maintenance and operation of the plant, quality assurance and reliability and finance and accounting. Developments of computer numerical control technology and computer based software for various manufacturing functions have opened up tremendous possibilities for increasing the productivity in each of the functional areas of manufacturing engineering. However, for world class manufacture with competitive edge it is not sufficient simply to employ new technological developments to improve individual functions independently or in an arbitrary manner.

CIM

What does computer integrated manufacturing mean? Is it just another computer based new technological element for manufacturing engineering, e.g. , CAD/CAM, MRP or JIT ? In broader sense, CIM is a concept or philosophy which advocates all the computer software/hardware based manufacturing engineering functions to be employed in an integrated and co-ordinated manner. As such CIM represents complete integration of all the manufacturing engineering functions including the human based activities and functions. Thus, CIM addresses the total information requirments of an enterprise as shown in Fig. 1, making it a business strategy rather than an automation or computerisation strategy. Its aim is the success of the company or enterprise as a business entity, not simply the improvement of the company's individual engineering, production and administrative functions. For integration to be effective, the business must be viewed as a whole and unified system, rather than as a collection of individual functions.

Does CIM imply that mere computerisation of information flow between individual functions will bring business success to a company? On the contrary, CIM suggests that such integrated system should lead to effective utilisation of the capacities of each individual functions through timely information flow on technical, managerial and administrative aspects. The point is that, for CIM to be successful in enhancing competitiveness, all the manufacturing related functions of a company need to be utilising the new technological developments.

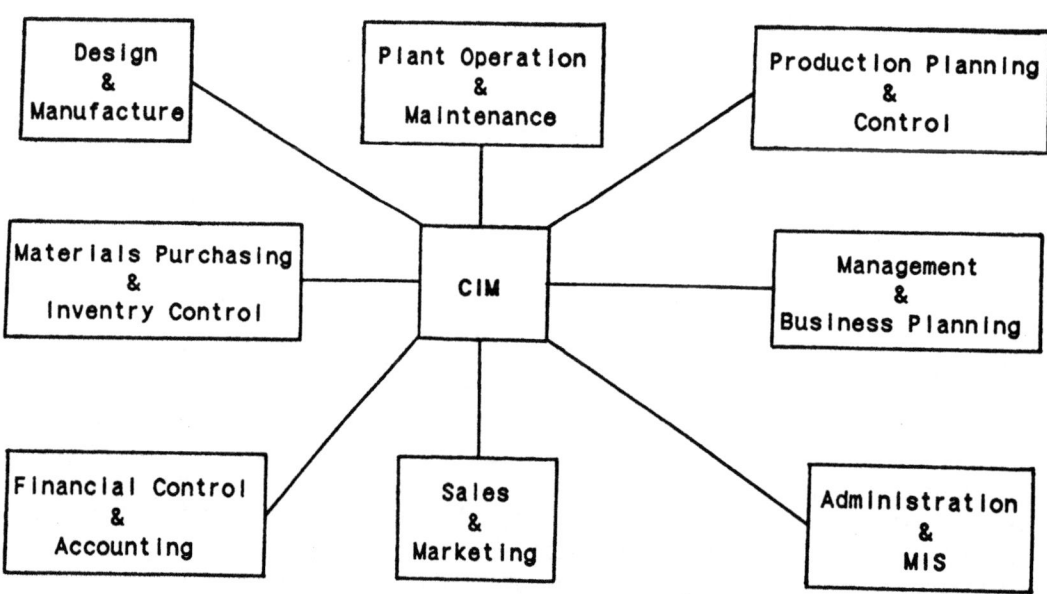

Fig. 1 CIM Elements in a Business Enterprise.

REQUISITES FOR CIM

Ideally, CIM strategy should cover all the possible CIM functions and be applicable to all types of industries. However, all the CIM functions may not be relevant to every industry. Thus, the global concept of CIM must be tailored to suit local application.

For CIM to be effective the environment internal as well as external to the company's domain must also be right. Very often it is found that application of CIM decreases the productivity of a company rather than enhancing it. This is mainly due to the ill conceived notion about CIM itself. In many cases CIM is thought to be the magic answer to improve productivity without realising that appropriate support system must be there for it to be of any good. This is similar to procuring an expensive motor car and then trying to run it with an inappropriate grade of fuel on a country lane full of pot-holes and yet expecting a smooth and comfortable ride.

There are a number of resources which need to be present before implementation of CIM should be considered. These resources could be listed as,
    (i)   human resources
   (ii)   technological hardware resources and
  (iii)   capital resources

Human Resources

This is the most important resource any enterprise can have. For both AMT and CIM to be successful appropriately trained professional are one of the major pre-requisites. Many AMT/CIM facilities fail to deliver due to the incorrect expectation of the users. The fact that CIM is a tool which has to be utilised expertly in order to get the most out of it is often not well understood. This is very often due to the lack of personnel in computer software engineering who have in-depth knowledge of the manufacturing processes and functions or in manufacturing engineering with in-depth knowledge of data processing, systems analysis and systems design. At management level, the decision makers need to be equally assertive in knowing how best to utilise the system they have invested a substantial amount of capital in. They must be aware of all the facilities a CIM strategy accords to them and make proper use for the decision making process towards long term enhancement of productivity. However, very often CIM support section is not manned at appropriate level and the system is used as hybrid between discrete and integrated system with unnecessary duplicity and ineffectiveness. The reason for such undermanning of the sytem could be twofold. Firstly, the management do not appreciate the need for them and secondly, there may not be sufficiently trained personnel available for recruitment due to the absence of a critical mass of trained professionals in this discipline.

Technological Hardware/Software

CIM is about integration of all the so-called computer aided functions within the business of manufacturing engineering. As such, it implies that the emphasis is on efficient communication between the discrete CIM elements. Problem in communication may easily result in the implementation of CIM to be impracticable. The situation would be synonimous to a discussion meeting between a number of persons each speaking a different language.

The reasons for such communication problem are many but the most frequently observed one is the lack of uniformity of computer hardware between each CIM element coupled with the lack of conformity between the software employed in these hardware. CIM is a natural development after AMT. Majority of manufacturing companies embarked on AMT, initially by investing into a single area of AMT and then expanding into another area and so on, thus building up their current level of investment in AMT. Invariably, in most cases, these discrete AMT facilities or CIM elements have different computer hardware and software systems. The development of such miss-match took place not because industry wanted it that way but for reasons that different systems were developed by different suppliers who are good at certain AMT aspect but not in others. Thus, the current hardware/software facilities dedicated to the discrete AMT elements pose serious problem for implementing CIM. Lack of effective intercommunication facilities between the discrete elements needs to be rectified before any progress can be made.

The other option is to start fresh. However, this option is impractical for any existing company which has already invested heavily in various AMT facilities. Re-investment of huge capital for implementing CIM is not a very attractive proposition to the Board of Directors of the company. Another difficulty towards the effective implementation of CIM is the fact that CIM system itself is undergoing change. Rapid developments in computer hardware and software technology is making existing system out of date. Thus, any existing system must be capable of incorporating new developments without causing serious inconvenience to the support personnel. Any new communication protocols would require to pay regard to the existing system. However, this is easier said than done and, hence, in reality new developments take place usually with less than 100 percent compatibility with the existing system. This leads to extra effort and resource to be spent to bring the existing system up-to-date.

Capital Resource

One of the main reasons for the low takeup of the advantages of AMT in general and CIM in particular is the level of capital investment required and the reluctance of the management to embark on such a large capital investment programme. Increase in the productivity is easily visible due to various AMT facilities which comprise the CIM system and the decision for capital investment in these various CIM elements (AMT facilities) is relatively easier to take. With CIM, the enhancement of productivity is a rather long-term process and it is not very easily identifiable for attribution to the implementation of CIM. In fact, more often, implementation of CIM brings disruption and short-term fall in productivity. Thus, unless company management and future business planners are absolutely convinced on the long-term benefits of CIM, they will not be prepared to invest considerable amount of capital.

Since CIM strategy is meant only to co-ordinate the productivities of different AMT elememts and CIM functions, its value is sometimes underestimated. Many companies would even shy away from CIM considering it to bring only marginal improvement in productivity for the disproportionately high level of investment requirement. Current evidence suggests that implementation of CIM has proved profitable only for about 20 percent of the participants in the U.K. it is perhaps not difficult to understand such misconceived notion about CIM.

MAKING CIM TO WORK

What can, therefore, be done in order that increasing number of companies continue to implement AMT in general and CIM in particular in order to enhance competitiveness in the world market. Many companies will not be involved in all the CIM functions and for them to implement any comprehensive CIM system often proves counter productive in terms of human, hardware and capital resource requirements. Ideally, CIM system should be optimised depending on the company's requirement and then implemented. Otherwise, the system is most likely to be counter productive and fail to meet the expectations of the company. The responsibility for implementing an optimised CIM system is on the suppliers/consultants. There may be incentives to supply a full and expensive CIM system to a company. However, it should be bourne in mind that a white elephant is always a white elephant and it can ruin both the company and the supplier in the long run.

It thus, transpires that enhancement of competitiveness through the application of CIM strategy requires the right environment both internal and external to the company. A single company may be in a position to control its internal environment and may in some cases be able to influence marginally its external environment but for most companies, especially the SME's, it is rather impossible to influence their external environment. In this context, external environment includes (i) extent of usage of AMT/CIM by other companies (ii) availability of appropriately trained professionals and workforce, (iii)

accessibility to continuing education and training at all
levels, (iv) financial aid/grant towards the
implementation of AMT/CIM systems and (v) financial
incentive towards long term capital investment in
connection with AMT/CIM implementation e.g., tax
exemption, etc. In national context this is where the
government can and must play a key role as depicted in
Fig. 2. In the context of the European community states
the EEC Commission has got a major responsibility to see
that the competitiveness of industries of the member
states, especially of the SME's is not only maintained but
enhanced in the world market scenario. The commission, in
concerted actions with the member states, can and should
initiate programmes not only for R & D and training but
also for direct financial aid or grant to enterprises
willing to invest in AMT/CIM.

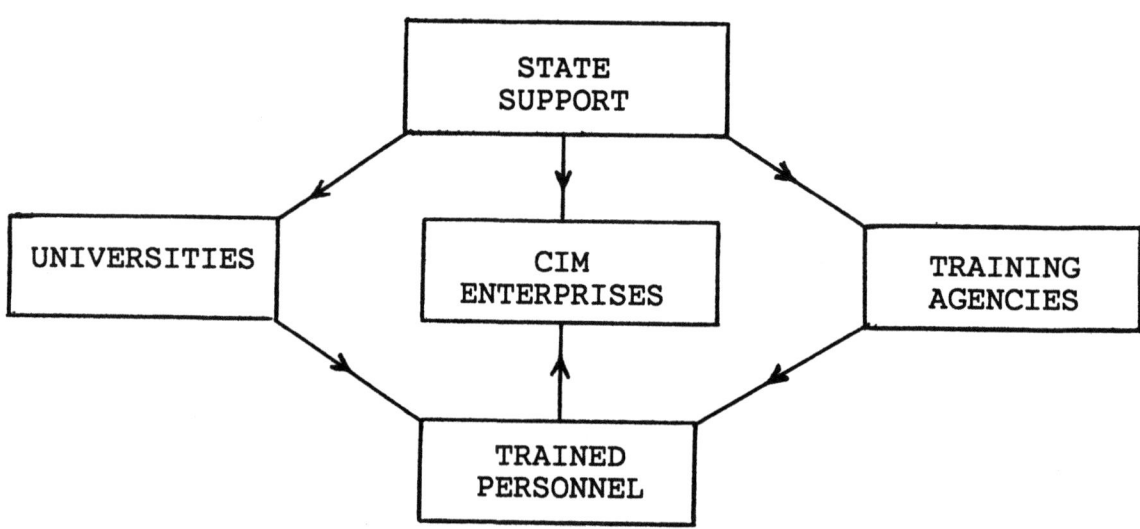

Fig. 2 Support System for CIM Enterprises

AMT IN IRISH COMPANIES

By international standards the size of indigenous Irish
companies is small. Currently, there are 6,000 small to
medium-sized enterprises in the country with employment
level between 20 and 150 people and annual turnover in the
range of 0.5m. to 5m. These companies are often
relatively young with high proportion of owner managers.
They supply products in the mechanical, electrical,
electronic, printing, packaging and other areas. Many are
short on management resources, with little middle
management structure between top management and shop
floor. Pressure on the tight management team prevents
strategic issues outside immediate financial and
production areas being confronted. The inability to
tackle strategic issues such as applying new technology
and product and service development can be a blockage to

the growth of such companies. For the long term strength of the indigenous Irish industrial base to develop, SMEs must be encouraged to grow through the use of appropriate advanced manufacturing techniques.

The Irish manufacturing industry has no option but to enhance competitiveness by facing the challenge of AMT if it is to survive and prosper. However, in times of high unemployment, use of AMT is seen by many as a threat to jobs. Implementation of AMT, by definition, does reduce the need for people. However, many of the tasks eliminated are unpleasant, tedious and sometimes dangerous. The positive effect which is most important for Ireland, of employing AMT is that it can enable companies not only to remain in business but also to expand; thereby creating more jobs in manufacture and associated services. Many of these new jobs are highly skilled. Most Irish companies are relatively new and in the process of developing. Most of these are also mainly involved in the so-called new technology products and services. Being new and small gives them flexibility to adapt to change.

For the majority of these companies automation has been a gradual process with a great deal more emphasis on improved manufacturing efficiency and productivity rather than on the reduction of workforce. Application of AMT is presenting a wide range of opportunities for providers of AMT systems and services. In recent years, a number of very important companies of overseas origin have established themselves in Ireland. In addition, there are a growing number of indigenous companies which have emerged. Many of these companies have developed to compete nationally as well as internationally. Until recent past almost all these companies started to take advantage of AMT in a piecemeal way. It is clear that a substantial proportion of the more progressive companies were well aware of AMT and many have made good progress with the introduction of new technology; some companies were still planning to introduce AMT. However, very few were in a position to consider implementation of CIM, purely due to the lack of sufficient CIM functions being present in computerised form.

THE INITIATIVES

The majority of the larger manufacturing industries in Ireland who use new technological developments towards enhancing productivity and competitiveness are of overseas origin. Majority of indigenous companies, however, are the suppliers of AMT systems and services. As these larger companies gradually implement computerised AMT/CIM technologies they would look for suppliers who can conform with their requirment with flexibility at the same time maintaining quality of highest standard.

The application of AMT in Irish industry started to take off at an increasing rate in late 80s. The main trend has been for companies to develop pockets of automation which gradually linked with automated materials handling and computerised materials control. This ongoing process lead most of the companies to endeavour to develop sufficient in-company capability or know-how to effectively utilise the technology and also to deal with the suppliers of automation related services and equipment. In this regard companies emphasised their need for multi-disciplinary engineers and technicians who can handle this type of work competently. Industries were keen to see that the Higher Education establishments took a vital role in relation to this type of training and also constitute an important source of specialised technical support to supplement their own resources. However, whilst Higher Education establishments were very active in various R & D projects in co-operation with a number of companies, it was difficult for the key individuals to contribute effectively as far as general training was concerned.

At this point in time (1986) it was felt that a concerted effort was necessary to be taken at National level to develop an indigenous capability for the provision of AMT systems and services. In this way all the companies would have access to such capabilities which in turn should promote implementation of new technology in an increasing number of companies. The target industries belonged primarily to three key areas, namely, (i) Electronics and other Manufacturing, (ii) Toolmaking and (iii) Plastics. Thus, decisions were made to support all these industries through new technology initiatives (see Fig. 3).

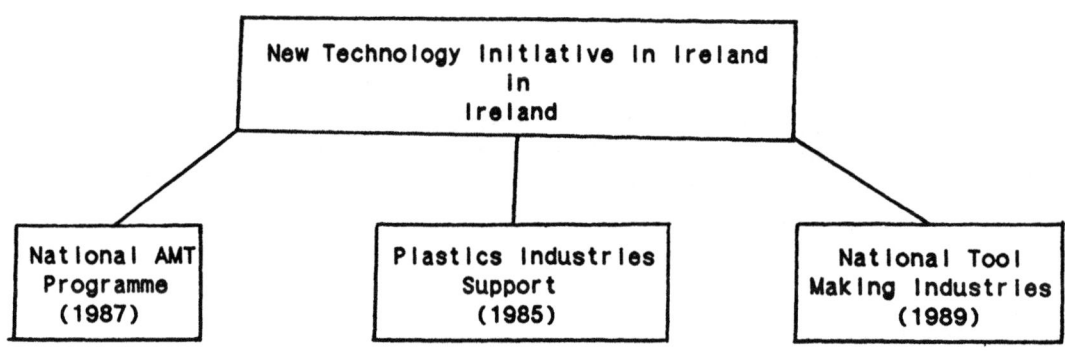

Fig. 3 Three Principal Initiatives taken in Ireland.

National AMT Programme

Following extensive discussions and consultation process between various authorities the National AMT Programme was initiated in 1987. The programme and its technical resources are co-ordinated by EOLAS, the Irish Science and

100

Technology Agency, and funded by the Department of Industry and Commerce. The primary object of this programme is to put Advanced Technology at the disposal of Irish manufacturing industries in order for these to be competitive through manufacturing excelience. The stipulated strategy was primarily

(i)   to educate companies in the potential, application and operation of AMT, and

(ii)   to transfer the technology to industry through centres of excellence in AMT and through the industrial support agencies.

Initially, four AMT research units were established, one each at University College, Dublin, Trinity College Dublin, University College Galway and University of Limerick (the then NIHEL) as shown in Fig. 4.

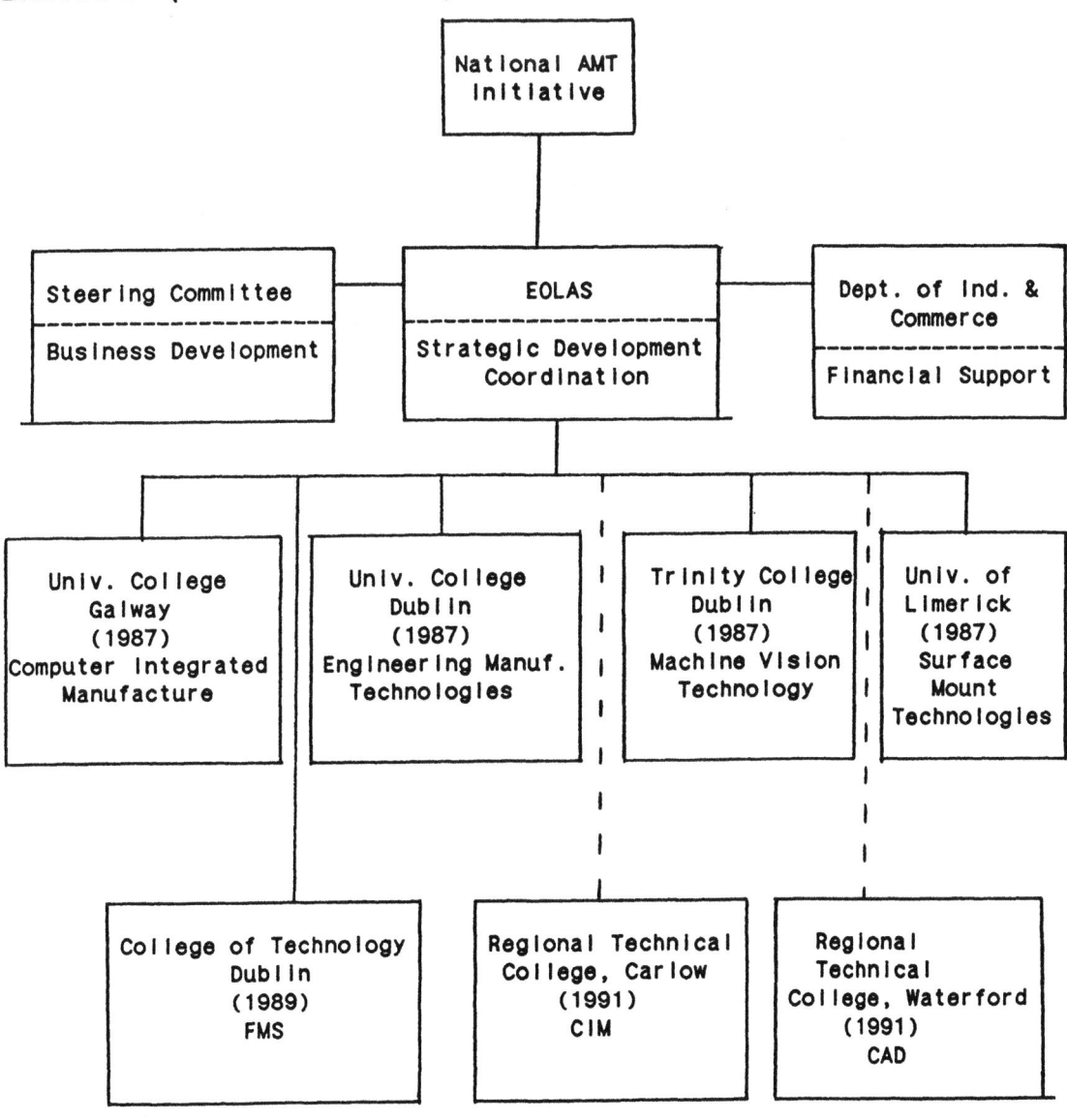

Fig. 4 AMT Programme and its constituent units.

The main reason for setting up these research units at these Universities is that relevant R & D work was already ongoing and thus no major expenditure in terms of space, manpower and capital equipment had to be incurred. Only modest support was needed to boost their activities and to give a more focussed direction.

It was intended that these AMT units would provide dual service of training professional engineers and of undertaking R & D work related to the application of AMT. Over the past few years the National AMT Programmes have been market driven and most of the efforts have been directed towards consultancy/contract research and AMT project activities. This resulted in training of limited number of engineering professionals in very narrow and specific aspects of AMT pertinent to the task they were undertaking. The desired pool of professionals trained in general aspects of AMT and its implementation has not materialised. This is mainly unintentional and can be attributed to inadequate resource being made available. Although the budget for the National AMT Programme has been increased by eightfold in just three years, much more is needed in order to make it more effective. Recently, another AMT Unit has been established and two further units are planned to be established in 1991 (See Fig. 4).

Despite the budgetary constraint the National AMT Programme, through its research units, has managed to make significant contribution in assisting a large number of manufacturing companies in connection with implementation of new technology. For example, in technology development and transfer some 25 multinational and 20 indigenous companies were helped. AMT units have also undertook consultancy activities during this period for around 70 multinational and 60 indigenous companies. It has to be stressed, however, that in majority of these cases the activity was comcerned with particular aspects of AMT, e.g. JIT, shopfloor control, workcell automation, automated inspection of PCB, surface mount technology, etc.

Through the CIM Research Unit (CIMRU) at Galway, the National AMT Programme has made some notable advance on software development including computer simulation. However, industrial involvement on the application of CIM strategy has continued to be few and far between, and limited to a handful of multinationals. Irish indigenous industries (SME's) are still catching up on the application of some form of AMT. At present most companies' involvement in AMT is such that CIM strategy would possibly show very little tangible result, at least that is the thinking of most SME's. However, things are changing very rapidly in Irish industrial scene and with increasing awareness programme and training support even smaller of the SME's will not hesitate from investing in AMT and CIM.

IDA Toolmaking Initiative

In order to foster use of new technology in tool-making and precision engineering industries a new initiative has been launched in late 1989. The primary objective of this initiative is to encourage implementation of AMT in tool-making industries by providing short term awareness and long term training programmes for personnel in these enterprises. The Centre is based at RTC, Sligo, to serve the tool-making industries which are mostly located in that region.

Apart from these specific AMT initiatives, successive Governments provided generous tax incentives to companies by way of encouraging them to invest in AMT related capital equipment. Most SME's have taken up the schemes to improve their productivity and, thereby, competitiveness. However, much more needs to be done if the companies are to be enticed into taking fullest advantage of AMT/CIM.

CONCLUSIONS

Vast majority of companies in Ireland are SME's, especially the indigenous ones. A few multinational companies are comparatively larger enterprises and most of these have made significant investment in AMT and CIM strategy. In recent years increasing number of SME's also have invested in various aspects of AMT.

The National AMT Progarmme has made substantial progress in terms of encouraging implementation of AMT in SME's in Ireland, both multinationals and indigenous ones, through the provision of limited training support, R & D, project management and consultancy services.

However, much more support is needed for training of engineering professionals in all aspects of AMT and CIM. Market-driven training and R & D may be satisfactory in the short term, but for meeting the long term needs of companies investing in AMT and CIM, a pool of trained personnel at all levels is vitally important.

State initiatives alone cannot be the solution. Only concerted and joint efforts by the State, industrial sectors and Higher Education establishments will make any such initiative really effective in promoting implementation of AMT/CIM systems towards enhanced competitiveness.

BIBLIOGRAPHY

1. O'Sullivan, D., `Manufacturing Systems Design in a CIM Environment', Proceedings Sixth IMC Conference on Advanced Manufacturing Technology, p.202, 1989.

2. Naim, M. and Schott, E. `The Implementation of FMS/FMC - The Vendor's Experience'. Proceedings Sixth IMC Conference on Advanced Manufacturing Technology, p.529, 1989.

3. Brodner, P. `Optioms for CIM: Unmanned Factory' Versus Skill based manufacturing; Computer Integrated Manufacturing Systems, Vol.1, No. 2, 1988.

4. McGuigan, K. and Yan, J. Optimisation of CIM Systems, from CIM to HIM to CIM; Proceedings Sixth IMC Conference on Advanced Manufacturing Technology, p.604, 1989.

5. Hutchin, C.E., `Manufacturing to the year 2000: we have the technology, do we have the people?' Proceedings of the Factory 2000 Conference. pp. 77-79, 1988.

6. Cuddy, J. `AMT Update', AMT (Ireland) Vol. 2, No. 13, p.21-21,1989

7. Burrow, G., `National AMT Programme Proposal', Internal Report of NBST (Ireland) 1985.

8. DTI (UK) Report 'Manufacturing into the late 1990's HMSO, 1989.

9. Carey, B. 'AMT - The Appliance of Science', Industry (Ireland), P. 16-18, January/February, 1990.

10. European Commission for Europe, Education of Training in the E.C. Guidelines for the Medium Term: 1989-1992, July 1989.

11. Rhumberg, M. and Alber, A. 'The Human Element: Its Impact on the Productivity of Advanced Batch Manufacturing Systems, J. Manuf. Syst., Vo. 1 P. 43-52 1982.

12. Kelly, T. 'MRPII The Price of Ignorance', Comput. Manuf., June 1989.

13. Bessant, J., 'The Integration Barrier: Problems, in the Implementation of Advanced Manufacturing Technology: Robolica, Vol. 3, No. 2, P. 97-103, 1985.

THE BENEFIT OR THE DOUBT

A Strategy for Knowledge Transfer on (ESPRIT-)CIM
directed at SME's

Henk Bolk & Martin van Manen

InterVisie Consultants in Management and Strategy,

1 Introduction

An important aspect of the policy of national governments
and the EC is to stimulate the development and dissemmi-
nation of (knowledge on) new technologies. The policy aims
in themselves may be clear, the effects and ways to reach
them are not always.
    By just spending money on projects that are techno-
logically impressive, accountable, sensible, reliable, etc
(when it comes to the technical expertise of the partners
and the likelyhood of reaching a result in terms of product
development) one it not _yet_ complying to the impressively
encompassing goal to "strengthen the national (or European)
industry"!
    A question that is connected to this has to do with
opinions about small to medium sized entreprises (SME's).
Since in most EC countries up to 80 % of the industry consists of SME's and only recently the budget going to SME-
participants in ESPRIT-projects rose to 20% (Zimmermann,
1989), one may wonder if _in fact_ something is the matter
with the convictions about SME's in circles of the EC and
of the larger IT-industy. Does one "grant the small entreprise some of the _benefits_ OR does one "_doubt_ about the
importance of the smaller entreprise as such", OR is something else the matter?
    The difficulties around technology transfer and know-
ledge transfer show in at least two ways:
1.
Knowledge and technology developed in funded projects, like
in ESPRIT's CIM-domain, is apparently only spread on a
piecemeal basis to a much wider audience. Not because partners are unwilling to inform others about their results,
but simply because there appears to be some sort of 'gap'
that is not understood well. A gap between partners and
other members of an 'inner project- or subject-circle' on
the one hand and the impressive number of organisations
that do not participate in national or EC-programmes, that
do not see the relevance of specific topics, that are
making a profit but not (yet?) with the help of the most
advanced technology, and that (for that matter) are perhaps
to be classified as 'late adapters'.
2.
The implementation of technology, advanced CIM-systems, appears to be an extremely complex issue. Project managers
recently questioned indicated that the organisational and
administrative obstacles often were greater than estimated.
Two kinds of obstacles arise:

- Decision processes are mostly based upon a conviction by the non-technicians that automation's main focuss and legitimation is that "it replaces routine-like labour". This results in unaccountable decisions and false expectations about impact and scope of projects.
- Although after careful designing the technology itself might appear to be clearcut, in having to implement it in a concrete social-organisational environment (always specific/unique <u>and</u> a complexity in its own) the problems were almost always leading to rising costs, and deadlines would be exceeded (see for instance Computable, 22 september 1989).

What is the matter?

In this paper we will offer some concept which allow us to better understand the situation.

Take a look at this picture and study it for a few minutes before you read on.

What is it that you see? An elderly woman? Or do you see a younger woman looking slightly to her right side? Or is it an elephant that you see? The picture can illustrate in two ways what this paper means to convey.
In the first place the picture means to illustrate that we are asking of the reader to change his point of view on organisations, to understand that more than one interpretation may be useful, that a 'radically' different point of view may also give an interesting interpretation.
In the second place the picture means to illustrate the mere fact that once people have made an interpretation about some fact, the same 'fact' may result in completely different definitions that, once established, are sometimes very hard to change. This will be used to illustrate that it is not always just one definition that is 'true'; several interpretations/definitions may be 'true' at the same time. This underlines the importance of reckognising that knowledge transfer is much more than just "telling the other person <u>how</u> the problem/the technology/etc looks like, <u>what</u> to do and <u>how</u> to do it".

2 "The" SME?

Of course larger companies in the IT-industry are capable of investing larger sums of money into R&D concerning CIM-technology than SME's. Also the investments involved with implementing (even small scale) CIM-technology are relatively easier to afford for larger companies. The possibility to appoint a group of people to a dedicated project is greater in larger companies.

Perhaps in this sense a major distinction between the larger and smaller companies is that the first are able and willing to broadly stretch their investment and their own R&D efforts, not only on product development but also on production development, while the latter simply are not.

We have seen many entrepreneurs in SME's and were always impressed by their involvement in running their (!) company. Perhaps we are allowed to draw some lessons from our observations.

In our view a major characteristic of many SME's is that they are driven by a small, dedicated group of people who are familiar to the maximum extent with the production process and the range of products being manufactured. Moreover, the general conviction in SME's is that upkeeping the (mostly person-to-person) relationship with the present range of customers is of utmost importance. Finally, perhaps, a general characteristic would be that management is not exactly driven by long term thinking. One of the most important elements of the strategy (if it is outspoken at all) is that the company should try to stabilize and only gradually extend the present range of customers since it is on its basis that the company became what it is to-day.

Willingness to take risks (in this respect!) is low and rightfully.

This is about all we can think of when trying to describe some general characteristics of SME's. Apart from these there are none, in our view. Markets appear to be unique for each company, internal social relationships are unique for each company, the technological state-of-the-art is unique, the knowledge-base of each company is unique, etc. If the reader does agree, then we must also realize that this is hardly a solid basis for developing a common and generally applicable strategy for knowledge transfer and technology transfer "for SME's". The policy denominators or anchor-points simply seem to be missing.

Unfortunately, however, the efforts of several national governments and sometimes the EC also, are directed at inputting money into already well organised institutions thereby hoping to increase the effectiveness of knowledge transfer and technology transfer without any concrete guarantees. Another strategy being applied, sometimes in conjunction with the previous one, aims at strenghthening only certain sectors of industry, while no strict dis-tinction can in fact be justified given the immense di-versities existing in the realm of SME's.

Key question then would be to find out what are the main forces 'driving' technological developments in SME's and elsewhere, and to decide which strategy would help in supporting technology awareness and technology transfer. Let us take a short detour to management science, in order to find some answers.

3 Does management science make any sense here?

Please give us the benefit of the doubt and let us explain why management science perhaps is able to contribute some valuable concepts to support the efforts to understand better what is blocking the awareness, disemmination and transfer of technology and knowledge.

3.1 Organisation: the old story

The practical use of organisational and management concepts in technically oriented industrial settings (and also elsewhere) can be summarized as follows:

1. Organisations are seen as goal-oriented 'thing-like' entities experiencing influences from different angles in their environment (Newman, 1987).

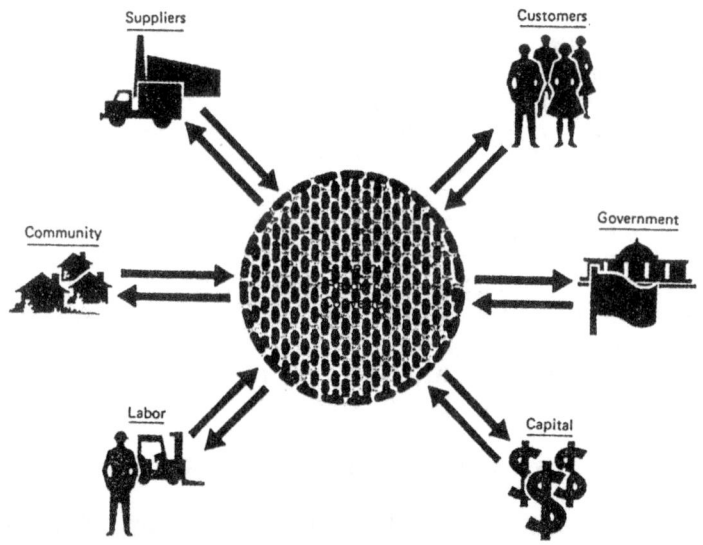

2. Departments and tasks are to be structured and predesigned to a large extent, for instance in tree-like images (Newman, 1987):

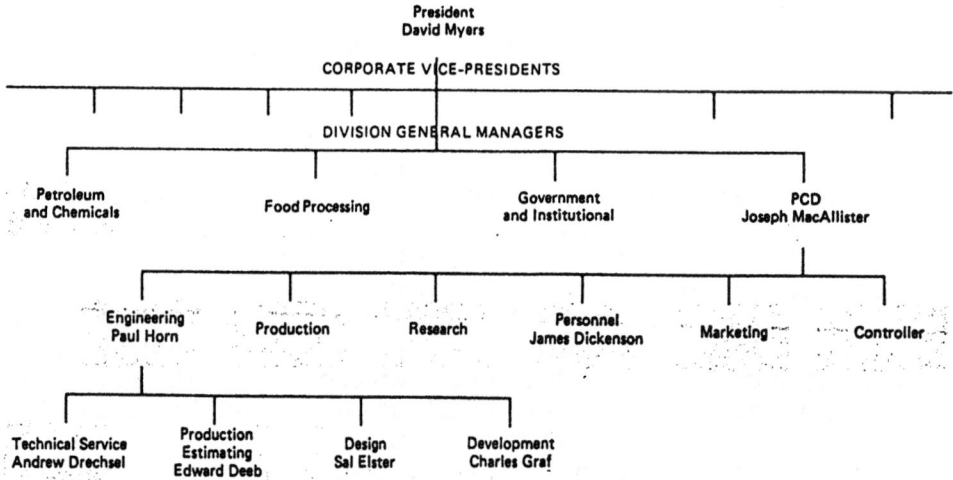

3. The optimal, one-best organisational build-up, fitting the demands of technical design, is looked for.

4. The optimization of throughput time, reducing stocks, minimising costs, increasing machine flexibility, etc., is seen as the prime motivator.

5. Control, detailed insight in and influence on every aspect of the organization's processes, are to be achieved.

All of this mostly based upon the conviction that for each situation or problem (defined (!) as technical) ultimately a best solution will be found.

3.2 Organizing: the new story

Why using the terms "old story" and "new story" (cf. Berry, 1985)?
    Well, one could bring foreward that the viewpoint described below simply is of later date than the one described above. Another reason might be that "old" and "new" distinguish between two distinct, discongruent and fundamentally incompatible viewpoints. The essential difference between the two is described below. And it is here that we ask of the reader to change from one interpretation to the other, exactly as when looking at the "ladies' picture" in the first section. The viewpoint below cannot be matched with (or, for that matter, made to be in line with, in extension of, brought in congruency with) the one described above!
    In the "new story" organisations are no longer perceived as unidimensional, uni-rational and goal-oriented structures that need to fit technological/commercial/etc objectives that in turn are decided upon in an abstract organisational region referred to as "the management".
    Organisations are seen much more like vehicles developed and choosen by individuals to gear the fulfillment of individual, diverse goals. The nature of the reality of the organisation, therefore, can never be objectively decided upon or fixed since there are always different individuals involved. There is just not one "true" definition of any organisation.
    Definitions of organisations and 'their' goals are always bounded to diverse individual interests, aims, objectives. When looking at a production firm, 'the' organisation is non-existant. What we do see within companies are mixed and mixing 'organizing processes'; processes that, moreover, may well cross the legal bariers of an entreprise (Weick, 1979; Bolk, 1989).
    Circles of management scientist nowadays tend to define these organizing frameworks as 'configurations', a term that simply fits the present managements scientists' needs. Also a term that came in handy as an alternative for worn-out and confusing concepts like 'group', 'network', etc. One of the authors of this paper defined configurations as "characterized by a matching between a relatively stable interaction pattern ('who') and shared cognitions ('what' or definitions of reality). The concept can therefore be considered as basic for the understanding of social processes" (Bolk, 1989). Let us translate this.
    People interact, and through this interaction people develop an understanding of what is going on outside them. It has been proven through numerous experiments that if one

does not interact with other people, one does not even notice certain objects, processes, incidents, etc. (Especially anthropology came with major contributions to support this insight).

Thus, interaction is vital for understanding and knowing. The scope of this paper does not allow us to go into detail here, therefore we advice the reader strongly to read Karl Weick on this (Weick, 1979). The knowledge that is developed through and in interacting with other people is always partial. It depends, of course, on whom one interacts with what the exact definition of reality will become like. Interaction with changing individuals mostly means changing content, changing objects, but most importantly changing particular (relevant) definitions of reality. Therefore, interaction does not only determine ones definitions, also the existing and adhered to definitions sort of predetermine the individuals with whom one will be interacting in the future. If relatively stable interaction patterns within or between organisations have resulted in definitions of the (organisational, societal, individual, family, etc) reality, <u>then</u> we speak of configurations.

Note the fundamental difference with 'group' or even 'organisation'. Configurations can cut through organisational boundaries simply because interaction patterns of individuals do not stick to legal boundaries. Also, an organisation may very well consist of more than just one configuration. Even, the formal organisational structure is likely to fall apart when looking into the details of the actual configurations that the organisation is comprised of. The formal, official structure could well be a definition of reality developed only by a staff department, not completely shared by others and completely overrun by (what we can now call communication <u>defined</u> as) 'informal' communication!

Mintzberg was one of the first authors to sketch possible 'configurations' within organisations that in his view formed the actual core to understand the structuring of work, production, etc. Official tree-like images were abandoned (Mintzberg, 1985).

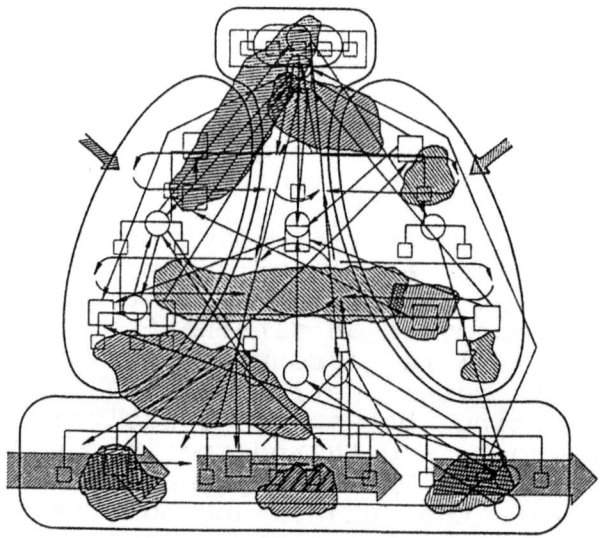

InterVisie, in cooperation with researchers from the Graduate School of Management, developed measurement instruments to thoroughly analyse the configurations within organisations. Earlier we reported on these instruments and there results (Bolk and Van Manen, 1989). Images like the one presented below are the result. Obviously the 'official' structure of the organisation can be seen to be falling apart. The instrument, meanwhile, has been applied to a great variety of organisational settings in order to get a deeper insight in underlying processes, for instance processes blocking further development of an organisation or blocking cultural changes.

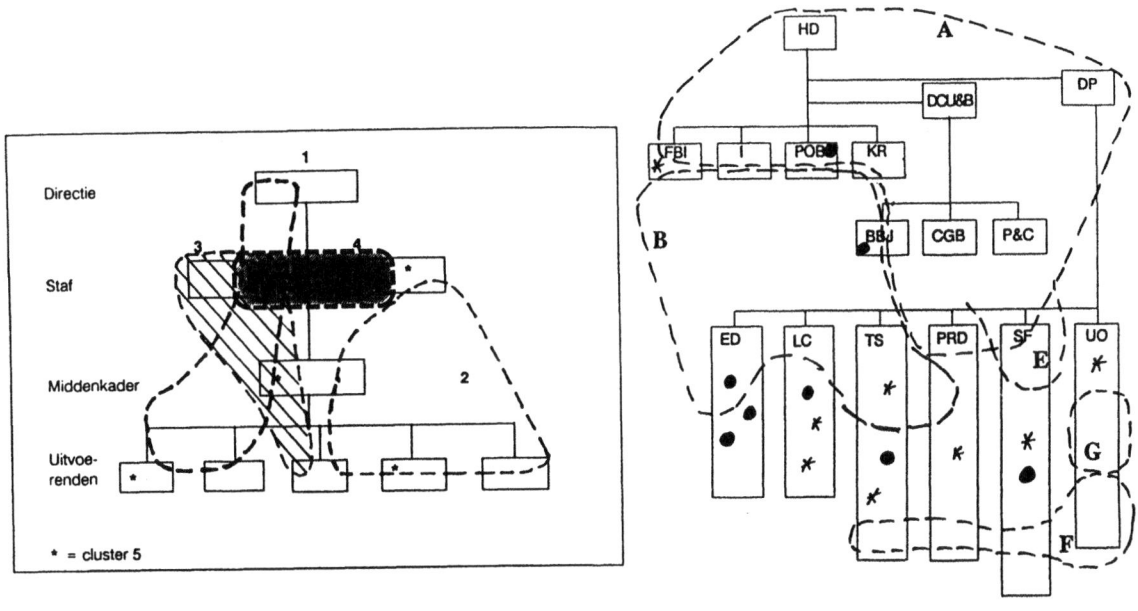

One of the authors of this paper continued the line of reasoning to the inter-organisational field and also tried to provide for longitudinal analysis and presentation of data. After having analysed data collected in organisational settings he came up with images like the one given below, sketching the gradual integration or desintegration of configurations (Bolk, 1989).

    At three succeeding moments in time data were collected on the basis of which the configurations (with respect to the subject under study) could be sketched. Below you will be able to get an impression on the sort of results that can be obtained. Especially the longitudinal changes within and between configurations can now be detected in a detailed way.

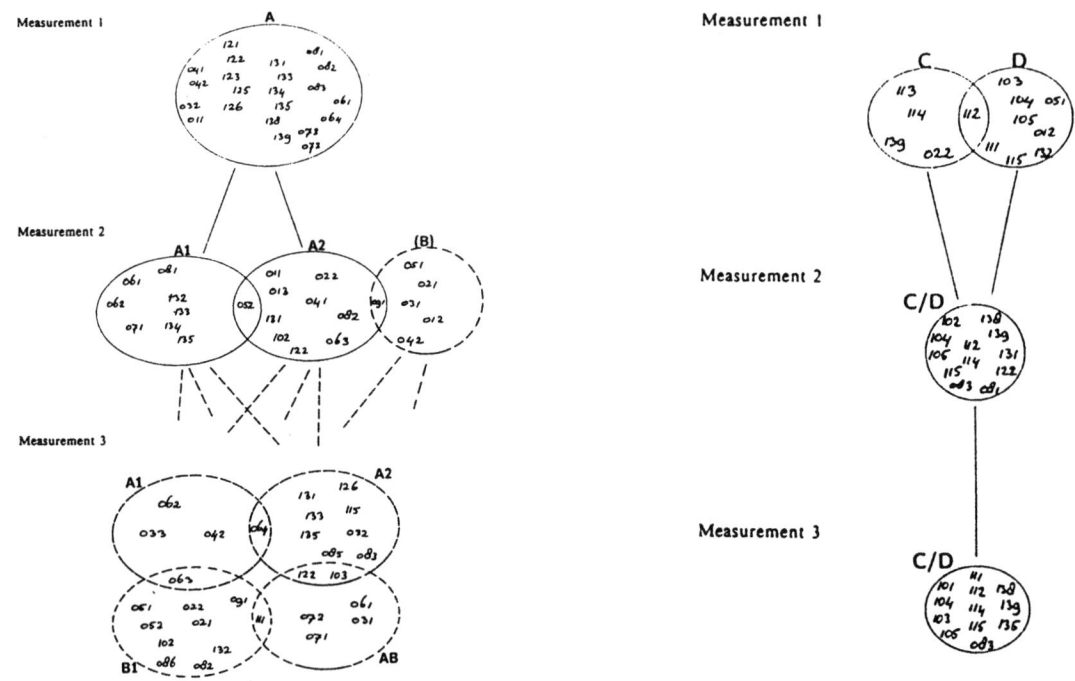

(The numbers in the diagrams indicate respondents codes, showing the inclusion of each respondent in configurations in the succeeding measurements. Note also that in the diagram nothing is indicated on the 'content' of the definitions of reality involved)

Letting loose an image of the organisation as an entity, that ultimately can be controlled and directed in all 'its' aspects, has some consequences. Management as the goal-oriented 'guide', 'expert' and 'controller' of the companies' course is being replaced by an image of management as the facilitator of organising processes becoming matched, intertwined, leading to new directions, new markets, innovations, etc. Management is suddenly to be seen as the binding element for various configurations to be related, to be matched, to be interacting, in order to develop a multiplicity of definitions to become a really effective organisation. Bolk studied the effects of both strategies is his book that was mentioned already.

MAMA-LOGICA: MAnagement by MAtching

PAPA-LOGICA: PAthology by PAtching

(Source: Broekstra, 1989)

112

The process from PAPA-logic to MAMA-logic is taking place at the moment. Due to growing insight into the functioning of organisations as seen in line with 'the new story', more and more questions are asked to consultants and researchers to provide for concrete instruments to facilitate management of change and technology management.

Not only research in the field of management science has come up with a broad range of proof underlining the validity of the 'new story'. Philosophical developments contribute to it (Bolk, 1989) and also the development of new instruments for research and analysis have resulted in more qualitative data underlining the value of the new story.

If the reader wants to know more about this new story he is advised to read the literature mentioned in the above paragraph. Let us no longer waste time on what might, at first glance, seem to be a purely academic issue and continue with what this recently described management theory has to say about knowledge and technology and see if it contributes to our understanding of the problem described in the beginning of our paper.

3.3 On knowledge and technology and their dissemmination

Seen in the light of the configuration-theory briefly discussed above, the meaning of 'technology' is changing fundamentally.

In line with the old story we were used to see technology as shapable but at the same time as indebatable matter. A byte is a byte, an open systems architecture is the most optimal standard for its purposes, a machining centre has its dimensions not for nothing, a programming language has been thoroughly selected, logistical problems can be solved only in objectively optimal fashion, etc.

Mostly the meaning of the word 'technology' is not defined in itself but its meaning is assumed to be derived from what technology 'does', e.g. in a factory. This whole line of reasoning is closely affiliated to the old story which approaches technology as something that is 'simply there' as an objective fact, to be discovered, to be invented. Technology, in other words, is seen as completely excluded from social (let alone psychological) factors.

Schwarz seriously questions and analyses how this idea may have become the dominant viewpoint today. "In our culture we live and think technology", "technical solutions define our problems, the advance of technology has become decisive for what we do. (...) Technology is a machine - an artefact that we can use or misuse. This image denies the ideological character of technological development" (Schwarz, 1988).

When Schwarz uses the term 'ideology' we could also use the term 'configuration'. Included in the use of technology should be the understanding of the configurational character of technology. Every knowledge, no matter what it is about, is the outcome of defining processes, of interactions between human beings; not of material, not of apparatus, not of the artefacts themselves. We would therefore like to say that technology represents nothing more or less than (more or less complex) appointments shared by a particular group on how to view and understand and also rule a specific reality. Matter is shaped into a form that

can contribute to control, to rule, to (in turn) shape itself. Therefore decisions, choices, are embedded in and
unbreakably part of technology. Technology, really, is the
result of interaction processes between people.

This angle of incidence allows us to understand
technical solutions and machinery as socially determined
and decided upon. It opens up our minds to the technological choices that we perhaps are no longer aware of;
to the 'prejudices', even, of the people who designed advanced applications, who are after all just interacting
human beings themselves, manipulating matter...

Why do we address such a seemingly fundamental issue?
Why do we want to rake up such a philosophical debate?
We want to make clear that effective knowledge transfer
about ESPRIT- and other R&D-projects involves a thorough
understanding of underlying configurations that are linked
to opinions, convictions, viewpoints of different 'parties'
involved. Also this understanding is necessary to grasp the
essence of differences in the degree of susceptibility for
new technologies in various sectors of industry and in
companies of different size. Let us sketch a diagram that
may illustrate the gap that much too often exists between
vendors and developers of new technology on the one side
and for instance SME's not involved in the developement on
the other.

```
  ╭────────────────────╮            ╭────────────────────╮
  │  configuration I    │            │  configuration II   │
  │        +            │            │        +            │
  │  its technology     │            │  its day-to-day reality │
  ╰────────────────────╯            ╰────────────────────╯
```

Configuration I may well represent an ESPRIT-project and
the partners involved. Configuration II may well represent
a particular small entreprise. The diagram illustrates the
gap between the knowledge base of both. There simply is no
interaction that can provide for the two realities to come
together. It therefore makes no sense to try to describe
the knowledge being developed in configuration I in an
'objectively understandable way' in order to cross the gap
to every person in the SME. The interaction needed to
incorporate the small entreprise is not organised properly.
However, it is still the most commonly tried policy to
'tell the world', to 'publish results', to 'convince with
the facts', to 'designate government funding for SME's to
develop their own R&D policy', etc. Roughly, the aim of
present knowledge transfer methodologies is therefore to
generalize project results and sort of spread the news and
socalled 'facts' all over the place. How unfortunate that
the greater part of SME's does not respond, does stick to
their present way of doing business, does cling to the
effectiveness of present client relationships, does not
seem to be interested in participating in ESPRIT-projects.

Sticking to the idea of configurations playing an important role here, we would like to say that perhaps what is missing is the understanding of the need for using intermediary configurations to gradually bridge the gap described. In a diagram this would look like this:

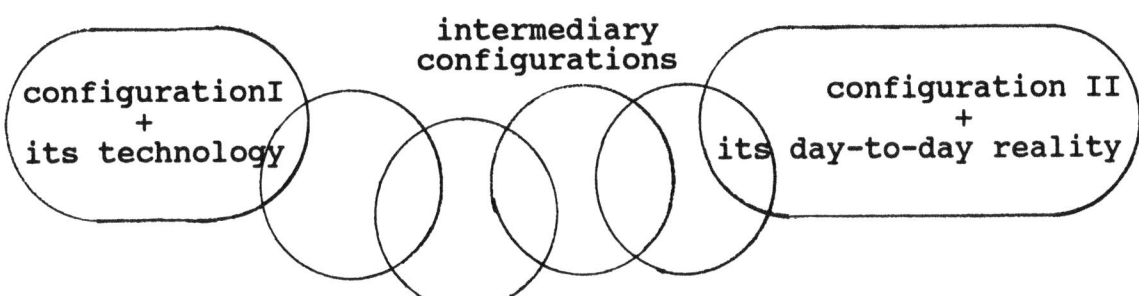

The diagram illustrates the sequence or chain of overlapping configurations needed to bridge the gap between both configurations. Put somewhat differently: one will have to transfer knowledge and insights and new technology by carefully stepping from one definition of reality to the other. In other words: by carefully reorganizing existing interaction processes through which redefinitions can take place. The gap cannot be bridged in just one step. Discongruency between the two realities is simply too great: worlds are too far apart. The overlappings in definitions existing between each sequential intermediary configuration must be understood and can be exploited in a fundamentally different view on technology transfer and knowledge transfer.

The implementation of advanced (CIM-)technology is not just a matter of learning people in SME's things they did not know before. Many vendors of systems and implementation consultants have experienced this in the past. Many vendors and consultants have also been able to massively neglect the issue by concentrating firstly on system design, hoping that organisational obstacles could be overcome once a working system could be presented to a customer or potential buyer.

Implementation, however, is really an issue of eroding and changing definitions of reality, of carefully constructed convictions based on dominant interactions with a particular (small) group of clients, employees and fellow entrepreneurs; also based on years of input, impulses, interactions on the basis of which and through which a particular organisation has become what it is. This idea of 'implementation' and also of knowledge transfer is something quite different than just 'system design' or 'reporting on project results'.

In the next paragraph we will briefly sketch the outline for an implementation tool that a group of researchers (in which InterVisie participates) wants to develop in line with the notions presented above. Also some strategic aims for the ESPRIT-programme and for partners within ESPRIT-projects are sketched.

4 Some practical implications

4.1 A tool for implementing CIM

In the diagram given below we present a sequence of steps to be taken in a proper implementation traject, incorporating the insights of the new story. The diagram is taken from our ESPRIT-proposal 5349, meant at the further development of a tool to facilitate optimal implementation of CIM in SME's. Several of the steps and tools of which it actually consists are a direct reference to the need to facilitate the interaction between and the integration of configurations. Configurations, sometimes discongruent, present within a company, and also different configurations present when looking at the relationship between the vendor/designers and the company. One cannot foresee which essential discongruencies will show. Therefore initial analysis should focuss on the actual organisational situation. One also cannot foresee which configurations will be relevant when actually entering the phase of system design and development. Therefore simulation techniques and knowledge based tools are used to experiment with alternatives amongst others in order to find out how in actual situations the various configurations match or clash.

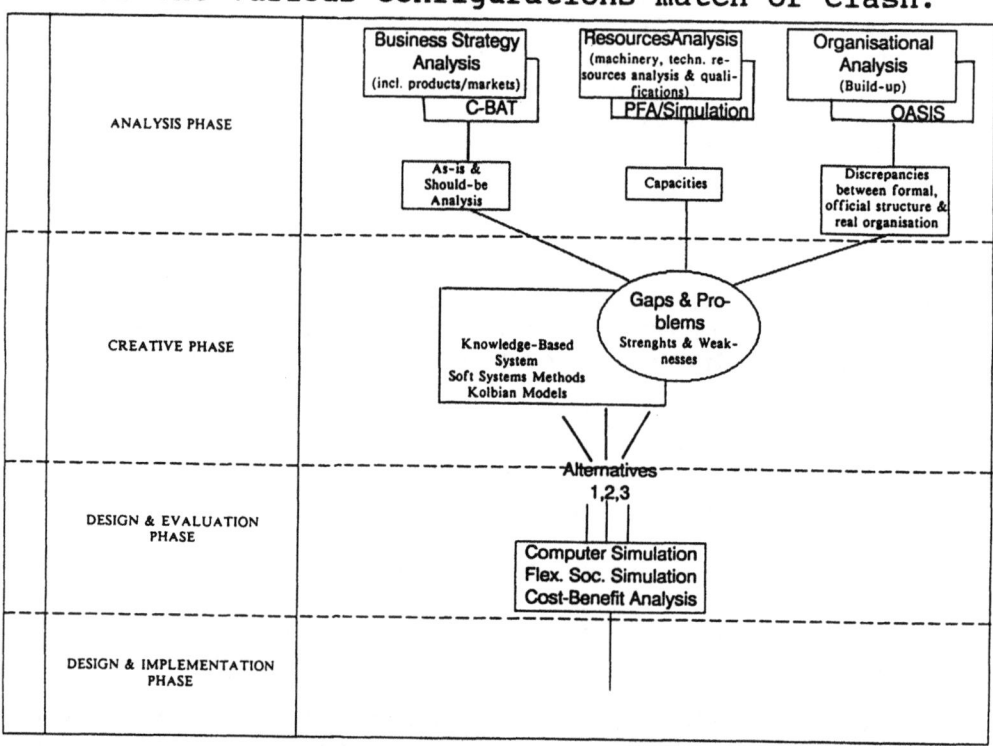

4.2 ESPRIT's and CIM Europe's strategic challenges

At the moment of writing this paper we could not give too much details on the further development of the strategy for ESPRIT and CIM-Europe, due to the fact that our work for CIM-Europe had just started. What we can do here is formulate some general guidelines on the basis of the insight from the new story sketched above. At the conference more details will be provided.

When looking at the strategic aims of the ESPRIT-programme, it is remarkable to see that, when presenting ones proposal, a great deal of attention is asked for the chapter "Industrial Relevance" and for the issue of "what will be done with project results". No doubt this part of a proposal is used to get an insight into the estimation by the partners of the importance of their work in order to have this compared with the general outline of ESPRIT's tasks.

However, when entering a project, the attention of the partners is completely absorbed by the work to be done, by the content of the project itself. It looks like one is hoping for the results to generate important spin-off without developing a matching _strategy_ for knowledge transfer and technology transfer. Such a strategy must naturally differ for each project and for each 'outsider' to be adressed. Why? Because one should be aware of many different configurations to be involved. Adressing 'the' SME, for instance, by writing reports or project and product documentation will not do. The gap described earlier will prevent knowledge from effectively spreading. It is necessary to involve intermediary configurations, so to say 'on the route to the market'. Operationally some of the elements to guarantee more effective knowledge transfer might be:
- The careful selection of relevant intermediary configurations requires specific attention and supervision from ESPRIT's project officers.
- Actually, it should perhaps be a specific work package in each project, designated to commercial or strategic experts to be involved from the start.
- The involvement of existing 'intermediary' organisations or networks should be studied (for instance innovation centres of national governments, branche-organisations, chambers of commerce, etc). Preliminary progress reports could be made to fit the language used in and needs of these organisations or networks, in other words: to fit the definitions of reality of these important external configurations. Further 'translation' into a language that might more effectively reach SME's (but also larger entreprises!) will then be organized.
- A great variety of media must be used. The standardisation of publications, and the choice for a relatively small range of particular means to spread knowledge (for reasons of effectiveness and overview at central ESPRIT level), might well be counterproductive. Susceptibility for 'new' knowledge is increased if one is confronted with it through channels hitherto used for 'confirmation' of ones definition of reality, through channels part of ones original configuration.

5 Conclusion

We have tried to give a first impression of changes taking place in the field of organisational and management science. These changes, in our view, have important consequences for the way technology and knowledge are (to be) transferred effectively from system designers and system developers to the environment, especially SME's. 'The new story' has been a means to convey these changes to the reader.

Consequences have been sketched for two fields: the field of the implementation of CIM and the field of the transfer of knowledge from ESPRIT-projects.

These consequences still need to be translated in more concrete strategic guidelines and methodological tools. InterVisie is working on this together with several partners. Amongst them: Cheshire Henbury Research and Consultancy, BIBA, EXTECH, The Danish University, and by expert-participation in the CIM-Europe team. You may expect from us that we will find the proper intermediary configurations to transfer the knowledge, instruments and tools to you.

Literature

Berry, T, as cited in Lincoln, Y.S., Organizational Theory and Inquiry: The Paradigm Revolution. Sage, Beverly Hills, 1985.

Bolk, H., Organizing, Changing, Simulating. Eburon, Delft, 1989.

Bolk, H., & M. Van Manen, Management Science's Role in Computer Integrated Manufacturing: Referee, Goal Keeper or First Aid? Paper presented at the 5th CIM-Europe Conference, Athens, 1989.

Broekstra, G., Het creeren van intelligente organisaties (The Creation of Intelligent Organisations). Inaugural speech, Eburon, Delft, 1989

Mintzberg, H., The Structuring of Organisations. 1985.

Newman, P., Organisation Theory, 1987.

Schwarz, M., Technology Policy and the Technological Culture. Paper presented at the 'Leergang technologie an verantwoordelijkheid' at the University of Nijmegen, 19 january 1988.

Weick, K.E., The Social Psychology of Organising. Addison Wesley, Reading (Mass.), 1979.

Zimmermann, H., Presentation at the ESPRIT Conference 1989, ESPRIT and SME's (session of 1 december).

Activity chains: A method for identifying and evaluating key areas of integration in SME's

Authors:

Jan Frick, MSc,   TESA as/ Aalborg University (AUC)

Jens O. Riis, Prof., PhD,   Aalborg University (AUC)

Empirical studies and our own experience indicate that most CIM technology applications are technically successful. However, only a few companies have succeeded in improving their competitive position. One explanation is that industrial enterprises have not been able to relate CIM technology to the corporate strategy. Another explanation is failure to manage the organizational learning process involved in the new manufacturing technology.

The paper will address these issues by introducing the concept of "activity chain" to denote a continuous chain of activities associated with the essential tasks of an industrial enterprise, such as product development, production flow, customer orders, quality, personnel. The method of using activity chains will be related to organizational learning through a discussion of problems of getting employees involved in a technological and organizational development process. An organizational learning approach will explain how to get employees involved in the planning and implementation process. We discuss how to visualize the benefit and basic idea of the CIM technology to key persons and employees who will be affected. The results obtained are methods for helping small and medium sized industrial enterprises find answers to the following questions: Where to integrate, to which extent, and at which pace?

The paper presents some of the experiences gained from applying the concept of activity chain in three industrial enterprises as part of establishing a CIM strategy.

PROBLEMS ENCOUNTERED AND THE APPROACH ADOPTED

The prospects for using CIM technologies for improving competitiveness in industrial enterprises are generally good. Empirical studies and our own experience from small and medium sized industrial enterprises mainly in the discrete part manufacturing indicate that most CIM technology applications are technically successful. However, only a few companies have succeeded in improving their competitive position, (cf. Voss, 1988)

One explanation is that industrial enterprises have not been able to relate CIM technology to the corporate strategy. Until recently, both in literature and in the practical planning and implementation of CIM systems, the notion of integration most often has been associated with connecting several computer systems. Little notice has been made of integrating goals, plans and activities. However, a greater awareness of a broader view of integration may be noticed. To an increasingly extent it has been realized that an essential element of CIM is integration across functional lines.

We shall adopt a broad view of CIM and define it as the tying together of production philosophy and logistics by means of information technology in order to increase the company revenue.

Following this definition we shall seek answers to the following questions: Where to integrate, to which extent, and at which pace? In this way it will be possible to relate CIM development plans to corporate strategy and central goals.

To provide a better background for our discussion we have identified the following characteristics of integrated manufacturing:

- ⊗ Integrated manufacturing requires the contributions and active participation of several departments of an industrial enterprise; thus, the operations cut across organizational boundaries.

- ⊗ Introduction of CIM technologies implies the changing of working modes, organizational structure, individual attitudes, and qualifications; hence, a complex organizational learning process must be supported and managed.

- ⊗ Several CIM modules are available, but their interplay is difficult to describe by traditional means for visualizing the effect of integration to the employees of an industrial enterprise is a new challenge.

Thus we need methods for analyzing the interplay between functions as a means for identifying key areas for integration. The methods should be able to relate activities to the overall corporate strategy and goals.

We shall deal with managerial aspects of introducing CIM technologies by taking into account both a technological and an organizational development point of view. This emphasizes the need to discuss organizational learning issues.

ACTIVITY CHAIN

In view of the importance of integration between functions and departments we shall concern ourselves with the horizontal flow of activities. We shall introduce the concept of "activity chain" to denote a continuous chain of activities associated with the dealing with an essential task of an industrial enterprise, such as product development, production flow, customer orders. An activity chain will cut across functions and departments, as illustrated above, as opposed to traditional vertical communication between a central management unit and functions and departments.

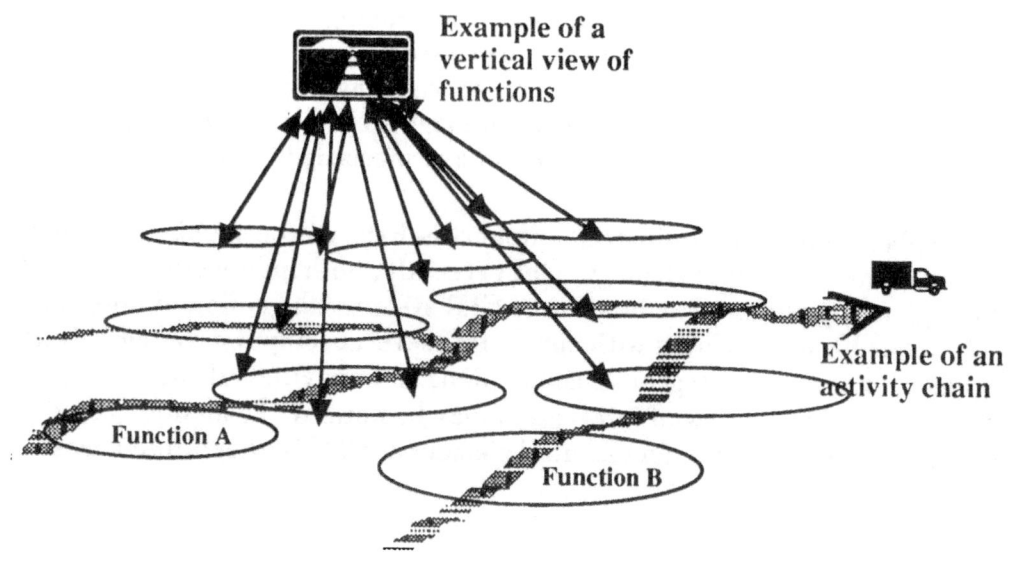

Example of a vertical view of functions

Example of an activity chain

Function A

Function B

An activity chain focuses on the horizontal flow and identifies key activity chains in a given industrial enterprise as a means for discussing where to integrate and to which extent. It is characterized by the following:

- it describes a continuous chain of activities, i.e. a flow
- it has a well defined start and end
- it will normally pass through more than one functional area of the company
- different activity chains may have the same start or the same end
- activity chains may have a tree structure
- activities in a chain may be either:
  - ⊗ processing or transportation of information ,
  - ⊗ processing or transportation of materials,
  - ⊗ decision making, or
  - ⊗ a combination of these.

By focusing on an activity chain it will be possible to disclose:

⇒ the overall through-put time for different types of orders, products etc., as well as the location of the main contributors to the through-put time

⇒ the overall consumption of resources, as well as the identification of activities using many resources

⇒ the degree of flexibility for different types of orders, products, etc., and the location in the chain of the constraining elements.

⇒ the bottlenecks along the activity chain

⇒ the need for competence development, e.g. individual and organizational learning along the activity chain

All of these items relate, in one way or the other, to the competitiveness of the company. Thus, we should be able to select the most important activity chains in view of the specific company's goals and strategy. At the same time, we obtain a means for identifying critical activities, individual or in a sequence.

Activity chains have two different main uses, namely

1. as a diagnostic tool for analyzing the present operations, strategy and goals, and

2. as a structuring tool for the development of CIM for increasing competitiveness.

In the following section we shall discuss how to work with activity chains.

STEPS IN WORKING WITH ACTIVITY CHAINS

Like any analysis, the process of describing activity chains may become quite time and resource consuming and yet not provide the desired understanding. In order to overcome this obstacle we shall suggest an adaptive approach which will allow the result of the initial analyses to be used to select the proceeding analyses.

Based on our experience we shall propose four steps in the process of working with activity chains. As may be seen from the chart below, the first rough diagnosis provides a basis for pointing to a few key activity chains to be analyzed and discussed in the second step. The third step represents a constructive phase in which new modes of planning and operations are developed and tested for key activity chains. A high degree of employee involvement is warranted in this step. Based on

121

the results of the third step, plans for implementing changes in key activity chains may be developed in step number four.

Process activities	Tools/ methods	Results
1. Rough diagnosis	Questionnaire and interviews Main characteristics of enterprise gives references and context	Selection of potential key activity chains
2. Detailed analysis and development of the key activity chains	° what is transferred to next element ° time elapsed ° use of resources, bottlenecks ° critical factors and decision points ° quality	Description of present status in key activity chains, and their connection to corporate strategy
3. Development and testing improvements in the performance of key activity chains	° group technology ° SMED ° simulation ° gaming	Suggestions for improvements in key activity chains, their feasibility tested and accepted by employees and in support of corporate strategy
4. Development of implementation plans for improving key activity chains	° project planning ° simulation ° group technology	Activity chain oriented plans for CIM development with plan resources allocated and ROI calculated

In the following we shall present four methods for carrying out the steps indicated.

1. A Priori Hypotheses of Key Activity Chains Based on Corporate Characteristics

Following the adaptive approach to analysis, it is essential quickly to be able to focus on key activity chains. The type of production mode may similarly point to important elements as a basis for selecting activity chains which should be subjected to detailed analyses.

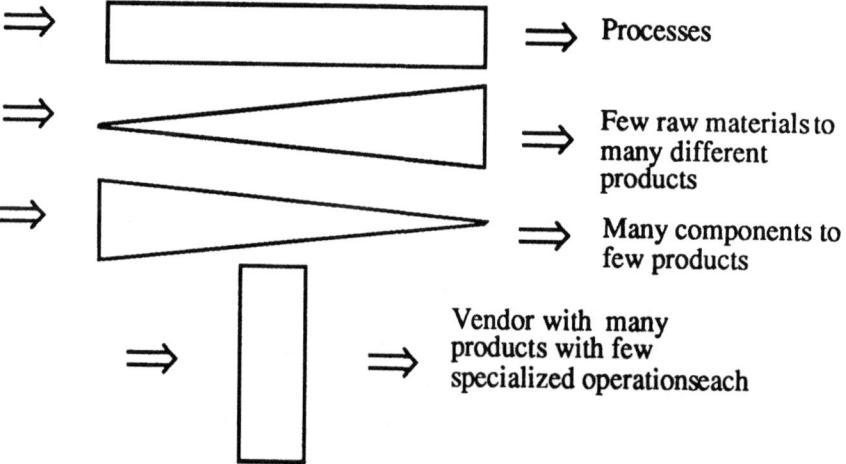

In practice, an industrial enterprise may use a combination of the above shown types. In the chart shown below we have identified some characteristics of an industrial enterprise and for each indicated potential key activity chains. A specific company may find that several of the characteristics are true, thus having several activity chains suggested as important.

Characteristic elements	Possible central activity chains
Producing enterprise	Materials flow, internal and external logistics
Information based enterprise, trading company	Information collection, Information structuring, and Information presentation
Order producing enterprise, order construction enterprise	Order information, order specification, production planning and control
Market with heavy seasonal variations	Prognoses development, master planning, stock control
Products with high costs, high demand on service, or much training involved	Making of documentation, updating of documentation
Products in large volume and low cost per piece	Prognoses development, master planning, stock control, distribution
High rate of new products introduced	Product development, production engineering, production planning and control

The method of formulating a priori hypothesis appears to be most relevant in the first step.

2. Integration Inside or Across Functions

Burbidge et al.(1987 and 89) discuss integration within or between management functions in industrial enterprises. A management function is defined as sets of closely related management tasks, which require similar skill for their efficient performance. Eight generic functions are introduced, which brings forth the suggestion that integration be discussed inside the functions as well as across functions. It is argued that efficient means exist for integrating the activities inside functions. However, the activities of one function will affect the conditions for the operations in other functions. This stress the need for focusing on the motivation to perform activities which are of benefit for the whole organization.

Four dimensions of integration across functional boundaries are identified:

Goals. There is a need for integration of goals, giving the same goals for all functions, in order to motivate employees indifferent functions to seek the attainment of company wide objectives, rather than sub optimal functional goals.

Main links. The main links between the functions cover those cases where the output from one function provides the input for another function; for example, parts list, being output from product design, is input to production planning.

Consultation links. The establishing of consultation links may improve the decisions made in one function; for example, purchasing consult with product design and production planning when new materials become available.

Cross effects of parameter changes. In traditional manufacturing systems, the specialists in each function were responsible for their own system outputs, and had their own parameters which they altered in order to control these outputs; for example, purchasing reduces the delivery frequency for a part to reduce transportation costs. This increase stocks which production control is in charge of, and reduces liquid capital (finance)

When analyzing an activity chain, the four dimensions of integration across functional boundaries may serve as a useful guideline for focusing on the relevant elements of the chain. The method appears to be most relevant in step 2.

3. Costs Connected to Key Activity Chains

Miller and Vollmann (1985) show that a main difference in the ROI of enterprises in USA and Japan is the size of the indirect costs. Their thesis is that most of the indirect costs may be attributed to activities connected to information, goods, and decisions. If this causes the majority of indirect costs, then the key to control the overheads in the factory is to control those activities which generate other activities. The chart below shows two example of the calculation of the total number of transactions per month. The example from Kverneland is based on a roughly estimate made by the author.

	Calculation sample Electronic part with 700 components ordered per month, shipped per week to store, transported in boxes from store.		Estimate for Kverneland Subassembly with 100 components with an average of 5 operations and 1 transportation from stock and 4 in between operation transportations Assumes 2000 subassemblies per month and 300 parts per batch in average (= a truckload)	
Ordering activities	700*4	2800	100*(5+1)*2000/300	4000
Receiving activities	700*4	2800	100*(5+1+4)*2000/300	4026
Goods activities (processing, in-out of inventory)	700*4*2	5600	100*(5+1)*4*2000/300	16000
Materials and components authorizations	700*4	2800	100*5*2000/300	3333
Total transactions per month		14000		27359

Miller and Vollmann estimate in their electronic example that simplification of the work routine and the product may reduce activities from 14000 to 8200 per month. Additional reduction may be obtained by analysis of the content and frequency of the information treated in the enterprise. This is quite in line with the preliminary estimation of the possibilities we have experienced at Kverneland.

4. Production Management Concept as a Reference

In the third step all the information gathered for the key activity chains should be compiled for a constructive effort to improve the performance of key activity chains. Many details are available, but it is often difficult to present an overall view of the way in which decisions should be made along an activity chain.

For activity chains related to the production management flow we have explored the idea of developing a Production Management Concept to give a coherent picture of the way in which production is to be managed (Riis, 1990). A production management concept includes mutual agreements between parties involved in production, such as sales, engineering design, production engineering, and the various production units. Hence, the production management concept focuses on an important issue with respect to the production flow activity chain, namely the interfaces between functions and sections along the chain.

A production management concept must be developed for each individual enterprise in order to grasp the specific nature of its market conditions and technology, as well as its specified goals. An example of a production management concept will be given in a subsequent section on "Experience gained".

ORGANIZATIONAL LEARNING

As pointed out in a previous section, analysis of an activity chain may disclose where learning actually takes place, individual as well as organizational. Traditionally, most of the organizational learning has taken place within departments and sections, leaving cross functional activities out of a systematic learning and experience gathering effort. The horizontal view adopted in the analysis of activity chains may bring to the attention of management what the need is for stimulating individual as well as organizational learning along activity chains.

Since the performance in key activity chains may be related to the competitive strength of the company, organizational learning along the activity chains becomes vital for its capability to compete.

Another aspect of organizational learning relates to the process of introducing CIM technologies for the improvement of the enterprise. It is essential to realize that in the initial stages of a project on CIM technology, typical the industrial organization is characterized by

• an unclear conception of what CIM technologies encompass and what it will do to the company, mainly because of lack of knowledge of CIM technology and its impact.

• little experience in carrying out a large and complex project, mainly because the production managers primarily are concerned with operations and not with development,

• inertia in the initial phase, because the project will cut across functional and departmental boundaries. (Riis, 1985)

This leads to the conclusion that organizational learning must play an important role in the planning and implementation of CIM technologies. For example, employees to be affected should be involved, and the project phases should be designed so as to allow for an adaptive process.

Important dimensions in organizational learning are included in a model called EIRAS (Andersson, 1983):

(E)	Element is the transformation of elements for example from idea to product
(I)	Individual is the employee participation in the process
(R)	Reference is the frame of references which gives basis for the actions in the process
(A)	Action is the actions done by individuals or groups in the process
(S)	Scenery is the organizational context where the action takes place

The EIRAS model has two extremes: Ambiguity Learning and Evolving Learning. Both involves these dimensions

	E	I	R	A	S
Ambiguity Learning	many elements	many individuals	many references	many actions	many Scenery
Evolving Learning	few elements	few individuals	few references	few actions	few Scenery

Ambiguity Learning is a vague and unclear learning process with a weak linkage between activity chains and their elements. Parallel events make learning difficult.

Evolving Learning is a well defined training in which the training activity can be seen clearly separated from other activities.

The development of a production management concept, discussed in the preceding section, has proven to stimulate an organizational learning process. It enables many employees to share a common view of where the company wants to go with respect to production management. Furthermore, the production management concept can be subjected to a common discussion and adjustment. A production management concept thus serves as a reference for the organizational learning.

In several instances, we have used role playing games as a means for creating a subtle understanding of the dynamics of activity chains. Especially, we have developed company specific games which enable the employees easily to accept the content of the game, because it resembled the core issues of their own company. Such games have also been used to demonstrate new production modes, experienced by the employees in a holistic way. As will be mentioned in a subsequent section, this has greatly stimulated the organizational learning process in one company to adopt a new production planning mode and to create enthusiasm for the improvement project.

EXPERIENCE GAINED

The concept of activity chain was developed and tested in the Spring of 1989 when a CIM strategy project was carried out in three Norwegian industrial enterprises. They belong to a group of 11 different companies called TESA with a total of 3500 employees in the south-western part of Norway. The main purpose of TESA is to initiate activities of common interest, such as purchasing agreements with major vendors, training courses, and development projects within manufacturing, engineering design, etc.

The three companies participating in the CIM strategy project may briefly be described as follows:

Company	No. of Empl.	Main products	Main market	% Exp.	87 Rev. M NOK	88 Rev. M NOK	89 Rev. M NOK
ABB Trallfa Robot as	200	Coating systems, Painting Robots	World-wide	90	106	90	120
Kverneland as	800	Ploughs	World-wide	80	382		
Øglænd DBS as	350	Bicycles, Bodygard equip.	Scandinavia	30	359		

Although the three companies in many respects are different, they followed the same steps of CIM strategy development, as suggested previously and indicated in the graph below.

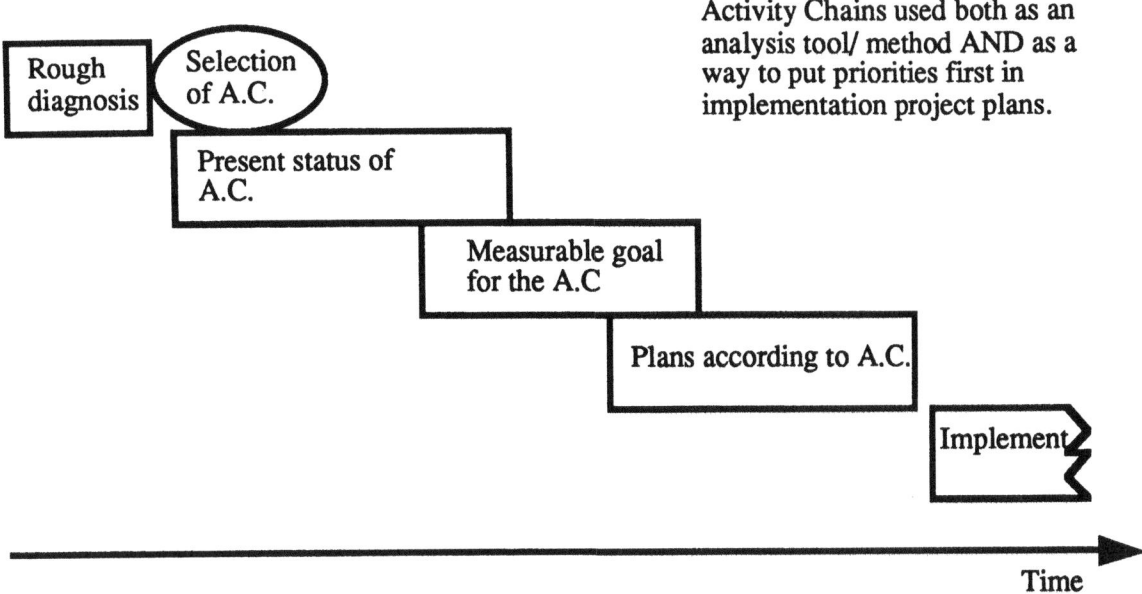

Activity Chains used both as an analysis tool/ method AND as a way to put priorities first in implementation project plans.

Rough diagnosis

Selection of A.C.

Present status of A.C.

Measurable goal for the A.C

Plans according to A.C.

Implement

Time

ABB Trallfa Robot as.

The rough diagnosis pointed to two key activity chains, namely product development and production flow. A detailed analysis of the first activity chain disclosed that the documentation phase of product development was a bottleneck. Solutions to overcome this were derived, and plans for their implementation developed.

With respect to production flow, it turned out that the computerized MRP system was of little help for handling the many disturbances in production planning and purchasing. It was difficult for the many employees involved in this activity chain to see the overall picture. Everybody was acting on the basis of his/her limited knowledge with "local common sense". Inspired by a seminar attended by about 20 employees from various sections, a role playing game was developed. It was designed as a simplified version of the company with the same nature of incoming orders, product structure and production processes. This allowed everybody to identify the game as a miniature model of the company, despite the fact that a part was symbolized by a colored piece of paper. A new mode of production was tried out similar to the ideas of continuous flow manufacturing. By playing for three hours including discussions, the participants experienced in a very realistic way how this new production management concept would work, and how it would affect their job. Hence, the game, aided by the production management concept, turned out to be instrumental for an organizational learning process which took place during the CIM strategy project.

Øglænd DBS as.

Employees were asked at a seminar to contribute to a diagnosis by supplying examples of mal-functioning and their conjectured causes. The detailed analysis focused on the production flow. A production management concept was developed and presented for the same group of employees, serving as a useful feed back to their suggestions. The concept spurred a constructive discussion about critical elements

127

along the activity chain, and formed a solid basis for preparing an implementation plan.

Kverneland as.

Several projects on introducing modern production technology had been carried out in recent years, or were in the planning phase, such as FMS cells for welding and machining, CAD, CAM, real time monitoring of work progress, etc. However, a detailed analysis of the production flow activity chain indicated that the projects to a large extent could be characterized as "islands of automation". The concept of activity chain thus made the production management aware of the need to form a coherent picture of the company's CIM development. Following the idea of production management concept (Riis 1990), a rough model of a idealized production flow was established in which high volume parts were separated from low volume parts. Furthermore, the plant was divided into homogeneous planning units, e.g. production groups. After some initial reserve and discussion, production management accepted the concept as a basis for further study. A Production Flow Analysis (Burbidge 1989) was carried out in the Fall of 1989 as the first step in the CIM implementation project, basically leading to a confirmation of the production management concept. See chart below.

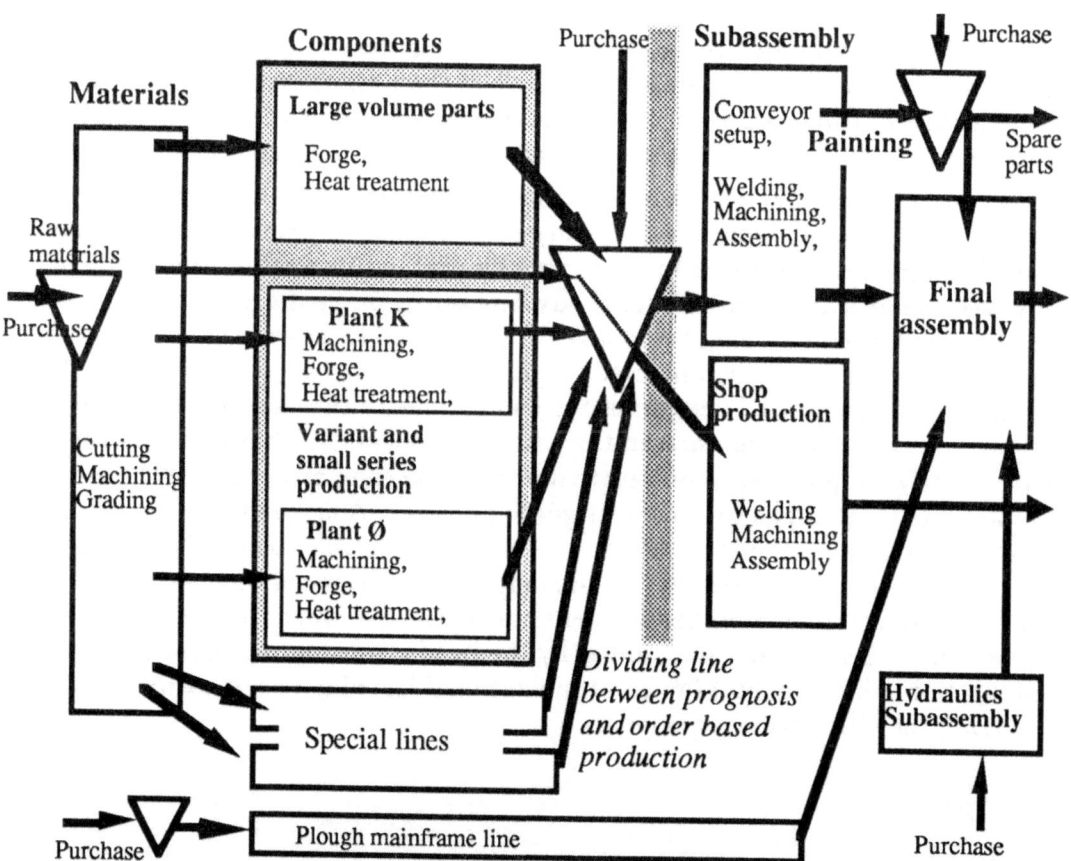

The computer network concept at Kverneland was also adjusted. It is based on a collection of different information activity chains such as the processing of customer order, production planning and control, product development and production engineering.See the chart below.

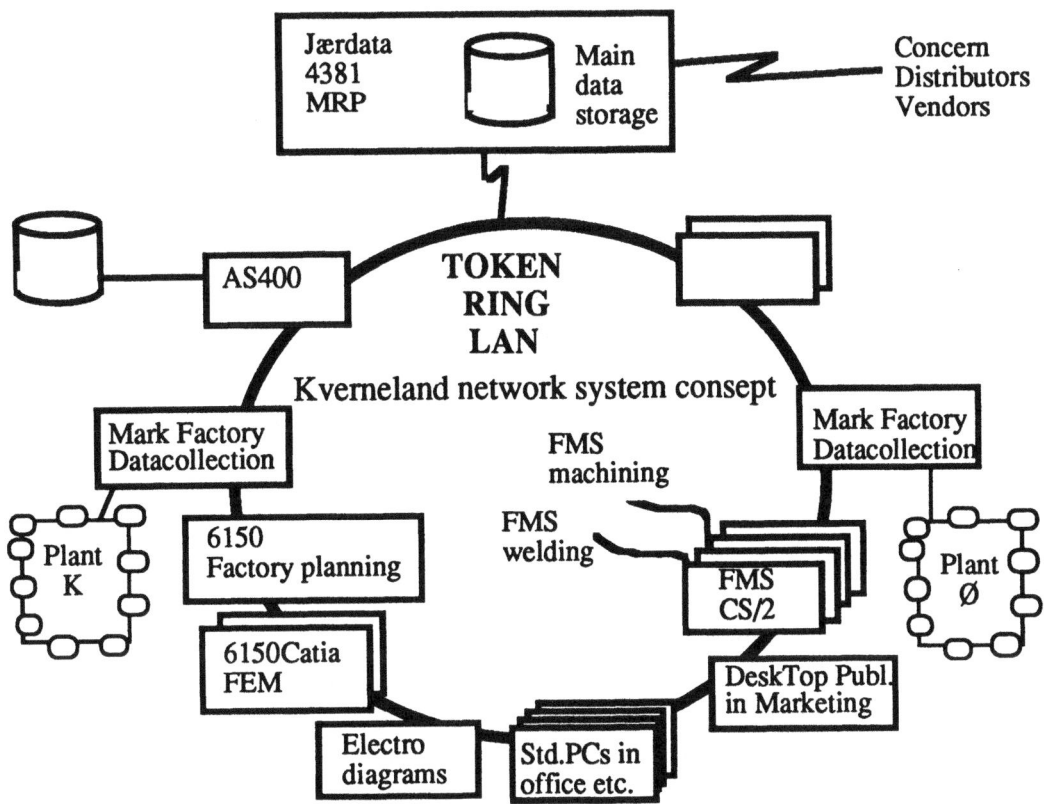

CONCLUSION

In view of the need to increase the rate of success in small and medium sized industrial enterprises of transforming CIM technologies into improved competitive strength, we have explored the idea of activity chain to denote a continouos chain of activities associated with an essential task of an industrial enterprise.

Analysis of key activity chains will pinpoint issues along the chains which are critical for the attainment of competitive objectives. Hence, focus on activity chains may be related to corporate strategy and competitive goals. Furthermore, analysis of key activity chains may create increased awareness among the employees of the relationship between their own work and the overall performance, thus stimulating an organizational learning process.

The concept of activity chain has been developed and tested in three industrial enterprises. On the basis of the experience gained in this context we may conclude:

1. that activity chains may help an industrial enterprise to realize how the introduction of CIM technologies will affect its competitive strength.e.g. by obtaining answers to the questions: Where to integrate, to which extent, and at which pace?

2. that activity chains tie elements together across functional boundaries to coherent pictures, thus stimulating organizational learning processes during CIM strategy development and implementation.

REFERENCES

Andersson 1983	Andersson, Kenneth : "Organisatorisk inlärning vid produktutveckling med exempel från förlagsbranschen" Studentlitteratur, Lund, Sweden, 1983
Burbidge 1987	Burbidge, John L.; Falster, Peter; Riis, Jens O;Svendsen, Ole M.: "Integration in Manufacturing", Computers in Industry, No 9 ,1987 p 297-305
Burbidge 1988	Burbidge, John L.; Falster, Peter; Riis, Jens O.: "Integration audit", Computer - Integrated Manufacturing Systems, Vol 2 No 3, 1988)
Burbidge 1989	Burbidge,John L. : "Production Flow Analysis, For Planning Group Technology" Oxford University Press 1989.
Miller, Vollmann 1985	Miller , J.G; Vollmann, T.E.:"The hidden factory", Harvard Business Review ,Sep- Oct 1985, p 142-150
Riis 1985	Riis, Jens O.: Seeking clarity in industrial organizational projects, Proceedings of Project Management - INTERNET 85, W. Vriethoff, J. Visser and H.H. Boerma (eds), North Holland, 1985
Riis 1990	Riis, Jens O.: The use of production management concepts in the design of production management systems, Production Planning & Control, Vol.1 No 1, p 45-52, 90
VIPS 1986-1988	the VIPS project, several reports in danish on different aspects of enterprise adapted production planning and management with cases, AUC, Aalborg University , Denmark, 1986-1988
Voss 1988	Voss, Christopher A. : "Success and failure in advanced manufacturing technology", International Journal of Technology Management, Vol 3 No 3, 1988

COMPUTER AIDED PLANNING

CIM-Elements for a Market Oriented Production Planning and Control

Dr. H. Ulrich
Institute for Operations Research, ETH, 8092 Zürich, Switzerland

INTRODUCTION

Today in manufacturing industry production planning and control is confronted with new market requirements. High flexibility in view of fast changing conditions is demanded. The principles of a JIT-production point out of the direction new developments have to follow.

For this purpose a research project has been initiated by the Institute of Operations Research of the Federal Institute of Technology, Zürich, together with a company of the manufacturing industry. This project is supported by a national research fund (KWF). Its topic is the development of an efficient computer support for a production management according to JIT-principles.

Whereas the manufacturing company had previous experiences with Japanese production control methods our Institute had already developed several successful computer supported planning tools for the manufacturing industry [1]. A co-operation offered therefore a promising perspective for the intended project. This research project has a duration of 3 years and will be finished at the end of 1991. Until now we have realized one planning tool for one production department as a prototype. Two more tools are realized as to their conception. We intend to develop a global planning proceeding for a JIT-production as well as at least four specific planning tools.

MANUFACTURING ENVIRONMENT

The main characteristics of the production process we deal with are the following:

- Products: An assortment about 120 electrical tools e.g. drilling machines, mechanical saws in a large variety of different completions. One product consists of a small electrical motor, a shell of synthetic material and mechanical parts dependent on the tool's purpose. In total, there are about 50-100 different parts per product.

- Market: The products are sold all over the world.

- Manufacturing

process: The production process consists in the manufacturing of parts in three departments (electrical, mechanical, synthetic material) on several production lines joining in a final assembly process. The material flow is organized according to flow shop principles.

The JIT-principles are part of Japanese manufacturing success [2]. However, these methods are not directly applicable under European conditions, a careful adaptation has to be performed first. Until now no sufficient computer support for this kind of production concept is available on the software market.

MARKET REQUIREMENTS

Today's manufacturing enterprise faces the challenge of keeping up with a fast technical progress and extended global markets with increasing competition. The resulting requirements for production planning and control are the following:

- High flexibility in changing production schedules and order specifications
- Short lead time
- Reliable delivery dates
- Low production cost

These requirements differ less by its content as by its ranking in priorities of former ones. The need of flexibility has taken the first position which implies a capability of reacting to market demands.

CURRENT PLANNING SYSTEM

The planning system in operation is in the tradition and follows MRP-principles [3]. The administrative aspects of material requirement are covered sufficiently whereas the management aspects are neglected. In a MRP-system the manufacturing of parts is performed "anonymously", which means a specific order for parts in the process has no direct link to its final schedule in the assembly process. Therefore we have in the production schedule in our case a "frost" zone of about 6 weeks in advance. In this period a change of production data is not allowed any more, because its consequences would not be controllable. In respect of the demand for high flexibility we expect better results of modern information processing facilities.

NEW CONCEPT

The basic elements of our new approach for a market oriented production management are:

- The anonymous part production is replaced by <u>internal production order management</u> which allocates customer orders to assembly orders. Each assembly order is on the other hand linked to all its respective orders for the manufacturing of the necessary parts (exception: Mass-production of frequently used small parts which are controlled according to a global demand-rate). The information basis for a flexible production management is established herewith.

Information links

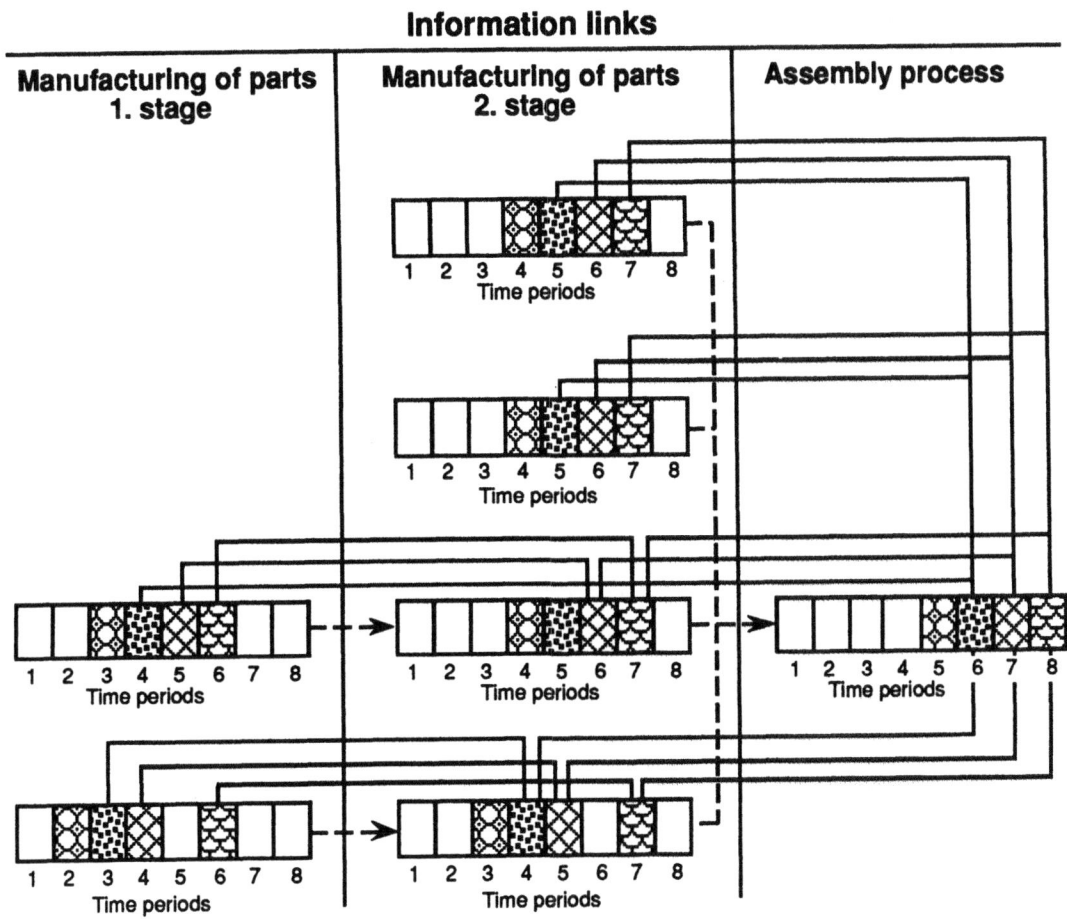

Information links between the manufacturing of 4 parts and the following assembly process (assumption: balanced lead times)

- MRP-systems use standard lead times to determine the timing of production. A standard lead time includes buffer-time to cover the normal time fluctuation within the production process. Short lead time are therefore not the main concern. To exploit the remaining potential for a shortening of lead times standard lead times have to be replaced by <u>individual lead times</u> per order. This individual lead time has to be determined with consideration of production capacity. Therefore, for every production step the use of production capacity is controlled with adequate accuracy by reserving in advance the necessary production capacity for every order. The <u>management of production capacity</u> is performed not only once in a master production schedule like in MRP-systems but permanently during the whole production process with updated production data.

- To simplify the management task of capacity reservation two main measures are taken.

  . For the production process the <u>most simple structure</u> is searched which just enables to control the production flow sufficiently. Herewith the basis for modelling the production process is established.

  . <u>Elementary orders</u> are determined. An Elementary order is an indivisible atomic element within the planning process with the following characteristics:

    - All parts within this order have the same specification.
    - All parts within this order have the same due-date on the assembly line.
    - The order quantity is optional but never exceeds the quantity of a daily production of the respective final product on the assembly line.

  The production schedule is now composed as a sequence of Elementary orders taking in account objectives as large lot sizes, low inventory (WIP) adjusted to the valid management targets.

Representation of work load

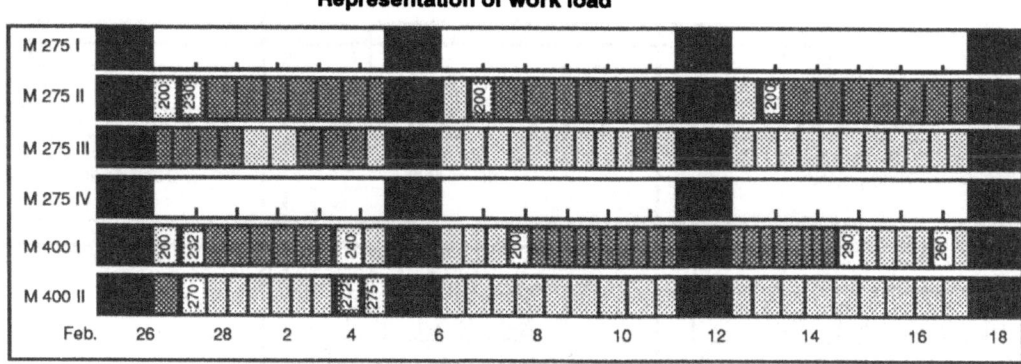

Elementary orders for 6 production lines over 3 weeks

- The administration of production data is performed by a <u>centralized</u> system e.g. one of the standard program packages available on the market. The decision process within this planning task on the contrary is allocated to several decision centers. Each decision center is destined for a well defined subproblem of the planning process e.h. short term planning of the department for mechanical parts or long term planning of the assembly process. As the whole decision process is in such a way distributed on several decision centers we have here a <u>decentralized</u> proceeding.

DESIGN OF A PLANNING TOOL

For each decision center a specific planning tool is realized on computer as a CIM Element. In the realization of such tools we base on the following main issues of former successful experience.

- The planner is enabled to solve this problems in a user-friendly man-machine dialogue with the computer.

- The computer provides him fast answers to his problem specific questions. All necessary information is updated permanently and presented in a easy conceivable way in view of global relations.

- The newest facilities of modern computer technology in particular visualization by computer graphics is taken advantage of. The computer programs are realized in a object oriented language (Smalltalk).

- A computer network providing an efficient data exchange integrates these CIM Elements together with the central data administration into the global system.

The user interface shall cope with the standard of the most advanced field in the CIM environment, the CAD applications. A hardcopy of the screen by our planning tool may illustrate our ideas more exactly (see diagram next page).

- The user operates by mouse-control, only few data he has to enter by keyboard.

- He can choose between predefined environments with graphic and alphanumeric data and windows he can open individually at will.

- The computer program is controlled by menus which can be activated by a click of a button on screen.

- The information is presented alphanumerically and graphically in color.

- All information in the system is permanently updated therefore all activities are performed on the latest system state.

- The response time is fast, the speed is adjusted to the planer's needs.

- A What-if-analysis is supported efficiently.

The requirements of the hardware as to its financial aspects are moderate. We need a workstation type computer running Smalltalk 2.5. The actual version was developed on a Macintosh IIx (8 MB memory) with a 19" color display. The development environment Smalltalk guaranties the portability to other systems, eg. IBM PS2/70, DECstation 3100, Sun-3.

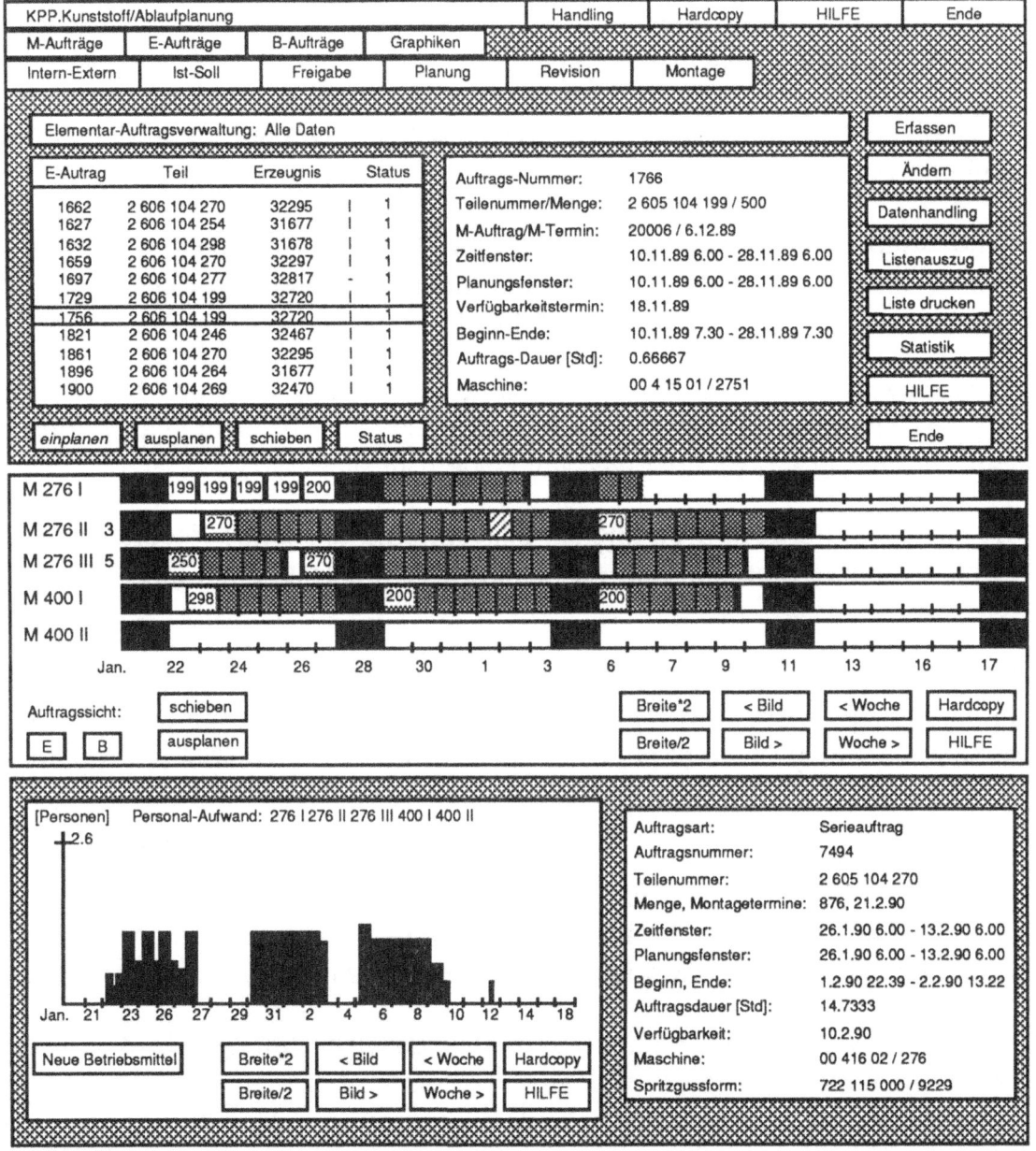

Planning Tool: Hardcopy of the screen

Our first experience with object oriented programming in the language Small-talk are very encouraging. We could reduce the expenditure of programming work at least by a factor of two and prospects of a reuse of programming code are favorable.

CONCLUSIONS

It is too early to speak about well-founded achievements and experiences in this project. But the following conclusions can already be stated:

- The results in respect to shorter lead time seem to be remarkable. As soon as the management of work load is performed with sufficient adequacy, a requirement we already have satisfied successfully in other applications, the "frost" zone is expected to be reduced to half the time it was before. The power for flexibility in production management is enlarged considerably therewith.

- The presented approach indicates a way of getting computer supported planning tools, which can be conceived specifically according to the needs of an individual planning task.

- The software development with an object oriented programming language allows to use the code of previous realizations to start with. The expense is in a dimension even smaller companies are able to cope with.

As an important perspective for the future a new segment on the software market emerge, individually designed planning tools for various specific planning tasks in the field of production planning and control. This will be an important challenge for production managers as well as for software engineers.

REFERENCES

[1] Ulrich, H.: A successful computer supported production planning in Swiss industry, Proceedings of international conference on systems science and engineering (ICSS88), Beijing, July 1988, 415-421

[2] Schonberger, R.J.: Japanese Manufacturing Techniques, The Free Press, New York, 1982

[3] Orlicky, J.,: Material Requirement Planning, McGraw-Hill, New York, 1974.

A CIM Approach to Scheduling and Material Handling: a case study[1]

M. Isabel Ribeiro
Carlos Bispo
João Sentieiro
Carlos Pinto Ferreira
CAPS/LRPI, Instituto Superior Técnico, Lisbon, Portugal
Luís Carvalho[2]
Ricardo Almeida[2]
José Nunes Ferreira
UMM, Lisbon, Portugal

1 INTRODUCTION

Many production plants have reached the state where important activities and processes have been automated, namely production planning and scheduling, machining, material handling and quality control. However, in general, these automated activities and processes have no dialogue in between them, behaving like isolated automation islands. The CIM approach to production management really addresses the problem of dialogue and synchronism between production activities.

This paper describes the research and development activities that were undertaken by a University/Industry joint group in the area of material handling systems and production scheduling. The project main goal was the development and implementation, at UMM - a portuguese industrial plant - of a flexible manufacturing scheme combining an automatic planner and scheduler with a transport system based on Automated Guided Vehicles (AGVs). The approach views the interaction of the two systems and aims in the future to use the production schedule generated by the scheduler to activate the material handling manager and to integrate other production sub-systems namely an automated warehouse, as shown in Figure 1. The paper presents the scheduler algorithm and the associated planning system, the main aspects of the project, test and implementation of an AGV prototype and the project main achievements.

For a particular production problem, the scheduler algorithm allocates the jobs to the set of existing machines by minimizing the total time of production. The schedule is generated by using, alternately, a forward and backward allocation strategy with a rule-based criterium. Sufficient conditions for optimality are stated and representative simulation results are presented. The interface between the industrial production needs and the scheduler is provided by a task planner. The planner finds the net production needs to accomplish each particular production order providing that information to the scheduler module. It also has a feedback

---

[1]Work partially supported by JNICT project #87361(MIC).
[2]This work was done on a schoolarship from JNICT.

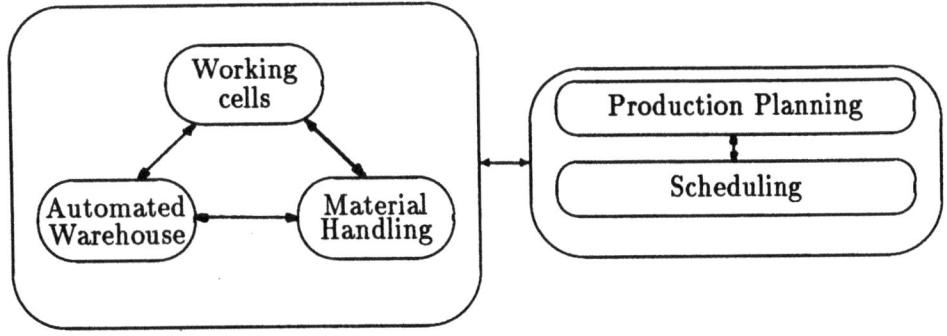

Figure 1: Integration of plant sub-systems

function, reactivating the scheduler when the production plan being executed is far from the projected one.

To satisfy the specific needs of material handling among the various work-cells of the industrial plant, an inductive steering AGV prototype was projected and implemented. The AGV is unidirectional with a tricycle drive configuration, achieved by a single drive and steering wheel. The mechanical architecture, the design of the steering antennae and its location within the chassi, all the vehicle control and supervision including safety procedures and the needs on battery power were studied. The project extends for the guidepath power generator and for the discrete communication system between the AGV and a central task and traffic supervisor system.

In section 2, the scheduler algorithm and the associated planning system are briefly described and representative simulation examples are presented. The project, test and implementation of the AGV prototype are discussed in section 3. Finally, section 4 concludes the paper, discussing the applicability of the planning and scheduling strategies to other environments and presenting directions for further developments within the project.

2 PLANNER AND SCHEDULER

In the production scheduling area an automatic **planner** and **scheduler** for manufacturing plants with m machines, n jobs and a *flow-shop* structure has been designed with the structure represented in Fig. 2.

2.1 Planner

The planner is responsible for the input data the scheduler receives, since it determines the net needs of production for each individual part.

For a given production order, the planner interacts with the system data base to convert it into a list of parts which are to be produced (or bougth) and assembled in order to form the desired products in the desired quantities. The data base has a complete description of all the products. That description includes information such as the product decomposition in subcomponents, the machines where the

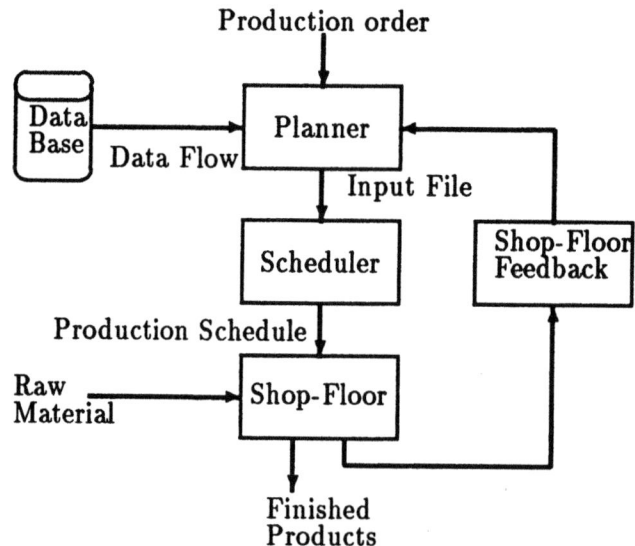

Figure 2: The overall architecture of the integrated environment for planning and scheduling.

parts are processed, the set-up times, the operation times, the parts produced in the plant, the parts which come from external supliers, etc.

As soon as the list of all the necessary components is available, the planner determines what has to be produced, taking into account the existent stock of raw material, the stock of components and the external supliers in terms of components and raw material. After this step, the planner finds the net production needs to accomplish that particular production order and builds a list of all the information needed to schedule that production order, [5,6].

The planner has also the capability of deciding if it is necessary to reactivate the scheduler when a production plan is beeing executed. That decision is based on the data colected from the shop-floor and can be taken when some significant differences are detected between the plan and its execution (feedback function).

The planner implemented for the integrated production planning and scheduling system described in this paper is therefore similar to a Material Requirements Planning (MRP) system, [3].

2.2 Scheduler

For the net needs of production defined by the planner, the scheduler algorithm allocates the jobs to the set of existing machines by minimizing the total makespan.

The scheduler is developed in an object oriented language allowing a uniform and user-friendly description of data and algorithms, [7]. Cases where different machines may have to work in different shift regimes happen very often in practice, in particular when the workload is not evenly shared by the machines. This additional complexity, usually avoided in most scheduling algoritms, [1,4], is considered here.

Notation. In what follows, relevant notation and definitions are introduced.

1. M is the set of machines $\{M_1, ..., M_m\}$.

2. J is the set of jobs $\{J_1, ..., J_n\}$, each job consisting of a set of identical parts.

3. C_i is the completion time instant of J_i.

4. p_{ij} denotes the processing time of J_i in M_j.

5. Let L_{ij} and R_{ij} be defined as:

$$L_{ij} = \sum_{k=1}^{j-1} p_{ik} \quad , \quad R_{ij} = \sum_{k=j+1}^{m} p_{ik}.$$

Assuming that there are no shifts, L_{ij} is a lower bound on the interval of time that J_i takes to get to M_j and R_{ij} is a lower bound on the time interval necessary to complete the job J_i after leaving the machine M_j.

6. CPI_j denotes the Critical Production Time Instant for M_j.

   Taking into account the existence of shift regimes, if one starts by allocating the job with minimal L_{ij} to the machines that precede M_j, and then allocates all operations on M_j and finaly allocates the job with minimal R_{ij} in the machines that follow M_j, then CPI_j will be the time instant when this last job is completed.

7. **Critical Machine** (M_c) is the one which exhibits the larger CPI_j, i.e., the machine with the highest workload.

8. **Left Ordered List** (LOL) is a list of all jobs processed in M_c, ordered in increasing order of L_{ic}. When two or more jobs compete for the same position in LOL, priority is given to the one that has the largest *complementary manufacturing time*[3], $(p_{ic} + R_{ic})$.

9. **Right Ordered List** (ROL) is a list of all jobs processed in M_c, ordered in increasing order of R_{ic}. When two or more jobs compete for the same position in ROL, priority is given to the one that has the largest *complementary manufacturing time*, $(p_{ic} + L_{ic})$.

10. **Minimal Time** (T_{min}) is an estimated value of the completion time. The first value assumed by T_{min} $(T_{min_{1st}})$ equals CPI_c.

Scheduling algorithm. For a particular problem and once identified the machine with higher workload, M_c, the algorithm tries to allocate the jobs in order to minimize the total time of production. As the UMM plant has a flow-shop structure the optimum is attained if the following conditions can be met simultaneously [4]:

---

[3]The term *complementary manufacturing time* is here understood as the sum, for each job, of the processing time in M_c and in machines that follow (precede) for LOL (ROL)

143

C1 The critical machine starts its operation at the earliest possible time.

C2 Once started, the critical machine operates continously.

C3 After the critical machine has exhausted its workload, the operations that remain to be performed[4] will be concluded in the shortest possible time.

The above are sufficient but not necessary conditions for the optimum. If they are not simultaneously verified, a worst case bound can be derived in order to evaluate the performance of the generated schedule (a bound on a "distance" measure to the optimum).

The algorithm generates a schedule using alternately a forward and backward allocation strategy, [2]. The allocation criterium is rule-based and the dispatch rules are defined based on conditions C_1 and C_3.

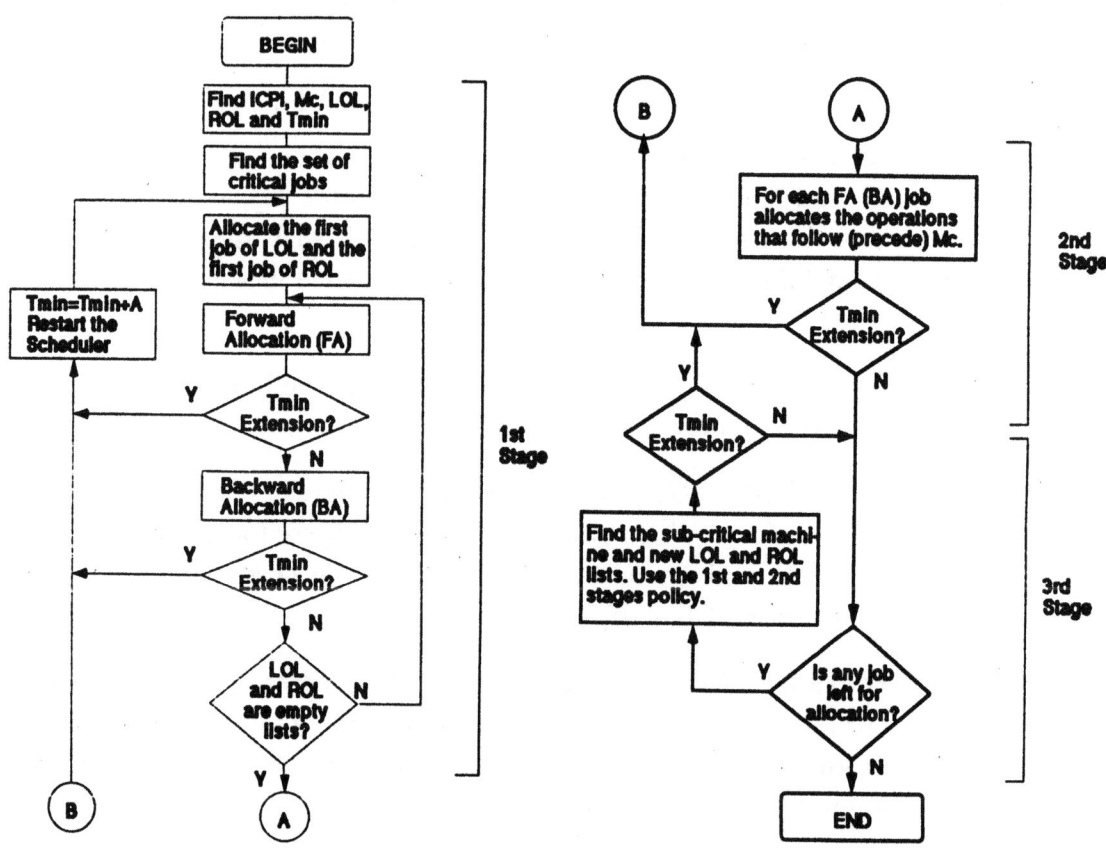

Figure 3: Flow Diagram for the Scheduling Algorithm.

The scheduler flow diagram is represented in Fig. 3 where the three main stages of the algorithm are displayed. In the first stage, and after the necessary initializations, the algorithm allocates the jobs which have one operation processed by the critical machine. This is done in two different and alternated modes: forward mode (FA) and backward mode (BA). In the forward mode the allocation rule is

---

[4]It is assumed that the production to be concluded refers only to the jobs that have one operation processed by the critical machine.

defined by LOL and the allocated operations are the ones that precede the critical machine. The backward mode uses ROL to allocate the operations that follow the critical machine.

After the first stage is completed all jobs that have one operation processed by M_c will have that operation allocated. Regarding the rest of the operations some jobs will have allocated the operations that precede M_c and the others will have allocated the operations that follow M_c. The second stage of the scheduler completes the allocation for the remaining operations of the jobs which are processed by the critical machine.

Finally, the third stage allocates the jobs which are not processed by the critical machine. For these, a new sub-critical machine is defined for which the allocation strategy is similar to the one used for M_c. Therefore, the third stage of the scheduler consists on recursive calls of the algorithm as long as there are jobs to allocate, see Fig. 3. For a detailed explanation of the scheduler, see [7].

If the scheduler generates a production plan whithin the $T_{min_{1st}}$ then conditions C_1 to C_3 are satisfied and the algorithm attains the optimum. Otherwise, a swapping heuristic induces an enlarged new estimate of T_{min} and a new schedule is generated[5]. If T_{min} is extended, the optimum cannot be insured, but the new (sub-optimum) T_{min} will eventually have a value close to CPI_c.

One of the interesting properties of this scheduling algorithm is the possibility of including in the model different shift regimes. This allows a uniform way of dealing with other situations. In fact, as the algorithm treats the inactivity periods due to shift regimes as black holes in terms of allocation, it is possible to make use of the same technique to deal with inactivity periods due to breakdowns or to the inclusion of new production orders into existent plans.

2.3   Results

In order to evaluate the behaviour of the scheduling algorithm presented in this paper (denoted by algorithm A), a comparative analysis with the well known Palmer's algorithm (denoted by algorithm B) is presented [4]. Using both algorithms, one hundred different problems were tested for a model with 10 machines and 25 jobs. The operation times were randomly generated with a uniform distribution.

In Figure 4 the distribution function of the deviations relative to the lower bound on the optimal ($T_A - T_{min}$ and $T_B - T_{min}$). The lower bound (T_{min}) was computed based on conditions C1, C2 and C3. As it can be seen, the results of algorithm A are closer to the computed lower bound than those obtained with algorithm B. Note that the deviations that can be red out of this figure do not correspond to an exact measure of the distance to the optimum but rather to an upper bound on it.

Figure 5 displays the distribution function of the difference between the total makespans for algorithms B and A, i.e., $(T_B - T_A)/T_{min}$. From this figure it is easy to see that algorithm A performs better than B in approximately 70% of the cases or, that it performs with a 15% gain in 20% of the cases.

---

[5]The different timings between shifts, make this system *time variant*.

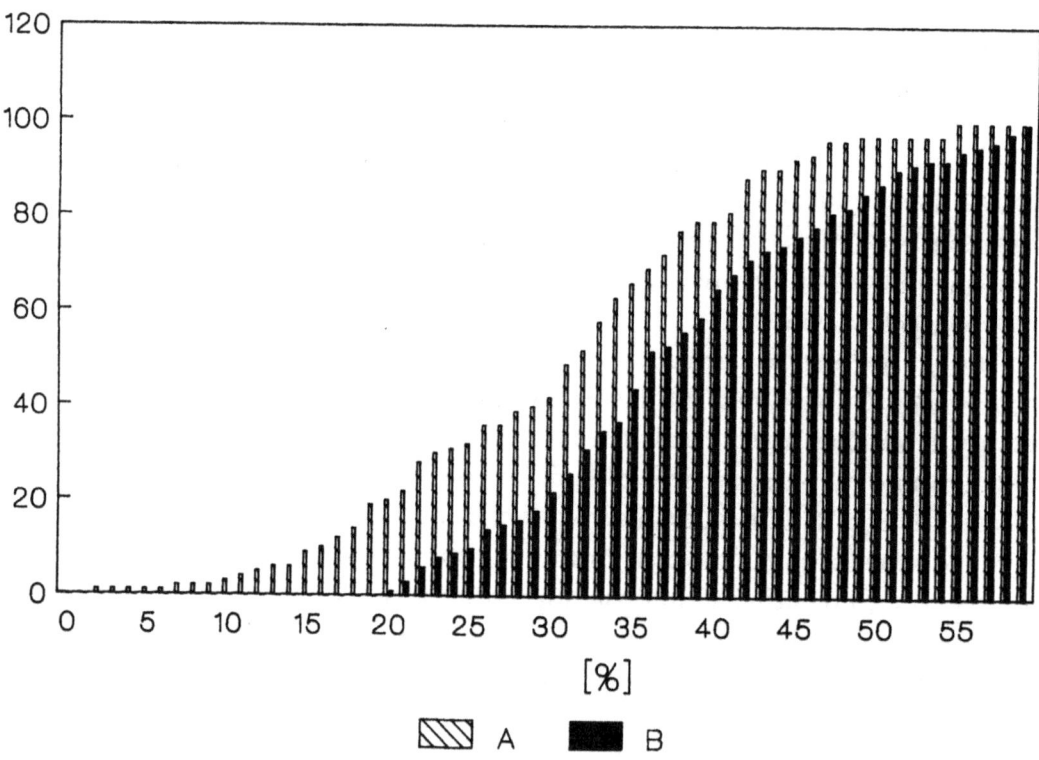

Figure 4: Distribution function of $(T_A - T_{min})/T_{min}$ and $(T_B - T_{min})/T_{min}$

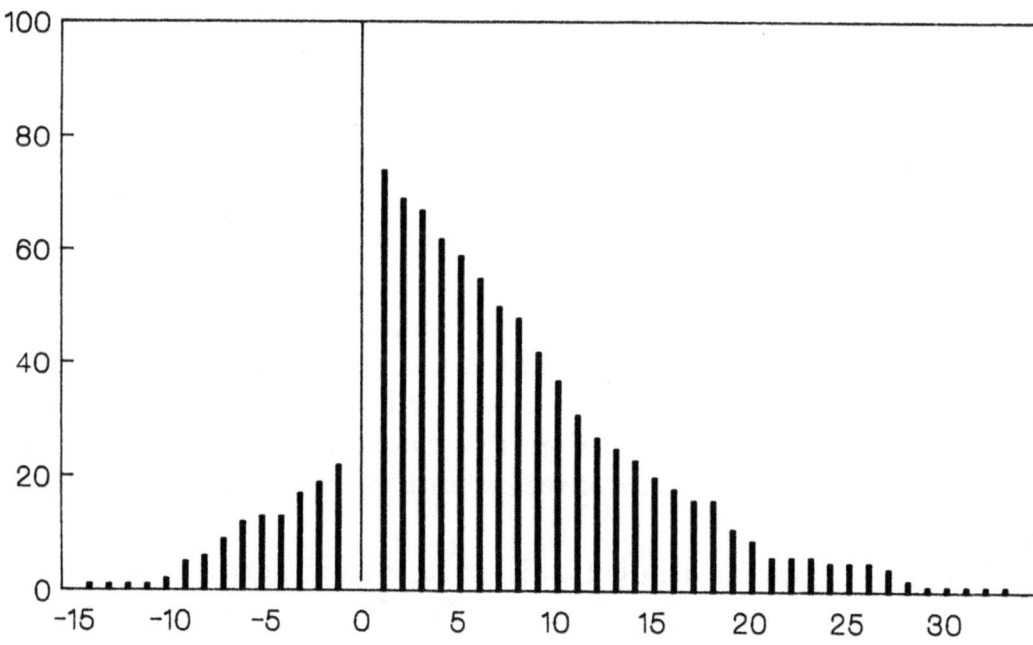

Figure 5: Distribution function of $(T_B - T_A)/T_{min}$

For the case of UMM plant, the algorithm achieves production schedules with a makespan that, in average, is 1.15 of the lower bound on the optimum. Palmer's algorithm does not apply to this real case since it does not deal with shift regime cases.

3 AGV PROTOTYPE

The material handling procedures in an industrial flexible environment may be achieved by an AGV System (AGVS) that combines a task planner, a path planner and a set of Automated Guided Vehicles.

The task planner optimizes the assignment of each material handling demand to a particular AGV. The allocated vehicle is driven between the loading/unloading working cells, through an optimal path defined by the path planner.

This section describes the project, test and implementation of an AGV industrial prototype considered as the basic component of an AGVS. The vehicle steering control, its identification and the analysis of the vehicle performance is addressed. All the details on the AGV project are presented in [9].

At the actual stage of the development, the task and the path planners, although projected, are not yet implemented at the plant.

AGV Specifications.

- Maximum load = 3000 Kg,

- Maximum speed = $1ms^{-1}$,

- Safety procedures, including bumpers, emergency stop/start button and obstacle proximity detectors,

- Loading/unloading transfer mechanism provided by a lift table,

- Electromagnetic brake,

- Manual control.

AGV Architecture. To fulfill the specifications, an unidirectional AGV was designed, with a tricycle configuration achieved by a single drive and steering wheel in the front and a pair of fixed wheels in the rear, as represented in Figure 6.

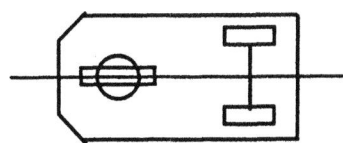

Figure 6: Tricycle configuration

The vehicle dimensions are $180 \times 120 \times 62.5$cm, including the bumpers. The 1KW driving motor and the 100W steering motor are powered by two 12V/220Ah batteries with an autonomy of five hours. A wire guidepath method with a guiding frequency of 8200Khz is used to steer the AGV. An electromagnetic brake, a stop/start emergency button and a manual control device were projected and implemented. The communication between the AGV and the path planner system is established, at discrete points along the path, using FSK modulation.

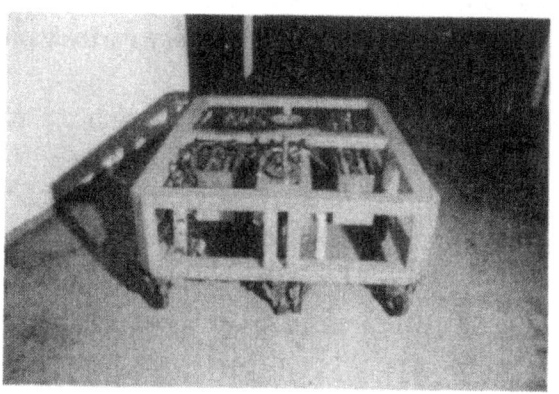

Figure 7: AGV prototype representation

The pictures in Figure 7 represent, in detail, the AGV and its steering device with two associated inductive sensors.

Steering Control. The steering control main objective is to drive the AGV along a wire buried in a slot in the floor. The low-voltage, low-current ac signal present in the wire generates an electromagnetic field which induces the voltages v_1 and v_2 in the two inductive sensors represented in Figure 8. The differential mode of v_1 and v_2, $v_d = v_1 - v_2$, acts as the reference for the control of the steering motor. The common mode, $v_c = (v_1 + v_2)/2$, is used to detect a vehicle misalignment greater than a certain threshold or a power failure on the guiding wire. Both situations lead to the execution of emergency procedures.

Consider Figure 9, where $\gamma(t)$ is the angle between the steering wheel and the

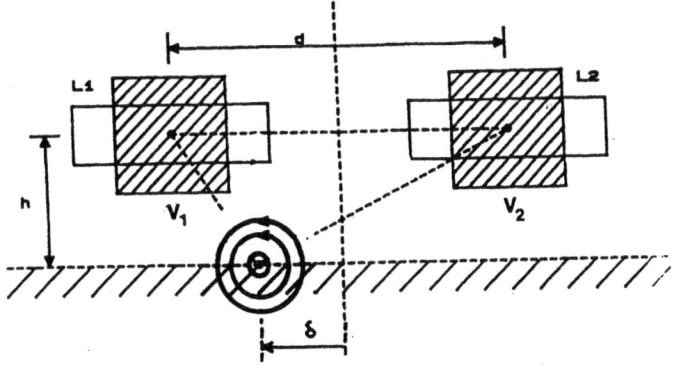

Figure 8: Inductive sensors and guide wire

vehicle axis, and $\theta(t)$ represents the AGV orientation reported to the x axis. Let $\xi(t)$ be the orientation in the $x - y$ plane of the tangent to the path defined by the buried wire and define,

$$\varphi(t) = \gamma(t) + \theta(t).$$

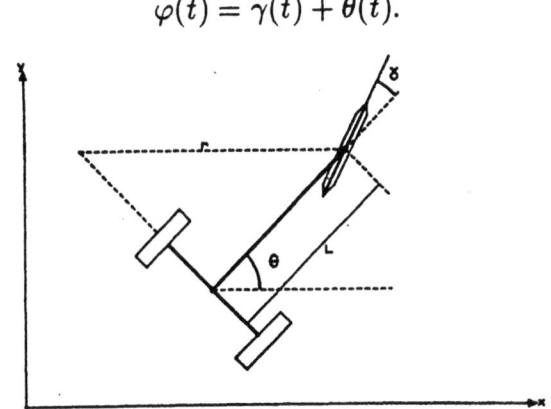

Figure 9: AGV and steering motor orientation

The AGV steering control block drives the vehicle in such a way that $\varphi(t)$ follows, as closer as possible, the path orientation $\xi(t)$. The AGV model, considered as a closed loop system, with $\xi(t)$ as the reference signal and $\varphi(t)$ as the controlled variable is represented in Figure 10. In this Figure, $M(s)$ and $P(s)$ are the steering motor and the controller transfer functions. The inductive sensors are not fixed to the chassis but are steadily linked with the steering block, at a distance d_r from the motor axis. This leads to an antecipative effect, modeled as a zero on the first block in the diagram of Figure 10. The optimum value of d_r results from a trade-off between the advantages of an early detection of a guiding path change of direction and the vehicle instability for inductive sensors placed far apart from the motor axis. An analytic study of this optimum and the dynamical relation between $\gamma(t)$ and $\varphi(t)$ are derived in [9]. See [8] for an exact kinematic AGV model used for steering control simulation purposes.

The steering motor model was identified as a first order system with transfer function $M(s) = 15/(s+8\pi)$. We adopted a velocity feedback control strategy and conclude that a PI controller satisfies the AGV performance requirements. The controller parameters were adjusted to provide a non oscillatory AGV response with a minimum establishing time, leading to the closed loop transfer function $G(s) = 9.6(s + 2.4)/(s + 4.8)^2$.

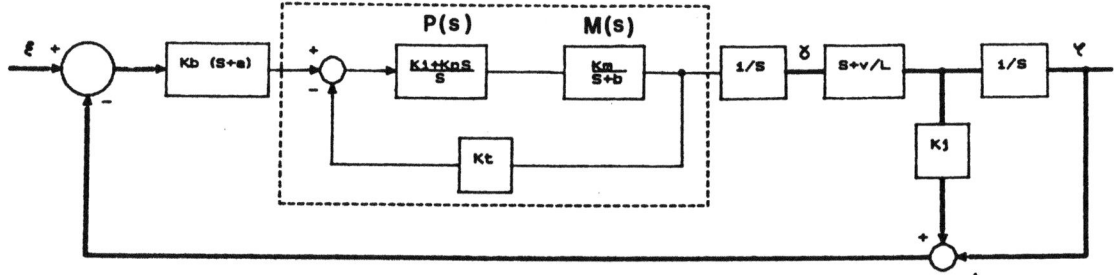

Figure 10: Diagram of the AGV steering control block

The performance of the AGV prototype, evaluated by its response to a step variation of 14° in $\xi(t)$ and a speed of $v = 1ms^{-1}$, is displayed in Figure 11. This input signal, $\xi(t)$, corresponds to the guidepath ramp evolution represented in the $x - y$ plane in Figure 11-a). The error $\xi(t) - \varphi(t)$ is displayed, as a function of t, in Figure 11-b). From this figure it is evident that the steady-state error is zero and that the AGV location error becomes negligible soon after the guidepath change of direction.

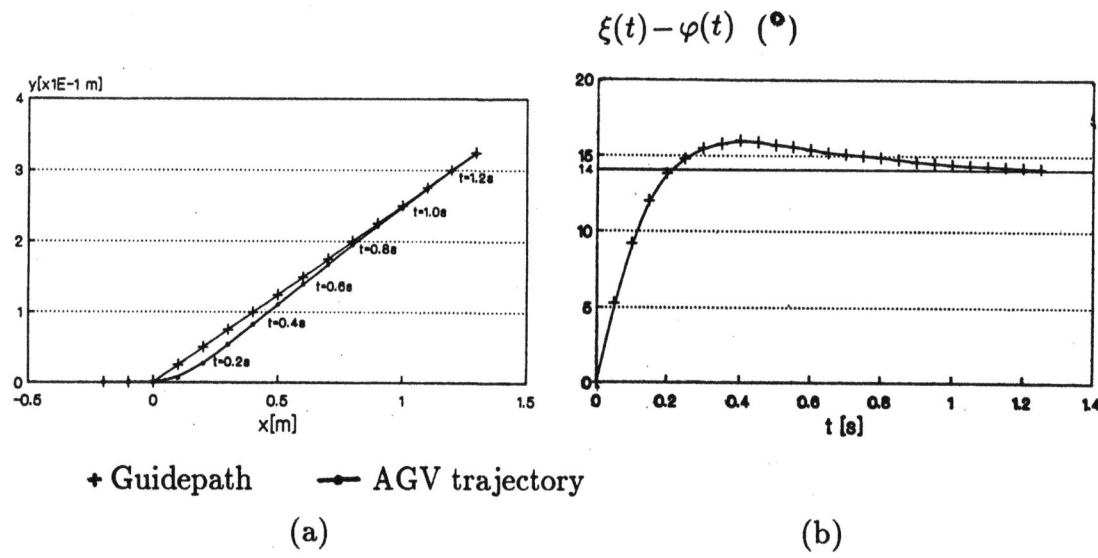

Figure 11: AGV performance test

4 CONCLUSIONS

This paper reports the results of an University/Industry joint project on the area of automation and production integration. A planning system and a scheduler algorithm, were developed and implemented at UMM, a portuguese car manufacturer, leading to improvements on the production time that amount to 30%. The performance of this scheduling algorithm is compared with another competing technique.

The planning and scheduling strategies, although developed for the specific case of the UMM plant, are general purpose in the sense that they can be successfully applied to other environments. In fact, the planning system defines the jobs to be executed based on a data base with no particular characteristic. The scheduling algorithm is mainly based on two assumptions: the flow-shop structure of the plant and an unbalanced distribution of the workload between the resources. Therefore, as long as these two assumptions are verified, which happens in a large number of industrial production processes, the scheduler algorithm proposed in this paper may be used. It also applies to industrial environments with or without shift regimes.

An unidirectional AGV, considered as the basic element of a material handling system, was designed and implemented. The mechanical architecture, the specification of the steering inductive sensors and its location within the chassis, all the vehicle control and supervision including safety procedures and the needs on battery power were studied. In the near future, the project will proceed towards the development of a more flexible and autonomous AGV.

References

[1] Baker, K. R., *Introduction to Scheduling and Sequencing*, John Wiley & Sons, Inc., 1974.

[2] Fox, M. S., "Constraint-Directed Search: a case study of Job-Shop Scheduling," Ph.D Thesis, Computer Science Department, CMU, December 1983.

[3] Aggarwal, S. C., "MRP, JIT, OPT, FMS?," Special Report, Harvard Business Review, September-October, 1985.

[4] French, S., *Sequencing and Scheduling - An Introduction to the Mathematics of the Job-Shop*, John Wiley & Sons, Inc., 1986.

[5] Baker, A. D., "Complete Manufacturing Control using a Contract Net: a Simulation Study," 1988 International Conference on CIM, Troy, N.Y., USA, May 1988.

[6] Brandimarte, P., Greco, C., "Integration of Knowledge-Based and Algorithmic Techniques for Production Scheduling," 1988 International Conference on CIM, Troy, N.Y., USA, May 1988.

[7] Carvalho, L., Almeida, R., Bispo, C., Sentieiro, J., "An Integrated Environment for Planning and Scheduling in Flow-Shop Manufacturing Plants," to be published in the Proceedings of the 2nd International Conference on CIM, RPI, Troy, N.Y., U.S.A., May 1990.

[8] Milacić, V., Putnik, G., "Steering Rules for AGV Based on Primitive Function and Elementary Movement Control," Robotics & Computer Integrated Manufacturing, Vol. 5, No. 2/3, pp. 249-254, 1989.

[9] Almeida, R., "Transporte Automático num Cenário de Fabricação Flexível: Veículos Guiados Automaticamente," Master Thesis, Instituto Superior Técnico, March 1990.

Dynamic production scheduling.

Stig Ulfsby
Noratom A/S
Norway.

1.1. Problem addressed.

Production scheduling is today most often done by traditional MRP II systems (Manufacturing Resource Planning). These systems have several severe problems:

- They are planning for unlimited resources. They only compute the resources needed and do not schedule jobs that need the same resource at different times.
- They do not take bottle-necks into account.
- They are most often planing on week basis.
- They do not take into account the status on the shop floor (machine breakdown, order status etc)
- They do not find the optimum sequence of jobs.
- They are not fit for consequence analyses.

These problems causes long lead times, delays in the delivery and large inventory.

The complex problem of optimized scheduling is too resource consuming for traditional computing. An example will illustrate this:

10 orders that all shall be scheduled on 4 machines gives $(10!)^4 = 10^{26}$ possible combinations to schedule the orders. A supercomputer that can compare 100 million combinations pr. second will use 10 billion years to find the optimum solution. The use of artificial intelligence will find a good (but not optimum) solution in few seconds.

1.2. Scope of system.

The purpose of the system is to do dynamic production scheduling.

The system is based on the OPT production philosophy (Optimized Production Technique) introduced by Eli Goldrath [Goldrath 1983]. The goal of the system is to create the optimum schedule regarding throughput and inventory, provided delivery within due time. It takes into account the production capacity at each workstation. Alternative workstations can be used for each operation. Artificial intelligence techniques are used to realise this.

By data capture changes in production capacity (like machine breakdown and absence of personnel) and status of each order is recorded. When necessary a new schedule is produced taking the actual status into consideration.

The data capture can also be used to record quantity produced, quantity scrapped, materials consumption, direct labour, stop times and stop reasons.

New orders, changes in quantity or due date can be entered any time.

The system has a powerful tool for analysing the cosequences of a schedule, which can be used for what-if-simulations. Different alternatives of production capacity and product mix can be simulated.

The product structure can be defined hierarchically with assemblies and subassemblies. Each assembly may have its own process plan, making it possible to parallel perform operations in different assemblies of a product.

The system has a modern graphical user interface utilising windowing and pop-up menus.

1.2.1. Advantages.
Achievments for the user companies will be:
- A schedule is always up-to-date and possible to follow, ensuring delivery at due time or at least have an early warning that delivery at due time is not possible.
- The total production capacity can be better utilised by focusing on the bottle-necks.
- The production flow will be better including reduced queues at bottleneck workstations, reducing work in progress.

In addition the data capture will give the basis for cost calculations, quality control and maintenance planning.

1.2.2. What the system doesn't do.
The system doesn't do stock management. The system will, however, calculate the materials needed by each operation of an order and give a warning when the material is not available on store.

The system doesn't do long-term planning. It assumes that the orders come from customers or are generated by prognoses (using a prognosing system).

The system doesn't do automatic data capture, but has functions to enter order status, machine stop and capacity changes.

The system doesn't do accounting, but can provide an accounting system with information like direct labour and materials, produced goods and scrap.

The system doesn't do process planning, but provides the user with a good tool to enter process plans into the system and change them.

1.3. Experiances from the ESPRIT project.

The system is developed within ESPRIT 2434 "Knowledge based CIM controllers for distributed factory supervision". The main goal of the project is to make modern production strategies like OPT, JIT and LOP operational on the factory floor using knowledge based software techniques. There are 18 partners and subcontractors from 7 countries with Philips as maincontractor. The project started in 1989 and lasts for three years with a possible extesion to five years.

The partners are all developing their own software systems. The cooperation is therefore on exchanging methodologies rather than trying to make a common software system. This has proved to be an efficiant way of cooperation giving results within short time. The authors experiance from other international projects is that trying to make several institutions with different motives develope a common software system, causes a lot of coordination roblems and no results.

2. BOTTLENECK SCHEDULING.

The main goal of a company is to be profitable for its investors. A widely used measure for the profit is return on investment:

Return on investment = (Income - Costs)/Invested capital.

This ratio may be increased by:
- Increasing the income.
- Reducing the costs.
- Reducing the capital needed.

For the factory this means:
- To produce more (if it can be sold).
- To reduce production costs.
- To reduce inventory (work in progress and stock).

The purpose of bottle-neck scheduling is to maximize the throughput (completed orders per time unit) and minimize the inventory.

The main idea is that the bottle-necks of the factory limits what the whole factory can produce. Therefore it is important to optimize the utilisation of the bottle-neck, while utilisation of the rest of the resources is of no interest. "One hour saved in the bottle-neck is one hour saved for the whole factory, while one hour saved in a non bottle-neck is an illusion" (Goldrath).

The way it works is that jobs on the bottle-neck are first scheduled to maximise the utilisation of it and hence the throughput. Then the rest of the jobs are synchronized with the bottle-neck to reduce the lead-times (the time from an order is released till it is completed). See fig. 1.

We have developed an algorithm for bottle-neck scheduling which is based partly on operational research and partly on artificial intelligence methods. It consists of three main parts:

- A bottle-neck scheduler which makes an optimum schedule for the bottle-neck.
- A forward scheduler which schedules the jobs after the bottle-neck.
- A backward scheduler which schedules the jobs before the bottle-neck.

The algorithm has been tested on several examples and compared with pure forward or backward scheduling. Results from a test with 25 orders representing about two months work on eight machines are given below. (See also fig. 2).

	Forward	Backward	Bottleneck
Computation time	12 sec.	81 sec.	32. sec.
Average lead time	12.3 days	11.5 days	7.8 days
Average delay	6.3 days	1.7 days	1.8 days

Table 1. Comparison of three scheduling techniques.

The forward and backward schedulers are implemented by heuristic search methods from OR using priority dispatching rules [Wang, 1983] [Guida, 1989]. In the tests above shortest processing time is used as priority rule.

The bottleneck scheduler is using a hill climbing method from AI described in [Hanan, 1972].

A more detailed description of the algorithms are described in [Kaldager, Ulfsby 1990].

3. SYSTEM PERSPECTIVE.

3.1. Input and output.

The main input to the system is production orders. The main output is workorders to the workstations, transportorders to transport units, materials requisitions to the store for each workstation, tools requisitions and requisitions of other common resources. This output controls the total production process according to the schedule.

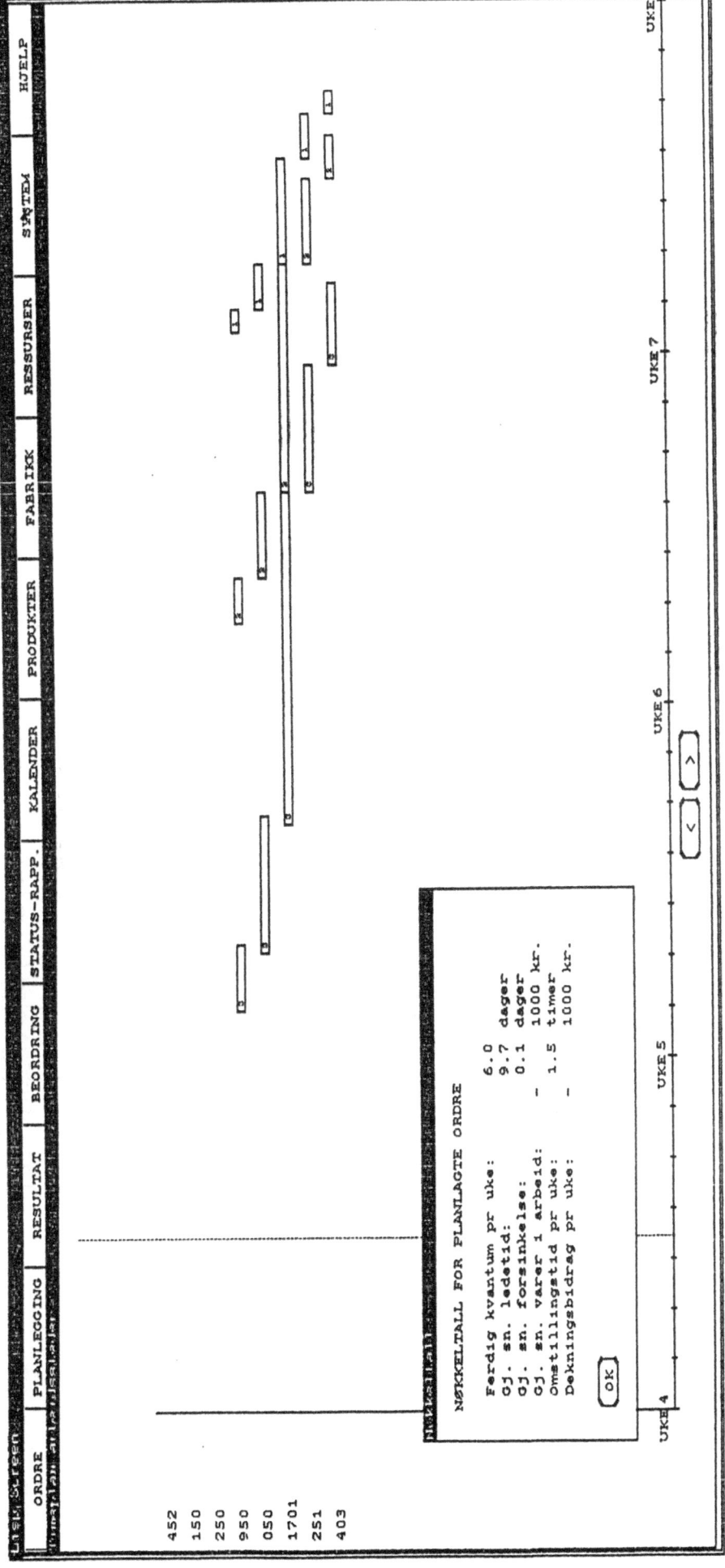

Fig. 1. BOTTLENECK SCHEDULING.
Three orders (1, 2 and 3) with five jobs each are scheduled. First the jobs on the bottleneck machine 1701 are scheduled. Then the jobs before the bottleneck are scheduled as late as possible, and the jobs after the bottleneck are scheduled as early as possible. In this case there is no waiting-time for any of the jobs.

157

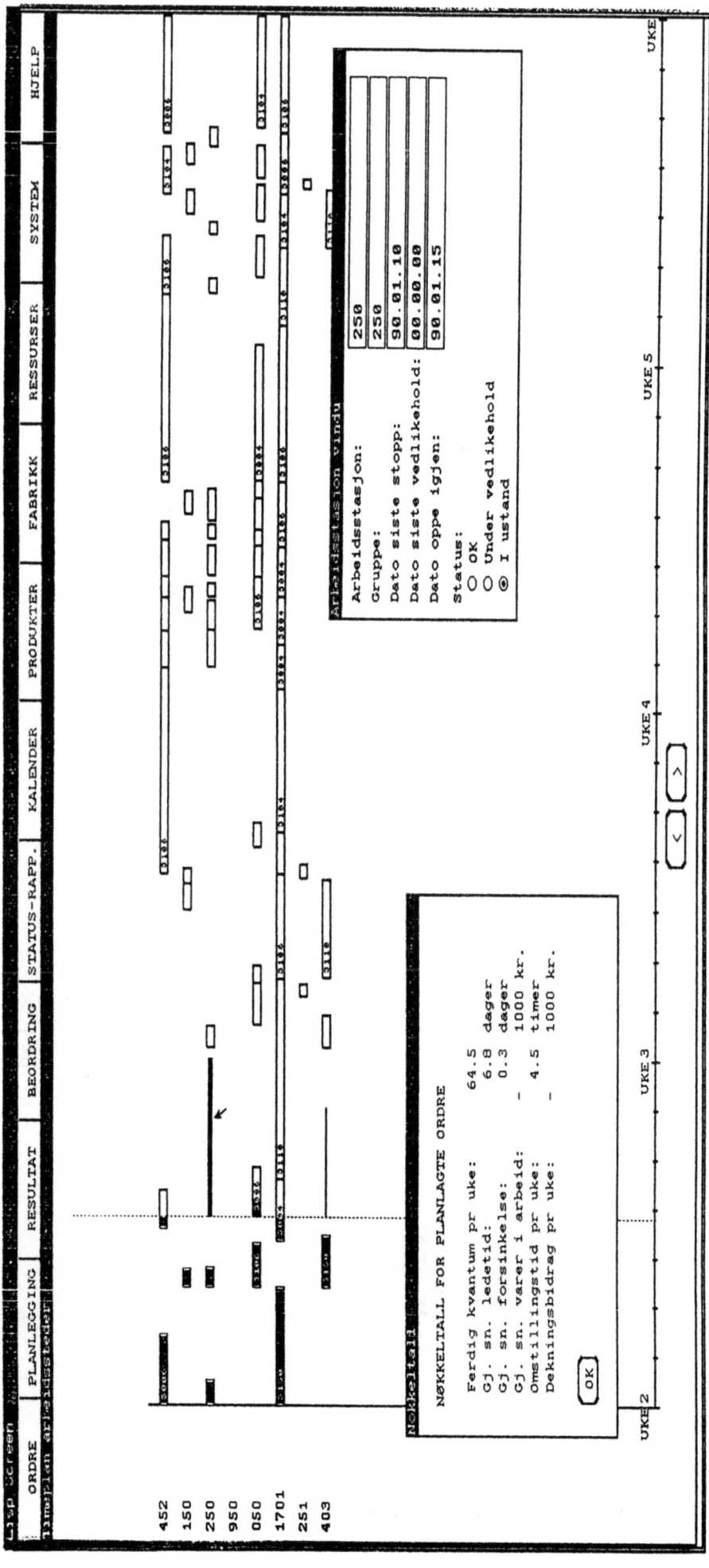

Fig. 2. A MORE COMPLEX EXAMPLE.
Here bottleneck scheduling is applied on a larger example. 29 orders with 5 to 7 jobs each are scheduled.
The average lead time is 6.8 days while it was 11.3 days for forward scheduling and 10.5 days for backward scheduling (both with shortest processing time as priority-rule).

The algorithm also takes into account jobs completed (black), jobs in process (partly black), stopped machines (thick line) and machines under maintenance (thin line).

158

To be able to produce this output the system also need a description of the products and their production processes, of the production resources (workstation capacities etc.) and of the factories calendar (shifts, holidays etc.).

In order to take the actual production capacity and order status into account when scheduling, this has to be reported to the system (by data capture). In addition quantity scrapped, materials consumption, direct labor, stop times and reasons can be reported.

As tools for analysing the consequences of a schedule, the system generates graphics presentations of load plans, schedules for orders, workstations and workcenters, and order status.

3.2. Scheduling in a CIM-environment.

The system can be used as a standalone system, but much of the input to the system can be taken from other systems, and much of the output can be given to other systems - putting the system into a CIM environment. (See figure 3).

For instance the production orders can come from a customer order management system or a long-term planning system. Product structure and materials requirement can be taken from a CAD-system or from a process planning system. From the latter the process plans can also be taken.

In an automated factory workorders, transportorders and materials requisitions can be sent to a plant controller that controls the production processes, transport and store according to the schedule.

3.3. Hardware.

The system is implemented on SUN and HP 9000 workstations. These are connected to Ethernet by TCP/IP protocols, making it possible to communicate with other computers.

The system is programmed in Lucid CommonLISP with Flavours and Windows. This is a standard programming language implemented on many computers.

3.4. Test site.

The test site for the system is EB Datakabel A/S (subsidiary of ASEA Brown Boweri) which is producing cable. The company has about 200 employees. It produces about 1000 types of cable and have at any time about 400 orders in production. Each order has about 10 operations. They have two plants with altogether 225 machines. This makes it a considerable scheduling problem.

Production scheduling in a CIM environment

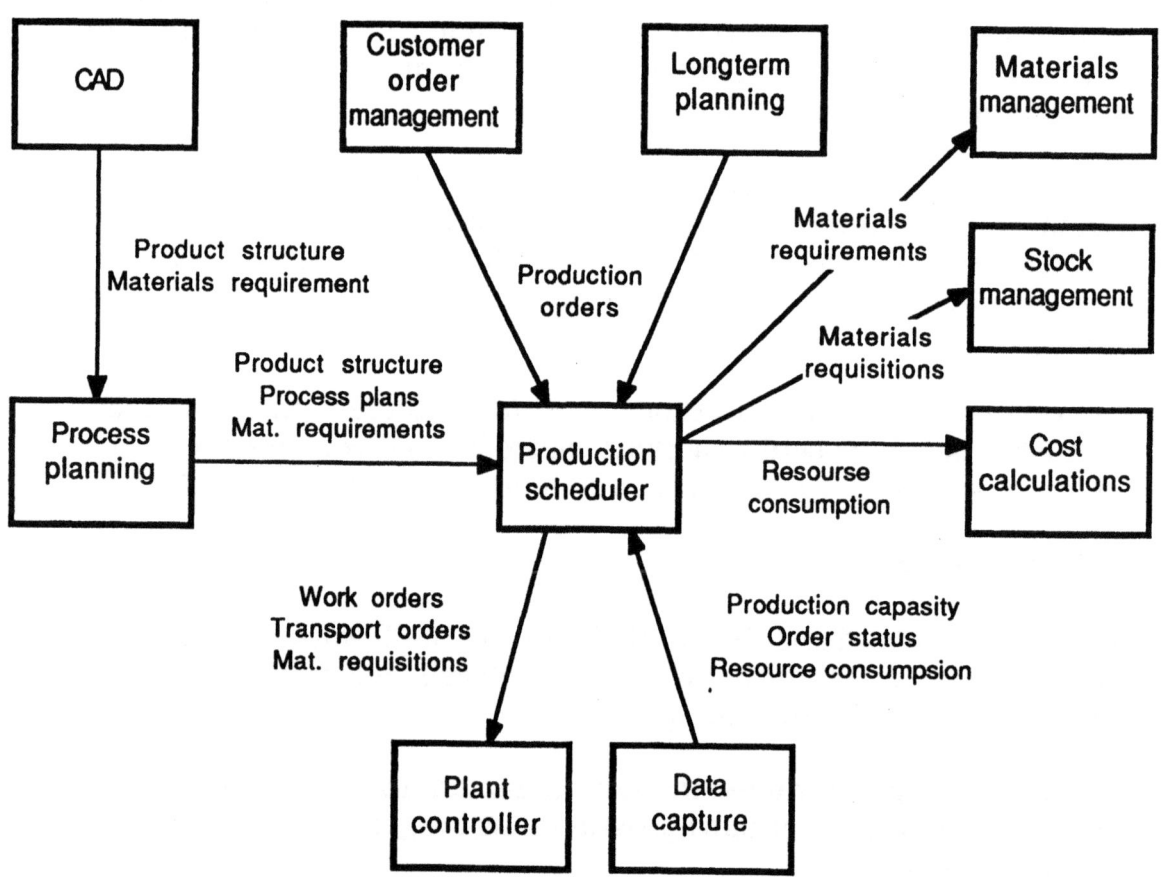

Fig. 3. Production scheduling in a CIM perspective.

At the test site the scheduling system is to be integrated with a material management system and an automatic data recording system. (See fig. 4).

4. SYSTEM FUNCTIONS.

This section gives a brief overview of the functions of the system. The system has functions for:

- Product and process description.
- Production resource description.
- Factory calendar description.
- Production order description.
- Scheduling and schedule analyses.
- Releasing orders for production.
- Data capture.
- Generating reports.

4.1. Product and process description.

The product and process description is a description of the product - what it consists of - and how it is produced. (See fig. 5).

A product can be defined in a hierarchical structure of assemblies and subassemblies. Each assembly has its own process plan, making it possible to perform operations in different assemblies in parallel.

The process plan is a list of operations. Each operation describes what is to be done, what and how much materials are used and what parts/assemblies are used. Furthermore it defines which alternative workstations can be used to perform the operation, and the time needed at each alternative.

4.2. Production resource description.

The main production resources are organised in workstations. These resources can be machines or employees. A workstation may only perform one operation on one order at a time.

The workstations are grouped in workcenters. All workstations within a workcenter must be able to perform the same operations and have the same production capacity. The machine capacity and labor capacity is given for each workstation. The workcenters are grouped into shops and shops into plants.

Common resources like tools, pallets and transport units that will be occupied for a period are also described.

Integration with longterm planning and data capture at pilot factory

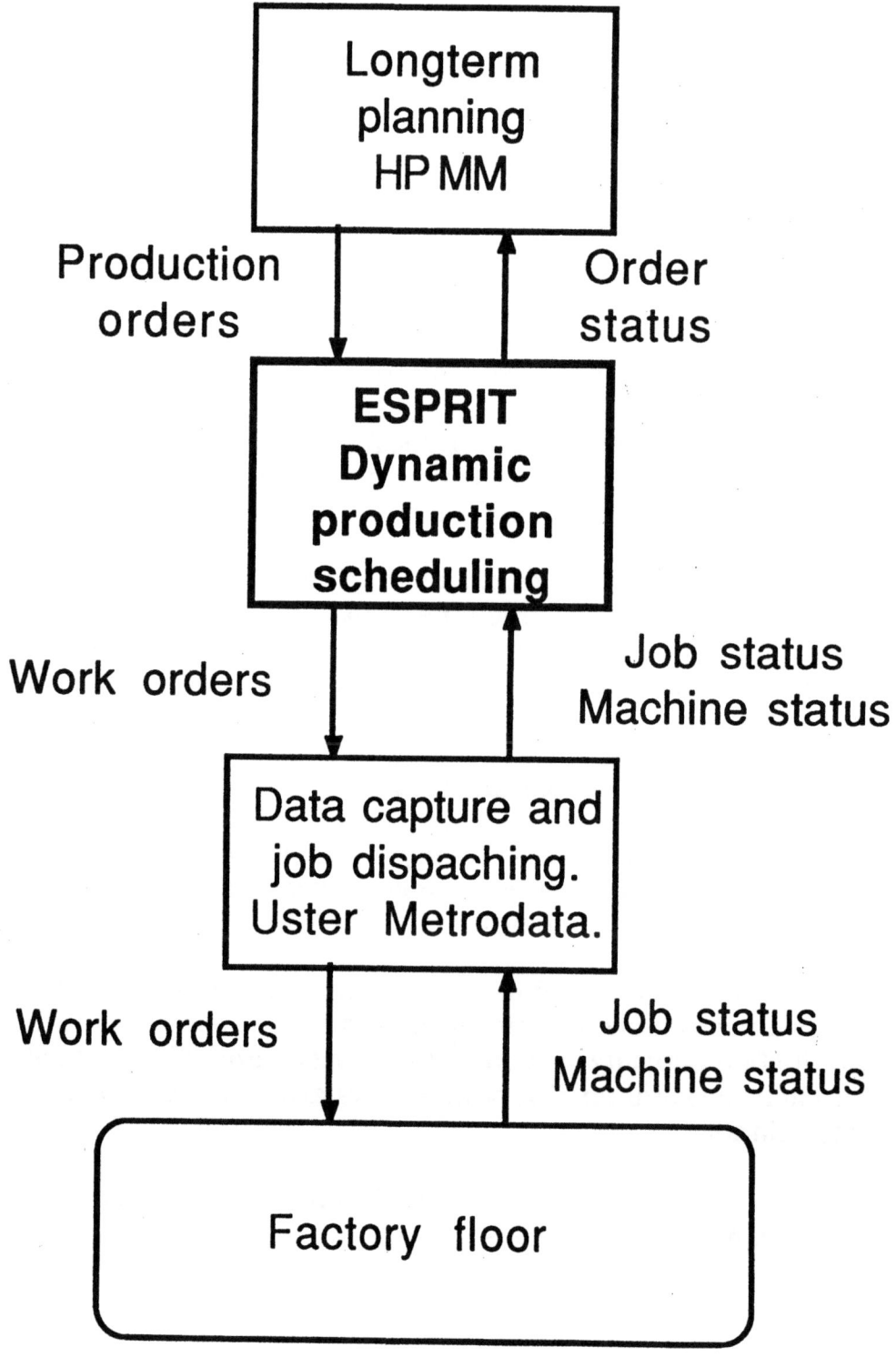

Fig. 4. The integrated solution at the test site.

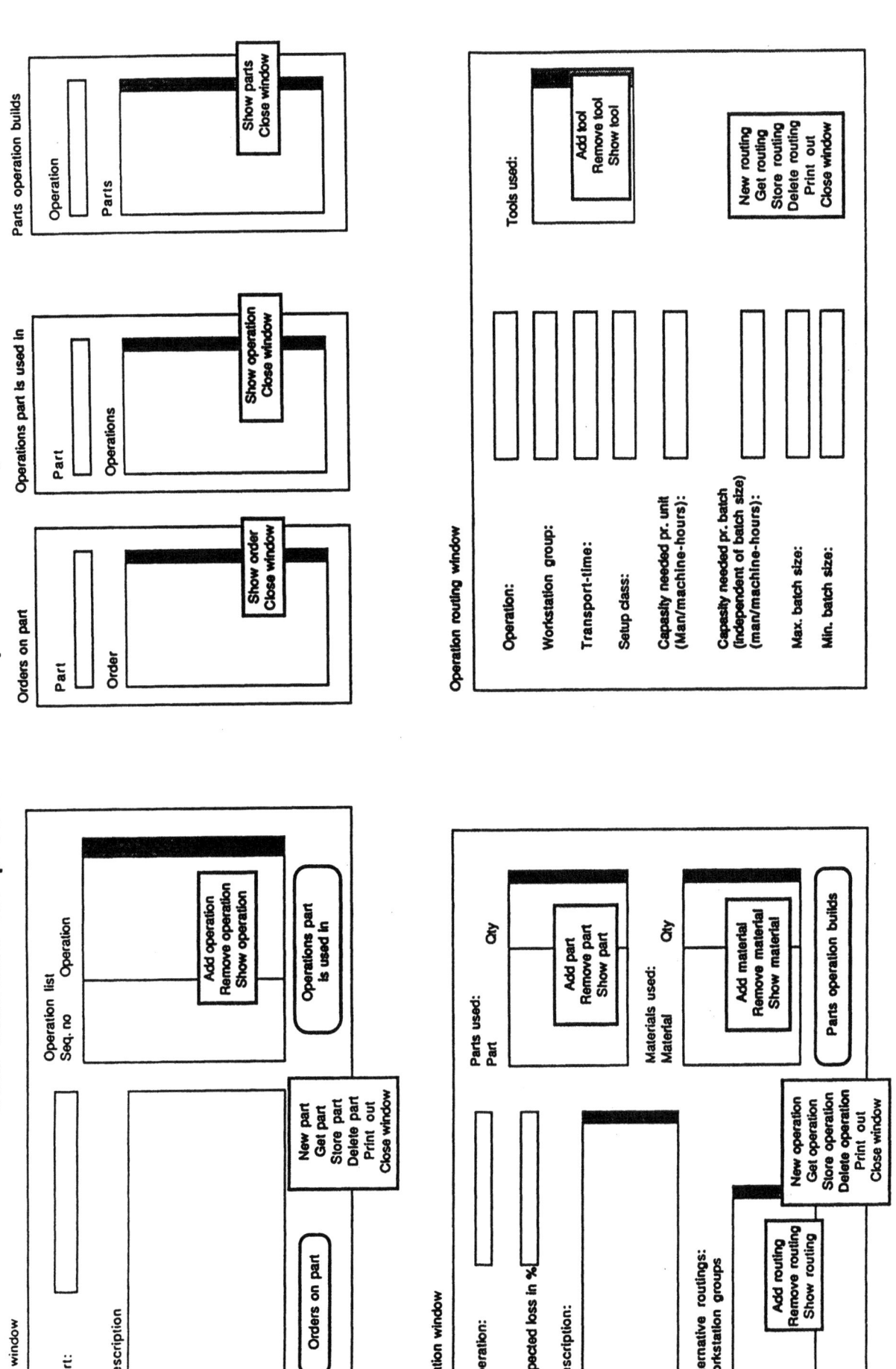

Fig. 5. Windows for product and process description.

4.3. Factory calendar description.

The factory calendar defines the available time for production for each workstation. (See fig. 6).

It defines the holidays and working-days, the number of shifts pr day, the length of each shift and the breaks (lunch-break etc.) within a shift.

The factory has a common calendar for all workstations, but it is possible to deviate from this calendar for a workcenter or a shop. This can be used to change the available production time at a workstation changing the capacity for a period.

4.4. Scheduling and schedule analyses.

4.4.1. Entering orders.
Each production order is entered with order identification, part identification, the volume to be produced and the due date. The system makes a working copy of the master part description and its process plan. The orders are then ready to be scheduled.

4.4.2. Scheduling.
The scheduling is based on OPT philosophy. The goal is to create the optimum schedule regarding throughput and inventory, provided delivery within due time. It will take into account the actual production capacity at each workstation. Alternative workstations can be used for each operation. The scheduler is handling parallel operations, overlapping operations and order split.

4.4.3. Schedule analyses.
The user is provided with powerful tools for analysing the consequences of a schedule regarding loads and lead-times (what if simulation). (See fig. 2, 7 and 8).

These tools are graphical presentations of:

- Schedule for each workstation.
- Schedule for one order.
- Rough schedule for several orders.
- Load-plan for a workstation or workcenter.
- Load for several workstations in one period.

4.5. Releasing orders for production.

When a schedule is completed, the orders can be released for production. The production information for the first operations are generated.

The workorders for the next operations are not released until they are needed. This will make it possible to do re-scheduling.

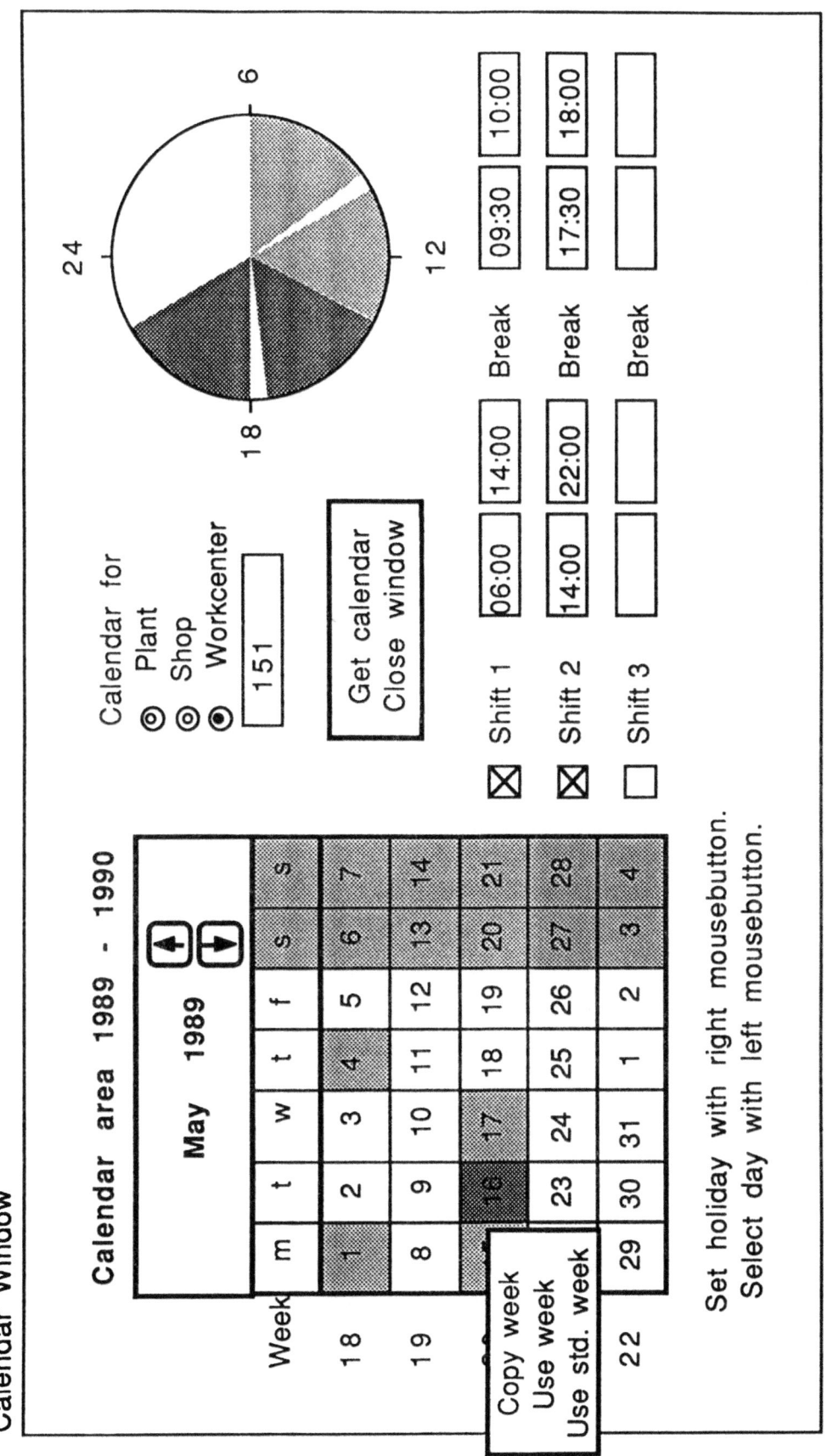

Fig. 6. Window for calendar description.

165

Order schedule
Order status

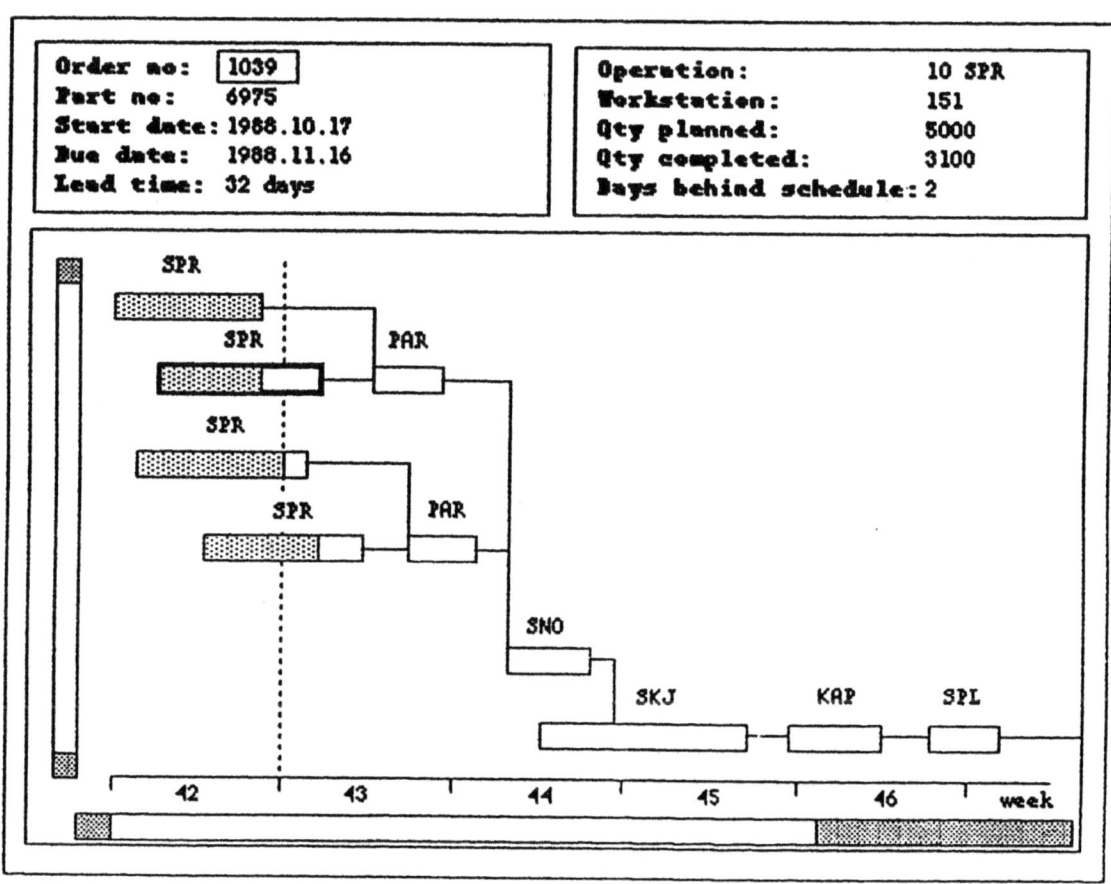

Fig. 7. Window showing the schedule for one order. Each box is a job (operation). The gray area shows the part of the job that is completed.

Workcenter
load

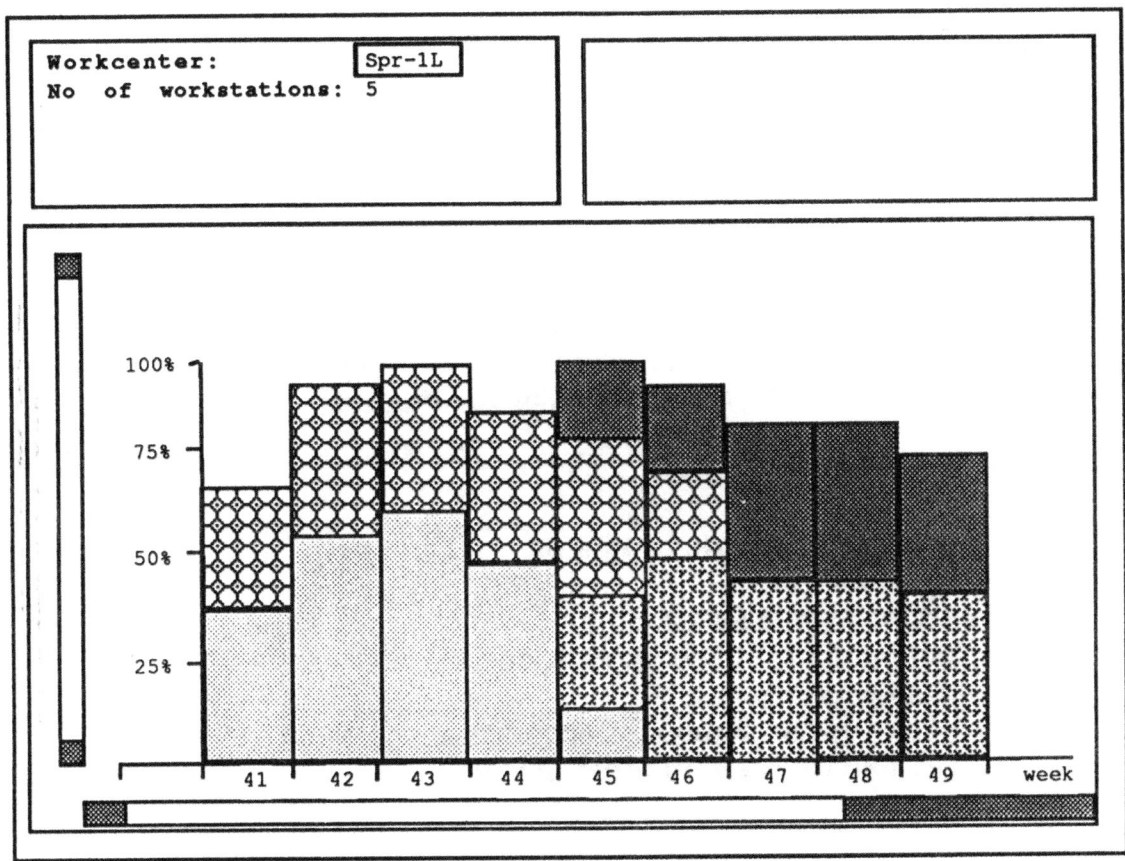

Fig. 8. Window showing the workcenter load for different categories of orders. The system will not generate a load of more than 100%.

4.6. Data capture.

Data capture is used to report order status and actual production capacity to the system.

4.6.1. Order status.
From each workstation the following can be reported:

- Order has arrived workstation (is in queue).
- Set-up is started.
- Processing is started.
- Quantum completed.
- Quantum scrapped.
- Quantum re-worked.
- Order completed at this workstation.

4.6.2. Actual production capacity.
The actual capacity at each workstation can be changed any time. Both machine capacity and labor capacity can be changed. A workstation can be reported to be down or under maintenance.

4.7. Reports.

The following reports can be produced:
- List of all parts.
- Operation lists for each part.
- Routing alternatives for each operation.
- List of all workcenters.
- List of all workstation for each workcenter.
- List of all orders with order status.

5. USER CHARACTERISTICS.

The users of the system will be production planners. They will usually have good knowledge about the production process and the production resources, but are not expected to have any knowledge about computer systems.

6. REFERENCES.

Goldratt, and J. Cox, <u>The Goal</u>, Scheduling Technology Group Limited, 1984.

Guida, L. Majocchi and E. Moglia, "State of the Art of O.R. and A.I. Techniques For Scheduling", Esprit 2434, 6 Months report, Supplement B1, pp. 2-45, 1989.

Hanan and J. M. Kurtzberg, "Placement Techniques", Chap. 5 in <u>Design Automation of Digital Systems: Theory and Techniques</u>, Vol. 1 (Ed. M. A. Breuer) Prentice-Hall, N.J. (1972) pp. 213-282.

M. Kaldager and S. Ulfsby, "Bottleneck Scheduling", ESPRIT 2434, 12 month report, 1990.

F. Wang and M. Rosenshine, "Scheduling For a Combination of Made-to-Stock and Made-to-Order Jobs in a Job Shop", *International Journal of Production Research*, Vol. 21, No. 5 (Sept/Oct 1983), pp. 607-616.

THE IMPACT AND INDUSTRIAL EXPERIENCE
OF CIM

Integrated Monitoring and Diagnostics in Modern Automated Manufacturing

J-M Le Veaux   M Fraile   G Mazzocchi
PSA-DITA   ENASA/PEGASO   MANDELLI

1. INTRODUCTION

During the past ten years the important changes which have taken place in the feature of the international market have imposed variation of equal importance in the way of producing to the manufacturing industries.

The necessity of improving their competitivity together with an easier access to the new software techniques have pushed these industries to acquire the capacity of reacting to these changes in the market in a much fast and more flexible way than in the past, by means of computerized automation. This real revolution, which is not yet over, has requested and still requires heavy capital investment a return of which can only be obtained through high productivity of the actual plants.

In order to continue improving the productivity, research and implementation of management strategy is necessary above all. This primarily signifies an improvement in quality and a greater availability, not only of the single machines but also of the whole system.

In order to reduce the effects of a fall in productivity which was not foreseen, several methods can be adopted the efficiency of which varies to the changing of the context in which they are applied.

The better known methods are as follows :

. over produce	- allowing for scrap or loss of availability
. produce for stock	- produce far in advance of due dates to buid up a stock in anticipation of problems
. install back-up devices	- to provide standby facilities or parallelism in production routes
. increase planned maintenance	- to reduce equipment failures
. re design plant	- to improve reliability and maintenability
. fault tolerant control	- give the control and management of the plant the capacity to maintain desired levels of productivity in the presence of fault conditions

The first two methods which are perhaps the most common due to their easier application, become unacceptable from an economical point of view when competition pushes companies towards a JIT type production. The same thing applies when a plant is provided with a very expensive back-up device which can seldom be accepted.

On the contrary, the increase of the preventive maintenance can only be efficient if withheld within specific limits beyond which the interference with the machinery can produce unacceptable periods of standstill. There is the risk then that the numerous interventions could produce much more faults than they can prevent.

The solution of introducing improvements to the design of the machine can radically eliminate most causes of the faults but this is not always possible for the existing plants (retrofitting) and it is often only applicable on new installations.

In any case, a good project can not certainly prevent all the "failures" many of which can derive from a bad maintenance, from a bad use of the machines or simply from their wear. Moreover, the machines are not the only possible source for plant productivity loss, indeed many problems can originate from other factors and circumstances which are typical of the working process on the whole.

The tools can be taken as a very good example : a wrong choice or a wrong way of using them, their progressive wear and possible breakage, all these factors greatly influence the quality of the final product and the standstill periods of the machines.

In conventional type production plants the impact of faults is today restricted due to specially trained operators which are always present nearby the machines and therefore always ready to operate in the best way.

However, in the automated plants this is not possible since one of the objectives is to be able to operate with less qualified personnel or even with almost the total absence of operators during the unmanned shifts. In this way the impact of faults, even those of little importance, can produce disastrous effects on the productivity such as long standstill periods of the plant or the production of a whole series of rejected parts.

For this type of plant a simple monitoring system which can draw the attention of the operators is therefore insufficient. What is instead required is a close integration of the monitoring and diagnostic functions in the system management loop so that a control can be also maintained during abnormal or fault conditions. Such a concept became known as "fault tolerant control".

2. FAULT TOLERANT CONTROL : CONCEPTS AND FUNCTIONAL ELEMENTS

The solution offered in the form of fault tolerant control seeks to address all of the faults which significantly impact on production. It therefore combines productivity monitoring, quality monitoring, process monitoring and machinery health monitoring.

The information from these different sources will be brought together to provide adequate diagnostics for the production activity as a whole. The results of diagnostics will be fed back into the control of the plant so that actions to minimize the effects of any disturbances can be planned and executed. This implies the integration of monitoring and diagnostics with plant control upwards from the lowest levels to establish an extended form of closed loop control, taking account of faults, and able to respond in a time frame appropriate to these real world events.

As already mentioned, in this way, the fault tolerant control simply appears as a further evolution of the real time close loop control (Fig.1). We seek to monitor the state of the manufacturing system, diagnose faults and degradation, then introduce controlling actions so as to minimize the disturbance to production.

In addition to the conventional functions the manufacturing control system must be capable of :

- status detection, this means the extraction of signals from the monitored process and the acquisition of data and parameters, which define the actual state of the object to be monitored
- status comparison, this means that the actual status must be matched with a predetermined desired status. The results of the status comparison are symptoms, which are input data for the subsequent diagnosis
- diagnosis, which analyses the symptoms on the basis of systematical strategies, localises the fault and determines its type and causes
- action planning, to select courses of action for recovery from a fault condition or for avoidance of predicted faults or to minimize the consequences of unavoidable failures. These actions may be :

  * issue alarms, warnings or safe shutdown commands
  * initiate repairs or preventive maintenance
  * reconfiguration of system hardware and software, eg. tool change, activate standby machine, execute further monitoring and diagnostics
  * replan the production schedule
  * modify process conditions
  * supply information to high level control for modification of machine design, product design, process plans or operational policy

The fault tolerant control concept, as per our description, was analysed and thoroughly examined during the ESPRIT 504 project phases (of which the ESPRIT 2349 project is the continuation) and its applicability was demonstrated with its implementation in a flexible machining cell test bed of this project.

Among other things, for this realization, we have operated on the control system both at the numerical control level of the machines involved and at the cell control level with specific hardware and software products expressly planned and realized for this aim.

More specifically for what concern the machines, a purpose built diagnostic monitoring computer (Data Acquisition and Analysis System - DAAS) was realized, functions of which are essentially those of acquiring signals, analysing signals and interpreting the health indicators to diagnose faults and select actions. In addition, DAAS incorporates facilities to enable it to communicate with the CNC and the cell controller; data acquisition and analysis are synchronized accordingly to provide real time status information.

For what concern the cell, a software module called "INTERPRETER" which can communicate with DAAS was realized. Once it has eleborated the DAAS reports and alarms, it is able to select the appropriate actions which will be passed to the cell control.

The main objective of the ESPRIT 2349 project is to further develop these concepts and realizations by demonstrating their validity in real production environments such as the PSA and ENASA/PEGASO plants which will be described here under.

3. POTENTIAL IMPACT OF INTEGRATED MONITORING AND DIAGNOSTICS IN AN AUTOMATED PLANT

In the previous paragraphs we learned how a strict integration of the monitoring and diagnostic features with control functions represents a convenient way of facing the problems of reliability and availability of modern manufacturing plants.

The potential impact on the way of producing deriving from the introduction of a system which implements the concepts of fault tolerant is estimated as being consistent. In particular we expect most of the effects on the following application areas :

(a) Tool Management

We have already underlined the critical aspects of the tools and their management in the flexible machining systems; a duration which is less than the one foreseen or an unexpected breakage could provoke damages to both the tools and the machines as well as the loss of productivity.

By means of a strict integration of the command functions of the cutting process and the informations collected from the appropriate probes (iron probes, vibration probes, etc.) it is possible to monitor the state of the tools in the different working conditions and according to specific reference parameters for each operation (synchronism with part program). This, for example, permits the recognition of premature wearing trends so as to plan the tools replacements by means of a predictive logic instead of preventive logic or of an intervention once the breakage has occurred.

This way of operating should therefore permit a reduction in the production costs which directly derive from the tools as it allows the following :

* Optimization in the use of the tools with replacements according to their real life and not to their estimated life
* Reduction of damages on the machines and on the tools due to unexpected breakages
* Reduction of standby periods due to the replacements
* Improvement in the global quality of the production

(b) **Maintenance Management**

The maintenance of the machine is an other crucial aspect in the functioning of an automated plant; if unforeseen stoppages of one or more machines occur during unmanned shifts, these could produce a considerable loss of productivity. Today this problem is faced in terms of preventive maintenance and therefore to statistic evaluations of the life of the various components, however, at times this technique leads to useless replacements or to standstill periods due to breakages which occur earlier than expected.

In this case also, an integration of the monitoring functions which permits the correlation of the various symptoms collected from the probing system, should permit the diagnostic functions to identify the trend of degradation of the devices and therefore to prevent most damages.

What we essentially expect from a system of this type is the possibility of switching from a preventive and/or curative maintenance to a predictive maintenance, with a substantial reduction in the managing costs of the plants by means of a reduction in the standstill periods which have not been planned.

(c) **System Control and Management**

A monitoring and diagnostic system which is integrated with the control, allows a wider general knowledge of the plant status. This means that the occurring of an event is known at the various control levels in different ways and according to the effects that this can produce. The automated management system is then able to decide the most suitable measures to take at the right moment.

For example, a dangerous fault on a machine could have as local answer its immediate stoppage. This fact propagates as a fault event for the production cell and the measures that the control can adopt is to no longer serve the stationary machine and if possible to divert its work load to other machines.

The evaluation of the original fault and an estimation of its duration, together with the effects that it can produce on the productivity in general, supplies information to the management system which is useful for deciding upon a possible radical rescheduling of the production.

The first result is an optimized plant management with degradated conditions.

(d) **System and Machine Design**

The scenario so far described inevitably affects the actual plant design. By means of a more sophisticated diagnostics of the existing plants, useful information and suggestions are obtained for the design of new machines and devices for automation. In this way, both the maps of the probes on board and their typology can be redesigned; moreover some causes of problems can be eliminated directly at their source by taking this in consideration during the design phase.

A consistant impact is also foreseen in the control systems where new functions are necessary and therefore there is a possibility that these are to make the proper use of artificial intelligence techniques.

4. THE INDUSTRIAL DEMONSTRATOR SITES

4.1 GENERAL CONSIDERATIONS

As already mentioned, the main objective of this project is the realization of a monitoring and diagnostic system, which is strictly integrated with the relevant control system (fault tolerant control), in three industrial demonstrator sites. This objective has widely influenced both the composition of the consortium and the project organization.

In the consortium, besides the four partners whose main task is Research and Development (STEWART HUGHES LTD, ADERSA, AMTRI and IKERLAN) there are also three industrial partners (PSA, ENASA/PEGASO and MANDELLI) who are involved in directing the research activity towards solutions to the real and not only theoretical needs.

In particular, while MANDELLI as constructor and utilizer of automated systems is to share its experience and to furnish informations about the market requirements by means of constant contact with its customers, PSA and ENASA/PEGASO are involved in supplying concrete examples, from their production reality, to which the concepts of Fault Tolerant Control are to be applied as discussed in the previous paragraphs.

For this aim, the project has been organized in such a way as to provide a level of identification for the possible demonstrator sites followed by an activity of Technical and Economical Audit for the identification of :

1. The economic justification and effectiveness for the application of diagnostic monitoring and fault tolerant control concepts. The objective was to highlight those methods which will be most cost effective and represent a good potential on investment.
2. How the machinery and systems presently operate and what sensors and control systems exist.
3. What techniques and methods would most suitably be added to achieve the degrees of fault tolerant control considered to be most effective.

This procedure which fundamentally is to act as a guide for the project phases which are to follow has been realized by means of a series of meetings with the site personnel whose complete collaboration and involvement has been essential.

The French Deputy Project Manager (DPM) ADERSA, has been in charge of this activity for the PSA sites while the Spanish DPM IKERLAN has been in charge for the PEGASO site. Support and evaluation came from STEWART HUGHES LTD, AMTRI and MANDELLI. An audit report was produced as a project deliverable. Details of the single demonstrator sites are as follows.

4.2 ENASA/PEGASO DEMONSTRATOR SITE

The PEGASO plant in BARCELLONA produces all geared axles and gearboxes for the truck assembly plants. The audit procedure has been undertaken in the machining area where the defferential gear housing are machined. In this area there are four MANDELLI machining centers and an IRSA vertical lathe.

The lathe is used to prepare references surfaces of the parts before entering the Mandelli Machines. Each of the machines is coping with approx. 20 different parts for the time being. Presently the activity has three manned shifts, where the operator's intervention is limited to loading/unloading parts, assembly within process, deburring and installation operation control.

The group of the four machining centers has been installed very recently and it is in the process of being updated to a fully automated flexible operation by Mandelli itself. This will give the opportunity for the introduction of fault tolerant control concepts working in close association with the operator and original vendor.

The expected economical impact due to the introduction of the 2349 technology in the PEGASO pilot plant is resumed as follows :

* Investment reduction by means of improved production rate and higher availability : 55 M.Ptas (= 1 Machining Centre)
* Deferred machines replacements investment :
  7 M.Ptas (1 year at 15% discount rate)
* Maintenance and breakdown costs reduction :
  5.12 M.Ptas (50%) per year
* Scrap and rework costs reduction :
  2.4 M.Ptas (50%) per year
* Tool investment reduction :
  1.47 M.Ptas (15%) per year
* Inventory reduction :
  1.56 M.Ptas per year

4.3 PSA DEMONSTRATOR SITE

The PEUGEOT plant is near MULHOUSE and produces the 205 car model. An audit procedure for the mechanical workshop where are machined and assembled the front suspension triangles for the 205 GTI and 309 GTI has been achieved.

There are four GRAFFENSTADEN machining centers which are overloaded. Three shifts per day work from Monday to Friday and one shift on Saturday. To machine the parts, 20 tools were initially used for drilling, boring, chamfering and face milling operations. There are 24 machining operations.

To increase productivity, composed tools have been set up to save the tool change time. These specific tools are very expensive. The present maintenance strategy is quite varied :

- planned preventive maintenance moving towards predictive maintenance through statistical process control mainly on oil analysis and consumption on some mechanical parts like ball bearings
- curative on the metal removing process

The objectives of the project are to address all the machine faults in an homogeneous way moving progressively towards predictive maintenance in order to attain the concept of fault tolerant machine. This implies a great effort in fault detection of the metal removing processes and on drive units monitoring.

The demonstrator machining centers will soon be equipped with a Data Acquisition System (DAAS) allowing a vibration analysis in the first part of 1990 and spindle monitoring. A hydrostatic spindle should be retrofitted for force measurements and fully implemented by mid 1991.

The potential economical benefit which results from the introduction of the 2349 technology on all the machines, is resumed as follows :

* Costs reduction for :
    Direct Labour   1%   (14 KFF)
    Scrap Allowance   10%   (24 KFF)
    Tools   10%   (70 KFF)
    Maintenance-Labour   5%   (60 KFF)
* If we consider the TU values of 80% (budget) and ~ 76% (achieved) it should be possible to achieve, or at least closely approach, the budget value using 2349 techniques and thereby avoid the need to operate the extra shift on Saturdays

4.4 PSA CITROEN DEMONSTRATOR SITE

The CITROEN plant is in ASNIERES near PARIS. This plant specializes in the machining of hydraulic and precise parts. We achieved an audit procedure of a workshop where the supervision regulator of the XM car model is machined.

There are six STEINEL machining centers which are overloaded due to the success of the XM. Three shifts per day work from Monday to Saturday. There are 45 machining operations per part. 34 tools are used for pointing, drilling tapping and face milling. These parts have been machined since the end of 1988. Initially the planned production was of 120 parts per day. Now they have to produce 420 parts per day. We do not have a significant historical account of the relevant facts. They have the same maintenance strategy as in MULHOUSE : measures, consumptions, oil analysis and cycle time.

The basic objectives of the ASNIERES demonstrator site are similar to those of MULHOUSE. However, the setting is different in the sense that the STEINEL machines are new without historical maintenance data; therefore, it is first necessary to acquire information, followup the elementary evolutions. As these machines constitute a production bottleneck, it is of primary importance to have a good mastery of their availability.

The potential economic benefit which results from the introduction of the 2349 technology is the following :

* Costs reduction for :
    Direct Labour   1%   (17 KFF)
    Scrap Allowance   10%   (4 KFF)
    Tools   10%   (32 KFF)
    Maintenance-Labour   15%   (38 KFF)
* If we consider performance against "Standard" performance, production was initially assessed as being 96 pieces per three shifts day per machine. Actual performance has been about 84. 2349 technology might be expected to reduce the "off-Standard" by up to 33%.

5. CONCLUSIONS

The economic success of highly automated manufacturing systems depends very much on overcoming the problems of their often poor availability and reliability.

We suggest that a correct approach to this problems is an integration of advanced diagnostic and performance monitoring with the normal control functions in such a way that the system maintains control and management capability under all conditions.

Success in this area will be a key factor in improvements in the economic justification and successful implementation of truly integrated manufacturing systems. It will also stimulate changes in the approach to machine and sensor design, machine controller design, system architectures and the real time management of functions such as production scheduling, tool management and maintenance management.

References

H.K. Tönshoff, H.J.J. Kals, W. König et al. (1988) Developments and Trends in Monitoring and Control of Machining Processes . Annals of the CIRP

R.D. Puetz, R. Eichhorn, K.P. Faehnrich (1987) Fault Diagnosis on CNC Machines. IXth Internat. Conference on Production Research, Cincinnati

A. Rault, J. Brunnet, A. De Viq et al. (1986) Modelling and Identification for Machine Tool Fault, Detection and Diagnosis. Conference on Decision and Control, Athens

Report Technical and Economical Audit ESPRIT 2349 Project

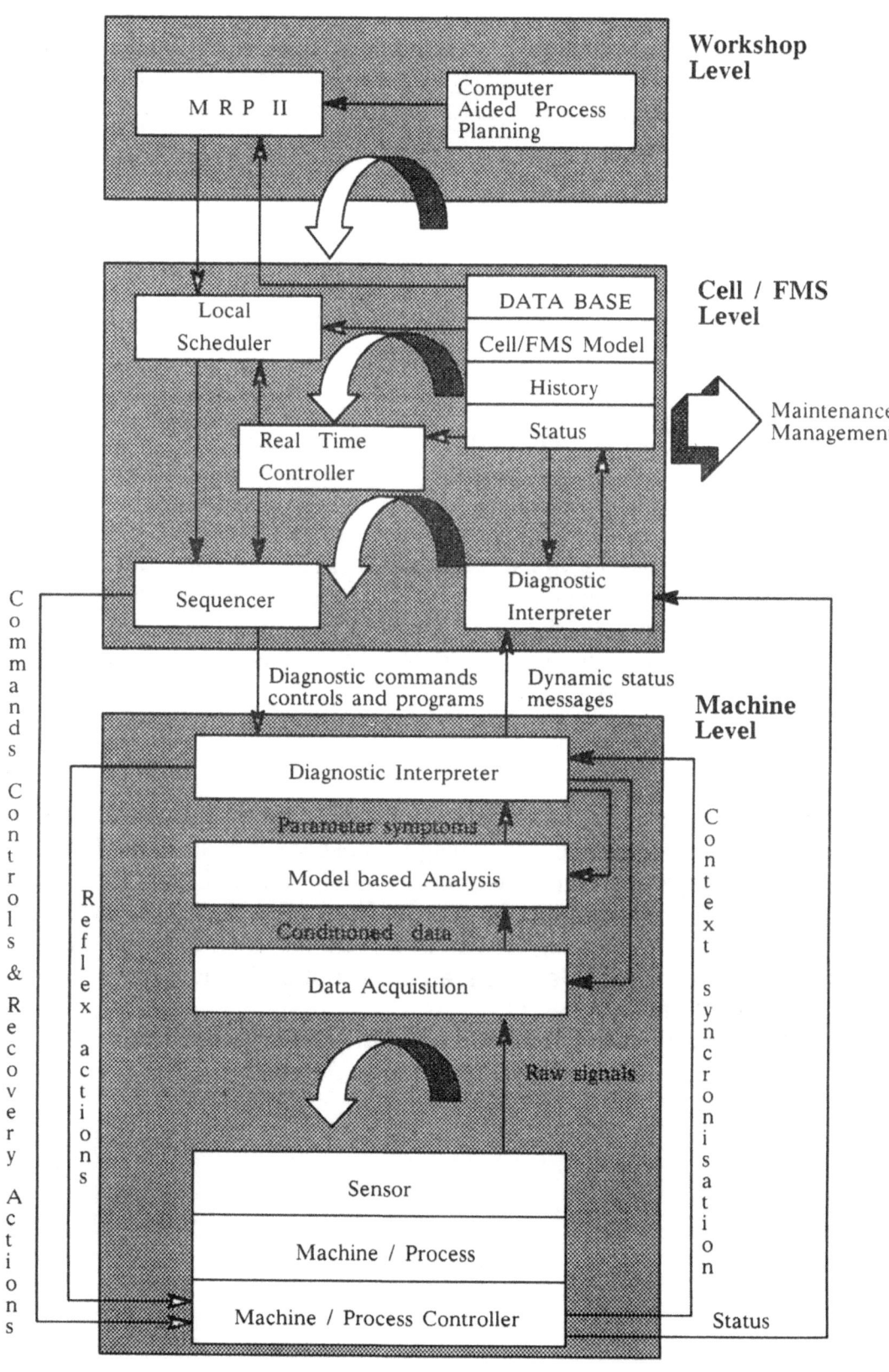

Fig. 1 Reference Architecture of Fault Tolerant Control in ESPRIT 2349

AN EXAMPLE OF A CIM REALISATION FOR AN ELECTROMECHANICAL MANUFACTURING FIRM

G.Ricottilli, R.Guiffrey, A.De Luca
Syntax Factory Automation, Turin (Italy)

1. INTRODUCTION

Factory automating has a complexity threshold beyond which, to obtain the best results, it becomes necessary implementing a full CIM system. Such a system, generally, implies a great complexity in all the main phases of the project: design, implementation, test, installation and start up. In this document we refer to such an experience trying to introduce clearly the problem and to describe the difficulties encountered during all work phases, together with the corresponding adopted solutions.

2. APPLIANCE ENVIRONMENT AND CUSTOMER'S REQUIREMENTS

2.1 Characteristics of the Target Factory

The factory under consideration operates in refrigerators production field. The production is almost entirely made by need plans for stock. These plans are developed by production management according to sale forecasts based upon information coming from the firm's world-wide trading network. Production reaches nearly 4000 end-products per day (1100000 per year) and is organized by recurrent lots. A high level of end-product differentiation is needed to effectively face the wide variety of requests coming from the world-wide market, and fashion evolution. Inside the firm's catalogue, then, a very large number of codes can be found, differing only by variations in components, even though only a limited number of basic models is actually produced.

From a production process point of view, we identify only two main process structures: one corresponding to the manual production line, the other to the automatized one. The technology employed is well proved and sound: for this reason the process structure is quite settled.

In the production process we can distinguish two main phases. A first one concerns the production of low characterized semi-manufactures: at this level the product flow is very high and setup of machines is a very time consuming operation, which causes high costs in increasing mix. The second phase makes use of the semi-manufactures to build the end-product; the product flow in this phase is lower and the production process is more complex and articulated on many different branches joining in assembly appointments. Such a production structure has some operative critical points due to the small size of existing interoperational buffers, which brings to a need of very precise and timely synchronizations among production steps. Moreover, the various working cycles differ on automation level, production flow, standard lot size and flexibility. For this reason a wide variety of operative organization strategies is needed: "push" management (activity triggered by components arrival), "pull" management (activity started by the control system upon checking components availability), with or without the possibility of rescheduling the production lots. Other requirements are the optimum balance of working load between parallel lines, a fair managing of a set of common resources, as tools or vehicles, and the timely adjusting of production orders upon unexpected events, as culls or machine failures. An important feature of the production plant is its topological and functional division in a number of autonomous different areas: this allows a modularized control and managing of the plant activities.

2.2 Customer's Goals and Requirements

The set of features of the product, its market and the production process outlined bring to some base constraints that, together with customer's requirements, circumscribe the possible solutions. The main one is that the size of the market requires a great skill to differentiate the end-product in terms of look, accessories and quality; for this reason a dynamic management of the component list and of the corresponding technological variations within each working step is a compulsory feature. Luckily, thanks to process technology, it's possible a static definition of the process structure. The main customer targets were:

Flexibility Target: the system had to be able to support a high mix level (variety of different end-products processed in the same day), and to adapt to the introduction of new models with substantial differences in component list, keeping a sufficient margin to satisfy possible order peaks; under this respect the goal was to be able to produce 360 different models per day and all the models in one week;
Minimum stock target: reduction to the minimum of raw materials and end-product stock level;
Quality target: the system had to be able to timely manage some anomalies confining the propagation of their effects and adapting the other involved processes to avoid wasting of semi-manufactures at the next appointment. The timeliness requirement is necessary to keep the wasting or temporary shelving of semi-manufactures at the lowest added value and highest reusability levels.
Quantity target: the system would have to be able to product about 1142000 codes in one year, with 272 working days per year and 3 day shift of 6 hours per day. This didn't represent a strong increase of productivity; in fact the customer was more interested in the other targets.

2.3 Context of the Problem

As already mentioned above, the problem has been faced in a CIM global perspective. The adopted approach has been to define first an organization of the information flow adequate to the foreseen targets; two distinct ways have been adopted: the use of flexible automation tools and the optimum integration of the main production managing functions. The solution core has been recognized in a production control system (PCS) acting on the whole plant and distributed on each area, the boundaries of this system being represented at a higher hierarchical level by an MRP system and at the lower hierarchical level by a layer of various process controllers.

The whole system has been framed in a CIM architecture providing the following hierarchical definitions:

Level 4 : System for General Production Planning and
          Materials Requirement Planning
Level 3 : System for master production scheduling and
          plant processes coordination
Level 2 : System for control of area activity
Level 1 : Numeric control of machines
Level 0 : Machines of the plant (shop floor)

The levels 3 and 2, subjects of our work, will be treated on the next chapter.

The level 4 system. This system is realized with the COPICS package in an IBM environment. It manages the acquisition of raw materials and maintains the lists of end-products and semi-manufactures composition with a synthetic visibility of the production process structure. The system communicates to level 3 the new lists and the corresponding material data, and waits for a producibility verifying outcome or an anomaly message; moreover, level 4 receives from level 3 the communication of materials stock status. Level 4 supports also the accounting management of all the materials known at every level; these data are updated upon the communications of level 3 about accounting evolution of materials in the plant.

Another main function of level 4 is the daily production planning; this plan has a three months horizon and is daily communicated to level 3 with a visibility window of five weeks. Level 3 at the end of the working day returns its corresponding production plan, after a replanning and an updating based upon the production of the day and the actual resources status.

The level 1. This level is comprehensive of all the machine/plant functions, i.e. the control and monitoring functions of actuators which make up the "shop floor": roller conveyors, AGV systems, robotized assembly cells, automatic painting lines, etc.. Obviously this level is strictly linked to the plant physical organization. This level interfaces with the upper one receiving specific orders related to its functionalities and returning the activity monitoring and if necessary some diagnostics about the plant status. As the adoption of flexible automating tools started with the design of the whole production control system, this level didn't represent a pre-existent constraint; however, this caused some problems in the last phases of the system development.

3. ADOPTED SOLUTION AND SYSTEM DESCRIPTION

3.1 Functional Description

As seen above, our goal was to design a CIM system able to meet four main targets: a flexibility target, a quality target, a quantity target and a minimum stock level target: the final goal of the firm management was actually to reach a kind of production closer to the make-to-order one. To meet these requirements, within the application context and according to the particular constraints discussed above, two different ways have been taken: on one hand the use of flexible automation tools, on the other hand the integration, as complete as possible, of the information system functions at the factory level, and the design of a PCS fully automatized and highly integrated in the CIM context, avoiding an unilateral approach to the problem (see Fig. 1). The roles related to the production management have been divided between level 3 and level 2 systems according to the following scheme:

1. level 3 communications functionalities:
   - acquisition, from level 4 system, of the production plan and of the list of raw materials waited for from suppliers
   - acquisition, from the level 4 system, of the updates in the lists of composition and material data
   - sending to level 4 updated production plans
   - sending to level 4 material accounting reports, stock and availability status, and new productions enabling
2. level 3 batch functionalities:
   - batch creation of the daily optimum production programs
   - batch creations of reports and data for level 4
3. level 3 operator transactional facilities:
   - operator definition of material and technical data
   - operator definition of all factory parameters
   - operator monitoring on production
4. level 3 real time functions:
   - real-time driving and optimum coordination of plant activities with updating of material and resource status
   - real-time managing of unexpected events like culls or machine failures and consequent timely adjusting of production orders
   - two way interfacing with level 2
5. level 2 real time functionalities:
   - real-time physical maps management
   - real-time production orders feasibility verification
   - real-time area global functionalities optimization
   - real-time data collection and recording for future processing
6. level 2 transactional functionalities:
   - area local functions and related procedures

In the design and implementation of a PCS able to warrant these functionalities, the following general requirements have been adopted:

- performance: the system is able to execute timely real time, interactive and batch functions; this must be possible under the specified load conditions and both in overload conditions or degraded situations;
- reliability: the system has to assure constantly the coherence between plant status and system plant image;
- application fault tolerance: the system can stop and restart in every state maintaining the coherence of data;
- networking: system architecture allows the system components communicate among each other in a fast and reliable way;
- expandability/flexibility: system to be designed must be easily expandable both in terms of computational power and data storing.

On the implementation side, some important guidelines have been adopted. The overall architecture had to be technologically and philosophically up to date. Moreover, the communication system had to be independent from all the elements (computers, PLC etc.) already present in factory; a substantial autonomy had to be reached between the various production areas, together with a high level of standardization in terms of operating systems, communication protocols and plant elements.

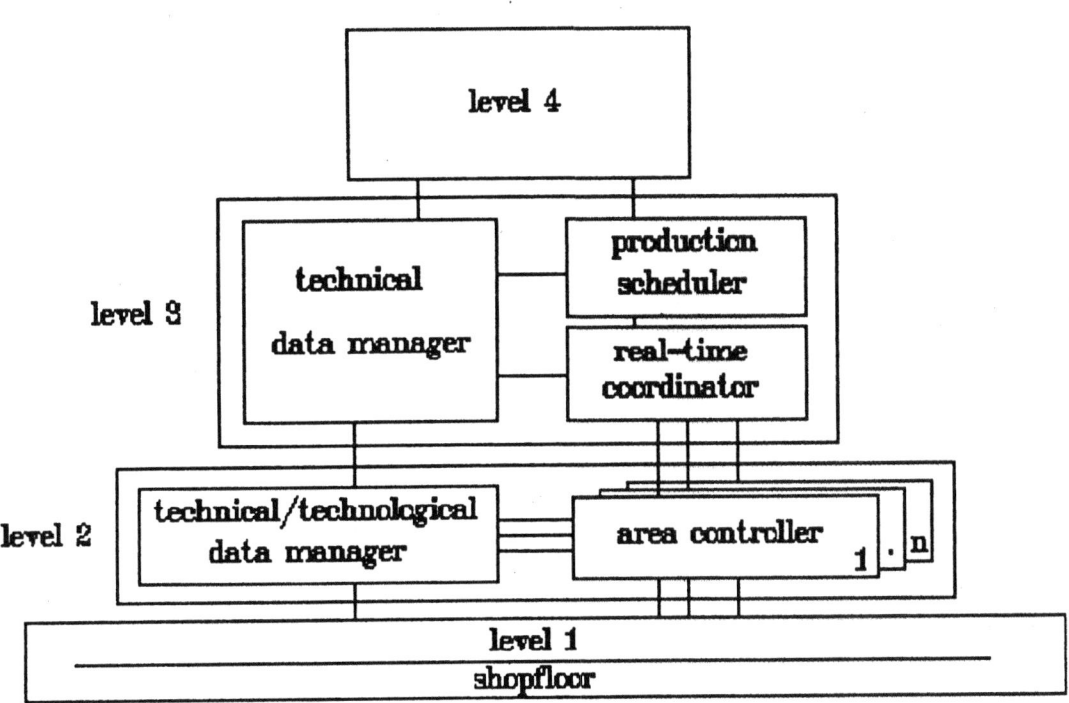

Fig. 1. Overview of the control system architecture

3.2 Fundamental Architectural Choices

The adopted solution came in a quite direct way from the main targets, the secondary targets and the above guidelines. In the remaining part of the chapter we describe the adopted solution in terms of the global system hardware and software architecture, that will be presented in a more detailed way in the next chapter.

Among the feasible choices a distribuited architecture has been chosen. At the plant control level we have placed a hardware able to manage the global data base: all the functionalities requiring access to this data base have been placed on the same machine, whereas all the other functionalities have been shared on other machines. Communication between machines have been implemented via files: these link files reside in the first machine, and accesses to them from other machines is obtained through the communication network. The area control functions have been divided into two different sets: standard control functions, placed on the same machine, and specific control functions, i.e. functions strictly related with the specific area layout, placed on the lower level local area controllers. All these area controllers see the higher level machine, supporting standard control functions, as completely dedicated to themselves. Lower level controller, such as cell/process controllers or shop floor plant elements, are linked with the area controller via local area network, while the other computers implementing higher level functions are linked via factory network. For the production areas a hierarchical distributed architecture has been chosen.

For the plant coordinator module an hardware fault-tolerant system was chosen to meet the required standards in terms of reliability: the chosen system was a CPS 32 XA 600 OLIVETTI/STRATUS. Some of the level 2 functionalities have been implemented on a similar system, a CPS 32 XA 400, others on programmable logic controllers, one for each production area: the functionalities have been shared between the two systems according to the main criteria already seen: however functions needing high computational power have been placed, in all cases, on the CPS 32. Using two CPS 32 systems to implement the main part of the PCS allowed us to deeply exploit a particular communication channel, STRATALINK, together with the features of the chosen operating system, VOS [4]. STRATALINK, a bus communication channel with a 2.8 Mbytes/s transmission rate, makes it possible to avoid access to the factory network for data interchanging between the CPS 32 systems.

The main criteria in choosing area controllers have been to standardize the models of PLC in the factory, both in terms of suppliers and elements, to standardize software development and documentation environment, to use the actually available know-how about a specific supplier, and, finally, to choose only one supplier also for other shop floor hardware elements. These criteria brought to choose SIEMENS as supplier. In particular we chose SIEMENS 150 U as high performance PLC, and SIEMENS 115 U as medium/low performance PLC. The same criterion, suppliers unification, has been adopted in choosing the Personal Computers for monitoring and diagnostics to be placed next to the PLC: we chose SIEMENS PC 16/11, an high performance PC, already in use in the factory but outside the current standards.

About the choice of the network, the following criteria have been adopted. About the higher level network, the factory network, the requirements were that the network had to be proven in other automation project of similar kind, support had to be warranted by important suppliers, the network had to be highly reliable, fully standardized and easy to install. For these reasons the Ethernet network, standard IEEE 802.3 was chosen. The same criterion of supplier unification has been adopted in choosing the area local network. The chosen network was the SIEMENS SINEC L1, a proprietary SIEMENS network able to connect up to 32 PLC for machine controlling. All shop floor typical elements are connected to these level 1 PLC.

4. SOFTWARE ARCHITECTURE OF THE INFORMATION SYSTEM

The software architecture had to meet the constraints imposed by the functional architecture in terms of performances, and to match the chosen hardware. At first sight it is clear that divisions among levels and internal modules were respected both for processes and programs; this was obtained keeping at the minimum possible level the coupling among them, limiting data interchanges to well defined shared archives, and managing all possible conflicts.

From a dynamic point of view we can distinguish among three kinds of processes: batch processes, real-time processes and user bound transactions.

The technical data manager module uses both batch processes and transactions to the user, depending on whether automatic computations on technical data or data acquisition from outside is required. To avoid possible couplings involving conflicts with the coordination module the data flow passes through archives and is managed by a dedicated procedure.

The production schedule module work in a different way due to the large amount of computation needed to optimize the production programs; for this reason it is implemented by a batch program working on a private "photograph" of the factory status extracted from the main shared data base. When computation is over (the module works typically during the night), the updated data base is merged with the main one and kept out of any possible conflict.

The third level 3 module, the production coordinator, has been realized mainly by real-time processes to timely and automatically manage the necessary synchronization between production phases and the control, testing and correction actions. All the inquiry and monitoring functions, interacting with the level 3 human operator, have been implemented in the transactional environment.

Looking at level 2, the area control functions have been implemented with transactional processes for what

concerns the technical and technological data manager
module, and with real-time and transactional processes for
the actual area control functions.

The software tools adopted were partly imposed by the
hardware choice: choosing a CPS 32 implied to work with
the VOS operating system. Many different working places,
as well as distributed data bases, are easily managed by
this system thank to its multitasking and virtual memory
management features. Other useful tools and utilities were
jointly provided, as a mask manager, the transactional
environment, a multiprocess debugger. Among these tools we
can remember the Forms Management System (FMS), the
Transaction Processing Facility (TPF), the Transaction
Handler System (THS), composed of a TP monitor and a Multi
Environment Handler (MEH). To help in the critical work of
analysis and performances optimization on the real-time
coordination module, a suitable software package has been
adopted for monitoring and statistical analysing software
performances: the TPS from Developer's Edge Ltd.[3].

A system to support comments inserting in the software
programs and automatic documentation production was also
used: this was the KeyOne from LPS [2].

5. IMPLEMENTATION AND INSTALLATION STRATEGIES

5.1 Implementation

Given the huge size of the project and the size of the
software programs to be written, a rigorous organization
of the design and testing work was found to be compulsory.
A programming standard was defined, setting the
directories organization as well as the files and archives
naming rules. The same standard provided also the
necessary rules for the best use of the adopted tools,
like the transaction manager, the user interface manager,
and the automatic documentation tool. Finally, a set of
programming rules was defined, together with rules for
errors and data access conflicts treatment.

Some specific activities were decided to deal with the
system test and validation phases. In particular the level
3 real time coordination module, because of its data
access concurrency and its cooperative functional
architecture, required a systematic use of the on-line
monitoring (already built for using in the final system),
the use of a multiprocess debugger and the development of
a synthetic simulator. This last tool was able to close
control loop, simulating level 2 interactive functions and
the evolution of production process (see also paragraph
5.3). These choices allowed us to reach the objective of a
very high software reliability and correctness, which
warranted a fast and safe installation.

5.2 Installation and Starting Strategies

A precise installation policy was adopted in order to
allow the autonomous installation of each area control
module, even without the coordination supplied by higher

level controller. Moreover, the installation of each area
control module has been done while normal production was
in progress. Contemporary has been adopted a double
gradual policy of installation for level 3 providing the
minimum risk of impact on lower levels and production.

The installation procedure started with the hardware
installation: the CPS 32 systems, the area and the
cell/machine controllers (PLCs), the factory network and
the area local networks; at this point began the
installation of the production control system software
components (level 3 and level 2).

Looking at level 2 system, we began installing the
area controller (PLC) software, so that production without
the level 2 standard functions on CPS 32 was possible.
Some problems arose in the areas containing complex
warehouses, not manageable by the area controller (PLC).
Here a suitable temporary strategy had to be added in
order to keep the production functions separate from the
warehouse management functions and make it possible to
install them separately. The level 2 modules had to be
provided with friendly interfaces to input production
orders and verify the production progress by operator. One
of the requests made by customer was to provide level 2
system with a mechanism to implement the automatic sending
of production orders among the various production areas:
although such a mechanism would have allowed autonomy from
the missing level 3 system, we avoided it, in order to
maintain the coherence between different level functions.

Looking at level 3 system, we began installing the
module for technical and material data management; this
allowed us to test the problems in interfacing with upper
and lower levels. The functional autonomy of the module
permitted its starting without the two other modules. The
second foreseen step would be the installation of the
module for production programming, followed by the final
step, i.e. the coordination module installation.
Currently the installation is waiting for a change in
progress on software due to a customer request about
obsolete management strategy of component stocks (see
chapter 6).

5.3 Level 2 Simulator (Tuning System)

As a support during test, integration and installation of
level 3 phases, the Tuning System [1] was employed. This
is a synthetic simulator of lower levels, developed "ad
hoc". This module implements ordinary communication
between level 3 and level 2, receiving production orders
from level 3, emulating events that may happen in the many
different working areas, and sending back to level 3 the
corresponding messages.

Simulator use. The need for using such a simulator can be
easily understood considering the complexity of the level
3 system to be tested. Level 3 comes to be a very
articulated system working on large amounts of data (the
whole manufacturing bill of materials, some information on
the plant, etc.). Moreover it is composed both of batch

modules that require a large amount of computational power and of real-time modules with very high efficiency and performances requirements. In such an environment, the integration of the various different components is a most important activity together with operational logics check. A synthetic production process simulator is very useful in this phase to close control loop, so providing the conditions for a correct system check. We can describe in more detail three phases in which the simulator has proved and still proves its usefulness:

1. test and function integration: given that the control system has a modular organization, connecting it to the simulator it was possible to activate some partial configurations, more and more complete as the work was performed, ending with the complete integration of the system. This was possible because the simulator was structured in such a way as to provide different performances dealing with different configurations of level 3.
2. control algorithms test: in this second phase the simulator was used to send to the outside (level 3) messages corresponding to the main events that could happen at the production level, to verify level 3 control algorithms.
3. installation support: once the control system has been checked, the simulator can be used also in the on-the-field installation phase. The installation can be performed in an incremental way, starting from an area and passing then to other areas; actually it can be possible to reconfigure the simulator to have it interacting with the control system only for those areas in which the true control system is not yet installed. This results in a less efficient operativity of the system, but it is very useful to reduce misfunctioning risks which can interfere with production activities during the phase of installation.

Functional architecture. The Tuning System has an internal control structure containing all information on the delays simulating production progress in the plant, taking into account all resources employed. Such a structure is flexible and detailed enough to cover the main events happening in the factory, as production reports, arrival of material in production area, scratch reports and so on. The resource model (a machine, a cell, a working area) is very simple and is made up of a set of parameters representing each production phase delay (stop waiting for an order, setting up time, execution time, time to transport the output item to the next area), process constraints (lots, mix etc.) and probability of random events (scratches, non-producibilities etc.). The delays can be pre-determined or random, depending on the chosen configuration.

Notes on software architecture. From the architectural point of view the simulator was conceived and built having its functions completely detached from the control system, and as much as possible non-interfering with its performances. This has been obtained making the Tuning System such as not to introduce delays in the control chain, and limiting the possibilities of conflict with the level 3 system in accessing the shared communication files. As all Tuning System parameters are contained in a table inside the data base, the system is also easy to reconfigure.

Advantages of the simulator. After test phase was completed, the costs/benefits ratio of the simulator appeared very good in terms of saved time, high reliability and correctness standard of software and assurance against the risks of the installation phase (risk to interfere with production process, to have long periods of degraded working and so on).

6. MAIN PROBLEMS ENCOUNTERED

During the various phases of the work many different kinds of problems have occurred. We are listing below some of the more interesting.

Among the difficulties linked in some way to the human factor we can mention a problem connected with the huge size of the project. It turned out to be difficult to manage the number of people involved in the implementation phase, especially given that part of them was provided by the customer and for this reason it was more difficult to control. To manage this problem a hierarchy was created in the project team, related to functional architecture of the system; moreover, very precise software development rules and standards were adopted (see paragraph 5.1).

Another problem met in the development and installation phases was represented by the requests by the factory personnel to provide level 2 (which was the level they mainly dealt with) with more powerful functions. This would have meant the introduction of control strategies often not fitting with level 3 functions.

Among the technical problems we can mention the small flexibility found in the already existing level 4, which led to serious difficulties in building an efficient interface to that system.

However the most important problem has involved a feature of the component substitution management: the automatic finishing of obsolete component stock. It arose because of a late realization by the customer about this matter. To add this feature to the system an analysis of the original design followed by software changes was needed. The work results easier and faster than expected due to the modularity features of the system.

It is worth to note that, as flexible automation machines were adopted on the production plant while design and implementation of the production control system was still in progress, a need arose to run-time modify specifications and already built systems, to match changes of the plant itself.

7. CONCLUSIONS AND ACKNOWLEDGEMENTS

The many critical features determined by the particular plant to be controlled required the use of a complex and very automatized solution. Such a solution has required a particular care to made it well accepted by the factory personnel, mainly because of psychological factors. A previous work of change in the plant outline, perhaps would have resulted in a reduction of the encountered critical points, simplifying the solution in terms of information system.

From the organization point of view the choice of paying a great attention on the test phase, acquiring or developing special instruments to monitor software performances and simulating the working system, proved to be a good one. Some different instruments for supporting the development and the documentation were adopted, together with the relative rules for software development.

Anyway, for its very nature this project resulted in a good increase of know-how to all participants at any level.

We like to thank Mr. Valerio Minero and Mr. Renzo Roveta for the suggestions they gave us. A particular thank also to Mr. Livio Pugliese for the precious help he gave us in the final writing of this document.

References

[1] V.Minero, C.Albera, G.Ricottilli "Simulazione dei processi produttivi finalizzata alla verifica ed ottimizzazione delle prestazioni e del comportamento dei sistemi di programmazione e controllo di fabbrica" Proc. Congr. Internaz. LA QUALITA' PER L'EUROPA, Nov. 1989, pp.247-256
[2] LPS, "KeyOne, manuale d'uso", LPS, Torino 1987
[3] Developer's Edge Ltd., "TPS Command Reference and User Guide",Developers Edge Ltd., Manchester NH 1988
[4]Stratus Computer Inc., "CPS 32, VOS User Guide", Stratus Computer Inc., Massachussets, 1986

Integration and Assembly in the Alvey "Design to Product" Demonstrator Project

Paul Fehrenbach
GEC-Marconi Research Centre, Chelmsford, UK

1. INTRODUCTION

The Alvey "Design to Product" large scale demonstrator project (DtoP) ended in March 1990 after five years and almost two hundred man years of effort from its eight collaborators. Part of the UK Government's Alvey Programme of research into advanced information technology, the project aimed to demonstrate computer based lifecycle support for a class of engineering products termed "light electro-mechanical devices". Beginning with design and going through manufacture (machining and assembly) to service life, the project aimed to use the techniques of artificial intelligence to maximise re-use of information in an integrated product support system for diesel fuel injection pumps. The project demonstrated its prototype system in February and March 1990 and has produced a specification of its approach to product support which is planned to be published as a book.

Throughout the project a key theme was the integration of product support functions into a unified system that makes best use of the information available within it on products and their design, manufacture and service. The project began without an overall system architecture, but during the first two and a half years drew on the experience of others to develop its own.

Part of the project's brief was to use and demonstrate relevant results from other Alvey projects and projects in other programmes such as the European Strategic Programme of Research and Development in Information Technology (ESPRIT). The project team built up close links with a number of other projects, often through becoming members of the consortia, and the final results of the project have drawn particularly on the Alvey Advanced Networked Systems Architecture (ANSA) project [ANSA 1987] and the ESPRIT project 688 AMICE [AMICE 1988] in their approach to computer integrated manufacturing (CIM) system architectures. Thus the project has an open, heterogeneous, distributed architecture that is suited to large and small systems, as required by the application.

2. SYSTEM INTEGRATION

The project was organised with each of several collaborators responsible for developing a part of the system to support a particular phase in the product lifecycle, such as design or assembly. Another collaborator had the task of integrating all the system components into a working whole. In the early years of the project the software developers working on the lifecycle support "tools" were given a great deal of autonomy by the integrators, who were also project managers. At that time there was no system architecture and the tool developers had some formidable problems to solve in their own domains without worrying too much about integration. This meant that the integration task, when it began in earnest half way through the project, was one of bringing together a disparate set of software packages developed using different languages, with different user interfaces and different or no system interfaces. At least they had all been developed on the same hardware: Sun 3 workstations running UNIX and the network extensible windowing system (NeWS).

While this may have been a less than ideal starting point for building a demonstration system, it fits closely with the project team's view of the real world. For new CIM systems to be successful in the future they must fit in with existing design office and factory systems and be capable of expansion in easy stages as more and more functions are added. Thus the integration or interfacing of existing software will be commonplace and by mirroring this requirement the project severely tested its own integration strategy for coping with it. Also, specifically as an example of interfacing to third party software, the DtoP system was interfaced to an Applicon Bravo mechanical analysis package. The subject of future CIM system requirements is returned to in section 4.

2.1 The System Architecture

Large information systems centre round two components: information and communications. The information is the reason for the system's existence and the communications is the means by which it is made available to those, whether people or computers, who want to use it. So it is with Design to Product. At the heart of the DtoP architecture are two basic elements around which the rest of the system is built:

- The Information Management System, or IMS
- The Tool Manager

The overall architecture is shown in figure 1. The way in which it operates reflects the origins of the different components and the project's view of the future of CIM systems. Thus the design tools, which may be many and varied and which include manufacturing planning and programming functions, all interact differently with the Tool Manager, and only with the Tool Manager at a functional level, though the communications protocol has been standardised. Their user interfaces have been functionally separated and, conforming as far as possible to a common style, communicate with the tools via the Tool Manager. This may sound complicated but is fairly simple to do using the NeWS environment [Sun 1988].

The manufacturing area controller supervises the operation of the shopfloor and communicates with the Tool Manager like any of the design tools. However, it is a VAX-based system running VMS in the demonstration system and has the sole distinction of dealing in repeat orders for products when design is not required. It is a different breed of tool.

2.2 The Information Management System

The IMS is the repository for all the useful and user accessible information in the system, which is of two kinds:

- Information about a particular product that the system is concerned with, held in a Product Description.
- Information about how to support a product during its life, such as design rules and manufacturing codes of practice, held in an engineering knowledge base.

The IMS provides all its own information management functions, such as version control, and is based on a commercial, object oriented software development toolkit called KERIS [Poulter et al 1989]. It is written in LISP.

2.2.1 Product Description. A Product Description (PD) is a set of related information objects in the IMS that contains all the information the system has about a particular product. When a new product is being designed its PD starts empty and is gradually built up to eventually include manufacturing and service information. A PD should contain the following:

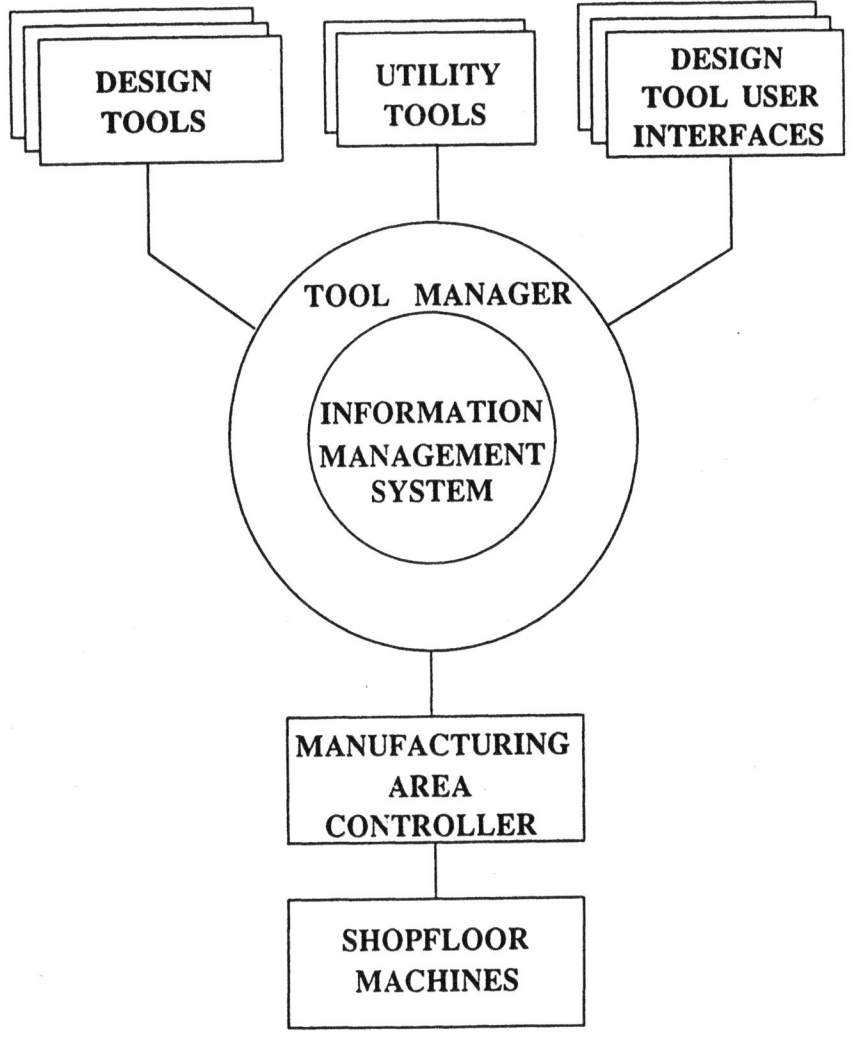

Figure 1. The overall Design to Product architecture

- Specification
- References to other designs investigated for similarity
- Design calculations
- Design decisions and their reasons
- Geometric models
- Parts list
- Component machining process plans
- Component machining routings
- Component machining bills of materials
- Component machining NC code
- Assembly operation plans
- Assembly process plans
- Assembly bill of materials
- Assembly cell program
- Installation documentation
- User documentation

However, none of these items is compulsory. The system relies upon the user to put things into the PD, but its whole point is seriously devalued if they do not, so conscientious users are careful to commit all their work to the PD in the IMS.

2.2.2 Engineering Knowledge Base. The information in a Product Description changes rapidly as a product evolves. The information in the engineering knowledge base (EKB), in contrast, is slow to change. An engineering knowledge base is built up over a long period and will be different for each system, containing information of particular relevance to the users. The structure of the EKB is determined by the contents and the application. Typical contents might include:

- Index to existing designs
- Design guidelines
- Design standards
- Machine tool capabilities
- Material properties
- Machining set ups
- Assembly robot capabilities
- Assembly grippers available
- Standard assembly operation times

2.3 The Tool Manager

The Tool Manager is the means by which the tools in the system communicate with each other and with the IMS. When the system is started up all the tools register their presence with the Tool Manager and declare their location, for the system could be distributed across several machines under NeWS and connected by Ethernet, and the functions that they can provide, in the form of calls to them with specified arguments. From then onwards the Tool Manager can put a tool requiring a function in touch with one that can provide it. The Tool Manager plays no direct part in the subsequent interaction, other than to ensure that the connection is cleanly broken when the called tool has responded. In many respects the Tool Manager is like a telephone exchange.

2.4 The User Interfaces

Perhaps more important than a system really being integrated is its appearing to be integrated to the users. The user interfaces to the system tools are thus an important integrating mechanism from the user's viewpoint. Separating them from the tools themselves, in the way that DtoP has, makes it easier to change their "look and feel" to be consistent with one another and to suit changing requirements.

All the major tools have window/icon/menu/pointer (WIMP) type interfaces and the project defined a style of appearance and of interaction to which they should all conform. Items in the style specification include:

- Size of icons
- Position of scrollbars on windows
- Font sizes and types
- Mechanism for activating a button
- Response of a button to activation
- Location of menus
- Maximum number of items on a menu
- Mechanism for exiting
- Background colour

The adoption of this common style has made it easier for users to change from using one tool to another and makes the system appear more like an integrated whole than a collection of connected tools.

2.4.1. Primary User Interface. As a further integrating feature, when a user logs on to the system they are first presented with a screen displaying the Primary User Interface (PUI), which is the means for accessing the tools in the system. The PUI shows all the tools that are available in a standard format and reinforces the impression of an integrated system. The PUI display is shown in figure 2. The user clicks the mouse over the name of a tool to call it into action.

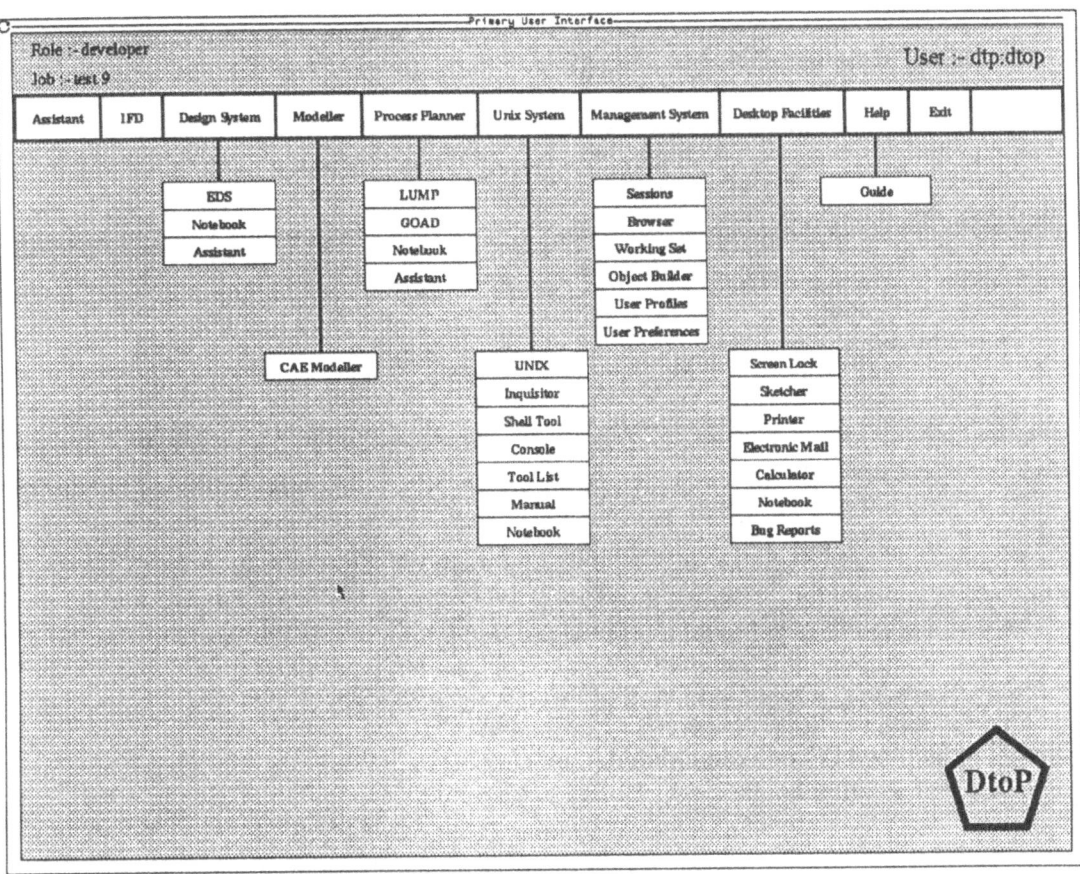

Figure 2. The primary user interface screen

2.5 Other Integrating Features

A number of other features of the system contribute towards its integration.

2.5.1. Infrastructure Tools. Several small software tools, small when compared with the application tools, aid the integration of the system. The most important ones are:

- Working set tool which allows a user to work with a subset of the IMS
- Object buffer tool which is a halfway house between tools and the IMS
- Browser tool for browsing in the IMS or elsewhere.

2.5.2 Socrates. The project has used the results of another Alvey project to provide a mechanism for linking in standard expert systems on IBM compatible PCs and other machines. Called Socrates [Corlett et al 88] the software allows DtoP to tap knowledge stored in existing expert systems and integrate more effectively with established computer support facilities. The process is one of conversion of the expert system rather than straightforward interfacing, but the outcome is more useful. Simple interfaces to PC based systems are also provided.

3. ASSEMBLY SUPPORT

The demonstration DtoP system supports the design, machining and assembly of a selection of components from a diesel fuel injection pump. There is a design tool for each of these functions and shopfloor machines for the latter two. A description of the support provided for product assembly will here serve to illustrate the use that the system tools make of the integrating infrastructure of the system described in the previous section, in addition to being worthy of attention in its own right.

3.1 The Generation of Assembly Data Tool

One of the software tools that registers its presence with the Tool Manager on system start up is the generation of assembly data (GoAD) tool. A tool call to this, usually from the Primary User Interface, with a reference to a Product Description, the one used by the current user session, causes it to perform all the planning necessary to convert the present design information on specified sub-assemblies of the product into the information and programs necessary to assemble the components on the system's robotic assembly cell.

The GoAD tool comprises:

- User interface
- GoAD shell
- Dynamic database
- Assembly knowledge base

Its structure is shown in figure 3. The assembly knowledge base contains modules of knowledge, known as "knodules", that are invocable by the GoAD shell.

The tool takes as input from the Product Description the following information:

- Shape of each component in constructive solid geometry
- Classification of each component, e.g. screw
- Relationships between components
- Weights of components

These inputs have been generated by another design tool, the Edinburgh Designer System [Smithers et al 1990] and stored in the PD.

The GoAD tool proceeds through the following sequence of actions.

1. *Load and review the design information,* as listed above, from the PD in the IMS selected by the user. GoAD checks the design for completeness and the user can abort the tool if the check fails.
2. *Select a lot* of sub-assemblies that are to be considered. The user selects the quantity of each sub-assembly to be planned for.
3. *Plan assembly operations* in the form of gross component assembly steps and ordering constraints. Operations are equipment independent and have attributes such as "moving", "fixed" and "possibly fixed" components and the directions in which components can be moved during an operation. The GoAD tool works by considering the disassembly of the product in a gravitational field and then reversing the sequence to yield an assembly operation sequence. All possible sequences are generated.
4. *Classify operations* according to their attributes. Categories include "get base" which has no fixed or possibly fixed components and "shaft in hole" which has only a single assembly direction.
5. *Generate processes* by applying assembly equipment constraints to the operations plan. Processes are equipment dependent and contain a parameterised robot program template, a list of robot positions, cell teaching information and a list of resources used. GoAD designs some simple jigs and feeders for the components so that they are presented to the equipment in the expected orientation.

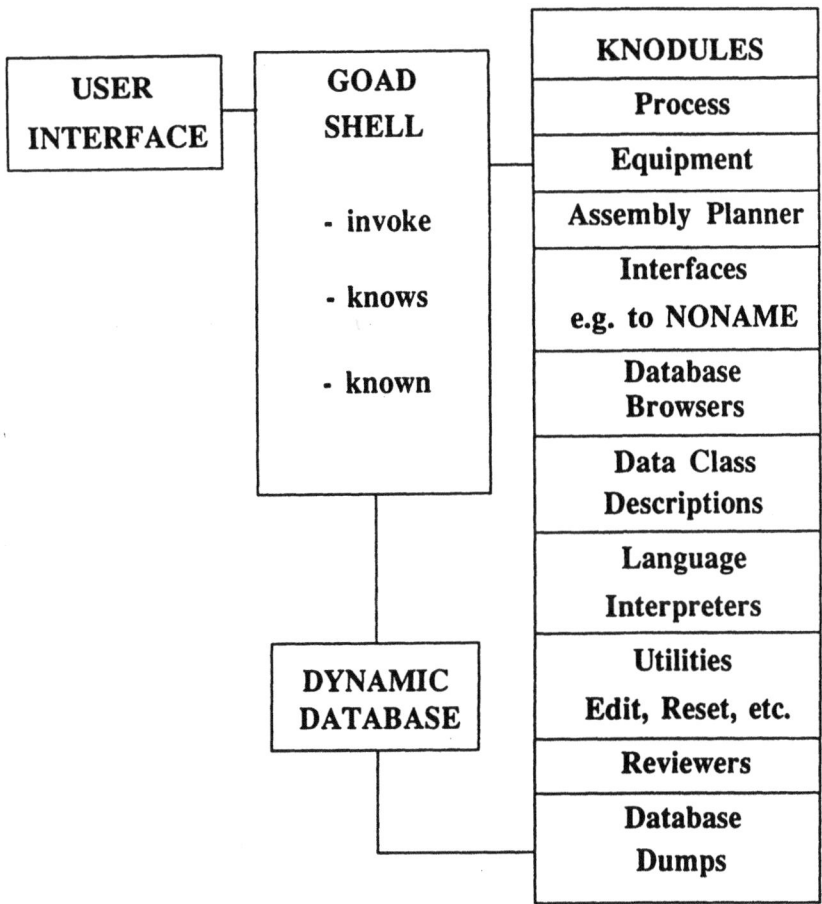

Figure 3. The structure of the generation of assembly data tool

6. *Review processes* to check that there is at least one operation plan that has a process for each operation. The user can abort GoAD if the check fails.
7. *Produce process plan* by first producing a general plan describing all possible sequences of all possible choices of process and then optimising it according to a cost function based on resource usage to arrive at a simple linear sequence.

The GoAD tool output is contained in four files which are committed to the Product Description in the IMS. They contain:

-   Assembly cell program in AR BASIC
-   Robot position definitions
-   Teach report in graphical and text form for use by the assembly cell operator
-   Details of resources required for assembly of the lot and other relevant information

The output of GoAD in the PD is later picked up by the manufacturing area controller and passed to the assembly cell controller for use in the cell.

The GoAD tool was developed under the POPLOG environment [Systems Designers 1987] on a Sun 3 machine and is largely written in PROLOG. It is a partly automated and partly interactive tool, a typical screen interface is shown in figure 4. During operation the GoAD tool makes extensive use of another DtoP design tool, the solid modeller, via the Tool Manager for geometric queries. Operation of the solid modeller is described in [Fehrenbach 1990].

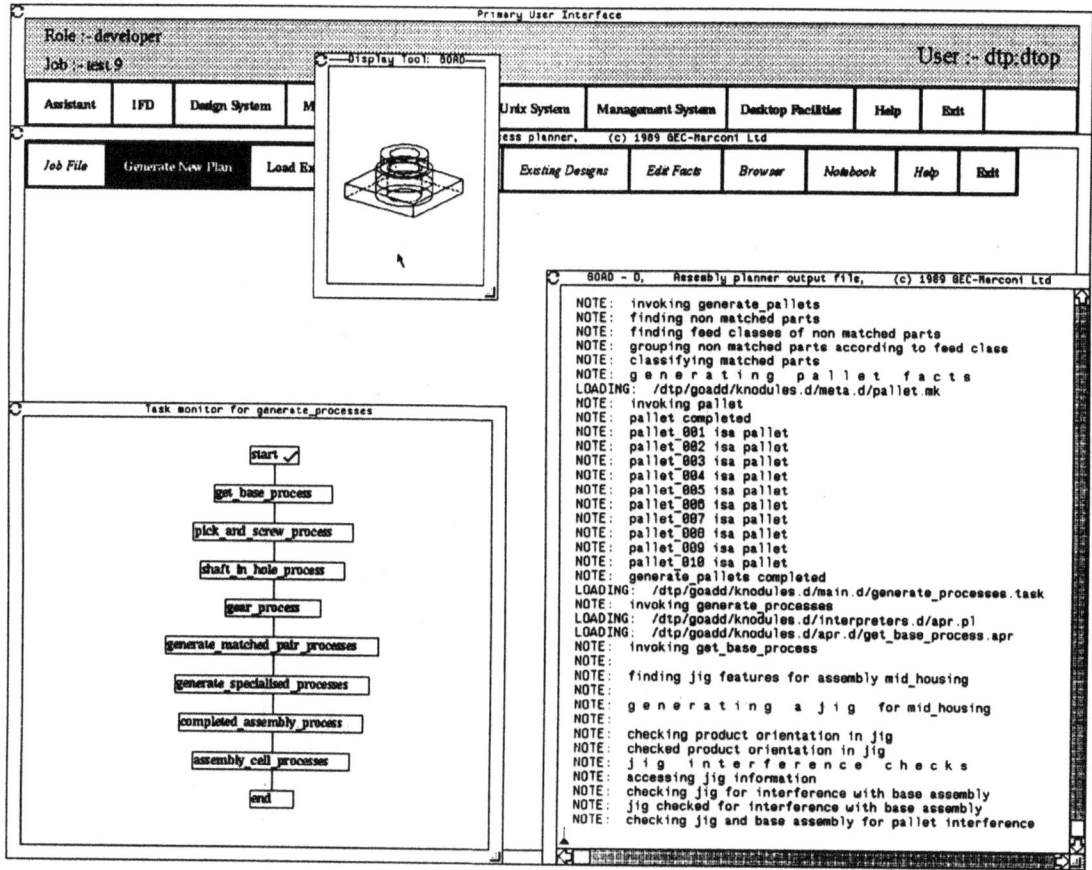

Figure 4. A typical generation of assembly data tool screen

3.2 The Assembly Cell

The assembly cell comprises:

- Six axis assembly robot
- Indexing table
- Three axis pallet transfer gantry robot
- Conveyor
- Pallet storage area

The cell has its own controller that is linked to the manufacturing area controller for downloading programs and sending back status information. A photograph of the cell is shown in figure 5.

The cell receives pallets on the conveyor in random order with one type of component per pallet. Each pallet carries an electronic tag that the cell reads to decide whether it requires to load the pallet from the conveyor. Pallets are loaded by the gantry robot into predetermined locations around the assembly robot. Tooling for the assembly robot is loaded in this way too. The robot has a gripper change unit and grippers for it are delivered on the conveyor. Unwanted pallets are allowed to pass by.

When the cell has all the components and tooling it needs, it assembles the required number of products and unloads them to the conveyor. The tags on the pallets for the completed assemblies are updated before going off along the conveyor by a tag read/write unit underneath. The used component pallets may then be unloaded from the cell if they are not required for the next lot or are empty.

Figure 5. The assembly cell

The assembly robot is a Rediffusion Reflex and requires to be manually taught new positions before it can move to them accurately. The teach reports and position definition files produced by the GoAD tool assist the operator in this task by providing instructions on the required points and their sequence. If the robot was capable of being off-line programmed without the need for teaching, the DtoP system would be able to program it totally.

The equipment in the assembly cell is all general purpose and can be used to assemble anything within its capabilities. Product specific tooling is delivered along with the components and can include gauging devices, power tools, suction grippers and sensors of various kinds. The main limitation on the capability of the cell is that the contents of a pallet must fit within a cube of side 300 mm, so no component, assembly or tool can be larger than this.

4. EFFECTS ON COMPETITIVENESS

The DtoP project was aimed at a class of products where customers are demanding smaller and smaller batches of ever increasing variety in shorter and shorter timescales. Typically these products come from small or medium sized companies that cannot afford large investments in CIM technology of the more conventional kind. They need systems that they can acquire in stages as funds permit and that they can introduce into their existing operations without major disruption.

DtoP allows all of these things. The only essential components are the IMS and the Tool Manager, onto which can be grafted existing systems that the company has, if they are suitable. There is no commitment to a single system supplier and the system can start small. As more information is put into the system it will become more useful and there will be greater incentive to add to it further. After a suitable period of building and learning, the system should enable the users to respond more rapidly to customer requests for new designs and to make maximum use of existing designs in new products, so reducing design and manufacturing costs and simplifying service.

The DtoP software is targetted at computers and operating systems that are or will be industry standards and that are set to become the personal computers of the future. In this respect also the system can be expanded in stages by networking machines together as required. The whole concept of a modular system is intended to allow particular installations to grow and fit their application at a manageable pace, that all sizes of company can afford and that brings rapid repayment of investment. Small companies in particular need to see rapid returns if they are to stay in business.

The DtoP demonstration system is only a prototype and even a fully developed version would not be a panacea for all manufacturers' problems. But the DtoP project team do believe that their ideas could give some companies the competitive edge they need by making best use of information to improve responsiveness. There are other benefits too. An integrated system should improve quality by incorporating feedback from manufacture to design and a well managed information base of the kind on which DtoP is based would be a valuable asset to any company.

5. CONCLUSIONS

This paper has described the integration mechanisms used in the DtoP system and the thinking behind them and has illustrated the way the system works with reference to support for product assembly. Finally there has been a discussion of the effects the adoption of DtoP concepts could have on the competitiveness of manufacturing industry.

DtoP will not solve everyone's problems but it is a new way forward. The project has produced only a specification and a demonstration prototype. There is much more work still to be done and fully engineered systems based on DtoP will not emerge for several years. The demonstration prototype will continue to be used by the project collaborators as a vehicle for stimulating industrial interest in the project's ideas and for exploring new applications.

6. ACKNOWLEDGEMENTS

The author wishes to thank all the collaborators in the DtoP project for permission to publish this paper: GEC Electrical Projects Ltd., GEC Avionics Ltd., GEC-Marconi Research Centre, Lucas Diesel Systems Ltd., University of Edinburgh, University of Leeds, University of Loughborough and HUSAT Research Centre. Most of the work on system integration described in section 2 was done by GEC Electrical Projects Ltd. and the assembly cell described in section 3.2 was designed and built by Lucas Systems and Engineering Ltd. for Lucas Diesel Systems Ltd. The generation of assembly data tool described in section 3.1 was developed at GEC-Marconi Research Centre but relies heavily on functions provided by the geometric modelling facility provided in the DtoP system by the University of Leeds.

The Alvey Programme, of which DtoP formed a part, was funded by the UK Department of Trade and Industry, Department of Education and Science and Ministry of Defence, as well as by the industrial participants.

References

AMICE project (1988) CIM-OSA Reference Architecture Specification. CIM-OSA/A-MICE, 489, Avenue Louise, Brussels.
ANSA project (1987) ANSA Reference Manual Release 00.03.ANSA, 24, Hills Road, Cambridge, UK.
Corlett, R. et al (1988) Socrates: A Flexible Toolkit for Building Logic-Based Expert Systems. Knowledge Based Systems, vol. 1, no. 3.
Fehrenbach, P.A. (ed) (1990) The Alvey 'Design to Product' Demonstrator Project - An Integrated, Knowledge Based Approach to Product Lifecycle Support. GEC-Marconi Research Centre, West Hanningfield Road, Chelmsford, UK.

Poulter, K.J. et al. (1989) The KERIS Reference Manual. GEC-Marconi Research Centre, West Hanningfield Road, Chelmsford, UK.

Smithers, T. et al (1990) Design as Intelligent Behaviour: An AI in Design Research Programme. Int. J. of AI in Engineering, forthcoming. Also available as DAI Research Report no. 426, Dept. of AI, University of Edinburgh.

Sun Microsystems (1988) NeWS 1.1 Manual. Sun Microsystems, 2550, Garcia Avenue, Mountain View, California.

Systems Designers (1987) POPLOG User Guide Issue 2.0. Integral Solutions, 3, Cambell Court, Basingstoke, UK.

EFFECTS AND CHAINS OF EFFECTS AT CIM-REALIZATIONS

HANS F. JACOBI

1   INTRODUCTION

The success of a company within a period of time is determined by the balance sheet
and the profit and loss calculations. The balance sheet provides information about the
assets and debts (value) of a company at a fixed date (deadline). The profit and loss
calculation provides information about the origin of the success by comparing profit
with expenses /1/.

Two units of evaluation to be analyzed not only regarding performance (income,
profits) but also respective costs (expenses, expenditures) are "**quantity**" and "**time**"
(time span). Examples of performance can be the amount of products sold or profits
from services sold per time unit - costs on the other hand are wages and salary (time
base), purchase and consumption of raw materials in the same period (quantity basis) as
well as expenses when processing raw materials, indirect materials and working
materials (quantity/time base).

By reducing the company-related success account to the basic units **time** and
quantity one opens up the possibility of using these units as aims (formal aims) of
action for the **entire** company divisions and in this context to determine the extent of
the attainment of the goal (increase, reduction etc.). As further determining aims there is
the **deadline** of the product delivery in relation to the purchases as well as the **quality**
of the product, whereby in the final analysis improvement of quality has a direct
influence on the quantity (quantity of goods) and the sales price. To sum up, the
following parameters are of fundamental importance for evaluation:

- o   QUANTITY
    - used, turned-over and delivered quantities
      according to length, weight, number etc.
    - number of processes, operations
    - number of stock

- o   TIME
    - work/machine-hours, etc.
    - production, swift, calender-time

o   VALUE
- types of costs
- calculation values
- stock values (warehouse values)
- turnover figures (profit)

o   QUALITY
[f (go quantity)]

o   SOCIAL ENVIRONMENT

o   WORK SAFETY/ENVIRONMENTAL PROTECTION

The above mentioned basic evaluation units can be transferred to the company structure, in particular regarding "quantification", and be used as aim factors for all divisions (Fig. 1).

When shaping the analysis of aim systems aim efficiency relations have to be considered which are made clear by the two basic related forms **aim complementing** and **aim competition.**

The principle of aims complementing each other is based on the assumption that when an aim of the company is attained, as many other aims as possible at the same and/or following level are at least partially realized at the same time.

Aim conflicts arise when the attainment of one aim leads to the non-attainment (or negative attainment) of another aim. These conflicts have to be recognized and solved, that is, for example, by eliminating, re-evaluating and/or modifying certain aims, or by changing to a superior, comprehensive aim level that includes sub aims /2/.

Known examples of aim conflicts are:

o growth versus profitability versus solvency

o high rate of delivery capability,

o high quality of material versus reasonable net prices versus low capital tie-up

o avoiding idle times of production equipment versus avoiding down times of products "dilemma of work sequencing".

Aim conflicts and complementing have to be determined not only for the division ifself (within and between the task areas) but also between the divisions (externally). Internal goals serve as a requirement for time and quantity consumption as well as for keeping

Company Policy

company purpose	company aims	company principles

• maintain and continually develop the company successfully
 	
• determine
  - type of products
  - techniques to be applied
  - consumer groups
  - social commitments
	• maximize profit
	• increase productiveness
	• increase share of market
	• guarantee solvency
		• independence
		• flexibility

Division Aims

production planning	materials management	quality assurance	production control	maintenance

• reduction
  - of production and indirect wages by appropriate deployment of work force
 	
  - equipment costs by deploying effectively and at an optimum
 	
  - of interest costs (capital tie-up for circulating material by reducing the production lead time
 	
  - of energy consumption and auxiliary material costs for production
 	
  - of space and area costs by planning a material flow appropriate workshop
 	
  - of quality costs
 	
  - of operating loss costs by reducing scrap, waste, re-work
 	
• keep deadlines
 	
• minimize production and production overhead
	• reduce capital tie-up
	• keep delivery deadlines
	
	• improve material quality
	
	• increase delivery capacity (procurement marketing)
	
	• minimize purchase prices (material)
	
	• minimize material and material overhead costs
		• reduce error costs
		• keep deadlines
		
		• maintain, increase share of market
		
		• minimize error prevention and test costs
			• guarantee capacity loading
			• maintain production lead time
			
			• maintain production final deadline, quantity
			
			• deliver product quality agreed upon
			
			• keep to planned production costs
				• reduce number of technical breaks in production as well as idle times
				
				• guarantee the functional capacity
				
				• maintain safety and environmental directives
				
				• minimize maintenance costs

Figure 1: Company and division aims

deadlines and guaranteeing quality when executing tasks in the divisions. The activities for attaining internal aims also influence other external divisions of the own company or an outside company. For example, the use of CAD in design/development influences the amount of time needed to make a drawing and the number of drawings (internal). On the other hand, less time consumption and the possibility of producing a large amount of drawings and/or increasing the complexity of drawings also has an influence on the fulfillment of tasks in production planning, materials management as well as production (external point of view). Moreover, the value of the equipment changes when using CAD. This operation influences the task accomplishment in finance as well as the accounting.

These effects should be analyzed using the evaluating units QUANTITY, TIME (DEADLINE), VALUE as well as QUALITY in connection with the structuring points of view "technology", "organization" and "employees" on the one hand as well as division related companys structures. For it is only when one knows which aims work in which way in the entire company that it is possible to go from a profit related basis to a need orientated CIM realization. Then, one can proceed to a preparation of the necessary planning, control, and monitoring information for the corresponding aim attainment.

2   RESULTS OF AIMS OF A SELECTED DIVISION

In the division "production control" ($\simeq$ PPS functions) the production plan which was defined by the production program and/or production planning for a period of time is to be translated into a detailed work program with clearly defined orders and deadlines. The tasks to be fulfilled in **work specific** production control, representing the PPS main functions, can be divided into:

      o production program planning
      o quantity planning
      o order forming
      o deadline and capacity planning
      o giving orders
      o monitoring orders.

Assuming that the aims are attained in the pre-production/assembly and/or accompanying divisions the following are basic assumptions in the sub-division "production control" - especially for serial production o in "Marketing/Sales" sales are forecast realistically,

o in "Development/Design" production and assembly appropriate
  products are designed within the framework of production equipment
  designreliable equipment/machines, as well as tools,

o the appropriate equipment/machine is invested in, the personnel and
  material demand is determined realistically,
  - the costs are calculated correctly,
  - the sequence (time, quantity, deadline) is optimized by the division
    "production planning"

o the right material has been made available according to the demand
  (deadline, quantity, quality) by the division "materials management".

This assumption leads to the conclusion that in the division "production planning" with the remaining degree of freedom "personnel deployment" the production/assembly can be "simply" taken care of. In the day-to-day reality however, in the other divisions the aims affecting the production sequences are hardly attained so that in "controlling production" one has to react first of all to the various aim deviations (time, deadline, quantity, quality) and the production process has to be harmonized in respect of the delivery quantity (yield) per time unit (how, day etc.).

Thereby in the division "production control" the aims of the production program and production planning affecting production and assembly are the basis that have to be realized by the "control" in production/assembly (figure 2).

In order to be able to attain these aims not only in planning but also in control of production execution it is necessary to employ appropriate methods and instruments within the framework of the above mentioned tasks (figure 3).

Aim Type

content	extent	time context
• time aims - capacity "equipment / machine" - capacity "personnel"	* do not fall below a certain value	
- processing times per operation+ (eng. office) - set up time (equipment/ machine/personnel - piece time - idle time - waiting time - transportation time - quality check time **• quality and value aims** - production and indirect wages - production equipment costs - interest costs (circulation assets) - energy consumption, indirect material costs - space and area costs - machining loss costs (waste, reject, rework) - costs of production control area (production overheads)	* do not exceed a certain value	* short, middle, long termed * for (from) certain time periods / certain time period
- yield quantity (yield)	* maintaining a certain minimum value * increase (by x%)	
• deadline aims - machining deadlines (eng. office) - production deadlines delivery deadlines **• quality aims** - quality values (product / machine)	* not exceeding / not falling below a certain value	

Figure 2: Aim overview "production control"

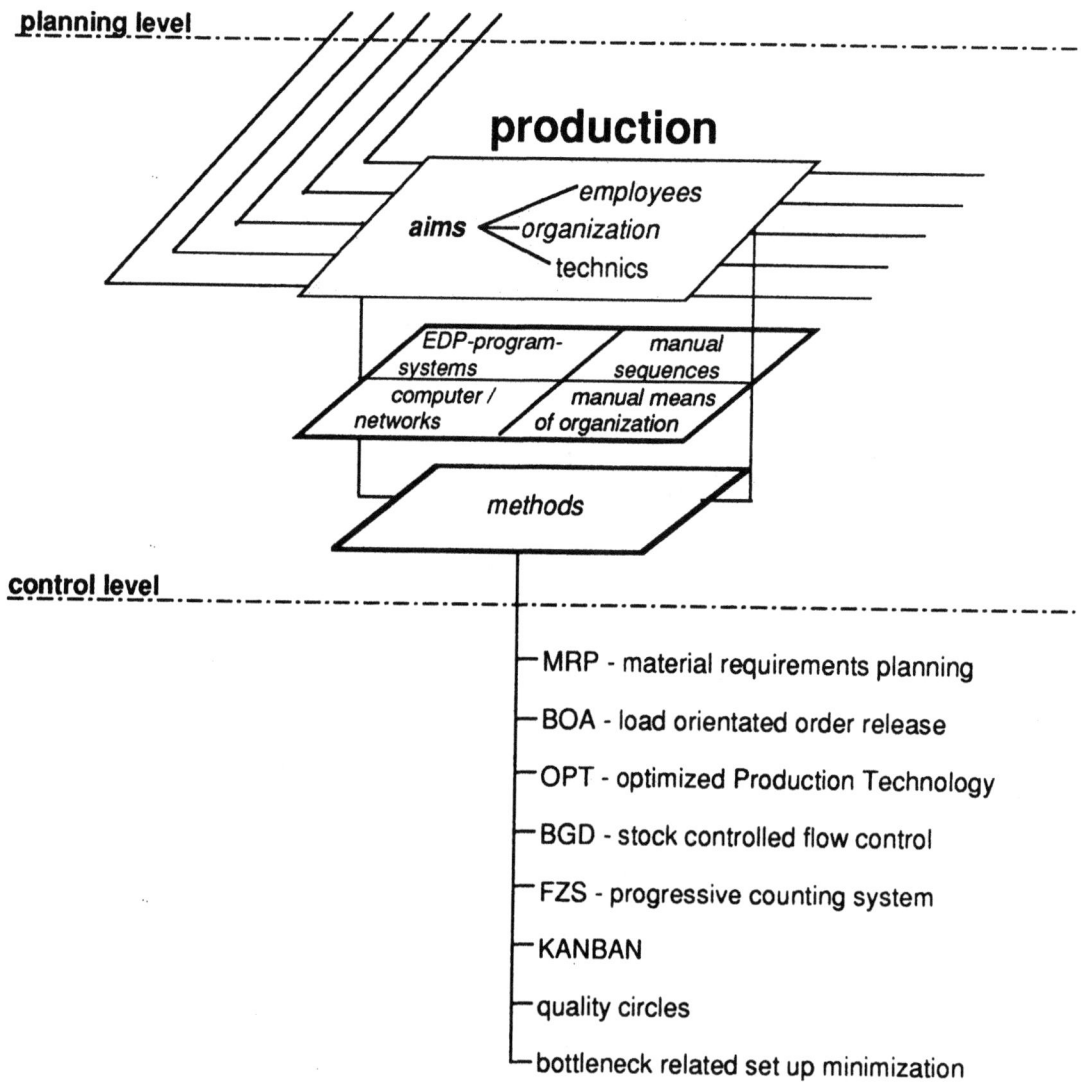

Figure 3: Relationship: Functions - Tasks - Methods - Instruments

The methods used as "means" to attain the aims have a specific range of effect. For example, as a function of the production type and principle they influence the sequence of order release and allocation to a great extent and thus the aim attainment "maintaining deadlines" and "minimizing stock".

Internally the established aims affect the machining times (eng. office), the qualification of production control workers, the production and indirect costs as well as the error costs (rework, rejects).

The external directional effects of the aims can be structured as follows (examples) (figure 4).

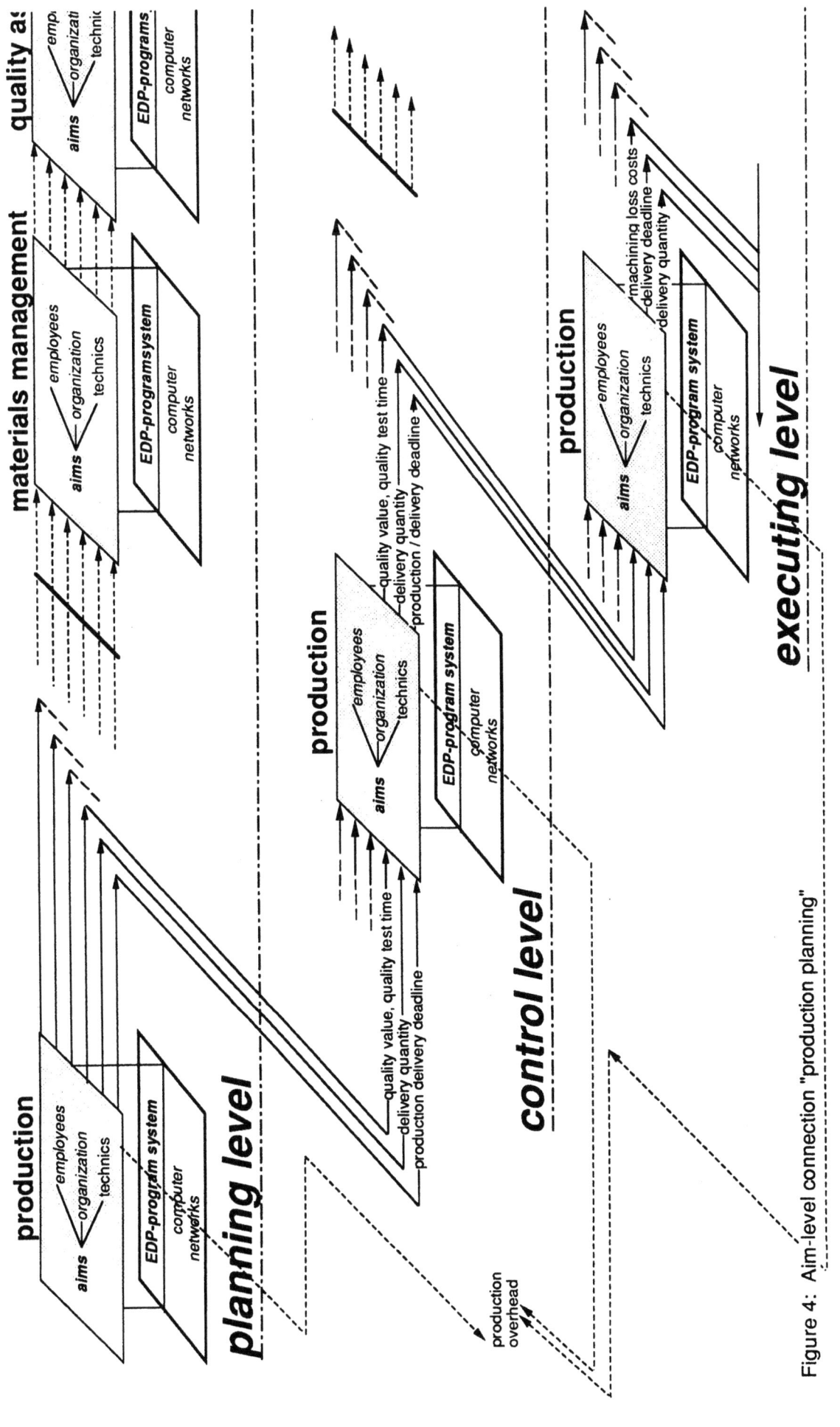

Figure 4: Aim-level connection "production planning"

3 CONCLUSION

The analysis of the operational aims/situations leads to the conclusion that primarily the division related interests (aims) are represented in the individual divisions today as a result of the hierarchical set-up and that the fulfillment of tasks is controlled accordingly. The differing aims and their effects pose an obstacle for CIM-realizations

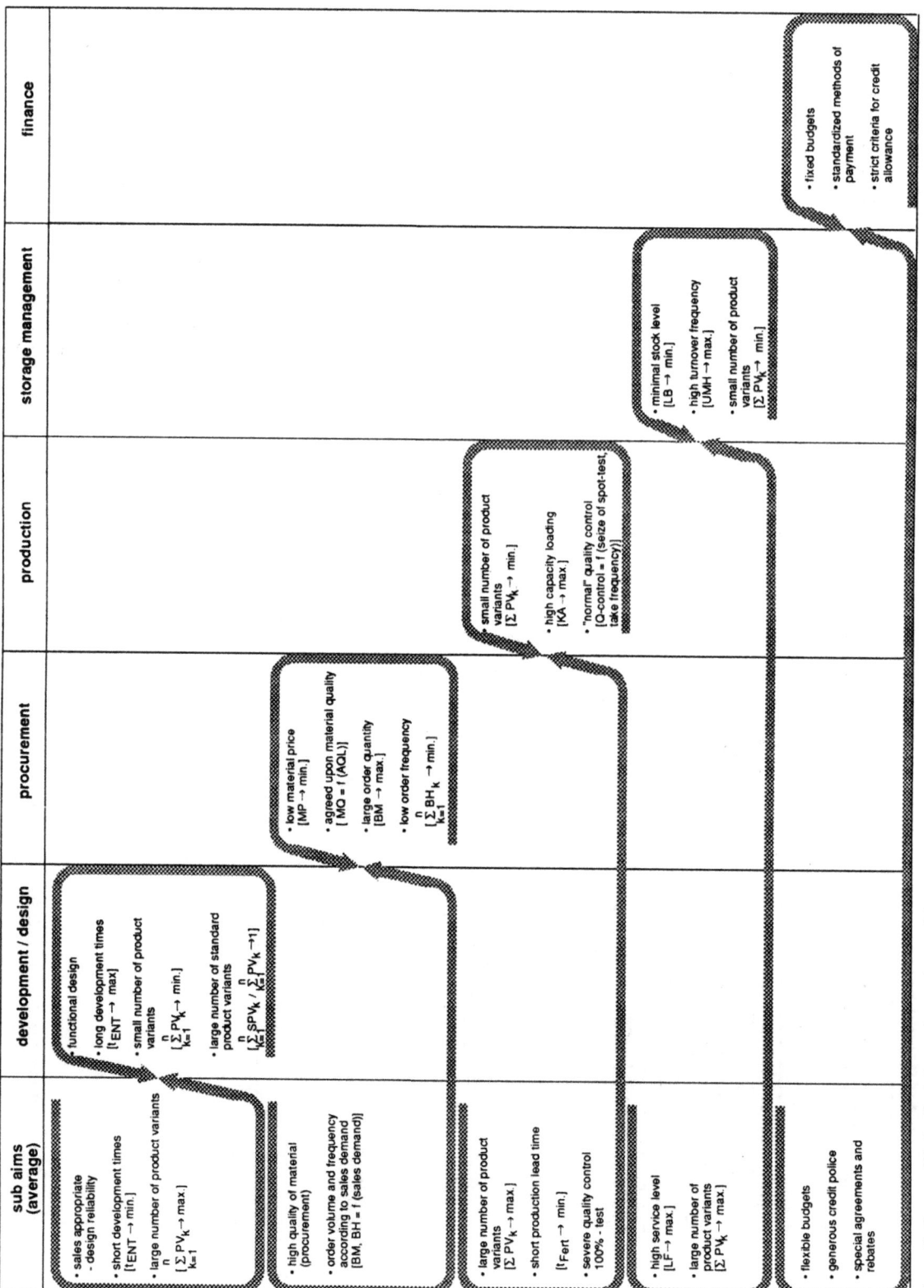

Figure 5: Aims and their effects in a company (example)

Working out the division-related aims and their effects is the prerequisite for conceptional and realizational decisions regarding "computer integrated production".

There is a way of demonstrating the usefulness of a CIM- component-usage in a quantified manner if one succeeds in proving that the company related aim attainment and/or aim conflict solution is assisted by using the CA and PPS-Systems. This realization is not new, however, in daily practice the knowledge of aim formation, attainment and conflict solution is still underdeveloped.

An approach is recommended that allows for the checking and/or development of CIM-concepts as well as CIM realizations regarding their aims:

1.   For each division the aims are to be formulated according to the factors time, deadline, quantitiy value and quality in conjunction with the company aims. The internal and external effects are to be elucidated as well as the aims conflicts and the complementaries (example materials management, figure 6).

Materials Management

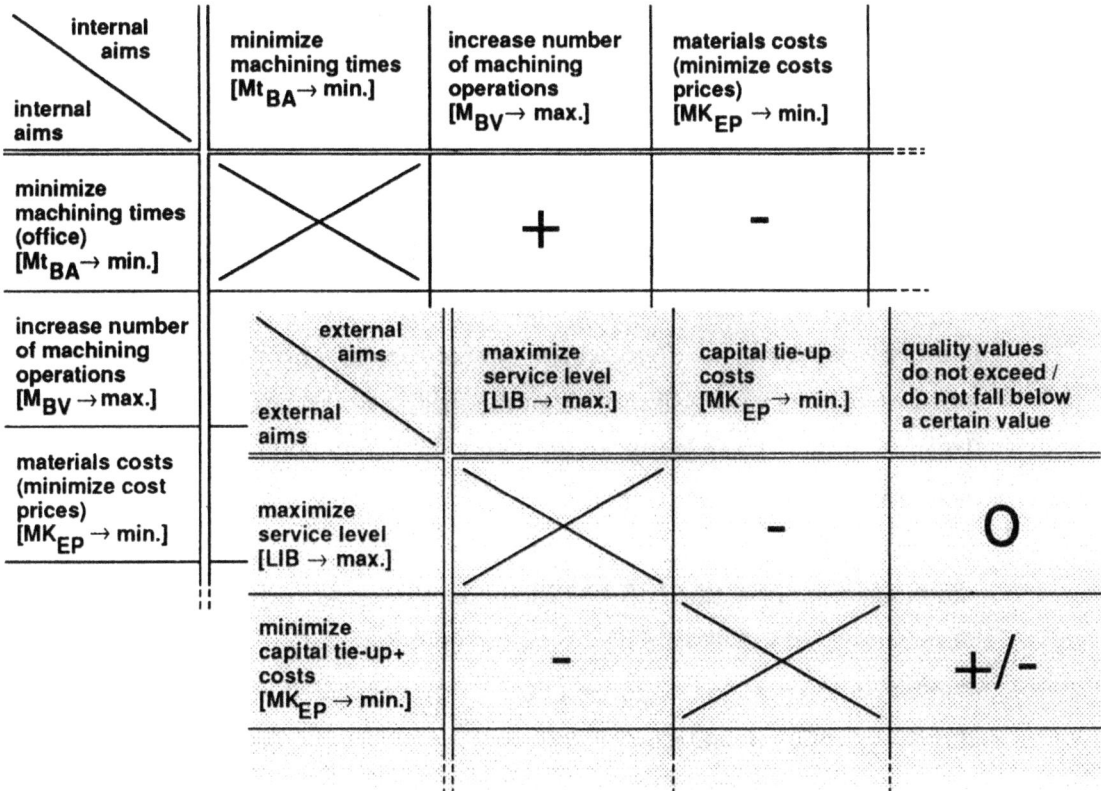

+ complementary effect; - rival effect; o indifference; +/-dependent on the aim characteristic

Figure 6: Division related aim matrices

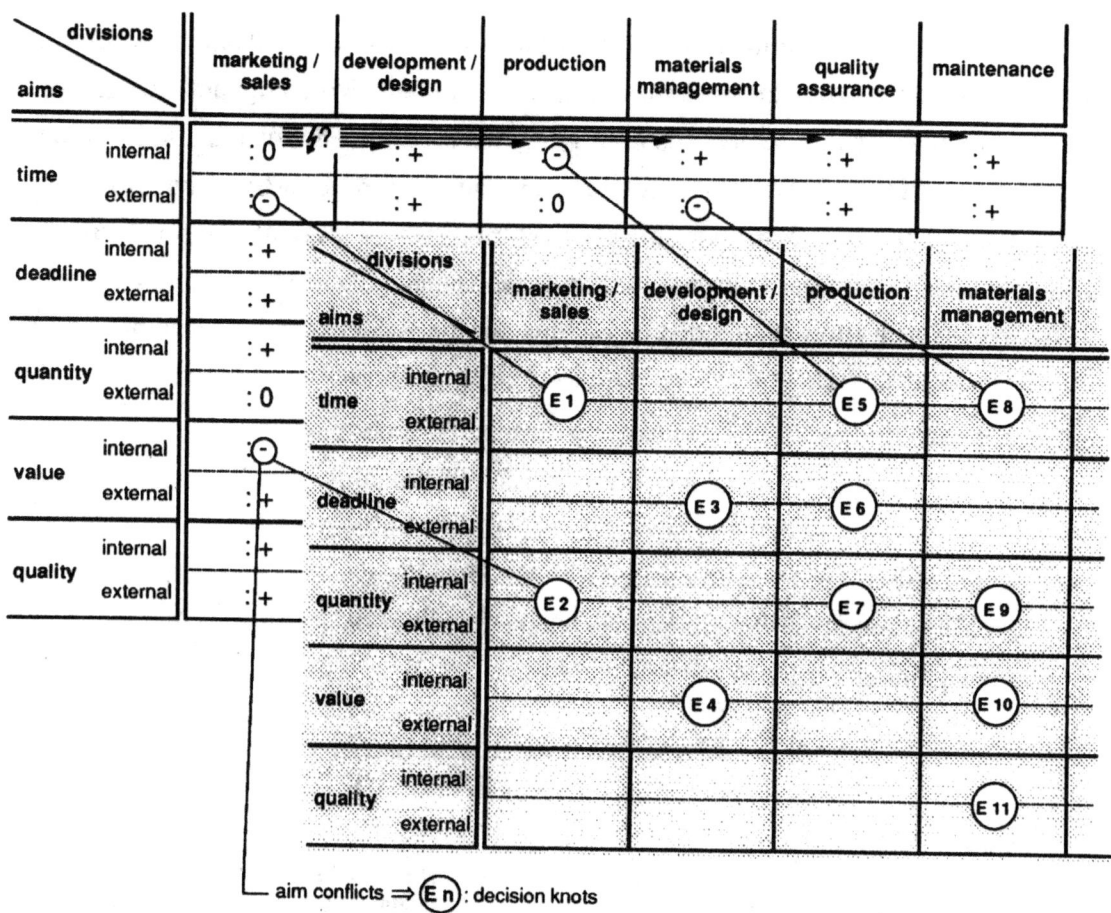

aim conflicts ⇒ (E n): decision knots

Figure 7: Company related aim matrix

2.   For the company as a whole aim matrices are to be set-up in which the supra-divisional aims, their effects and the aim conflicts can be determined (figure 12).

3.   In order to solve aim conflicts a decision between at least two alternatives has to be made. Starting from the localized aim conflicts level specific decision structures can be set-up.

   In addition to the

   - automated data handling (office automation) and

   - automated use of methods (FFM forecast calculation etc.)

      the

   - automated preparation of information for support when making company aims orientated decisions is definable

      as "useful" CIM component.

   For example: A summarized representation of the P, Q and M aims and their effects are given in figure 8. These effects of aims are only quantitative and have yet to be evaluated qualitatively.

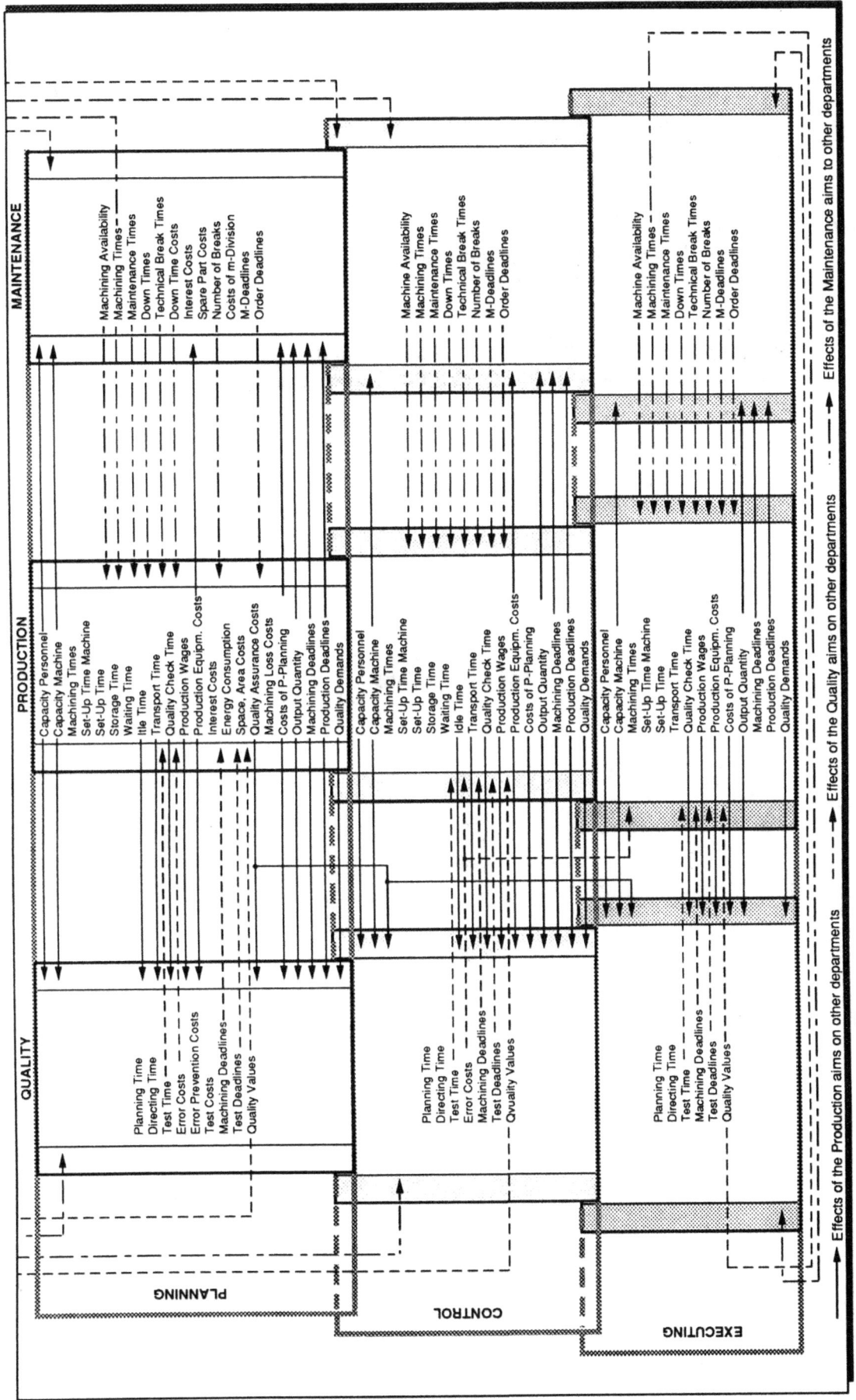

Figure 8: P, Q, and M aims and their effects

217

LITERATURE

/1/   Horvath, P.:
        Controlling. München: Vahlen 1986.

/2/   Hünerberg, R.:
        Marketing. München, Wien: Oldenbourg, 1984.

/3/   VDI 2235:
        Wirtschaftliche Entscheidung beim Konstruieren.
        Düsseldorf: VDI-Verlag 1987.

/4/   Jacobi, H.F.; Wincheringer, W.;
        Production Aims and Chains of Reaction in the Field of
        Maintenance. ESPRIT 2434, 12 Months Report, Task 2,
        1989.

ENABLING TECHNOLOGIES – PART I

FLEXIBLE ASSEMBLY CELL: AN IMPLEMENTATION EXPERIMENT USING THE ESTELLE FDT*

V. B. Mazzola[1], J. P. Courtiat[1], M. Diaz[1], A. M. Druilhe[2], J. F. Lenotre[2], P. Michaud[2]

[1]LAAS/CNRS   [2]MARBEN

1. INTRODUCTION

Flexible manufacturing systems are composed of several interacting heterogeneous devices which present several synchronization and resource allocation problems. Designing the driving system of a flexible assembly cell is then rather complex, and a well defined methodology is therefore required.

Estelle is a Formal Description Technique standardized by ISO [Estelle 88] which presents a powerful set of features for describing communication and parallel behaviours. Although Estelle has been mainly devoted to the formal description of OSI protocols and services, it may also be applied to the formal specification and implementation of complex distributed systems. In addition, this technique is supported by a consistent set of tools which are useful along the life cycle of the distributed software.

Within the European framework, the most representative studies on Estelle have been conducted within the ESPRIT program, and in particular within the SEDOS [Diaz 89a] and SEDOS Estelle Demonstrator projects [Ayache 89, Diaz 89b]. Starting from the results yielded by SEDOS, the aim of the SEDOS Estelle Demonstrator project was twofold:
- the development of an Estelle WorkStation (EWS)
- the evaluation of EWS (as well as of the Estelle methodology) on several applications characteristic of different application areas (computer networks, telecommunications, spatial communication and industrial systems).

The paper presents the experience gained in applying Estelle to the design and implementation of the programming system of a flexible assembly cell (FAC) and shows the usefulness of the Estelle tools for this purpose.

Section 2 summarizes the main concepts of the Estelle Formal Description Technique and describes the main facilities provided by its support tools. Section 3 presents the specification approach used for the FAC programming system by detailing in particular the modelling of an assembly task and the architecture of the programming system. Finally, Section 4 demonstrates how the previous specification approach, as well as the Estelle support tools, have been applied to a particular assembly task, which has been validated and implemented on the experimental cell available at LAAS/CNRS.

Finally Section 5 draws some conclusions about the results obtained in this study.

2. THE FORMAL DESCRIPTION TECHNIQUE ESTELLE

2.1. Overview and Main Concepts of Estelle

An Estelle specification [Estelle 88], [Budkowski 87], [Tenney 88] describes a hierarchically structured system of non-deterministic sequential components (*module instances*) interchanging messages (called *interactions*) through bidirectional links between

---

* This work was supported by the CEC, within the framework of ESPRIT project 1265 Sedos Estelle Demonstrator.

their ports (called *interaction points*). Both the hierarchy of modules and the structure of their links may change over time, making thereby the system a dynamic one. A module is defined by a *module_header_definition* and a *module_body_definition* associated with this header. Several instances of a module may be created and simultaneously be present during the execution of an Estelle specification. Each one of them has the same external visibility characterized by the module_header_definition and the same internal behaviour defined by its module_body_definition.

From an external point of view, a module instance is seen as a "black box". Access in and out of that box is made via finite sets of interaction points. Each interaction point of a module instance x has an associated unbounded FIFO queue which receives and stores interactions sent to x through that interaction point. A module instance may also send interactions to other module instances through its own interaction points. The set of interactions (together with their parameters) which may be received and sent through a specific interaction point is determined by the *channel_definition* associated with the interaction point declaration.

The visible results of a module instance behaviour, in the form of the interactions it sends, are the effect of a module instance internal activity described within an associated module_body_definition. The internal behaviour of a single module instance may be characterized in terms of a non-deterministic extended state machine [Estelle 88]. Informally, each local state of a module instance has a complex structure characterized by the following components:

• a control part represented by a value which characterizes the control state (also called major state) of a module instance;

• an input_environment part represented by the contents of all the input queues associated with the module instance interaction points;

• a data part represented by the values of the variables declared either in the module_body_definition or the module_header_definition;

• a "delay" part represented by the values of the dynamic firing time intervals associated with the (enabled) transitions.

The **declaration_part** of a module_body_definition includes the declaration of constants, the definition of data types (among them the module_header and the module_body definitions of the child modules), the definition of channel types, Pascal procedures and functions, and finally the declaration of the major states and local variables.

The **initialization_part** of a module_body_definition specifies an initial control state and the way variables are initialized; it further defines an initial hierarchy and interconnection structure of the descendant module instances (if any).

The next_state relation of a module instance is specified within the transition part of a module_body_definition by transitions. Each transition is made up of a guard and an action which may be characterized as follows :

• the guard may contain the clauses "from", "when", "provided", "delay" and "priority" where :
   – the clauses "from", "when" and "provided" determine the enabling condition of a transition by testing, respectively, the major state of a module instance, the existence of an interaction in head of the FIFO associated with an interaction point, and, finally, by evaluating a Pascal predicate. A transition is said to be enabled iff all the previous clauses are satisfied, thereby enabling (but not requiring) the firing of the transition;
   – the clauses "delay" and "priority" characterize the firing condition of an enabled transition; in particular, the clause "delay(tmin,tmax)" defines the minimum time tmin (respectively the maximum time tmax) during which an enabled transition must (respectively may) be delayed; the clause "priority" defines a priority relationship among fireable transitions of the same module instance.

• the action is made up of the clause "to" and of a transition_block where :
   – the clause "to" specifies the next value of the control state;

222

– the interpretation of a transition_block defines an operation to be executed by the transition; the transition_block is given by a Pascal-like compound statement with some specific restrictions (no "goto" statement for instance) and extensions of Pascal (specific Estelle statements like "init" (for creating a new module instance), "release" (for releasing a module instance), "connect" (for creating a link between two interaction points of two distinct module instances), "disconnect" (for releasing a link between two interaction points), "output" for sending interactions, ...); the operation may therefore change the module instance state, modify the configuration of its descendant module instances and, finally, output interactions to the environment.

Execution of a **transition** by a module instance is further considered to be an *atomic* (indivisible) operation. Once a transition execution is started, it cannot be interrupted, and, conceptually, one cannot view intermediate results regardless of how "large" a transition is (i.e. how many statements its transition_block is defined with).

2.2. Support Tools

The use of a Formal Description Technique requires a consistent set of tools to be applied at the different stages of the design trajectory of a complex application. It was one of the main goals of the European projects SEDOS and SEDOS/Estelle/Demonstrator to make such tools available. Two tools have been used for the flexible assembly cell application, which are respectively **ESTIM** – Estelle SimulaTor based on an Interpretative Machine – for simulation and verification purposes and **EWS** – Estelle WorkStation – for simulation and automatic code generation purposes.

ESTIM [Saqui 89a], [Saqui 89b], has been developed at LAAS within the framework of the SEDOS project . It supports a version of Estelle enhanced by a Rendez-Vous mechanism [Courtiat 87], [Courtiat 88].

EWS [Ayache 88] is the environment developed within the SEDOS Estelle Demonstrator project. It consists of an integrated set of tools for edition (syntax oriented editor) and automatic C code generation purposes. The generated C code may run either in a simulation environment (simulation motor) or on some target run-time environment (implementation motor).

3. THE FAC PROGRAMMING SYSTEM - MODELLING AND SPECIFICATION

In order to ensure the genericity of the design being made, the programming of an assembly task has to rely on an abstract model. This model is intended to formalize both the interacting cell components and the specificities of the assembly task to be performed on this cell. Several concepts, detailed in [Chochon 86], [Chochon 87], have been used for this purpose and are introduced in the sequel.

3.1. Modelling an Assembly Task

An assembly task is a set of operations to be applied on input parts with the purpose of yielding output parts (or final products). An assembly task operation performs some transformation on the input parts, changing therefore the identity of the parts. Each operation may consist of several actions performed by the cell components. The following notation will be used for modelling the operations specific to an assembly task, where the symbols X, Y, XY, $X_1...X_N$ represent part identities, and the symbol $!$ represents an unknown part (for some given assembly task):

Feeding of part X:		$\rightarrow$	X
Identification:	$?(X_1,...,X_N\ !)$	$\rightarrow$	$X_1\ /\ ...\ /\ X_N\ /\ !$
Technologic operation :	X	$\rightarrow$	$X*$
Assembling of parts X and Y:	$X\ \&\ Y$	$\rightarrow$	$X\ Y$
Inspection of part X:	X	$\rightarrow$	$+X\ /\ -X$
Unloading of part X:	X	$\rightarrow$	

3.2. Modelling a Flexible Assembly Cell

Two kinds of information have to be modelled: the function carried out by each cell component (functional modelling) and the spatial resources of the cell (spatial modelling).

3.2.1. Functional Modelling. The concepts of actor and agent are introduced in order to specify in an abstract way (i.e. not directly related to a specific implementation) the function of each cell component.

Actor: an actor is a cell component which manipulates parts. For instance, a robot gripper, a camera, a screw machine are actors of a flexible assembly cell. Five classes of actors may be defined: Effectors (for moving parts), Holding devices (for holding a part in a stable position), Sensors (for providing informations about part identities within the cell, Specialized machines (for transforming part identities) and Feed/unload systems (for feeding and unloading parts in the cell)

Agent: an agent is a set of actors which either modify the part identities or move the parts inside the cell. For instance, a robot which performs an assembly operation is an agent composed of two actors: an arm (effector) for moving parts and a gripper (holding device) for holding them.

3.2.2. Spatial Modelling. In order to manage the allocation of the cell resources, and to have an abstract representation of the spatial constraints of the cell, the concepts of site and places are introduced.

Site: a site is an area within the cell, which can hold a single part (two assembled parts will be considered as a single part). In general, sites are associated with actors (for instance grippers, cameras, conveyors, etc...) and constitute shareable spatial resources. The space region where two or more sites can interact is defined as a *working area*.

Equivalent sites: two or more sites are said to be equivalent iff they have the same functions (functional equivalence) and they access the same working areas (spatial equivalence).

Place: a place is the equivalence class of sites.

3.3. Architecture of the FAC Programming System

The FAC programming system refers to the concepts introduced so far in order to monitor and control the operation of an assembly task. Functionally, the FAC programming system may be structured into three different control levels:
 • the Task level, which defines a set of actions (also called a task plan) which ensures that the assembly task makes progress from its current state;
 • the Action level, which manages the execution of each action specified in the task plan, in order to allow a maximum concurrency among the components of the cell, as well as to synchronize these components when and where required (e.g. to avoid resource allocation conflicts);
 • the Function level, which interfaces the Action level to the run-time environment by sending requests/receiving indications to/from the cell actors.

3.3.1. Task Level. At the Task level, the state of the assembly task is characterized by the location of the parts in the cell (e.g. "part X located in place Y"). The task plan defines, starting from the current state of the assembly task, all possible evolutions of the parts within the cell. It is modelled as a set of partially ordered actions which will be managed by the lower functional level (i.e. the Action level).

The module of the programming system which implements the Task level is called the *Task Planner*. It is in charge of the task plan generation from the current state of the assembly task.

3.3.2. Action Level. At the Action level, the state of the assembly task is defined into more details by taking into account the location of the sites and effectors within the cell. Starting from the task plan generated by the upper level, the actions are synchronized with respect to the spatial state of the assembly cell, always with the purpose of providing as much as concurrency as possible among the cell components. The module of the programming system which implements the Action level is called the *Driving System*. It is in charge of the allocation of the spatial resources, and consequently of the synchronization of the actions specified in the task plan. For that purpose, the synchronized actions are translated into functional primitives to be performed by the lower level (i.e. the Function level).

3.3.3. Function Level. At the Function level, the state of the assembly task is characterized by the state of the actors present in the cell. The functional primitives, as requested by the upper level, are implemented by the actors. The module of the programming system which implements the Function level is called the *Control System*. At this level, the functional primitives, sent by the driving system, are translated into requests which will be performed by the robotic components of the cell.

Figure 1 shows the architecture of the FAC programming system. The *Supervision System* module is intended to provide on-line error detection and management facilities. It can be associated with any of the three levels of the programming system.

3.4. Specification approach

The hierarchical structuring of the FAC programming system, as introduced in paragraph 3.3, allows further evolution of the cell without having to rewrite the whole programming system software. The required evolution may consist in programming new assembly tasks for the same cell, as well as introducing/removing components of the cell. The problems involved in designing the software of the FAC programming system are quite difficult to solve, mainly at the Task and Action levels. The methodology followed for these two levels is detailed in the following paragraphs.

3.4.1. Task Level Specification: the Task Planner Module. The task planner module is intended to generate the task plan by starting from the current state of the assembly task. Operational research and artificial intelligence techniques are often used for this purpose in order to control the large number of possible cases. In this study, however, we have adopted the methodology initially introduced in [Chochon 86] where the task plan is obtained from a model, established off-line, of the assembly task. In our approach, this model of the assembly task is described as a set of communicating extended state machines, and its formal description in Estelle is therefore straightforward.

Each extended state machine corresponds to an Estelle module which describes the behaviour of one or several agents of the cell with respect to the assembly task to be performed (more precisely each action of an agent is modelled by a transition of the associated state machine). Starting from the current state of the assembly task, the transition firing sequences of the communicating state machines will specify the set of (possibly concurrent) actions which will constitute the task plan.

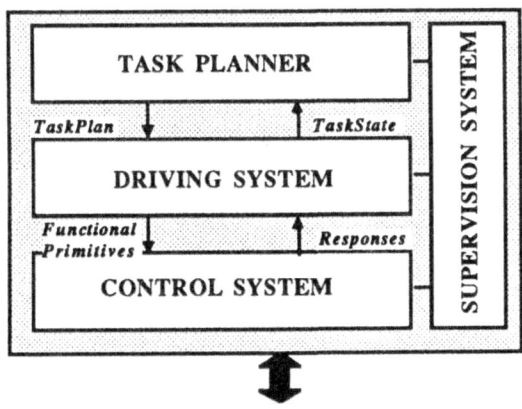

Figure 1 - Architecture of the Programming System

225

The task plan process can functionally be divided into two main phases: the definition of the actions to be performed and the establishment of the synchronization conditions among these actions. The definition of the actions is, as previously mentioned, mainly based on the (static) knowledge of the assembly task (as modelled by communicating state machines) and on the (dynamic) knowledge of the assembly task state. The establishment of the synchronization conditions among actions is performed for each action of the task plan, and these conditions will be used at the Action level in order to guarantee an assignment of the actions to the agents, which ensures a maximum concurrency among the agents of the cell.

As illustrated in Figure 2, the task planner module is composed of two types of module: the action manager module described as an extended state machine which is in charge of generating the actions for an agent (or a set of agents) and the plan manager module which is in charge of the establishment of the synchronization conditions. An action manager module may interact with the plan manager module or with other action manager modules. The plan manager module interacts with all the action manager modules composing the task planner and it furthermore implements the interface with the Action level (implemented by the driving system module).

For associating the action manager modules with the cell agents, some simple rules have been observed:
— an action manager module is associated with each "complex" agent of the cell; agents which can perform several action sequences will be considered "complex" agents (e.g. robots);
— an action manager module is associated with each agent presenting synchronization problems with others agents in the cell (e.g. a turn table which may be accessed by two robots);
— the functionalities of agents which perform one single action sequence (e.g. cameras, conveyors, etc.) may be represented within the action manager module associated with a complex agent.

In the Estelle formal description of an action manager module, two classes of transitions are defined: control transitions and action generation transitions. A control transition is activated by the occurrence of an interaction and its firing will enable action generation transitions (either of the same module, or of others communicating modules).

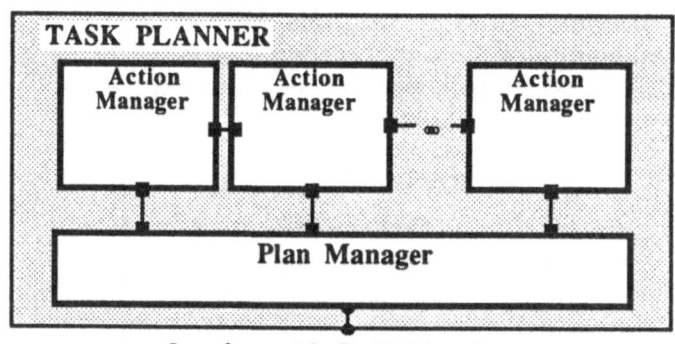

Interface with the Driving System

Figure 2 - General architecture of the Task Planner module

An action generation transition will send an action identifier to the plan manager module, allowing it to compose the task plan. The text below shows an example of the Estelle description of an action generation transition.

```
FROM State_A TO State_B
    PROVIDED (TaskState[place1] = Task_state1) AND ... (TaskState[placeN] = Task_stateN)
        begin
        {       Task State changes      }
        {       other instructions      }
            OUTPUT IP_x.ActionMessage(Action_ID);
        end;
```

The rules below are helpful for describing the behaviour of an action manager module:
– the states of the state machine must be associated with logical states of the agent within the assembly task context (for instance, a robot can be represented by the states Assembling, Loading, Unloading, etc.);
– the initial state is the Idle state which represents that no actions are feasible in the current task state;
– at the end of any action sequence generation, the state machine must be in the Idle state;
– the set of the action generation transitions in all the action manager modules must represent all the possible actions of the considered assembly task.

The plan manager module is described by a quite simple state machine whose main function is to receive each action sent by the action manager modules, to add it to the task plan and to establish the synchronization conditions for all actions composing the task plan. Once the generation is completed, the plan manager module is in charge of sending the task plan to the Driving System module, which will be described in the next paragraph.

3.4.2. Action Level Specification: the Driving System Module. Remember that the driving system module, which implements the Action level introduced in paragraph 3.3.2., is functionally in charge of:
– performing the synchronization of the actions;
– allocating the spatial resources of the cell;
– requesting the execution of the functional primitives which compose each action to be performed.

The goal intended, when designing the driving system module, was to define a general modular approach being as independent as possible of both the physical configuration of the cell and of the particular assembly task to be performed. This led us to design a cell recursively as a set of sub-cells, the bottom sub-cells in this hierarchy being *agents* which cannot be further refined. Well defined criteria may be determined for identifying the sub-cells (e.g. functional independence, autonomous set of components dealing with the assembly of a part, ...). These sub-cells, said to be of level (N-1) with respect to the upper (N) level cell, are in charge of performing at level (N-1) the (N) actions (i.e. the actions defined at level (N)). As (N-1) cells may have spatial or physical resources in common, allocation conflicts of these resources have to be dealt with at level (N). Any (N) action is therefore analyzed at level (N) in order to be assigned to one of the (N-1) sub-cells. If the resources required for performing the (N) action are local to the (N-1) sub-cell in charge of implementing the considered (N) action, then this action may be directly assigned to this (N-1) sub-cell. If, however, the resources required are common to several (N-1) sub-cells, then these resources have first to be allocated at level (N) before the action be assigned to the considered (N-1) sub-cell.

Starting from the modular approach introduced so far, the general architecture of the Driving System module is based on four types of module which are the following:

• The synchronizer module – This module, which appears only at the top level, implements the interface of the driving system module with the task planner module; it

227

therefore manages the concurrency and synchronization among the actions specified in the task plan. It receives from the lower level the responses related to the execution of the requested actions and then updates, when all actions have been completed, the logical state of the assembly task in order to request a new task plan.

   • The sorter module – This module, which appears within any cell (or sub-cell) refined into sub-cells, deals with the assignment of each (N) action to a (N-1) sub-cell.

   • The manager module – This module, which appears within any cell (or sub-cell) refined into sub-cells, deals with the resource management. A manager module may be associated with one or more resources of the cell. This module must perform the resource management with the aim of ensuring a high degree of concurrency among the cell components. The approach consists of decomposing a (N) action into several (N-1) sub-actions, when some required resource is not yet available.

   • The agent module – This module, which appears only at the bottom level, implements the interface of the driving system module with the control system module (i.e. with the actors composing the agent). It translates the (N) actions into a set of functional primitives to be performed by the actors.

4. AN APPLICATION EXAMPLE

The specific assembly task considered in this paper is a significant example which illustrates how synchronization and resource allocation problems may be solved. The cell considered here is actually available at LAAS/CNRS where this experiment has been carried out. The approach described in the previous section has been successfully applied to this particular example in order to derive an Estelle formal specification of the cell programming system. Starting from the formal specification, the Estelle support tools have been used for:
   – debugging the Estelle Formal Specification;
   – testing the C code automatically generated from the formal specification;
   – and implementing it on the actual flexible assembly cell.

4.1 Informal Specification of the Assembly Task

The assembly task consists in assembling three different parts (called respectively A, B and C) in order to produce a final part ($ABCC$). This assembly task is made up of the following set of operations:

Feeding of part A:		→	A
Feeding of part B:		→	B
Feeding of part C:		→	C
Assembling of parts A and B:	A & B	→	AB
Assembling of parts AB and C:	AB & C	→	ABC
Assembling of parts ABC and C:	ABC & C	→	$ABCC$
Inspection of $ABCC$:	$ABCC$	→	$+ABCC$ / $-ABCC$
Unloading of $+ABCC$:	$+ABCC$	→	
Unloading of $-ABCC$:	$-ABCC$	→	

The flexible assembly cell carrying out the execution of this assembly task is depicted in Figure 3. It consists of the followings robotic elements:
   – a robot, RightRobot, for moving and assembling A and B parts; this robot is also in charge of moving the assembled parts ($+ABCC$ and $-ABCC$);
   – a table, RightTable, for feeding A and B parts;
   – a turn table, TurnTable, for moving AB and $ABCC$ parts;
   – a robot, LeftRobot, for assembling a AB part with C parts;
   – a table, LeftTable for feeding C parts;
   – a camera, Camera, for inspecting assembled parts;
   – a conveyor, Conveyor, for unloading correct assembled parts ($+ABCC$);
   – a garbage, Garbage, for unloading incorrect parts ($-ABCC$).

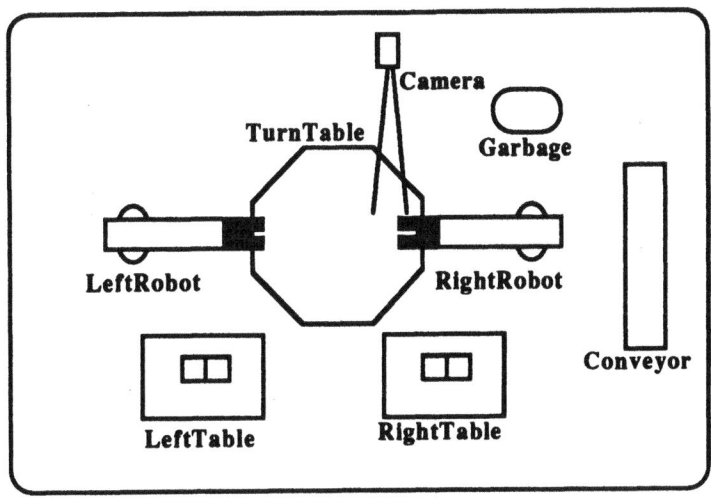

Figure 3 - The Cell Configuration

The normal behaviour of the cell, when performing this assembly task, is as follows:
   – *A* and *B* parts lie on *RightTable*;
   – *RightRobot* picks up a *A* part and puts it on the right side of *TurnTable*; then, it picks a *B* part and assembles it in order to get a *AB* part;
   – *TurnTable* then performs a half-rotation, moving the *AB* part to the left side of the cell;
   – *LeftRobot* performs twice the assembling of *C* parts (which are located on *LeftTable*) in order to get a *ABCC* part; meanwhile, *RightRobot* may generate another *AB* part on the right side of *TurnTable*;
   – after a new half-rotation of *TurnTable*, the *ABCC* part is on the right side of the cell, where *Camera* will inspect it; once inspected, the assembled part will be moved by *RightRobot*, in order to unload it either on *Conveyor* or into *Garbage* depending on the result of the inspection (either *+ABCC* or *-ABCC*).

4.2. Programming System Specification

The architecture of the task planner and driving system modules is defined by applying the specification methodology introduced in section 3.4. Both the task planner and the driving system modules have been specified in Estelle.

4.2.1. The TaskPlanner Module. Figure 4 shows the architecture of the task planner module. Action Manager modules are associated with the three main agents available within the cell – RightRobot, LeftRobot and TurnTable. In the RightRobot module, the actions associated to the Camera and Conveyor agents are also dealt with. As these agents are rather simple, there was no need for defining for them a specific Action Manager module.

4.2.2. The DrivingSystem Module. Figure 5 shows the architecture of the Driving System module. This module is only refined into two levels; the upper level is associated with the cell itself, whereas the lower level is related to the agents of the cell. The modules suffixed by "m" are manager modules and the modules interfacing the control system module are agent modules. Each manager module is identified by the place associated to the managed resource.

229

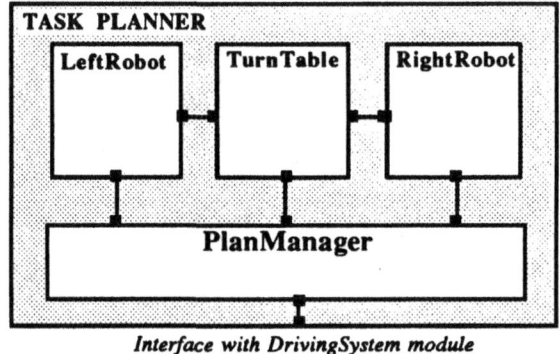

Figure 4 - Architecture of the Task Planner module.

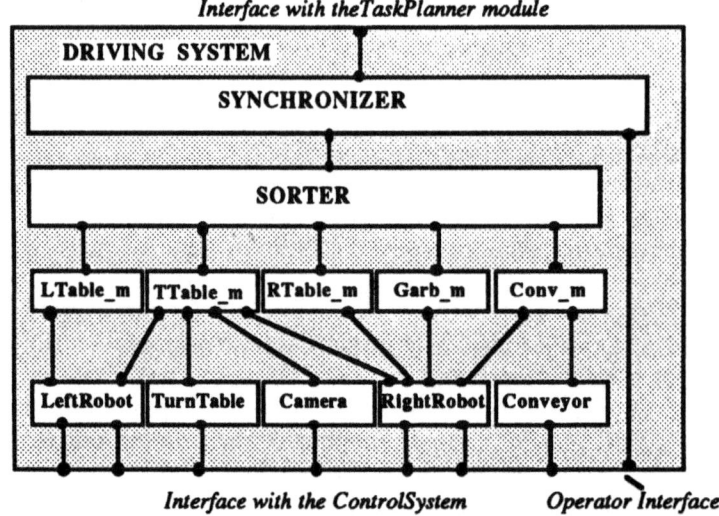

Figure 5 – Architecture of the Driving System module

4.3. Validation of the Programming System

A main advantage in using a Formal Description Technique for specifying the cell programming system is to ensure a good level of confidence at the design stage by validating the formal specification before generating the implementation. Using the Estelle support tools available at LAAS, a simulation of the programming system has been carried out, following three main objectives:

- an initial debugging of the formal specification,
- a validation of the generated C code within a simulation environment,
- a graphic validation of the generated C code within an implementation environment.

4.3.1. Debugging of the Formal Specification. This debugging phase has been carried out by using the ESTIM tool [Saqui 89a], [Saqui 89b] which offers a set of nice capabilities for monitoring the run-time behaviour of the formal specification by providing a symbolic access to all the Estelle objects. A step by step simulation permits the analysis of the specification by observing the way variables and interaction queues are modified after each transition firing. A non deterministic transition firing mode permits furthermore to analyze the global behaviour of the specification; this random simulation has been very useful for detecting several potential unspecified receptions as well as deadlock states in the early versions of the specification. ESTIM provides therefore a rapid prototyping environment for debugging and simulating Estelle specifications, as it interprets directly the Estelle statements and does not require to generate the C code associated with the specification. Extra Estelle "mirror" modules have been introduced in the specification in order to represent the behaviour of the actors of the cell.

4.3.2. Validation of the Generated C Code within a Simulation Environment. Following the previous phase, the EWS simulator functionalities [Ayache 88] have been used in order to simulate the C code generated from the Estelle specification. In particular modular and integration tests have been carried out. Modular testing has consisted in verifying separately the correctness of each module composing the Driving System, whereas integration testing has ensured the global consistency of the Driving System (i.e. consistency of the module configuration, of the interfaces between modules, etc...). EWS provides therefore facilities similar to the ones of ESTIM but which are applied to the C code generated and not directly to the Estelle statements.

4.3.3. Graphic Validation of the Generated C Code within an Implementation Environment. A graphic simulation of the flexible assembly cell has been developed for validating the integration of the generated C code with the implementation motor of EWS. This FAC simulator has been written using the functions of the SunCore graphic package available on Sun workstations. A graphic window representing the flexible cell configuration (as depicted in Figure 3) shows the behaviour of the cell components and therefore makes it possible to observe how concurrency and synchronization are managed by the software generated from the Estelle formal specification.

The C code considered at this phase presents a high degree of reliability and only few adaptations are required in order to integrate it with the actual run-time environment supporting the robotic components.

4.4. Implementation on the Cell

The cell available at LAAS is composed of the following components: two robots (SCEMI 6P01), a vision system (VISIOMAT from Allen-Bradley) supporting several cameras, a conveyor, a turn table, several sensors and pneumatic grippers, several UNIX-based computers (Sun3 workstations). The workstations are connected to an Ethernet LAN, and BRIDGE CS/100 communication servers link the robotic components to them.

The software generated from the Estelle specification (task planner and driving system) is integrated with four distributed control systems interfacing the robotic components (implementation of the control system module). These four control systems correspond respectively to the control of the two SCEMI robots, the peripherical control (conveyor, turn table and robot grippers) and the vision system control. The whole software consists of seven UNIX processes implemented on a Sun3 workstation: the Estelle application process implementing the task planner and the driving system, the four communication processes to access the remote control systems, the operator interface process and the initialization process. The compiled C code of the Estelle application process is linked to the EWS library which performs time and buffers management. The operator interface process allows the cell operator to notify the feeding of the parts and to stop the assembly task operation at any time (for instance, in the case of an emergency ...). The initialization process manages the other processes (creation, start of execution and release).

The software which is not generated from the Estelle specification represents less than 10% of the whole software (without taking into account the dedicated software associated to the four remote control systems of the robotic components which was already available). The size of the C code generated from the Estelle specification (task planner and driving system modules) represent about 7000 lines of C, whereas the size of the Estelle specification is less than 2500 lines.

5. CONCLUSION

An experience in using the Estelle Formal Description Technique for specifying, validating and implementing the programming system of a flexible assembly cell has been presented in this paper. The results obtained are much better than the ones initially expected and the usefulness of the Estelle approach has been demonstrated for an area of applications for which it has not especially designed for.

Considering the possibility of using these results on large industrial scale, the following comments may be drawn:

• By using the Estelle approach and its support tools, it has been possible to produce a highly reliable software, as well as to shorten considerably the time required for its development (compared with previous experiences at LAAS/CNRS);

• The modular approach proposed in the paper ensures both the genericity and the flexibility of the design, which results in particular to have only very few modules to adapt if it is required to modify the assembling task to be performed on the cell or even the configuration of the cell; this modular approach ensures therefore the reusability of the design, which is a strong requirement for large scale industrial applications;

• Although ESTIM and EWS may not (yet) be considered as CASE Software products, they present already a high degree of maturity (functionalities, performance and user interface), allowing a complete validation of the design.

6. REFERENCES

[Ayache 88] J. M. Ayache et al. "EWS - An Integrated Workstation for the Design and the Automatic Generation", 1st International Conference on Formal Description Techniques, Stirling, September 1988.

[Ayache 89] J. M. Ayache et al. "Presentation of the SEDOS Estelle Demonstrator Project", The Formal Description Technique ESTELLE, Results of the ESPRIT/SEDOS Project, p. 423-434, North-Holland, 1989.

[Budkowski 87] S. Budkowski, P. Dembinski. "An Introduction to Estelle: A Specification Language for Distributed Systems", Computer Networks and ISDN Systems, Vol. 14, 1987.

[Chochon 86] H. Chochon, R. Alami. "NNS, A Knowledge-based on-line system for an assembly workcell", IEEE International Conference on Robotics, San Francisco 1986.

[Chochon 87] H. Chochon, R. Alami. "A knowledge-based system for programming and execution control of multi-robot assembly cells", 3rd International Conference on Advanced Robotics, p. 431-442, Versailles 1987.

[Courtiat 87] J. P. Courtiat. "Contribution à la Description Formelle de Protocoles", Thèse de Docteur d'Etat de l'Université Paul-Sabatier, LAAS - Toulouse, December 1987.

[Courtiat 88] J. P. Courtiat. "Estelle*: a Powerful Dialect of Estelle for OSI Protocol Description", Proceedings of the 8th IFIP Symposium on Protocol Specification, Testing and Verification, Atlantic City, June 1988.

[Diaz 89a] M. Diaz, C. A. Vissers. "SEDOS: Designing Open Distributed Systems", IEEE Software, pp. 24-33, November 1989.

[Diaz 89b] M. Diaz, J. Dufau, R. Groz. "Experiences Using ESTELLE within SEDOS Estelle Demonstrator", 2nd International Conference on Formal Description Techniques, Vancouver, December 1989.

[Estelle 88] ISO IS 9074. "Estelle - A Formal Description Technique Based on a Extended State Transition Model". International Standardization Organization, November 1988.

[Saqui 89a] P. de Saqui-Sannes, J. P. Courtiat. "Rapid Prototyping of the ESTELLE Simulator: ESTIM", The Formal Description Technique ESTELLE, Results of the ESPRIT/SEDOS Project, p. 325-352, North-Holland, 1989.

[Saqui 89b] P. de Saqui-Sannes, J. P. Courtiat. "From the Simulation to the Verification of Estelle Specifications", 2nd International Conference on Formal Description Techniques, Vancouver, December 1989.

[Tenney 88] R. L. Tenney. "A Tutorial Introduction to Estelle", Invited Paper at the 1st International Conference on Formal Description Techniques, Stirling, September 1988.

Quality Assurance as a Dynamical Production Process Guide - Control Elements Supported by Dedicated Knowledge Based Systems

Mina-Jaqueline Schachter-Radig, Diederich Wermser

NTE NeuTech Entwicklungsgesellschaft
Munich, FRG

1. INTRODUCTION

Quality assurance is relevant for all steps in the lifecycle of a product, ranging from requirements definition, design, manufacturing and post-production control of each part manufactured to sales and maintenance. Quality management thus requires and enables a *semantic* integration of manufacturing. Not only data on lots of products to be manufactured, tools to be applied etc. have to be communicated, but rather higher level informations like tendencies in deviations (which e.g. may be used to control readjustment of corresponding tools), critical steps in manufacturing (which e.g. may recommend a less sensitive design of future products) etc. As this kind of integration obviously does not allow to predefine behaviour of respective systems procedurally, knowledge based systems are required which allow for a flexible and situation adaptive reaction of various components as well as the complete system. This paper describes current work and experiences of the authors concerning definition, development, introduction and usage of generic modules which each support single quality elements and will be connected to form a complete quality framework, which including more and more modules will integrate manufacturing through high level distributed quality control.

Due to extremely decreased life-time of products the post-hoc approach of quality control is no longer usable to ensure quality requirements. There is a need for quality assurance during the whole life cycle of a product, i.e. a holistic approach considering all phases and activities that influence the quality of a product during its formation. Within each state of planning and realisation (and usage) quality elements (QE) can be formulated, which are relevant to keep quality requirements for the corresponding phase.

The relations and interdependencies between the QEs are demonstrated in the quality loop [DIN 55350] (see fig. 1). Each phase, respectively its achievable contribution to the quality of a product, depends on the results of the preceding QE, i.e. it is not possible to make amends for not accomplished quality of a antecedent phase.

For some QEs there exist already computer aided support like editing facilities for inspection planning or systems for inspection data collection and evaluation. But in most cases these "CAQ" systems are stand-alone solutions without connection to existing CIM-applications i.e. there exists no common data model, no use is made of available geometrical informations (CAD), technological knowledge etc.

Basic aspects characterizing NTE´s approach to CAQ are:
- Quality assurance has to cover all phases in the lifecycle of a product ranging from requirements definition and construction, post-production control of each part manufactured to sales and maintenance. Quality assurance thus requires and enables a *semantic* integration of manufacturing as not only data on lots of products to be manufactured, tools to be applied etc. have to be communicated, but rather higher level informations like tendencies in deviations (which e.g. may be used to control readjustment of corresponding tools), critical steps in manufacturing (which e.g. may recommend a less sensitive construction of future products), etc. As this kind of integration obviously does not allow to predefine behaviour of respective systems procedurally, knowledge based systems are required which allow for a flexible and situation adaptive reaction of various components as well as the complete system.
- Fig. 3 shows five different levels of maturity of manufacturing processes with respect to quality assurance. Only manufacturing processes having reached at least the level

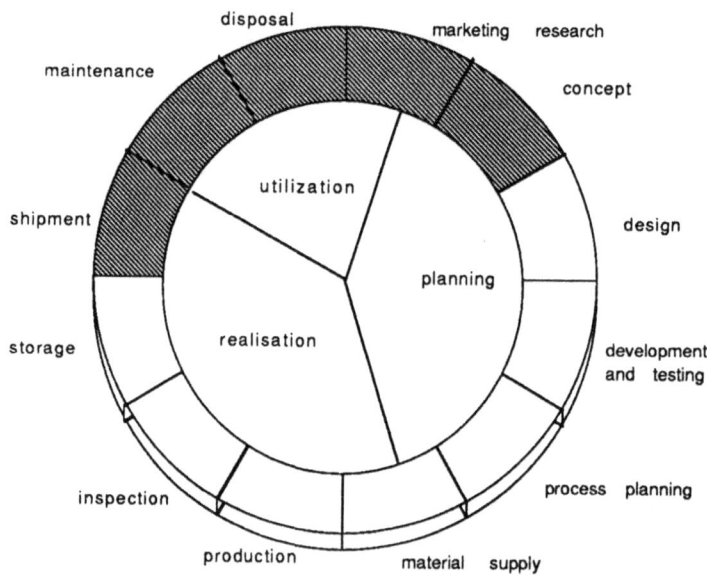

Fig. 1: Quality Loop with QEs

"controllable" fulfil the basic needs of quality assurance.
- Integration of CAQ components into the whole CIM environment is a crucial issue for the success of CAQ. Fig. 4 outlines important interoperations with other relevant areas in a CIM environment.
- Quality assurance is and will more and more become a regulative process rather than a post hoc control mechanism separating the parts fulfilling the quality requirements from those, which do not. Fig. 2 demonstrates this idea [Bläsing 87]. This approach is true as well for a micro view, where e.g. the parameters of a galvanic process can get controlled depending on the quality of the parts manufactured, as for a macro view, where complaints of the field service will influence definition and design of new versions of products resp. new products. This will further increase the need for interoperation of components in a CAQ scenario.
- A sound methodological approach is necessary as well for the development of an integrable and extendable CAQ concept as for the development of the various components of this scenario.

On this basis NTE has developed the CAQ concept partly introduced in this paper. Chapter 2 gives an overview of the CAQ scenario. In chapter 3 subsequently two example components are described. Chapter 4 explains the methodological basis and approach applied to develop the knowledge based components.

2. SYSTEM COMPONENTS OF A CAQ SCENARIO

An overview of components of a CAQ scenario integrated into a CIM environment is given in fig. 4. Referring to the quality loop given in fig. 1., the elements of the CAQ scenario are beyond the actual manufacturing and inspection (CAM) related particularly to product definition and design (CAE), process planning (CAP) and materials supply.

Typical functional elements of a knowledge based system in a CAQ scenario are outlined in figure 2.2 [Schachter 87]. Characteristic elements of many knowledge based systems in an integrated CAQ scenario are particularly sensor and effector interfaces including the knowledge necessary to interpret and evaluate sensor data, like e.g. visual information as well as process manufacturing interfaces, which directly interconnects with the respective processing machinery. These facilities enable a higher degree of automation in that the actual judgement of results can be performed by the knowledge based system itself.

Depending on the dedicated role of the corresponding knowledge based CAQ system the various functional elements get more or less emphasis or do even vanish. Different from the aspects mentioned above, the IPG system (Inspection Plan Generation) outlined in chapter 3 e.g. has a strong emphasis on man - machine interaction and no sensor/effector interface.

3. ILLUSTRATION OF TWO EXAMPLE COMPONENTS

Subsequently two typical components of an CAQ scenario as outlined before are explained in order to make the ideas of the scenario more concrete and demonstrate typical functionalities:

- The Inspection Plan Generation system (IPG) has to be associated with the phases process planning and materials supply in the quality loop as given in fig. 1. This system supports the elaboration of inspection plans for parts manufactured in the respective company or delivered by suppliers. Knowledge on classes of products and materials as well as classes of production processes is comprised. The responsible engineers get suggestions for all elements of an inspection plan like aspects to be checked, means for the inspection steps etc., which are generated by the IPG system on the basis of general rules for establishment of inspection plans, characteristics of materials used and analogies with similar kinds of parts, where valid inspection plans exist.

- The system for quality control in manufacturing processes by monitoring and preventive maintenance (MPM) as outlined in subchapter 3.2 is an example for a CAQ system which continously monitors and controls process quality in a regulative process as depicted in fig 2. Besides continous regulation of quality relevant parameters this kind of system has a serious economic benefit in that degradations of process machinery are detected early. This enables a significant improvement of the average up-time of complex systems. The MPM system relates to the production phase of the quality loop given in fig. 1.

Like a number of other components relating to other areas of the quality loop resp. other areas of the CAQ scenario as outlined in these systems have been developed as far as possible as reusable systems. Dedicated systems are developed for and introduced for industries like medical equipment manufacturing and printing industry currently.

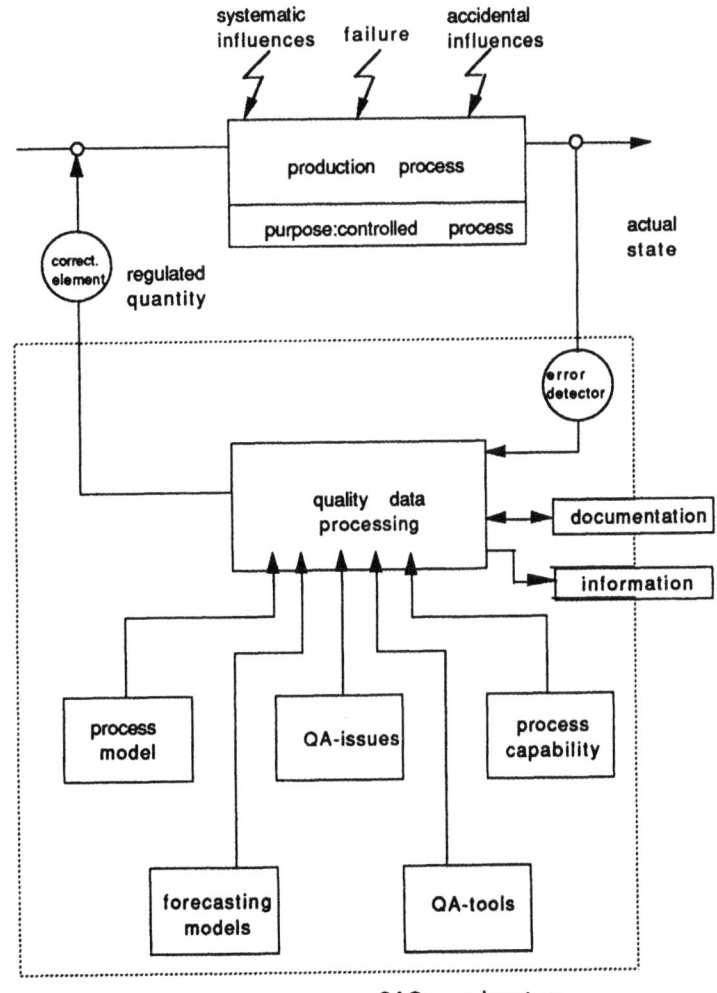

Fig. 2: CAQ: Process Layer

235

3.1 INSPECTION PLAN GENERATION (IPG)

One of the first elements a quality information and control system has to support is the realisation phase of a product (as shown in fig. 1) where the first element is the material supply needed for the manufacturing of products. A basic component of the material supply is the inspection planning in the receipt of goods department where the elaboration of inspection instructions is the main task. The dedicated IPG system for the receipt of goods department supports this task within the CAQ scenario. Because of the similarity of elements of the inspection planning the next proposed extension of the system will support the inspection plan generation for the manufacturing process as well as the final product inspection.

Reasons that make inspection plan generation for the receipt of goods department particularly relevant in the first phase of implementation of the CAQ scenario are:

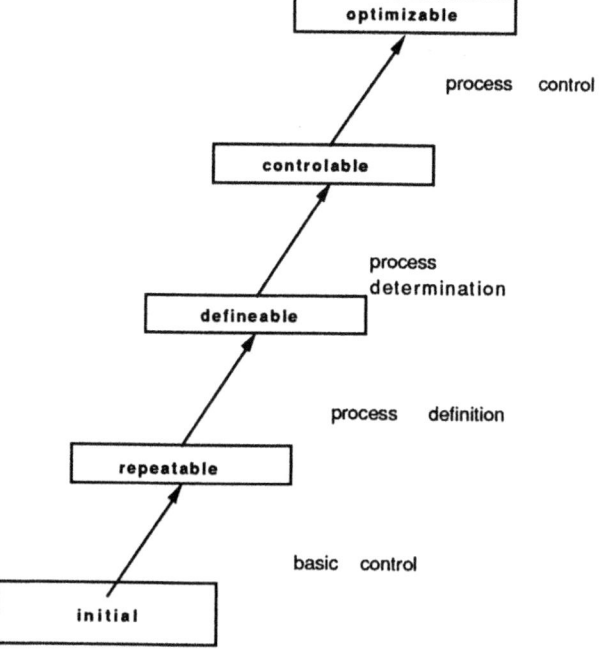

Fig. 3: Process Evolution

- IPG is the starting point of all following activities within the quality inspection and the evaluation of the inspection results will have strong impacts on the estimation of the corresponding suppliers
- Integration of this component to the customers environment can be done without major organizational restructuring of the customers working process
- The knowledge representation structures (e.g. taxonomies of materials and their char-

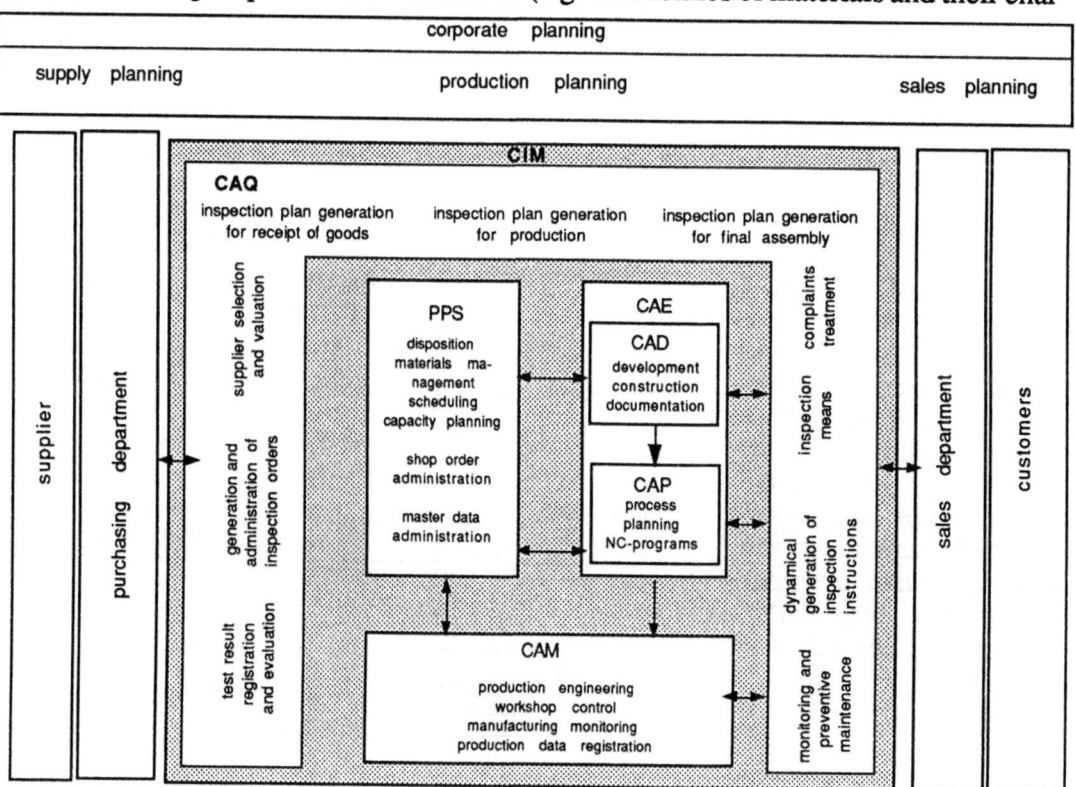

Fig. 4: CAQ as an Integrating Factor within CIM

acteristics, taxonomies of parts) can be reused in other CAQ systems.

- Many of the knowledge sources for the problem solving model of the IPG system can be reused for other systems of the CAQ scenario.

The main goal of this module is support for the development of inspection strategies for the inspection planning staff, i.e. reducing the dependence from written specifications. Besides this, the improvement of the quality and similarity of the inspection instructions and the dissemination of all included knowledge concerning the elaboration of inspection instructions has been focused on.

The inference structure in Fig. 6 represents an abstracted view of the problem solver of the IPG. The idea of an inference structure [Breuker 87] is to specify knowledge sources (elipsoids), that comprise bunches of knowledge available for performing a problem solving task. The metaclasses (rectangles) indicate, which input is necessary for a

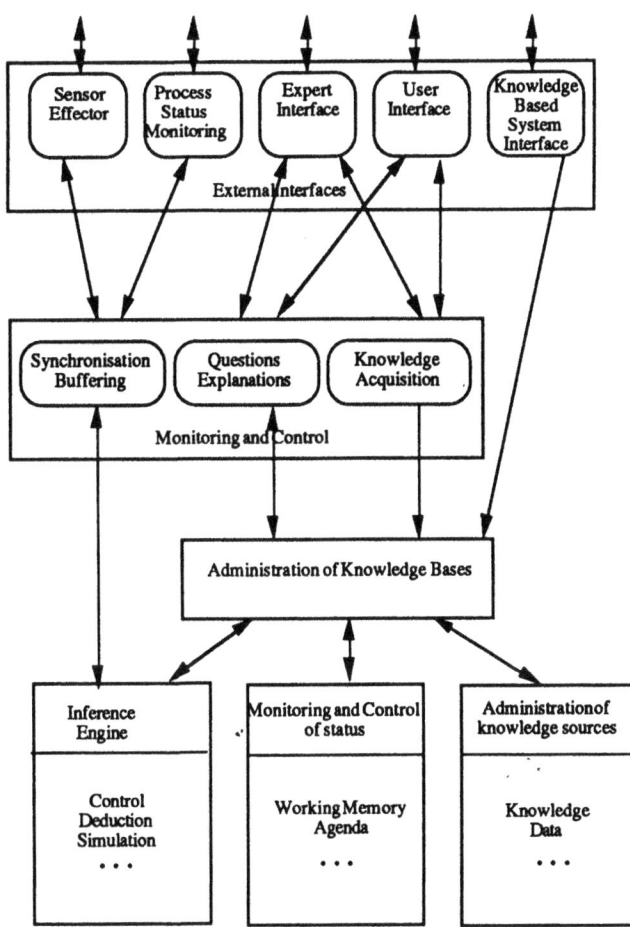

Fig. 5: Architecture of a KBS in the Production Process

knowledge source and what can be determined by a knowledge source. An inference structure does not specify which knowledge source to apply in a particular situation of the problem solving process but simply which knowledge sources are available. Strategies how to use the knowledge sources are represented in different layers of a model of the knowledge necessary.

A description of the inference structure of the IPG system, respectively its Metaclasses and Knowledge Sources is given below.

Description of Metaclasses

System model
> encloses the part of the domain knowledge necessary for inspection planning. It is a description of concepts and relations in the production world, especially taxonomies of products, manufacturing processes, materials, regulations, standards, inspection characteristics, inspection means and all their related attributes.

Part data
> contains information about e.g. the supplier or producer of the part to be inspected, his quality system management, kind of material, applied manufacturing technology and the area where the part will be used (medical equipment, aerospace etc.)

Parameter
> enable the final instantiation of inspection specifications on the basis of geometrical information about the part like desired values and their acceptable tolerance or limiting values.

Inspection part
> is the class of inspection part which is identified

Inspection characteristics
> is the set of relevant classes of inspection characteristics

237

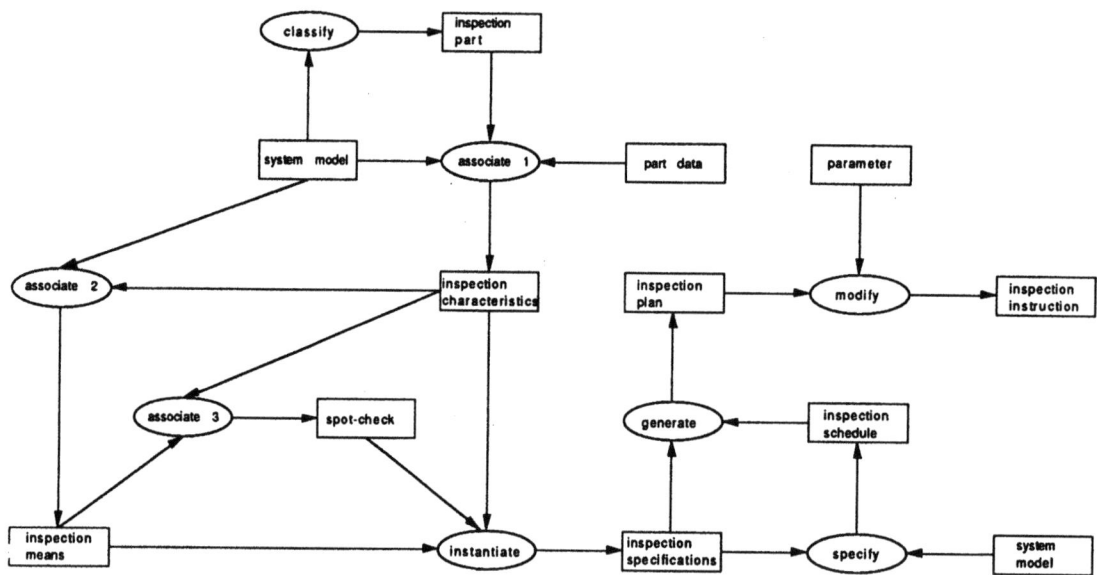

Fig. 6: Inference Structure for the IPG System

Inspection means
  is the set of relevant classes of inspection means
Spot-check
  is the description of the spot-check parameters determined
Inspection specification
  is the set of relevant inspection classifications for all inspection characteristics identified
Inspection schedule
  comprises adapted sequence of instantiated inspection specifications
Inspection plan
  is the set of inspection specifications in the selected sequence
Inspection instruction
  final result for use in the inspection department

Description of Knowledge Sources
Classify
  This knowledge source will identify the class of inspection part based on a part taxonomy in the system model by evaluating user inputs.
Associate I
  This knowledge source will associate to the class of the inspection part by using additional information concerning the part and the input of system model (taxonomies) all relevant inspection characteristics.
Associate II
  This knowledge source will associate to all inspection characteristics the adequate inspection means by utilizing links between inspection characteristics and inspection means taxonomies .
Associate III
  This knowledge source will determine for each inspection characteristic and the according inspection means the spot-check parameters.
Instantiate
  This knowledge source will instantiate all inspection specifications based on inspection characteristics and the associated inspection means and spot-check parameters.
Specify
  This knowledge source will specify the inspection schedule which gives the correct sequence of inspection specifications using possible dependencies between inspection characteristics etc.
Generate
  This knowledge source will generate an inspection plan including all inspection specifications in the adequate order .

Modify

        This knowledge source will modify the inspection plan by associating tolerance or limiting values to each inspection specification.

3.2. Quality Control in Manufacturing Processes (MPM)

This system focuses on continuous monitoring and controlling of process quality (fig. 2 "CAQ: Process Layer") and preventive maintenance in case of a detected degeneration of the process. Central tasks of such kind of KBS for industrial process control are :
- continuous regulation of quality relevant parameters
- diagnosis of the cause of observed degenerations of product quality
- changing parameters influencing the respective characteristics of product quality and in case this does not achieve adequate improvement of product quality,
- recommend actions for preventive maintenance

Within the analysis an approach was to develop quite generic, re-usable models of knowledge for these tasks, being specialized e.g. to a printing domain [Chon 90], [Busche 89].

Fig. 7 presents a top view on the inference structure of the system. Monitoring the production line is followed by an assessment of the further evolution of the production process: this is considered as the main critical conclusion from which usually necessary actions are derived.

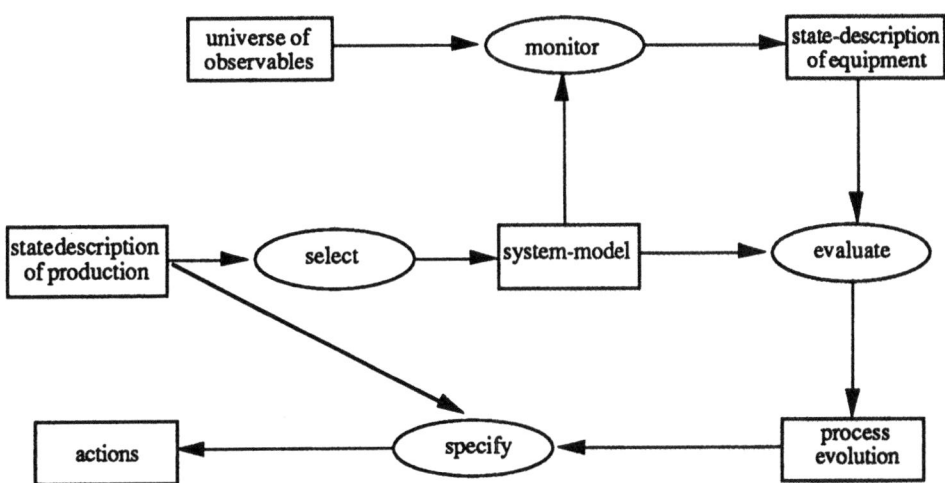

Fig. 7: Inference Structure for Quality Control in Manufacturing Processes

Description of Metaclasses:

The inference structure in Fig. 7 represents the top view only; "monitor" as well as "evaluate" are described by own inference structures at two more detailed levels, which are not described in this paper.

Systemmodel

        Systemmodel encloses the part of the knowledge base necessary for monitoring and evaluation. It is a description of a normally functioning production process and equipment. It comprises knowledge in the domain concerned with the set of relevant parameters and their associated values and processes.

State description of production

        This concerns the current status of the production and is a description of, e.g.:
-     the kind of production / the production capacity, etc.
-     phases and subphases of production: e.g. starting and endphases, normal production phase, periods in which materials are changed etc.

-       the status of job performance, e.g. production schedules and deviation of schedules are continuously updated.

State description of equipment
Description of the present status of plant, concerning machinery etc. It comprises:
-       parameters with values out of norm and tendencies as well as
-       a characterization of this observed degeneration

Process evolution
Consequences of the deviations from normal state observed in monitoring on the further behaviour of the production process

Actions
Correction of the production process by regulations of quality parameters and recommendations of appropriate measures to prevent damage to the production process. These regulations and recommendations are directed to effectors (see fig. 5), like valves, electrical switches, etc.

Universe of observables
All quantities of the products manufactured and the respective production machine which are measured by sensors or the values of which can be provided by the operators. Examples for typical sensors in a production environment are: cameras, laser measurement equipment, temperature sensors, force sensors, etc.

Description of Knowledge Sources

Select
This Knowledge Source fixes the set of monitoring parameters relevant for the next monitoring cycle as well as knowledge concerning these parameters (norm_ranges etc.) required for the monitoring subtask. The system model depends on the status quo of the production. In general however it does not vary continuously with time but discretely when the production operation changes phases.

Monitor
Quality of products manufactured, the production process and the manufacturing equipment are continuously monitored. Deviations from desired values and normal state are registered and classified according to their implications on the quality and the production process.

Evaluate
Consequences of the detected deviations of current parameter values from norm ranges and the detected tendencies on the further status of the production process are inferred.

Specify
Actions are recommended to the operator (or directly performed by the system) as appropriate countermeasures against the degenerations which are described by the predicted process evolution

4. METHODOLOGICAL APPROACH

The methodological approach for the development of CAQ scenario as well as its knowledge based component systems has to be seen in a context of the general methodological framework developed and applied by NTE. Fig 8 gives an overview of the areas of activity and the relations with respect to usage and further development of the methodological framework. Basic aspects of this methodological framework, which will be outlined in more detail below, are in particular

- reusability at all intermediate stages of system development, in particular on the conceptual level.
- risk driven system development approach, which allows for an early identification of critical areas and thus results in a more reliable anticipative metrication of projects, i.e. in particular more reliable effort estimation. Moreover this is a basis for a participative approach already in early stages of system development.
- NTE's methodological framework is the basis for all its projects . Vice versa the experiences gained when applying the elements of this framework for projects influence as a feedback the ongoing improvement of the methodological framework.

The risk driven approach mentioned above is shown in fig. 9. The LCM is of particular importance for project managers whose decisions should be strongly supported.

"Risk management" is the appropriate term referring to the importance of focussing attention on those aspects of a project which most likely will cause trouble and to determine the appropriate sequencing of performing project activities. Risk management encloses [Boehm-88] assessment of risks (risk identification, analysis and priorization) and risk handling (risk management planning, execution, monitoring and control).

The software process is considered as a sequence of development cycles starting always with a - risk dependent - identification of objectives to be elaborated in one cycle followed by a step of analysing and evaluating the risks. Further steps within each cycle include the execution of the objective, the validation of the results and a plan for the next cycle. Finally there will be a new review taking place. The basic idea of this approach is that progression in developing portions of a software system at specific levels of elaboration always addresses a similar sequence of steps which is represented by the project model.

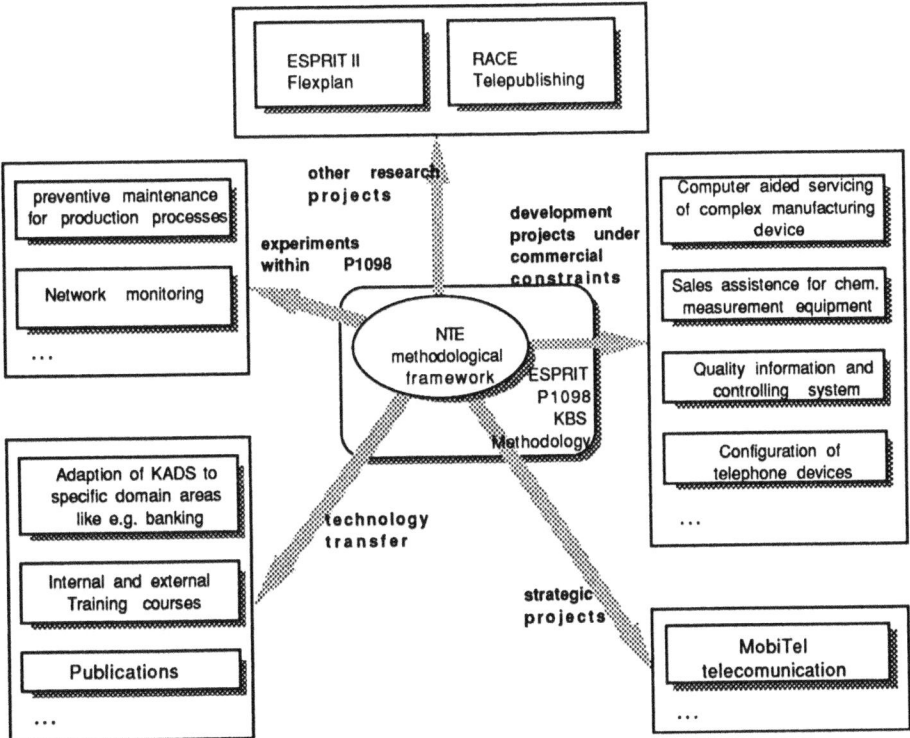

Fig. 8: Methodological Framework

4.1. The Project Model

The purpose of the spiral presented in Fig. 9 is to model important strategic issues driving KBS evolution [Dorbes 89]. The crucial feature of KBS development is that the strategies have to be planned quite frequently and at various points of the KBS life-cycle.

241

The four quadrants in Fig.9 are associated with different high level development activities. The radius indicates the cumulative cost of the project. The separation chosen in the figure is mainly guided by the intensity of involvement of the different groups. The left upper quadrant represents those activities carried out during the review. This is where normally all groups participate and also where the spiral is initiated. The second quadrant contains the crucial planning activities. A major part of planning is concerned with an analysis and evaluation of the uncertainties and risks associated with the project (a preliminary list of possible sources of risks in KBS development can be found in section 3.2). Quadrant 3 contains the development activities to derive required results specified by the objectives fixed in the planning stage and quadrant 4 activities (mainly by the development team) necessary to prepare the next review: an assessment of the previous results (validate, verify, test), an assessment of the remaining risks and a preliminary plan for the next cycle.

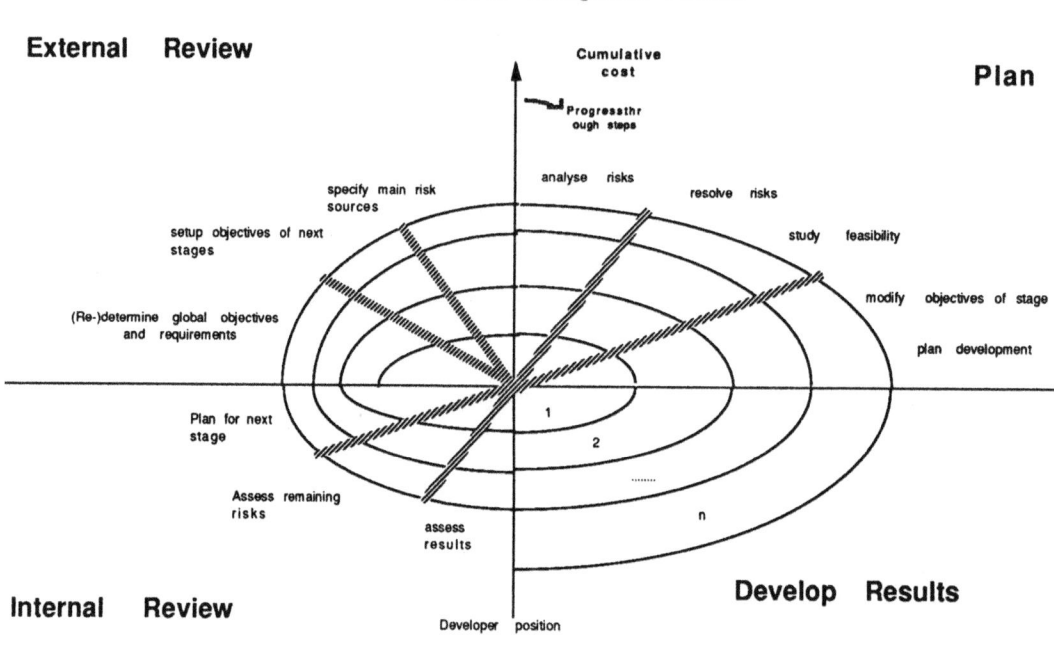

Fig. 9: Model of system development process (risk driven approach)

Note that the number of cycles (= number of reviews, etc) depends strongly on the complexity and size of the project and may be very large.

Each cycle of the spiral is engaged in establishing a well-defined and unambiguous result on some predefined aspects of the prospective system. It is unavoidable to focus on more or less accurately specified objects / models representing meaningful aspects of the prospective KBS in order to identify clearly results to be derived in a given cycle. Furthermore concentrating on results in terms of objects / models provides the essential conceptual simplicity and allows for instance to
- specify possible parallel development by different members of the development team
- decide which objects should be developed together in one cycle (perhaps after incomplete versions of these objects have been elaborated in earlier cycles) because of their dependencies.
- reuse of components from previous projects
- apply sound metrication methods [Readdie 89]

4.2. Object Model

Fig. 10 gives a view on KBS evolution from a point of view putting the intermediate objects during development of a KBS and their logical relations into focus. It is described in more detail in [Schachter 89]. For each of the intermediate results graphical representation languages are available in the methodological framework, which give a basis for sound documentation, allow for a more participative development and are of particular use for the metrication methods developed. The relations of the object model and the project model are indicated in fig. 11.

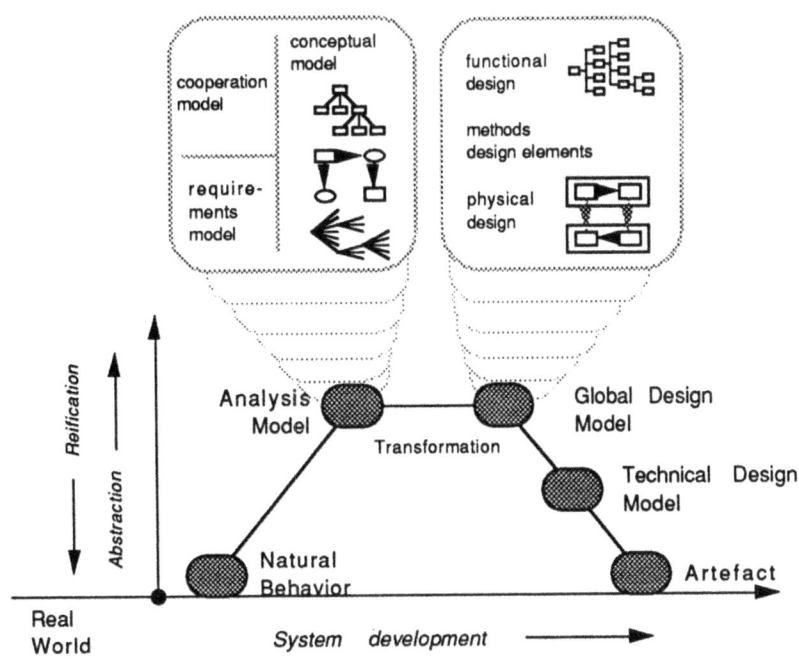

5. CONCLUSION

This paper describes a CAQ scenario and knowledge based systems implementing two components of this scenario. The development was influenced by the needs of different kinds of industries like manufacturing of medical equipment, printing industry, electronics industry and others. The methodological framework introduced is a basis guaranteeing reliable, extendable and reusable systems. Particular emphasis is put on reusability of results on a conceptual as well as an implementational level.

Next steps concerning the development of CAQ systems will in particular focus on an integration with other CIM components, like the knowledge based process plan generation which is developed in the context of the FLEXPLAN project (ESPRIT project 2457) [FLEXPLAN89].

The experiences made using the CAQ systems in the shop floor in the regular production have shown, that systems, which because of a lack of integration and sensors do need extensive inputs from humans, do not find sufficient acceptance. To increase the acceptance and thus the actual gain of introducing such systems in industrial usage, a more tight integration with production machinery, existing databases etc. as well as high level sensors like imaging devices will be a main focus of work in the next term.

Fig 11: Relationships of project model and object

243

6. ACKNOWLEDGEMENTS

The authors would like to thank Stephan Becker, Heinz Knödelseder and Andreas Pagel, who have developed the IPG system as well as Yoon-Chae Chon, who developes the MPM system. Karl-Heinz Streng has given support concerning the life cycle model. Jörg Winkler of Drägerwerk AG gave many useful contributions to the development of the CAQ scenario and supported the industrial evaluation especially of the IPG system.

7. REFERENCES

/Bläsing 87/
Bläsing, J.P.: Rechnerintegrierte Qualitätssicherung.
CIM-Handbuch - Wirtschaftlichkeit durch Integration, Hrsg. Uwe W. Genner, Vieweg&Sohn, Braunschweig (1987)

/Boehm 88/
Boehm, B.: A Spiral Model of Software Development and Enhancement. IEEE COMPUTER, Vol. 21, No 5, May 1988

/Breuker 87/
Breuker, J.A. (editor): Model Driven Knowledge Acquisition, Interpretation Models, Deliverable D1, Esprit Project P 1098, 1987.

/Busche 89/
Busche, R.; Krickhahn, R.: Modellgestützte Entwicklung eines Wissensbasierten Systems für die Fehlerdiagnose komplexer Industrieanlagen. KI 3/89, Oldenbourg Verlag, München, Sept. 1989

/Chon 90/
Chon, Y.-Ch.; Streng, K.-H.: Process Monitoring and Preventive Maintenance - Evaluation Report, Esprit Project 1098, Deliverable D10.2 of Task F10, January 1990

/DIN 55350/
DIN 55350 Part 11: Begriffe der Qualitätssicherung und Statistik, Hrsg. Deutsches Institut für Normung e.V., Ausg. 1987

/Dorbes 89/
Dorbes, G.; Hickman, F.; Porter, M.; Streng, K.-H.; Tansley, S., Taylor, R.: System Evolution - Principles and Methods (The Lifecycle Model). Draft Deliverable of Task G9, Esprit Project 1098, Nov89.

/Flexplan 89/
Deliverable 1 - Analysis Report, Esprit Project 2457 (Flexplan) June 1989

/Readdie 89/
Readdie, M.; Streng, K.-H.; Wermser, D.: KBS Project Development Metrication Methods. Draft Deliverable of Task G10, Esprit Project 1098, Nov89

/Schachter 87/
Schachter-Radig, M.-J.: Expertensysteme. In: CIM-Handbuch - Wirtschaftlichkeit durch Integration, Hrsg. Uwe W. Genner, Vieweg&Sohn, Braunschweig (1987)

/Schachter 89/
Schachter-Radig, M.-J. et al.:
KBSM - Structures and Models: Basis for the Reusable Development of Knowledge Based Systems.
Systems 89, Springer Verlag, Berlin 1989

The adress of the authors is:
NTE NeuTech Entwicklungsgesellschaft
Dachauerstr. 44
D - 8000 München 2
Germany

KNOWLEDGE-BASED SUPPORT OF QUALITY IN DESIGN

Inga With, Computer Resources International A/S,
Chris Irgens, Paisley College of Technology
Herman Goedman, HCS Industrial Automation BV

1. INTRODUCTION

The application of expert system technologies within the CIM area is
growing, just as the interest, not least for SME's, in Quality Assurance and the application of dedicated philosophies in that field.
The present paper describes how support of quality in design and
manufacturing can be approached by means of knowledge-based techniques. This approach can be seen as an exploitation of known software
technology in new ways, hereby widening the application area of knowledge-based systems. An expected technology development outcome of
the approach is the integration of knowledge-based techniques with
other computer applications. Also the focussing on quality support is
a new achievement in the field of knowledge-based systems in manufacturing.

2. THE APPLICATION AREA

The area of design and manufacturing as such is very broad. In the
present context the application area is highly automated production
facilities, focussing on mechanical mass-production, with the automotive industry as a typical example.

The design of such products is often based on a well established
product structure, defining the parts and subassemblies of the product
and their functionalities and geometry. In the design and development
of each new product this structure is adapted to meet the requirements
to the specific product. The design process could thus be said to
take place as a modification or restyling of existing products. However, the process of designing the product so that it fulfills the
customer's requirements often requires several iterations of design,
prototyping and testing, all in all consuming a considerable amount
of time.

Traditionally the quality activities in manufacturing companies
have been concentrating on control of processes, procedures and products. However, there is a growing awareness that on-line sampling
and quality inspection is not alone sufficient, and that there is a
need to focus also on off-line activities, such as product design
and development. This tendency is also seen in the change from quality

activities being handled by few, specific persons to a situation where all personnel is involved with quality, with "total quality management" as the goal.

Some important factors influencing the quality of products are accordingly related to and depending on the accumulated knowledge and experience in the design and production team. Often a designer specialises in developing or designing one specific part or subassembly of the product, thus maybe there is only one person who has got the expertise on that specific part.

2.1.   Problems Encountered in the Application Area

One reason why several iterations of design and prototyping is needed before a satisfactory quality is reached could be a difficulty in communication between designers and production engineers. It means that sometimes designers can design a product requiring specific manufacturing performance (tolerances of parts etc.) which cannot be met by the production environment, meaning that a perfect design ends up as a product with an unsatisfactory quality. The lack of designers' awareness of process capabilities can cause such problems.

Another reason is the functional complexity of the products, which often means that it is impossible, or at least practically out of reach, to calculate and predict the functionality of the product before it has been constructed. It is necessary to build and test a prototype in order to verify whether a given design will meet the requirements from the user.

A general fact which can cause problems for an organisation is that experience is built up in people rather than in organisations. This means a loss of experience if people leave the organisation. In the present situation it means that the expertise, maybe built up over a long period of time, regarding development of specific products and parts and the factors influencing the quality of these, is present only in the minds of the developers or other people in the production staff.

In order not to be too sensitive to the presence of certain persons it is an advantage for the companies to have their specific experience documented in the company, so that it is usable also by other members of the staff.

3.   PRESENTATION OF PROTOTYPE

The knowledge-based support of quality in design is established by means of a prototype being developed in Esprit II project no 2178, "Revision Advisor - an Integrated Quality Support Environment" (RA-IQSE).

The prototype is meant to support the designer, hereby intending to cover the first phase in the product life cycle. A possible extension of the project would focus on supporting the production planning and machining, hereby expanding the focus point from CAD applications to covering also CAPP and CAM.

The support provided by the prototype is limited to the above mentioned application area of mechanical mass-production with structurally well-defined products, meaning that invention of quite new products or product technology is not supported.

3.1. Appearance to the User

The prototype is developed and intended to run on SUN 3/60 work-stations, and it works together with the Medusa CAD system.

The system provides a user interface, which presents the various functionalities provided by the system as mentioned in section 3.2. The CAD system appears to the user in a specific window in the RA-IQSE user interface. A screen example is shown in fig. 1.

The design of a product is conceptually done by the designer as a process of "restyling", inasmuch as the product structure, which is known to the system, forms the basis for the design. The system shows the "building blocks" which must or can be used for a specific product, and it prompts for the features and attributes which have to be instantiated, such as geometric values, material etc. The designer builds the product from the parts defined in the structure, hereby "instantiating" the generic product structure. In the system the product is represented as a tree-structure, with parts, subparts, and their related features. The visualisation of the product structure in the user interface reflects this tree structure, so that the user sees a conceptual picture of the product, rather than a traditional drawing.

Anyhow, there is also a need for viewing the product as it is going to appear. This is required because the designer needs to see what the product looks like, to avoid the feeling of working in the dark. Also the CAD drawing may show some obvious errors. Another reason for integrating the CAD system with the RA-IQSE is to make the distance to the traditional design process smaller by using a well-known design tool such as CAD, whereby the RA-IQSE becomes more familiar to the user.

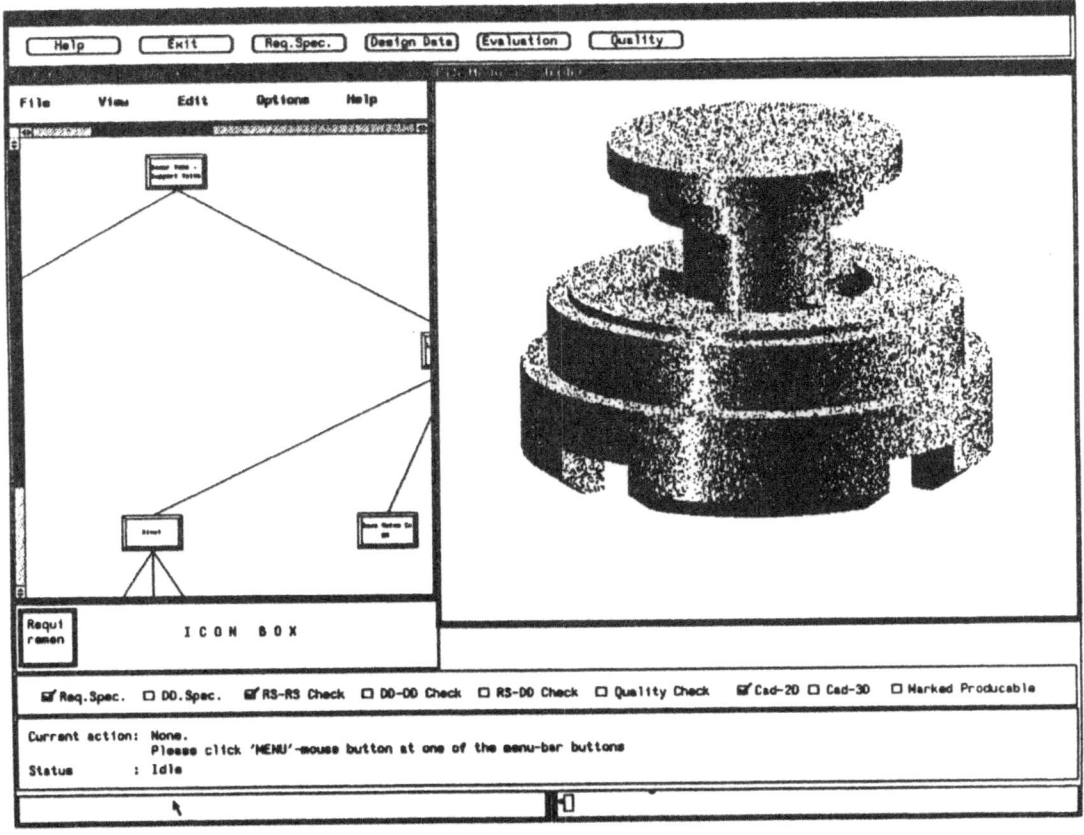

Figure 1. Screen example from the RA-IQSE prototype.

Whenever the user wants a visualisation of the product as a CAD drawing, this is made in the CAD system window of the prototype. This is possible, because the system contains a CAD driver which translates the RA-IQSE internal representation of parts and features to the representation adequate for the CAD system. I addition to display upon request, it is foreseen that automatic display after performing certain kinds of design actions will be possible too. The current layout of the user interface is presented in figure 1. This user interface stems from a mock-up prototype of the system, which has been made for demonstrations, so as to illustrate the foreseen functionalities. The actual looks of the prototype user interface may change when the real prototype has been developed.

The users of the system would be designers and design engineers, each on their level of responsability. The designer is the lower level, having access to a specific set of parts and features for the products he is working on. He would use the system for filling in requirements specifications, for retrieval of matching designs, for designing the product, and for evaluating requirements specifications and designs, in order to predict quality and get support for experimentation.

The design engineer would be responsible for the work of a number of designers and would typically be in charge of maintaining and updating the system: the database with available parts and features; the various designers' access to different data; CAD macro library (CAD driver); and the knowledge in the knowledge bases.

It should be noted that the system is not a tool for automatic design. It provides decision support for the designer in order to facilitate and speed up his work. It is at any time the user who actually makes the decision.

3.2. Functionalities Provided

The prototype provides the following functionalities: search for successful designs of the past with matching requirements specifications; consistency checking of more or less complete designs; prediction of product quality based on knowledge of the capabilities of the production environment; support of dedicated experimentation based on the Taguchi quality philosophy; general access to the knowledge present in the system; and addition and updating of knowledge in the system. In the following some of these are elaborated.

Search for Matching Requirements Specification. The requirements specification from the customer is the starting point for a new product. The customer requirements specification goes through a first interpretation and is entered to the system. Based on this, and on the previous designs which have been made, the system can find existing successful designs with more or less the same requirements as the new one, and which could be candidates for reuse of the design or parts of the design.

This facility means that if parts of an existing design can be reused, only the other parts will have to be re-designed by the designer.

Consistency Checking. This kind of control can be performed on the requirements specification, telling if contradictory requirements have been made. Partly or fully instantiated designs can be checked, so that for instance incompatible geometric definitions can be exposed. The connection between the requirements specification and the design of a product can be controlled, covering purely geometric aspects and, as far as possible, the mapping from geometry to functionality.

If inconsistencies are found, the system will give advice on suitable actions to be taken in order to revise the design or the requirements specification so that it will be consistent with the knowledge in the system.

Prediction of Product Quality. Each product is manufactured through a set of manufacturing processes, which influence the quality of the final product in various ways. The product quality can be predicted, based on a model of the product, identifying the quality characteristics which are connected to the different manufacturing processes, and on knowledge about the capabilities of the various processes.

The process capabilities are achieved by production measurements. If there is sufficient information available concerning the performance of the manufacturing processes, the design may be simulated, giving as result a prediction of the product quality, comparable to the interpreted outcome of a test of the real product. The quality prediction could be denoted a stochastic simulation of the product quality, seeing the product as a static entity. The functionality of the product is not simulated.

It is obvious that the mentioned prediction of product quality can only be of interest if a relevant quality metric can be established for the product, and if the model of how the various processes influence the quality can be established.

Support of Taguchi Quality Philosophy. A design problem may be encountered by the designer. In such cases the prototype can give advice on how to perform experimentation dedicated to solving the problem, and on how to analyse and utilise the experimentation results. Experiments can be performed as part of product prototyping or they can be simulated.

This facility builds on and supports the use of the Taguchi quality philosophy, which is gaining more and more interest in European industry these years. The main idea in the philosophy is to make design robust to noise, ie making the product less sensitive to those factors which can cause bad quality and which are difficult, expensive, or impossible to control. Such factors can be in the product, as eg material, or they can be found in the manufacturing processes, as eg available machines and tools.

The Taguchi quality philosophy indicates how statistical techniques can be used in order to focus the experimentation. It shows a way to select experiments, so that only a limited subset of the possible combinations of factor values need to be tested, and it provides techniques to analyse and interpret the results of the performed experiments, so that important factors which are robust to noise can be identified.

By using knowledge of the various factors influencing the product quality, the described prototype gives advice on which experiments to perform in order to get the best understanding of the specific problem. Furthermore the prototype provides means for analysing the experiment results in order to be able to give advice on which changes to make in the design in order to solve the problem.

If experiments have been performed in the past for a similar problem, the results of such experiment may be reused, whereby time and money can be saved by reducing the number of necessary experiments.

If the experiments can be related to specific processes and the capabilities of these processes are known, then the experiments may even be simulated.

3.3. Test Environment for the Prototype

The described approach will be tested by installing the prototype in a live manufacturing environment. This is a shock absorber factory, which has already in the current routines well described quality procedures. It is a factory with high quality products and running` massproduction, working constantly round the clock and through the week. This factory provides a test environment where it can be verified how and to which extent the prototype adds functionality to and supports a manufacturing environment characterised by an up-to-date quality policy. The factory at present accomplishes its high product quality more by expertise in manufacturing than by specific design specialisation: It could be said to be a manufacturing rather than a designing company. This means that the design support provided by the RA-IQSE prototype is expected to strengthen and improve the design process in the company.

The notion of testing the prototype in a real life environment has several advantages to the project. It gives a chance to verify and validate the applicability of the approach and it gives experience with the tasks of collection and analysis of knowledge (knowledge acquisition) which have to be encountered when the system should be installed in another environment, and of organisational problems which must be taken into account. This altogether gives a better platform for an industrial exploitation of the work for the partners, in the sense of developing the prototype into a product in a later stage.

3.4. Genericity in the Prototype Work

However, the usefulness of the prototype work for further exploitation is heavily dependent on genericity in the prototype design. This characteristic is achieved by making loosely coupled interfaces to the different other applications which are involved. This is in the first place the CAD system, where the interfacing is made, as mentioned, by means of a CAD driver which transforms the internal representation of product data into a CAD-relevant representation. When using a different CAD system, only a new CAD driver would have to be made.

Also the use of support tools such as statistical packages, simulation packages etc. will need a "clear-cut" interface, so that the prototype will not forever depend on a specific package.

Special attention has been paid to the development of general data models with which it is possible to describe products with parts and features, product functionality, quality aspects, processes etc., for in principle any manufacturing environment and product type.

The genericity of the knowledge bases is a special problem. Some of the knowledge in the system indeed has to be specific to the manufacturing environment and the product, in order for the system to be able to provide any useful support. However, parts of the knowledge needed will be relevant for a number of different manufacturing environments, if they belong to the same field of production, such as eg car manufacturing.

This problem cannot be eliminated, but it can be restricted by separating the knowledge into a number of knowledge bases, depending on the subject the knowledge concerns, and on the genericity of the knowledge. This would mean that only the more specific knowledge bases would have to be exchanged, if the system should be installed in another environment. The knowledge acquisition for these knowledge bases is, anyhow, not a trivial task, and an extensive load of work would have to be done in each case.

4. ACHIEVING QUALITY SUPPORT

4.1. Quality Terminology

The meaning of the term quality is not easily defined. The quality of a certain product cannot be judged without taking into account the requirements to the product, including requirements to price. The ball-pen example can be mentioned here: a cheap "BIC" ball-pen may reach requirements just as well as the expensive "Parker", depending on the requirements; if you need a ball-pen which you can loose any time without being sorry, the BIC is the better, while if you want to show style and class the Parker might be more adequate.

For producers the quality of their production is related to the profit it provides, which means that costs related to production is expected to be minimised. In this respect the direct cost is money, but also indirect costs in terms of bad working environment, pollution etc. should be considered.

The definition of quality, used in the RA-IQSE project is a broad and simple definition covering the above, and it describes high quality as "meeting the requirements of the customer at the lowest possible cost".

This definition is not far from the viewpoint of Taguchi. His definition of quality is the following: "The quality of a product is the (minimum) loss imparted by the product to the society from the time the product is shipped". Taguchi sees loss in a very broad sense, ie loss to society, related to every product that reaches the consumer's hand. This could be such as consumer's dissatisfaction, added warranty cost to the producer, or loss due to a company having a bad reputation and losing market share in the long run. [Byrne & Taguchi 1986]. However, these examples are so broad that they are hardly operational. When concentrating on losses caused by a product's functional characteristic deviating from its desired target value, the concept becomes more applicable. According to Taguchi a loss function, defining the loss as proportional to the square of the deviation from target, is valid for many instances.

Different products have different quality aspects, specific to the product. However, a number of general quality aspects can be described, such as durability, safety and reliability. These aspects indicate that quality cannot be seen as a static feature, but must be viewed over the lifetime of the product. Reliability is the probability for the product to maintain the quality throughtout the product lifetime. It is foreseen that the quality of a product at a fixed point in time (eg shipping time) can be predicted more precisely than it is possible to forecast reliability.

4.2. RA-IQSE Quality Model

The RA-IQSE prototype relates to a quality model, which is used to describe the quality procedures and policy of the organisation, where the prototype is installed. This quality model covers aspects such as design and production environment, personnel and product, with

251

the heavy emphasis on the product. The model is established in order
to provide a structure for the quality information pertaining to
design and manufacturing processes, product design, manufacturing
operation, and customers. As such it can also be used to support the
process of knowledge acquisition for future installations of the RA-
IQSE system. The quality model of the product incorporates the various
quality characteristics which can be identified for the product, and
indicates which factors influence these quality characteristics. Such
factors could for instance be related to the manufacturing processes,
or to design decisions such as selection among possible varieties of
a part.

Another purpose of the quality model is to use it as a means for
mapping the local quality procedures of the company to various quality
standards, such as ISO 9000. This area becomes more and more important
for the manufacturing companies, as more and more customers require
certification and as the open market approaches. The intention is
that the quality model will facilitate the process of certifying, by
providing to the company a means for mapping their own procedures to
the standards, helping them a part of the way to certification.

4.3. The RA-IQSE Contributions to Higher Quality

In the following is outlined some of the ways in which the RA-IQSE
prototype is expected to provide quality improvements to the manufacturing environment, ie how quality problems are solved and how quality
in general supported. The results mentioned could also be seen as a
list of results generally expected from the project.

The main advantage is that errors in design will be found before
expensive product prototypes are manufactured, meaning that product
development costs are lowered. Also the possibility of simulating
experiments will decrease cost. This conforms to the definition of
quality as equal to meeting customer's requirements at lowest cost,
inasmuch as a satisfactory quality will be reached sooner, and therefore cheaper, with the aid of the prototype.

Concentrating the quality assurance effort on the design phase
rather than on quality control of products means that scrap may be
decreased.

Implicit knowledge, which can be personnel's experience or such
as knowledge which is potentially available, but at present invisible,
hidden in existing data from production and product test, becomes
explicit and easier to use by more people. This happens by the accumu-
lation of staff experience in the knowledge-base and by the system's
and the users' interpretation of production and test data. Also the
reuse of specific experiments is related to this, as this now becomes
explicitly available and not only present as staff remembrance and
experience.

The Taguchi quality philosophy is supported, meaning that the use
of this approach, which is indeed not trivial, is facilitated. The
staff is assisted in the process of learning to use the methodology.

The reuse of designs or part of designs, which have proven their
success and high quality, has two positive effects. The new design
is made faster, as some of it is already at hand, and parts of the
final design has already proven its quality.

The designers' working situation is improved, as they can advance
more quickly to the interesting and technically challenging parts of
the design process and concentrate on them, instead of perhaps spending time on large calculations, or relying on "best guess" which is
not always satisfying. Also the reduction of the "response time" by

catching errors and simulating experiments instead of building prototypes is attractive to the designer.

It has been discussed whether the system would take the creative
part of the work away from the designer. As the concept of "automatic
design" is not the issue, and the user, the human being, will be in
full control of the system, it is not considered to be a problem.
The prototype aims at taking away the trivial part of the designers
work, making him work conceptually instead of just using sophisticated
tools like CAD for traditional design work. It could be compared to
using a text processor integrated with database- and spreadsheet-
programs for production of documents, merging of letters, etc., in
stead of just for writing and editing text.

The needs for practical support have been expressed by the test-
site where the prototype will be installed. Two examples of this shall
be referred to: There are cases where there is a number of options
available when selecting a specific part, ie where different types
of the same structural part can be used. In such cases it would be
helpful if the system supports the choice by indicating the suitability, and quality implications, of each of the possible selections.
This could be expressed as a validation of design decisions with
respect to product quality.

The mapping from design to functionality of the product is another
issue, which would be very useful for the designers. This mapping is
at present not practically available, or only to a certain extent.
It is needed because the complexity of the relationship between design
(geometry) and product functionality is one, if not the main, reason
for the long process of prototyping which in the current situation
is necessary before the customers' functional requirements can be
been achieved.

5.   KNOWLEDGE-BASED APPROACH

5.1.   Different Kinds of Knowledge

The support functions provided by the prototype are based on a variety
of knowledge, of which the most important is related to the product,
the manufacturing environment, and to the quality philosophy applied.
The following description of the main categories of knowledge is of
course closely related to the description of the functionalities of
the prototype, and some repetitions cannot be avoided.

Product Structure. The general structure of the products is known to
the system in terms of geometry and functionality. The product is
described and represented as a tree structure, containing the parts
and features constituting the product. Features are geometric, such
as holes, chamfers etc. Parts as well as features have attributes,
such as size, angle, diameter, material etc. Each design is an instantiation of the general product structure. The system has access to
these existing designs, and parts of them may be used in future
designs.

Besides the conceptual representation of the product, the system
contains knowlegde about the internal relations and constraints between the various parts and features of the general product structure,
so that possible conflicts in a design, which could cause quality
problems, can be identified. Connected to the knowledge of parts'
and features' relations is knowledge about how possible conflicts
can be resolved. This knowledge is the background for revision advice,

ie feedback to the user on how the conflicts can be avoided or errors removed.

Taguchi Quality Support. The prototype contains the necessary knowledge and algorithms to provide the experiment support and analysis according to the Taguchi quality philosophy. This knowledge is partly related to the method in general, and partly to the application of Taguchi's philosophy on the specific product and manufacturing environment. Concerning the latter, this knowledge comprises such as the following relationships:

1. The relationship between a product's parts and features and the known or accepted design procedures used to design the product. This relationship can be viewed as a set of "available" design routes and as such presents the "variety" in the design procedure.
2. The relationship between a product's component parts and features and the product's functions. This kind of relationship provides the designer and the RA-IQSE with the "variety" in the product structure that may provide a suitable solution to the stated requirements specification.
3. Quality is dependent on the manufacturing and assembly processes employed. The RA-IQSE must therefore couple the product model with a model of manufacturing resources and the historic performance. It is envisaged that each part or form feature is related to its manufacturing process sequences.

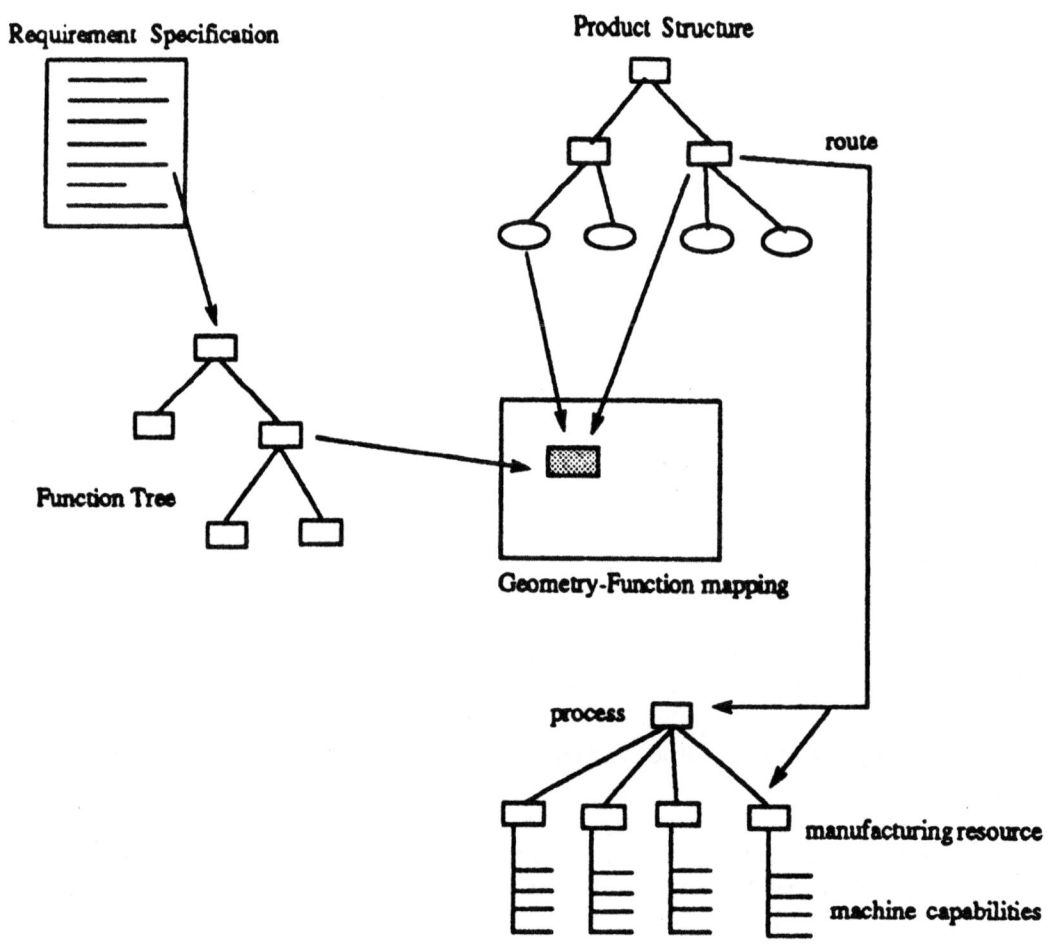

Figure 2. RA-IQSE Data Models

254

The relationships between parts/features and function gives rise to the need for the decomposition of the product's functionality. In the requirements specification, functionality is normally expressed as "composite" requirements such as rigidity, physical size, damping factor, operating temperature ranges etc. It is considered desirable to decompose such requirements into a requirement hierarchy, so that a mapping between simple requirements and parts/feature types can be effected. This hierarchy of functional properties can be viewed as fairly static product knowledge and may be updated as required with knowledge derived from prototype and field testing and customer feedback.

Manufacturing Environment. The manufacturing environment is known to the system in terms of knowledge of the manufacturing processes' capabilities and a model of how the various processes influence the quality of the product. This is related to the above mentioned mapping from geometry to functionality, in the sense that tolerances etc. of machined parts which are provided by the manufacturing processes are related to the geometry, and through the geometry-functionality mapping also to the functional quality of the product. The basis for a stochastic simulation is interpretations of measurements made on production and product test. The measured data are continually captured and entered to the system, and if possible interpreted. This means that the quality prediction is based on actual feedback from production and product test. Also rules may be derived from the analysed data.

5.2. Meta-knowledge Aspects

Meta-knowledge Overview. The term meta-knowledge is often used in a limited sense, covering control structures within the knowledge base, or knowledge embedded for control purposes in the inference process. With a broader definition of the term it can be used for expressing knowledge about the quality characteristics of a knowledge-based system. The quality of a knowledge-based system is influenced by three factors: the quality of the input data and information; of the knowledge embedded in the knowledge base; and of the reasoning mechanism applied, ie how the system uses the knowledge.

The structure as well as the contents of a knowledge-base can be validated in various ways. A number of relevant qualitative issues could be considered:

Content plausibility, which is concerned with the qualities and characteristics of the knowledge in the knowledge base; whether it can be considered true and genuine.

Correctness, which is concerned with the general validity of the systems output when applied to cases of interest.

Managing of uncertainty requires knowledge of how to deduce the consequences of variation and other types of uncertainty during the inference process; knowledge of to which degree the knowledge-based system conforms to the assumptions made in the methods applied in the system; and knowledge of the bias of consequence when applying the knowledge on borderline cases.

Robustness of the knowledge base is concerned with "the extent to which the system reacts to non-standard input and stochastic variation". Investigations on robustness can clarify weak spots of the knowledge model.

5.3. Meta-knowledge Concepts in the RA-IQSE

One conception of meta-knowledge in the RA-IQSE prototype is the separation of control structures in the knowledge-bases from the knowledge parts. Also the structuring of the knowledge according to subject and level of genericity is a conception of meta-knowledge.

The application of a meta-knowledge strategy throughout the development of the RA-IQSE knowledge-based system is considered a useful vehicle for ensuring and documenting the quality of the system in the end. Furthermore, this gives better possibilities for the project to provide guidelines for knowledge acquisition and validation when adapting the system to other manufacturing environments in the future, an aspect which is of importance for the genericity of the project results.

Knowledge Acquisition. The knowledge acquisition activities in the project work have been concerned with analysis of the product which is manufactured in the selected test environment, and selection of a subassembly within that product. Also the design and product development processes in the manufacturing environment concerned have been analysed, through interviews with people working with the processes, and by studying the documentation used during these processes. General technical literature related to the product technology has also been a subject for study.

The work with extracting information from the manufacturing company has primarily been taken care of by a partner in the project, which in several ways is close to the manufacturing company: situated geographically very close, speaking the same tongue, with a mutual confidence between the two companies, knowing each other and having worked together in the past. This combination, with a partner, fully integrated in the project, of course knowing and understanding the needs of the project, and at the same time having a good and close contact to the test-environment has been very fruitful for the project.

Knowledge Validation. The knowledge validation of the RA-IQSE will be approached from two sides. The prototype will be tested in the manufacturing environment, and its capabilities compared to the solutions which would have been found using the traditional ways of problem solving. For this purpose a bench-mark is established by collecting test-data from the design and manufacturing processes which will be involved by the prototype. This test relates to the the correctness of the system. Also the knowledge comprised in the system should be validated more specifically. If for instance the knowledge is represented as rules, people from the manufacturing environment should look into these rules and see if they are conforming to their understanding of the manufacturing environment. The purpose would then be to see if the persons doing the knowledge acquisition have understood the information correctly. This aspect of validation is related to the content plausibility of the knowledge-based system.

A special problem occurs as new knowledge will have to be added during the use of the system. This is necessary, as the prototype partly builds on feedback which comes while the system is in use. The system will have to be able to validate such new knowledge. This functionality is expected to be incorporated in the process which allows the user to update the system.

6. CONCLUDING REMARKS

The prototype presented here is different from a usual, stand-alone expert system, developed in a certain expert system shell, and not interfacing other computer applications. The knowledge-based aspects of it appears as a part of a system, which also builds on developments in other computer application areas. The most important of these are the CAD system and Man-Machine Interface (MMI). The representation of CAD related information and the connection to the CAD system, as well as the MMI, are programmed with "traditional" programming, using eg procedural languages and window systems. The "kernel" of the system, which provides the knowledge-based support, ie gives access to the knowledge, is programmed in a language, or by means of a tool which is adequate for this purpose. The main points are that the various modules have to work together and that they form one single system towards the user with one single man-machine interface.

The evaluation of the RA-IQSE prototype at the end of the project will indicate whether the approach of using simulation in order to predict quality in the way it has been described here is possible and relevant.

It would be interesting to predict different design desicions' influence on the quality in order to give advice to the designer. It is as yet not clear whether this will be possible within the scope of the prototype.

At present the project, which started in January 1989, has come through the phase of problem definition, prototype delimitation and selection of tools to use for the implementation. The products and processes in question have been analysed and described for providing the knowledge to be contained in the system. The architecture for the prototype and the included data models have been agreed upon and described, and the functionality determined. The decision on functionality included a process of limiting the prototype to a set of functionalities which could be implemented in the limited time-scale of the project. Techniques for off-line quality control have been investigated, especially the Taguchi methodology. In addition, work has been done on the quality model, including its use in connection with the target environment. Concerning the Man-Machine Interface and the knowledge-based part, applicable techniques have been studied. In relation with the work on the prototype architecture and the MMI a mock-up prototype has been built with the SunView toolkit and integrated with the Medusa CAD system. A part of the work with the knowledge-based support has been an evaluation of tools for implementation of this part of the system. At the moment it is foreseen that Lisp-Works will be selected. In addition a strategy for developing the knowledge-based part has been determined.

The current main activities are the specification of quaity metrics, the design of the MMI and the knowledge-based part, the implementation of the prototype, and the establishment of a bench-mark for testing the prototype. In spring 1991 the running prototype should be installed in the manufacturing environment, leaving until January 1992 to test and evaluate the system.

Acknowledgements. The research described in this paper is being carried out under ESPRIT Project no. 2178, RA-IQSE, Revision Advisor - An Integrated Quality Support Environment, and is funded by the Commission of the European Communities. The project consortium comprises: Paisley College of Technology, UK, HCS Industrial Automation BV,

Holland, Computer Technologies Company Ltd, Greece, Asociacion de la Industria Navarra, Spain, and Computer Resources International A/S, Denmark. The authors gratefully acknowledge the contribution of all members of the consortium.

REFERENCES

BYRNE, D.M., TAGUCHI, S. "The Taguchi Approach to Parameter Design". 40th Annual Quality Congress Transactions. American Society for Quality Control 1987

ROSS, P.J. "Taguchi Techniques for Quality Engineering". McGraw-Hill, New York 1988

BENDELL, A., DISNEY, J., PRIDMORE, W.A. (ed.) "Taguchi Methods: Applications in World Industry". IFS Publications, UK 1989.

MACARTHUR, E.W., LEES, B. "Quality Support through Knowledge Engineering". Proceedings of the 5th International Conference on Computer Aided Production Engineering, Edinburgh 1989.

"Description of the manufacturing location to be used as target environment for the RA-IQSE". Deliverable 1.1 from Esprit project no 2178, RA-IQSE, AIN Pamplona 1989.

"Description of the Integrated Architecture of the RA-IQSE", Vol. I-II. Deliverable 6.1 from Esprit project no 2178, RA-IQSE, CRI Copenhagen 1989.

BILALIS, N., LEES, B., POTZE, R., STAUSGAARD, J., WITH, I. "Survey of methods for modelling and simulation". Report 5.A from Esprit project no 2178, RA-IQSE, CRI Copenhagen 1989.

METHODS AND TOOLS FOR INTEGRATING CIM ELEMENTS – PART I

MAGIC: A NEW APPROACH TO GROUP TECHNOLOGY

PETER SCHOONJANS

WTCM-CRIF MACHINEBOUW

Group Technology (GT) is the concept of identifying and bringing together similar parts in order to obtain advantages of their similarity for design and manufacturing purposes. To put Group Technology into practice a coding and classification system is used. Although Group Technology as a concept is considered to be most valuable, a lot of companies have encountered serious problems trying to implement it. Efforts to lower the barriers to Group Technology have resulted in the ESPRIT project 2623 MAGIC. This paper will outline the main items of the new Group Technology approach adopted in MAGIC. It will also be pointed out how this approach provides some key elements for the integration of computerised islands in the manufacturing environment: CAD, CAPP and CAM.

1. THE MAGIC PROJECT.

1.1 Objectives.

The MAGIC project aims to improve and facilitate the use of Group Technology methods and tools in small and medium sized factories, by automating these methods and by integrating the tools in existing CAD/CAM systems.

1.2 Partners.

Industrial partners as well as research centres are involved in the project MAGIC.

WTCM is the prime contractor of the project. WTCM is a research centre of the Belgian metalworking industry. The department Mechanical Engineering has a research team of 28 engineers, working in the fields of CAD/CAM and FMS.

CETIM is a Technical Centre of the Mechanical Industry in France. The Production Engineering department (37 people) directs the research in company management, process planning, machining techniques and Group Technology.

N.V. MICHEL VAN DE WIELE is a medium sized Belgian company (700 employees) manufacturing weaving machines. It has a manufacturing experience of more than 100 years in textile machinery engineering for carpet and velvet weaving machines.

LVD, a Belgium machine tool manufacturer has been involved in mechanical engineering for more than 30 years. The extensive production program covers not only standard machine tools but also special manufacturing systems.

C.M. MARES is an injection mouldmaker for plastics materials since 1952. The 90% of their turnover is directed at the automotive industry. They are specialized in big tools until 40 tons, with complex surfaces.

Eigner & Partner are German system consultants for the solution of problems in the technical field.

CAP SESA Industry is a French software house that works on a very large range of different fields of Computer Integrated Manufacturing, covering CAD/CAM, Flexible Manufacturing Systems, networks, data management systems,...

1.3. Project Status.

ESPRIT-2623-MAGIC is a three year project, presently in its second year. In the first year of the project, two prototypes have been made. These prototypes were intended primarily for orientation and evaluation purposes. After the evaluation of the prototypes, a functional specification of the MAGIC system has been established. During the second year, the structure of the relational database used for the representation will be elaborated.

2. THE MAGIC PHILOSOPHY.

2.1 The Classical Group Technology Systems.

Group Technology systems are tools that make it possible to identify and group similar parts. This implies some sort of classification. In the classical GT systems like Opitz, Miclass and Cetim-pmg the classification is realized by giving each part a codenumber. The digits of this codenumber depend on the properties of the part. Although these systems can be used successfully they have certain disadvantages:

1. Manual coding. The manual coding of the parts is rather timeconsuming and can lead to ambiguities when performed by different persons.

2. Information Degeneration. Due to the use of a codenumber a lot of information about the part is lost during coding.

3. Rigid classification. The use of a codenumber leads to a rigid classification. This classification allows only one point of view. Therefore most classic GT systems are oriented towards one application: design or manufacturing.

MAGIC wants to solve these problems and to provide some extra advantages.

2.2 Comparing MAGIC with a Classical GT system.

A GT system can be divided into three main functional blocks: generation, representation and retrieval (Fig. 1).

	GENERATION $\Rightarrow$	REPRESENTATION $\Rightarrow$	RETRIEVAL
CLASSIC GT	MANUAL COCIFICATION	THE DIGITS OF A CODENUMBER	SELECT RANGES FOR THE DIGITS
MAGIC	AUTOMATED GENERATION OF THE REPRESENTATION	STRUCTURED REPRESENTATION IN A RELATIONAL DATABASE	MISCELLANEOUS COMPLEX SPECIFICATIONS

Fig. 1 MAGIC compared with a classic GT.

1. Generation. The representation, to be used in the GT system has to be generated for all new parts. With the classical GT, the codification of the part is done manually or interactively with questions and answers. In MAGIC we want tot automate the generation of the representation of the part as much as possible.

2. Representation. A good representation is very important since it is the limiting factor for the application possibilities of the GT system. Instead of a multi-digit code, MAGIC uses a structured representation in a relational database. This representation is based on the "feature" concept and far more complete than the classical codenumber.

3. Retrieval With the classical systems retrieval is performed by setting ranges for the different digits of the code. MAGIC allows complex specifications to be formulated. One of the goals is to retrieve drawings based on a rough sketch on the CAD system. .

How are the classical GT problems solved in the MAGIC system?

1. Manual Coding. By automating the generation, codification work is reduced and ambiguities are excluded.

2. Information Degeneration. The information on the part that is saved is much more complete than a classical codenumber. Moreover instead of dividing variables into classes using the digit of a codenumber, the real values are used.

3. Rigid Classification. The relational database structure, which is used for the MAGIC representation, allows to make the representation more like the real part. The representation is therefore a better model of the part than a codenumber, and

is independent of the application. This independence between representation and application allows a flexible classification and a wider application field.

2.3 Extra Features and Advantages.

<u>Simple parts and assemblies</u>. The classical GT systems are only suited for simple parts. MAGIC also deals with assemblies.

<u>Functional Properties</u>. In the design department, the function of a part can be an important retrieval criterion. MAGIC deals with these functional properties.

<u>Integrated with CAD Design</u>. MAGIC is integrated with CAD design. This integration is double. First, the information that is introduced in the CAD system is used to generate the MAGIC representation. This means that coding is no longer a separate process. It is completely integrated with the design task on the CAD system. Secondly, the CAD-user interface can be used to generate retrieval specifications. For example, the rough sketch of a part on a CAD system serves as the retrieval specification for a wanted part.

<u>Feature Based Design</u>. Instead of using lines, arcs and text to represent parts in CAD drawings one can use features like cylinders, holes, engineering information etc... MAGIC makes use of feature based design.

<u>Efficient Retrieval Strategy</u>. To retrieve a drawing with a GT system one has to formulate a specification for retrieval. MAGIC will loosen or tighten the specification to make sure that a predetermined number of drawings is found. From these drawings, the user can then select the appropriate one.

3. THE MAGIC SYSTEM.

MAGIC is a classification and retrieval tool. The items to be classified are parts and drawings. A part can be a simple part or an assembly. Like all classification systems MAGIC can be divided into three main functional blocks: generation, representation and retrieval (Fig. 2).

The MAGIC representation is the heart of the system. Instead of the classical GT code, a relational database is used to store all the data necessary for the representation of parts and their drawings. Several information sources are used to generate the MAGIC representation. The degree of automation of the generation depends on the information sources used. The retrieval combines the data from the MAGIC representation and a user specification to retrieve similar parts and their drawings, or to find the appropriate family for a given part.

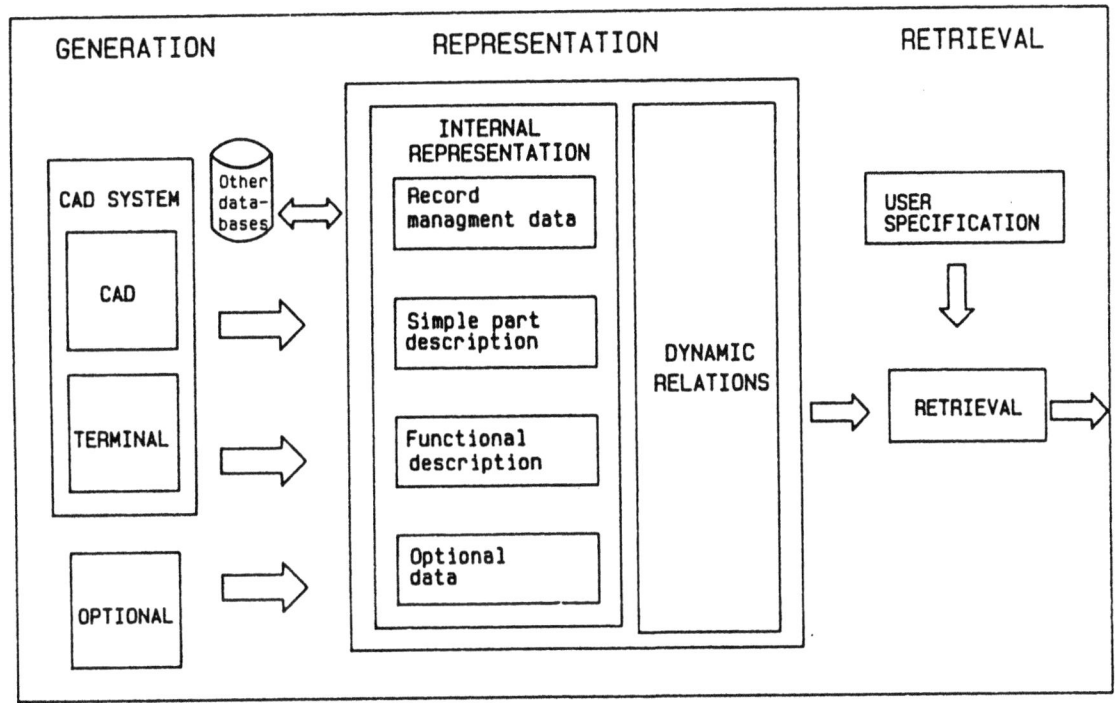

Fig. 2 General diagram.

3.1 Generation.

It is the aim of MAGIC to generate a part representation during the design on the CAD system, so that there is no need for separate coding. The information sources for this generation during the CAD design are CAD-Models and terminal input (Fig. 2). Other information sources are the other company databases and the optional non-design department information sources.

1. <u>CAD-Models</u>. CAD-Models are internal representations of parts in the CAD system. They contain a lot of useful information for the MAGIC representation. Using this information makes separate manual coding superfluous. There is however a problem of information level. The representation of a part in a CAD system depends on the model used: 3D constructive solid geometry, 3D boundary representation, 3D wireframe or 2D. At present most mechanical CAD design is 2D. The low level information used in the CAD model (especially the lines, arcs, and text in 2D and 3D wireframe models) is not suited for GT-applications. Therefore the CAD-Model is transformed into the MAGIC representation, by the MAGIC CAD-Model interpretation system (Fig. 3). To make this transformation possible the CAD-Model interpretation system makes use of a combination of different techniques: Design with features, Intelligent drawing interpretation and Automatic-interactive input.

Design with features. A set of macros on the CAD system enables the designer to draw higher level entities like cylinders, holes and slots, instead of lower level entities (lines, arcs, and text), used by the CAD system. To these low level entities in the CAD-Model, attribute information is added. These attributes facilitate the interpretation of the low level entities.

Intelligent drawing interpretation. The interpretation of low level information on drawings is a work that requires human interaction. Automatic interpretation however is possible for limited applications.

Automatic-interactive. For those entities in the CAD-Model that can not be interpreted by the intelligent drawing interpretation the system will ask the designer for more information. The information here comes from both the CAD drawing and terminal input (see below).

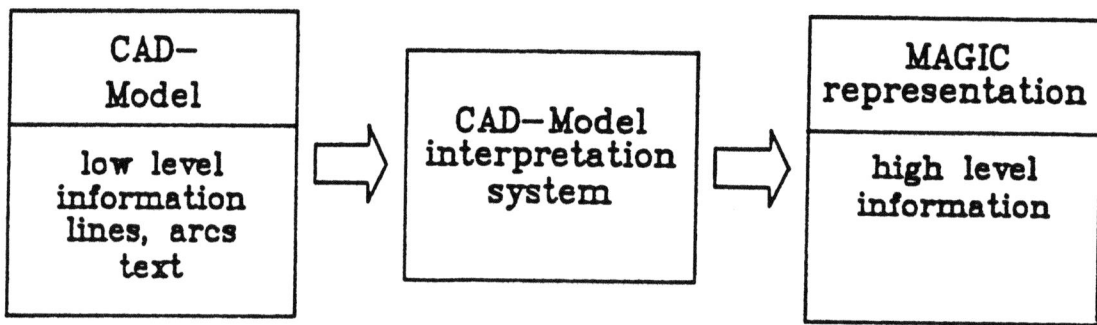

Fig. 3 CAD-Model interpretation system.

In the original MAGIC concept the idea was to use design with features for all the drawing work, and to use the attribute information for the generation. Disadvantages of this method are:

1. Generation is not possible for old CAD-Models (Made before the implementation of the feature based design).
2. Design with features can not cover all the necessary drawing activities.

Therefore the three methods mentioned above are used for the MAGIC CAD-Model interpretation system. This drawing interpretation takes place when a drawing is approved.

2. Terminal Input. There is some information about parts that can not be found in the CAD-Model. This is the case for those entities in the CAD-Model that can not be interpreted by the drawing interpretation system and for functional information.

Functional information is usually not present in the CAD Model. Even if the data are present in the CAD-Model extra information is necessary. E.g. , in the case of a functional dimension, it has to be specified that the dimension is functional, and what the functionality of the dimension is.

The terminal input is semi-automatic. This means that the information-gathering procedure is directed by the computer, and is integrated in the design process with its retrieval and approval procedures. For the functional information one of the techniques that can be used is, to keep track of previous retrieval specifications. For example, if the designer has tried to retrieve a piston with a maximum pressure of 160 bar, it is likely that an adapted drawing approved after this retrieval represents such a piston. The computer will however ask for confirmation.

3. Other Databases. A lot of information, useful for GT purposes can be found in other databases. Examples are the bill of materials and the material specifications.

The ideal situation would be that the MAGIC system could access information that is physically located in another database, and use it as if it would belong to the MAGIC database. In most cases however, practical problems make this impossible. This means that a lot of information has to be copied from one database to another. The information that is stored, where and how it is stored, and where it is introduced, are strongly company dependent.

4. Optional Information sources. This information does not come from the design department. Examples are the use of tools, coming from the manufacturing department, or cost information

This information can be obtained through interfaces with the application programs of other departments (e.g. process planning programs).

The information will not necessarily be physically located in the MAGIC database (See above: other databases).

3.2 Representation.

The MAGIC representation can be divided into the MAGIC internal representation and the dynamic relations (Fig. 2). The MAGIC internal representation is a collection of data representing parts independently of the application in mind. The dynamic relations are application oriented. They establish a link between the MAGIC representation and the retrieval. The dynamic relations can be divided into application oriented Definitions and supplementary data.

3.3 The MAGIC Internal Representation.

The internal representation should be able to contain all possible useful information and should avoid Information Degradation. It should also be independent of the application, being suitable for design as well as for manufacturing. This is achieved by using a relational database structure to bring the real entity[1] structure in the representation model. The internal representation contains different types of data. These data can be classified under Record management data, Simple part description, Functional description and Optional data (Fig. 2).

1. Record Management Data. The Record management data is used for simple parts and assemblies. The input for the Record management data comes from CAD-Models and other Databases. It contains administrative data on parts and their representations (CAD-Models and CAD drawings). Examples of administrative data are the partname, the physical location of a drawing and the name of the person who made the drawing.

---

(1) Parts and related entities like CAD-Models and drawings.

2. Simple Part Description. The Simple part description is used for simple parts only. It contains the information that can be found on the detail drawings of simple parts. The input for the simple part description comes from CAD-Models and other Databases. The simple part description can be divided into:

1. Material data:
    - material specifications.
    - raw part dimensions.
2. Engineering data:
    - surface quality specifications.
    - special treatments like heat treatments and coatings.
3. Geometry: This is a geometrical description of the part, based on Form features. Further investigation will point out how complete this description can and should be made. The possibilities are to have:
    - The outer contours (For rough sketch retrieval).
    - Production oriented features like holes and slots.
    - Surfaces to be machined
4. Tolerances and relationships: This description deals with the form and place tolerances on the detail drawing of the part

The geometry of the simple part in Figure 4 is an ideal example to illustrate the structure of the MAGIC internal representation. The example shows how features are used to represent a part. The part, in this case a piston, is composed out of form features. In this very simple case three tables associated with as many types of form features are used to represent the part.

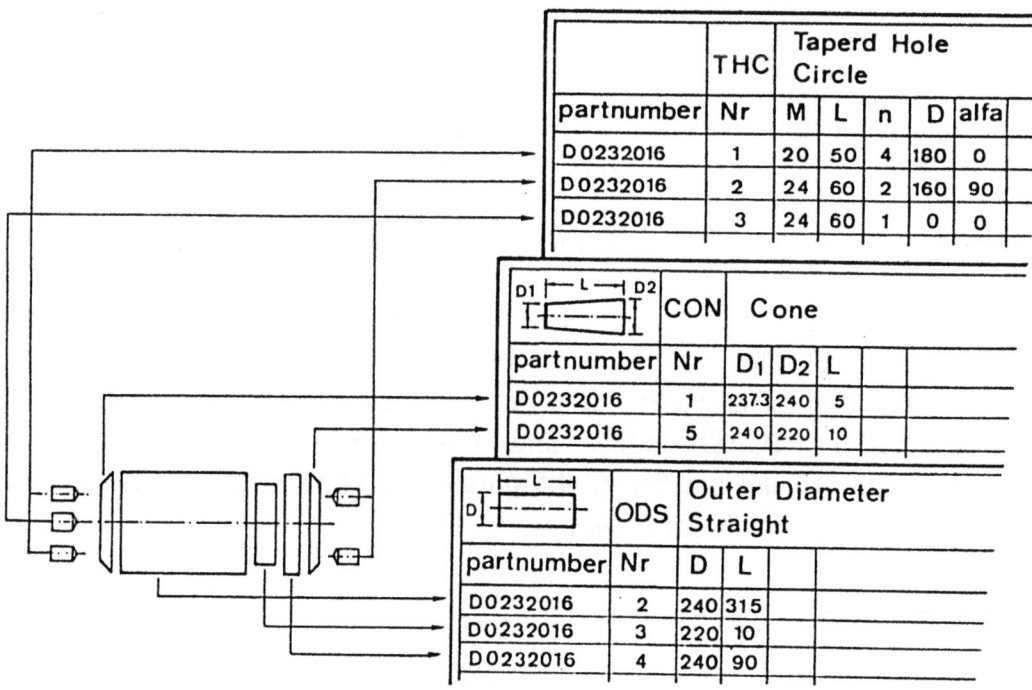

	THC	Taperd Hole Circle				
partnumber	Nr	M	L	n	D	alfa
D0232016	1	20	50	4	180	0
D0232016	2	24	60	2	160	90
D0232016	3	24	60	1	0	0

	CON	Cone		
partnumber	Nr	D_1	D_2	L
D0232016	1	237.3	240	5
D0232016	5	240	220	10

	ODS	Outer Diameter Straight	
partnumber	Nr	D	L
D0232016	2	240	315
D0232016	3	220	10
D0232016	4	240	90

Fig. 4 The geometric description of a part with feature tables

268

Surely other parts need more tables to make a representation. These tables are common for all the parts stored in the MAGIC internal representation. This implies that for every occurrence of a form feature in a table, the part to which it belongs should be identified. This is done by means of the partnumber.

3. Functional Description. The functional description is used for simple parts and assemblies. It is generated trough terminal input. The functional description of a part contains information about the use of this part. This information is very useful to the design department, where the design problems are often presented as finding a part that can perform a certain function. It is clear that this functional description is very much related to the company's know-how.

The functional properties of parts are used in one of the prototypes (The Functional Data Model). The piston family has been chosen to be used in the prototype. A study showed that all important functional properties of the pistons can be described with a limited number of attributes in one table. The possibilities for extending the prototype by incorporating more families of parts have been investigated. A possible approach would be to have a table for every family, each table containing the relevant attributes for that family.

Being the straightforward extension of the prototype, this "one family, one table" approach however results in a rigid classification of functional families. This is opposed to one of the aims of MAGIC, namely having a flexible classification. By analogy with the simple part description a more complex data structure, that does not have the previously mentioned disadvantage should be worked out.

Compared to the simple part description the data structure for the functional description is much more company dependent. This is due to the fact that the functional description is very much a reflection of the company's know-how, and its design and production methods. Therefore it will not be possible to work out a data structure applicable for a range of companies. The idea is however to work out some sort of basic frame and to provide adapted tools to work out and maintain a data structure for the functional description.

4. Optional Data. The information coming from the optional information sources is found in the optional data. This part of the MAGIC internal representation is not dealt with in this project.

3.4 The Dynamic Relations.

The internal representation, used in MAGIC, does not limit the possible applications, like the classic GT code. However due to its rather complex structure, and the large amount of data in it, retrieval can be difficult. Creating the retrieval specifications can be complicated. Another problem is that with large amounts of data and complex retrieval specifications, the retrieval speed decreases. Therefore the dynamic relations act as an interface between representation and retrieval. The dynamic relations can be divided into Definitions and Data.

The Definitions of the dynamic relations are application oriented. They help to create complex retrieval criteria. The data in the dynamic relations contain redundant information that allows faster retrieval.

The dynamic relations are also used to categorise the parts into groups of similar parts: families. Since the dynamic relations can be changed, it is possible for the user to create new families and to change or delete existing families.

3.5 Retrieval.

The MAGIC retrieval system combines the MAGIC representation and the user specification to give the answer to 2 kinds of questions:

<u>Type A</u>. Given a user specification, find the parts that comply with it. The user specification can be:

1. A direct query on the tables in the internal representation.
2. Using Definitions from the dynamic relations.
3. A combination of both.

<u>Type B</u>. Given a part, find the appropriate family for this part. These

4. MAGIC AS AN INTEGRATING ELEMENT.

The relational database used to store the MAGIC Representation, can be considered as the core of the MAGIC system. This database could serve as a common technical database for a number of applications in design and manufacturing, hereby acting as an integrating element. The intention is to provide links to computerised applications allowing input and retrieval of information (Fig. 6). Due to the use of features, resulting in a high information level, the MAGIC representation provides information that can be interpreted by computerised applications.

4.1. Design.

The CAD system is the main input source for the MAGIC representation. At approval time the CAD-Model is transformed into the MAGIC representation by the CAD-Model interpretation system. However, the CAD system is not only used for input, but also for retrieval purposes. Retrieval specifications can graphically be generated on the CAD system (e.g. A rough sketch of a part serves as retrieval specification). The integration with the CAD system both for input and retrieval purposes is considered to be a main innovation in MAGIC compared to classical Group Technology systems. Therefore the ESPRIT-2623-MAGIC project concentrates on the link with the CAD system.

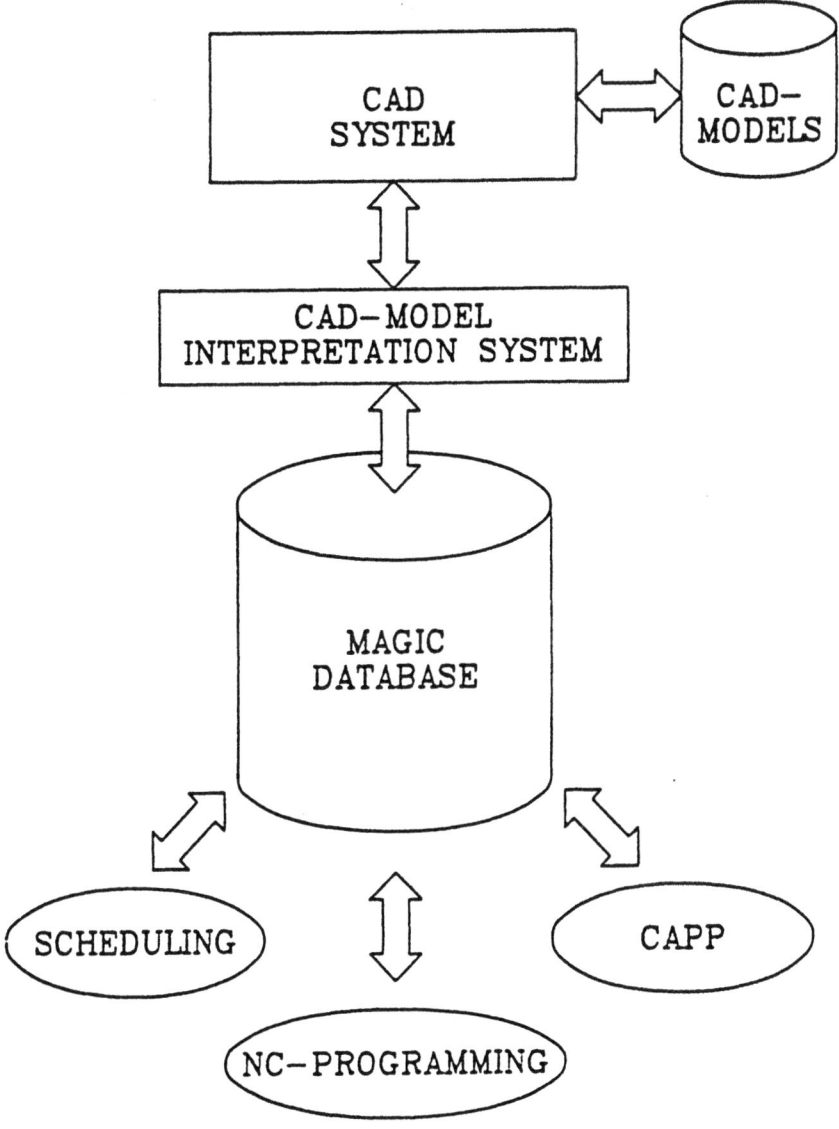

Fig. 5. The MAGIC database as an integrating element.

4.2. Manufacturing.

The use of Group Technology in computer aided process planning, CAPP, is not new. Variant CAPP systems use Group Technology as a foundation. The parts are grouped into families, and for every family a standard process plan is established. The codenumber of the part, used in the classic GT systems determines the family to which a part belongs, and therefore also the standard process plan. Adapting the standard process plan to the particular part results in the process plan of the part. Since the information in the MAGIC system is more complete than a classic codenumber, it will be possible to produce process plans that are better adapted to the particular part than standard process plans. The feature based representation can provide the necessary input for an expert system to generate process plans (A generative CAPP system). The MAGIC Representation can also serve as an input source for other applications. In Scheduling similar parts can be grouped and produced together avoiding setup and tool changes. Parametric NC-programs can be elaborated, reducing the NC-programmers effort to the introduction of a reduced number of parameters. On the other hand, these applications can also provide input to the MAGIC Representation. An example here is the use of tools. The tools used to manufacture a part, determined during NC-programming can be stored in the MAGIC representation. This information can than be used for Scheduling.

5. CONCLUSION

The MAGIC project started off with some innovating ideas about Group Technology, explained in chapter 2: The MAGIC Philosophy. Since these ideas needed clarification it was decided to build simple prototypes, primarily intended for evaluation. The conclusions of this evaluation lead to the specification explained in chapter 3: The MAGIC System. The next step to be completed by September 1990 is the elaboration of data structures and the conceptual design of the final prototype software. The third year of the project is reserved for the establishment of the final prototypes. It can be expected that the collaboration of different industrial partners as well as research centres and software houses is the best method to produce a general usable system.

6. BIBLIOGRAPHY:

Mikell P. Groover and Emory W. Zimmers,
"CAD/CAM", Blz 275-302,
Prentice-Hall International, Inc, London, 1984

Esfandiar Kamvar and Michel A. Melkanoff,
"Automatic Generation of GT Codes for Rotational Parts from Cad Drawings",
UCLA, California USA

Hsu-Pin Wang and Heng Chang,
"Automated Classification and Coding Based on Extracted Surface Features in a CAD Data Base",
Int. J. Adv. Manuf. Technol. Vol. 2. No. 1

Peter W. Chevalier,
"Group Technology: The Connecting Link to Integration of CAD and CAM",
OIR Europe Inc, The Netherlands.

V. Van De Steen,
"Methods for advanced Group Technology integrated with CAD/CAM",
Proceedings of the 6th Annual ESPRIT Conference, Brussels, November 27-December 1, 1989.

CIM SYSTEM PLANNING TOOLBOX: CIM-PLATO

R. Bernhardt, V. Katschinski, G. Schreck

Fraunhofer-Institute for Production Systems and Design Technologie;
Head: o. Prof. Dr.-Ing. Drs. h.c. G. Spur
Department of Robot Systems Technology, Director: o. Prof. Dr.-Ing. G. Duelen
Pascalstr. 8-9, D-1000 Berlin 10

1. Introduction

The increased performance requirements for manufacturing control demands an improvement of the a priori planning process. This means an entire manufacturing system has to be planned down to the level of application programms of all components before the real system is build up. Additionally the planning lead time and costs have to be drastically reduced. In general it can be stated that a poor a priori planning restricts the performance of the order related manufacturing control and increases the control complexity. Examination of the state-of-the-art methods and procedures of a priori planning revealed a siginificant absence of integrated computer-aided tools to support these tasks.

For that reason a R&D project (ESPRIT 2202, CIM-PLATO) has been launched to realize tools in the areas manufacturing system planning, process execution planning and information integration.

The first area includes the manufacturing task analysis and suitable description, the planning of operation sequences and material flows, the component selection, as well as the planning and optimization of the layout.

Based on these results a detailed process execution planning, down to the generation of application programms has to be realized and supported by tools in the second area. This includes event and motion-oriented planning for all components of the manufactuing system as well as their co-operation. Additionally a verification and optimization phase have to be considered in this planning process. Furthermore a linkage to the execution area as well as procedures and tools for the handling of exceptional situations have to be realized.

The third area -information integration- concerns the integration of the manufacturing system planning and process execution planning tools as well as their integration into a CIM environment. In some ways, the task to be done in this area can be regarded as an integration of all project activities and is therefore an important part of the project.

In chapter 2 an overview is given and the aproaches are described. Chapters 3 and 4 deal exemplatory with specific tools to be developed in the project.

2. Survey and Approaches

The overall objective of the CIM-PLATO project is the development of an industrial toolbox prototype of computer-based procedures and tools which support the design, planning and installation of FMS and FAS systems in a CIM environment. To reach this goal, three closely interrelated fields of R&D can be stated. These are the manufacturing system planning, the execution process planning and the provision of all necessary information to fulfill these tasks as well as the integration into a factory information system. In fig. 1 a functional reference model is presented which shows how the manufacturing system design tasks are embedded in a CIM environment.

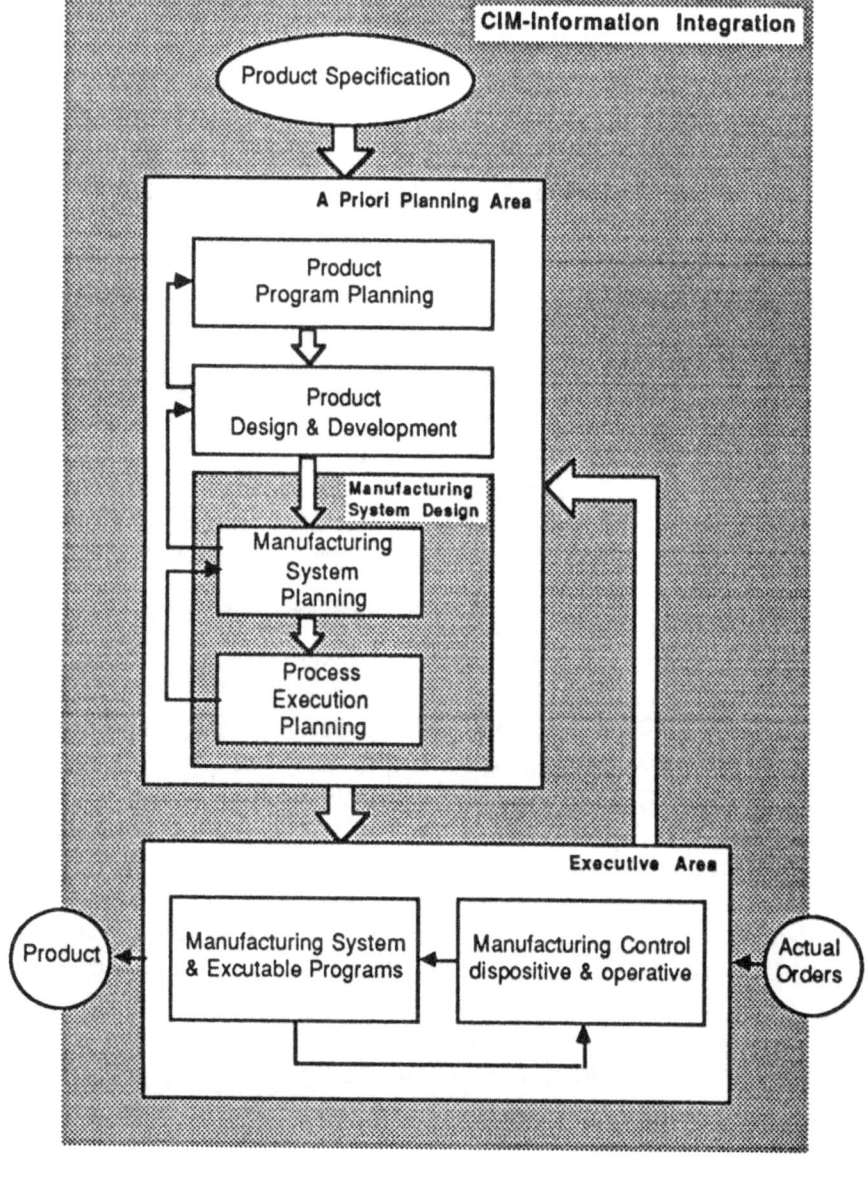

ce/01/PRMS/WB

Figure 1: Functional Reference Model

The manufacturing system design is part of the a priori planning area. Its output is the design of the manufacturing system, the control design, the generation of all required executable programs and strategies for the manufacturing control valid for the class of products envisaged to be produced in the plant. This represents the base for the manufacturing control process in the executive area which has to control actual order-related production of the required products in an optimal manner.

In the case where events occur in the executive area which have not been considered during the a priori planning process and which cannot be handled by the means of the manufacturing control, a feedback to the planning area is required, and a planning process taking these situations into accounts has to be realized.

The a priori planning area can be regarded as a sequence of planning subtasks which are fairly non-interacting in the forward direction. This does not exclude iterations between sequential planning steps. The main inputs for the manufacturing system design are coming from the product program planning and the product design and development area: In a first step the manufacturing system design can be further decomposed into the subtasks

- manufacturing system planning and
- process execution planning.

Planning of a manufacturing system requires initially a system configuration which can be subdivided into a selection of components and the planning of layouts. As a second step of the planning process, a scheduling has to be performed which is an assignment of resources and times to manufacturing operations. For the completion of the planning process a verification has to be carried out. The resultant information is a formal representation of the layout as well as a description of the production task and this constitutes the input for the following more detailed planning phase in the overall goal of getting a manufacturing system into operation.

Process execution planning is based on the delivered information produced by manufacturing system planning. Using additional technological and geometrical information the process execution planning as well as the exceptional case planning has to be carried out. Process execution planning contains event and motion oriented planning, technological process planning as well as application oriented verification, simulation and test procedures. Furthermore optimization procedures related to task sequencing and trajectories have to be considered. Another important item is related to the conversion and transfer of application programs to the real manufacturing system.

The integration of manufacturing system planning and process execution planning for manufacturing system design purposes is mainly a problem of information integration. Therefore the information flow within both systems, information exchange between both and their integration into a CIM structure have to be considered. This requires the elaboration of an information system architecture, the development of suitable (and possibly general valid) knowledge/information models, the use of standards for modelling and information exchange as well as a 'normalization' of user interfaces. This includes to guarantee a safe data handling and to reach a most flexible system structure related to the integration and management of the tools as well as to a quick adaption to user needs.

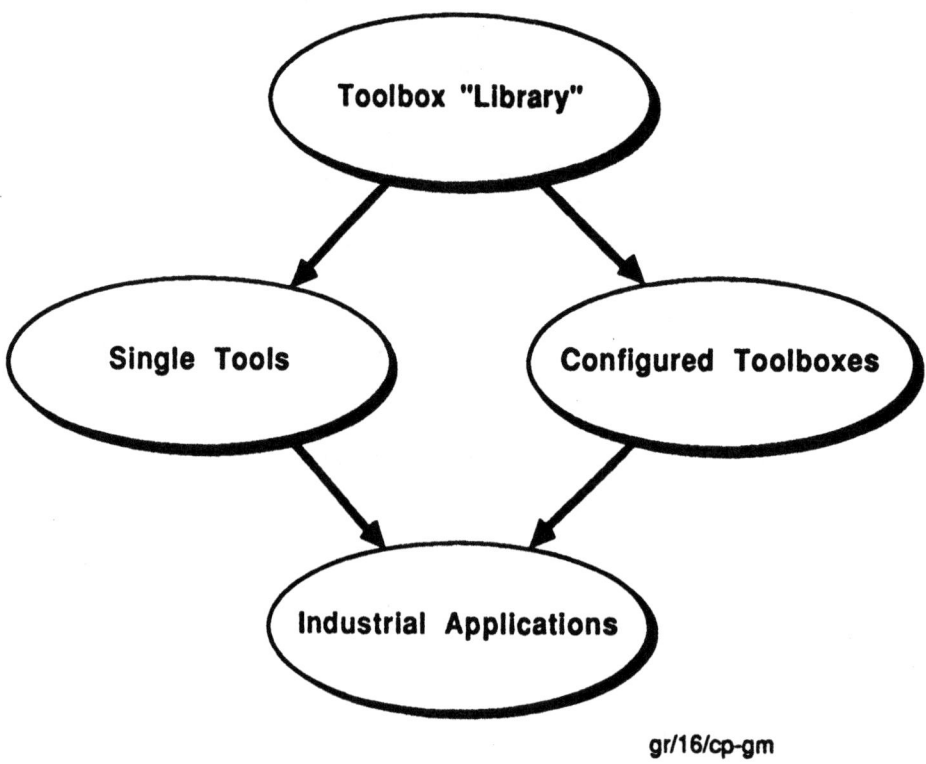

Figure 2: Principle Approach

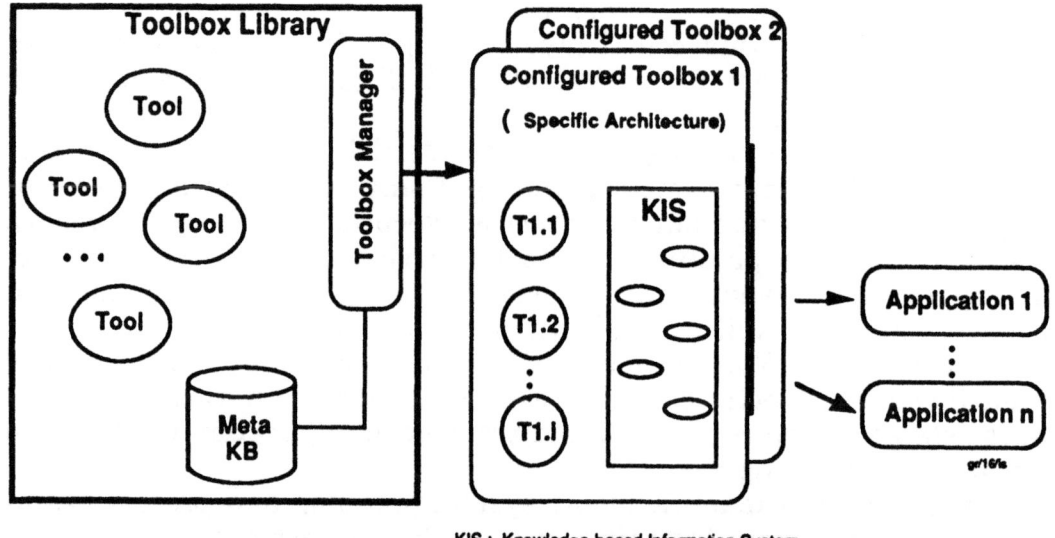

KIS : Knowledge-based Information System

Figure 3: Basic Project Structure and Principle Tasks

Within the frame of the project tools will be realized or already existing tools will be adapted and integrated to a toolbox 'library'. Tools are designed as far as possible and required to be applicable as 'stand-alone units' (single tools) for specific industrial applications. Examples of such tools are described in detail in chapter 3 and 4 of this paper. Additionally

276

specific toolboxes can be configured from the toolbox library for applications requiring the cooperation of different tools, e.g. planning and programming of a robotized assembly cell. This principle approach adopted in the project is outlined in fig. 2 and 3.

To ensure an effective cooperation between partners, a subgroup structure has been installed. There are two subgroups for the configuration of exemplary toolboxes to evaluate and to proof the functionality and effectivity of the tools to be developed as well as to demonstrate the benefits and advantages reached through integration of tools to a toolbox. Additionally a subgroup has been installed working in the areas of toolbox management system and information integration.

3. Tool: 'Technological Process Planning'

The application of innovative technologies increasingly requires the integration of experts into the planning process. The expert knowledge includes technical parameters and the requirements for the execution of specific manufacturing procedures.

In the frame of the CIM-PLATO project a tool for the planning of arc welding applications is being developed. The realization is done in close cooperation of partners from the industry and research institutes. The critical point in this tool development is the modelling of the technological process.

The result of the welding process is a welding seam on the workpiece, which may have different profiles and different qualities, depending on a number of parameters that can be manipulated by the welding expert. These parameters are:

- the composition of the gas shielding,
- diameter of the wire,
- speed of the wire,
- voltage,
- torch distance,
- torch position in relation to the workpiece,
- thickness of the workpiece,
- form and width of the gap between the two parts of the workpiece that are welded together,
- welding position and
- welding speed.

The information about the influence of the adjustable parameters on the resulting parameters (concerning the welding seam) is partly encoded in welding data tables and partly in rules. An entry in a welding data table is an n-tuple of values, one for each of the adjustable and resulting parameters. Thus, a welding data table can be conceived as a partial mapping from the adjustable to the resulting parameters. However, the domain of this mapping consists only of a finite set of single points in the m-dimensional space of adjustable parameters, and the gaps between thes points ara more or less large. The rules on the other side, are given verbally and are qualitative in nature, because in general they dont't relate exact values of parameters /1/.

The welding expert is able to process this welding information and also to estimate parameters between known parameter sets. Due to the above mentioned facts an expert system approach will be used for this tool realization. The term expert system refers to a computer program that can perform tasks in the domain of application, similar to a human expert /2/.

An important realization aspect is the knowledge acquisition. This has to be done together with experts from the arc welding domain. To support this strong cooperation the project realization is divided into two phases (Figure 4).

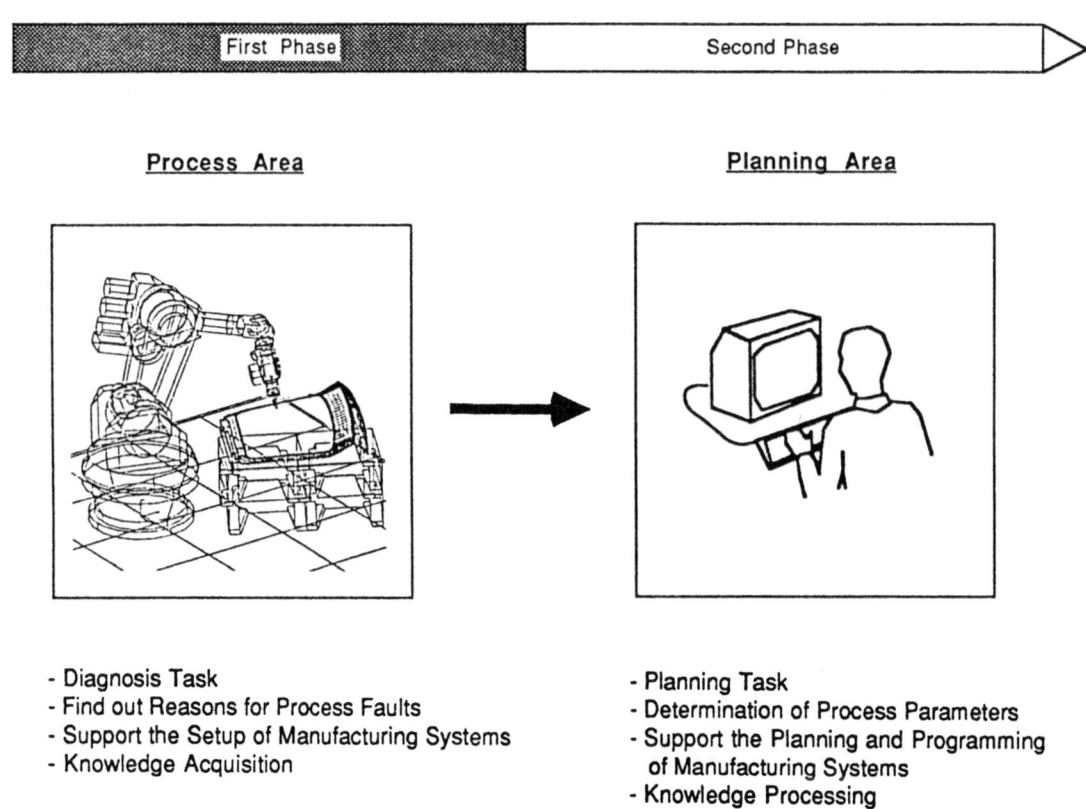

Figure 4: Principle Realization Approach

In the first phase an expert system for the process area will be developed. It supports the human operator to find out the reasons for faults in the welding seam (diagnosis task). The application of different prototypes with enlarged knowledge bases and improved inference strategies will lead to a comprehensive conceptional model of the technological process. In the second phase the present experience will be used to design and implement an expert system for the planning area. This system supports the determination of process parameters and requirements for the execution of seam welding tasks.

In the following the approach for the development of the tool for the process area is described in more detail.

The prototype of the diagnosis system is restricted to the diagnosis of pore faults in welding seams. By further development the system is going to be enlarged for the diagnosis of other faults. For the prototype, the following test scenario is planned: The prototype shall be able to diagnose reasons for the occurrance of pores during the welding process and propose measures for their avoidance. In order to achieve this, in the first step objects and rules of the welding process which in the welding expert's view are a cause of pore occurance will be included in the knowlegde base. The knowledge base contains:

- the objects of the welding cell configuration
- objects specific for the welding process
- fault type objects and
- rules which define the relationship between the objects in form of predicates and relations.

Objects are represented by object identifiers and a set of characterizing attribute-value-tuple. There are two different object hierarchies known to the system:

- a 'is-a'/'instance-of'-hierarchy for the classification of objects and
- a 'has-component'-hierarchy for the definition of the functional interaction of the objects during the welding process

The 'has-component'-hierarchy decomposes all welding process objects relevant to the fault diagnosis into components and sub-components. The functionality of the objects of this hierarchy with respect to the welding process is characterized by a slot 'state'. The state of an object is the mapping of the object characterization onto the value set W = (ok, fault, not defined). Admissable object descriptions are verbal expert descriptions contained in the object characterization. For the object 'welding tip' the descriptions 'spatter below', 'spatter inside', 'no spatters', 'clean' and 'dirty' are admissable. The state of one component is given by its description by an expert or is infered by rules evaluating the state of the sub-components. By the 'parts of' relation, the influence of the object's state on the functioning of the whole welding configuration is inferred.

The connection of fault description and object states is formed by a rules system. This rules system shall emulate the heuristic approach of a welding expert. Up to this point, three classes of rules can be discerned:

- rules which describe the relationship of objects or slots with each other (object rules),
- rules which connect a fault specification directly or via intermediate goals with the state of objects (fault rules),
- rules which propose measures for the elimination of faulty component states (measure rules).

In Fig. 5 the procedure of fault diagnosis is represented. At the start of the diagnosis process the user provides information about the fault specification and the state of the welding configuration. These information is optional. If the information provided by the user does not suffice, the system tries to gain additional knowledge by posing questions.

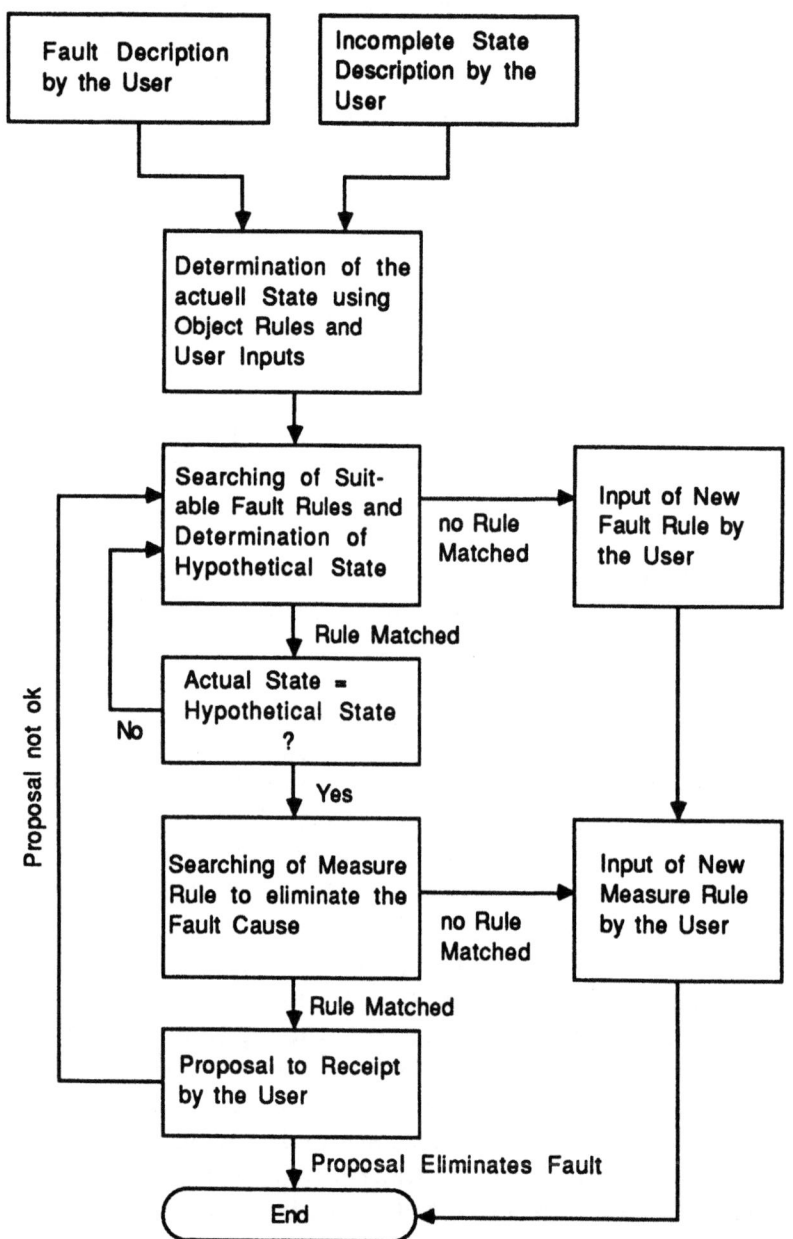

Figure 5: Diagnosis Procedure

The system tries to find a description of the state which corresponds to the one which has been input - and partly inferred through object rules - by applying fault rules. In the knowledge base such a description is related by measure rules with proposals for the elimination of fault causes. After the discussion of the first prototye it must be decided whether the inference component must be enlarged by a mechanism which can process probabilities and hypothetical knowledge.

For the prototype development a rudimentary dialogue interface is planned. The components of the welding system are directly implemented as objects of the knowledge base and cannot be input via the dialogue interface.

For further development of the system the user interface of the start dialogue is initialized according to the diagnosed fault. Hereby the number of questions to the user should be limited as far as possible.

Further activities should be directed towards the realization of an explanation component and an interfaces for the enlargement and modification of the knowledge base. The latter should provide the following functions to the user:

- Input/alteration of welding cell objects
- Input/alteration of a whole configuration
- Input of fault types
- Input of fault rules
- Input of object rules
- Input of measure rules

The prototype development is done on a VAX computer using the language PROLOG. For applications on site the system will be transferred to a portable PC.

4. Tool 'Process Execution Simulation'

A Simulation System is in development which allows a test of the executability and practicability of the created programs. Based on a general valid and formal description of detailed motions and processes a simulation of manufacturing processes is possible.

An important aspect of simulation systems is the quality of their simulation results, i.e. to what extend the simulation fits reality /3/. This depends on the computer internal modelling of the behaviour of all work cell components. Further the user interface plays an important role,i.e. which knowledge and experience related to manufacturing technology as well as system operation is required by the user and what is executed by the system automatically.

The practicability test contains testing of the defined trajectories related to positions, orientations, velocities and accelerations. Furthermore end-effector commands and interactions with peripherals have to be checked.

The simulator requires computer internal representations of all components of the work cell and their behavior according to predefined criteria. This repesentations are called simulation models and are structured in the following way /4/:

- control models which describe the motion behavior of components;
- kinematic models which contain the frame relations of the different links;
- shape models which describe the graphical representation of components.

Figure 6 shows the functional connection of these models for the example of a robot motion simulation. The robot application program is loaded into the control model of the IR. This control model interprets the application program and supplies the kinematic model with the joint values of the links. Within the kinematic model the frames describing position and orientation of each robot part are calculated. The connection of these frames to the relevant shape models enables the visualization of the robot motion on a graphic system.

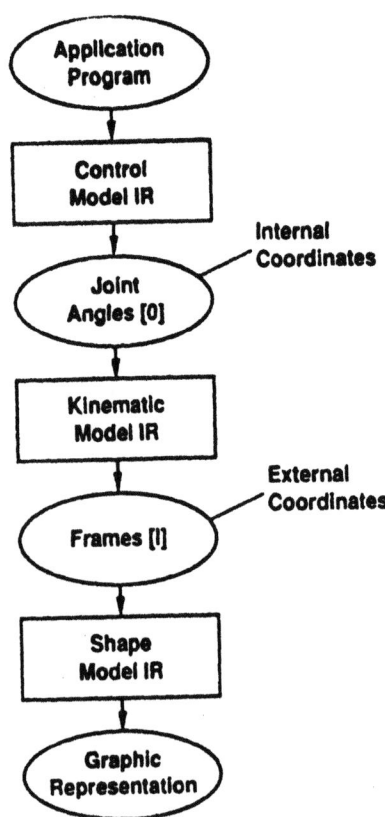

Figure 6: The Concept of Motion Execution Simulation

Gererally the above described procedure is also applicable for other components. To simulate the task execution of an entire work cell, models of all components are required.A lot of production tasks e.g. assembly tasks need the handling of time variant kinematics.

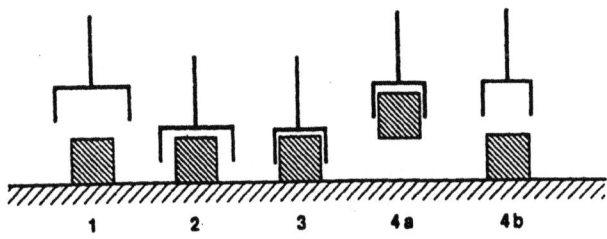

Figure 7: Results of the Process of Gripping a Part

The example of gripping a part is shown in figure 7. The gripper is positioned above the part (State 1), approaches it (State 2) and then closes (State 3). At this stage, two different cases can occur: Either the part is linked to the gripper and both will move upwards together (State 4a) - which is what we would expect - or the part is not linked to the gripper and consequently will remain in position (State 4b). In reality the laws of physics decide which case occurs; in a simulation this decision is made by a program.

This decision is often made by including some special statements to relink the simulated parts within the application program which controls the gripper´s movements. However, in a simulation this may form an obstacle for error detection. If such a statement is executed, the part will be linked to the gripper regardless of the gripper´s state or position.To avoid such fixed link definition the realization concept is based on the following approach:

The simulation contains declarative statements which define the linkstate of parts and model physics in an abstract way. A suitable formalisation of such statements is predicate logic. The predicates are dual and thus make it possible to decide whether a link is valid or not; their values depend on the state of the simulation. The level of detail can be chosen by the programmer, but it must be highly abstract in order to achieve a real-time behaviour of the simulation.

The applied kinematic model is based on kinematic chains modelled by the method of Denavit-Hartenberg /5/ which provides rotational and translatory links. It was extended by fixed links and branches which results in kinematic tree structures. Such a tree describes the actual kinematic of a simulation state. The concept of defining several links for a joint and determining a currently valid one by predicates leads to a structure of a partially ordered graph which is the graph of possible trees. Figure 8 shows such a graph (a) and its trees (b,c).

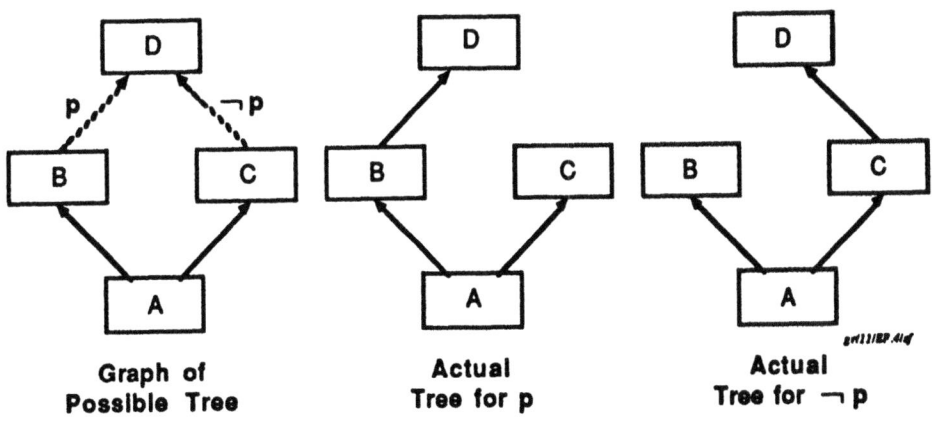

Graph of
Possible Tree

Actual
Tree for p

Actual
Tree for ¬ p

Figure 8: Graph of Possible Trees and Actual Trees

The links are denoted by blocks (A,B,C,D), the joints are denoted by arrows. The continuous lined arrows between links AB and AC denote time invariant joints, the dotted arrows between the links BD and CD denote time variant joints. The validity of the joints is determined by the predicates "p" and "¬p". The actual tree for the validity of p is shown in Figure 3b, the tree for the validity of ¬p is shown in Figure 3c.

The formulation of such predicates can be done by using elementary functions like AND, OR, ADD, DISTANCE. These operations build a function network which describes the link states.

The data structure representing the controls, kinematic and the communication model, which forms the connection of control and kinematic, build a network called simulation network.

The Simulation Network is the complete set of data which are loaded into the simulator to execute a simulation task. A Simulation Language Interpreter interprets text files written in a specific simulation language and builds the Simulation Network.

KERNEL

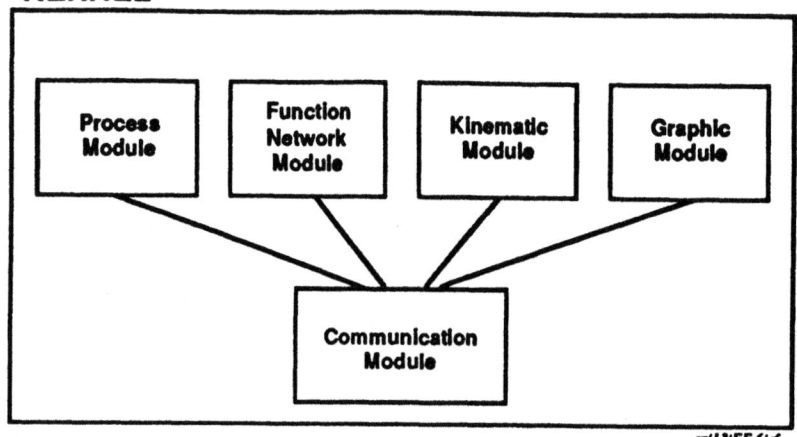

Figure 9: Modulare Structure of the Simulator´s Kernel

Figure 9 shows the modular structure of the simulator´s Kernel. The Communication Module provides several types of input and output buffers e. g. real, bool, integer, frame. It is the definite interface for the information interchange of the other Kernel modules. The Process Module includes the modelling of controls and physical aspects. The Function Network Module provides fundamental operations. It allows to define functions like relinking predicates and other physical aspects. The Kinematic Module provides ports and joints of several types e. g. rotational, translatory, fixed. It contains the actual and possible kinematic trees. The Graphic System Module is the definite interface for the graphic monitor. It initialize the graphic system, loads the graphical description of the parts and transmits the actual position and orientation of the parts. Based on this modular concept and the flexible configuration of the simulator using a Simulation Language the process execution simulation of FMS/FAS Systems is possible.

5.Conclusion

The paper reports about the ESPRIT CIM-PLATO project which has been started in May 1989. It is a four years project whereby the first year is a low budget definition phase which aims to the elaboration of functional specifications of all tools. Within the remaining three years the tasks to be fulfilled are, principly spoken, twofold: On the one hand tools in the areas system planning, process execution planning and information integration tools have to be realized and integrated into a toolbox library. This includes also the realization of a toolbox manager.

On the other hand two toolboxes for specific industrial application are to be configured out of the toolbox library to demonstrate the advantages and benefits of the adopted approach.

Important and common aspects for both of these general tasks are related to the information system and information integration which includes knowledge/information modelling, information exchange and management, safe data handling, flexible system structures to enable an easy tool integration as well as a quick adaption to user needs.

This requires the use of already available standards and to consider international standardization activities.

6. References

/1/ Dilger, Espen, Schuck: Process Modeling and Simulation with PEPS.
EXPS´87, Berichte des German Chapter of the ACM, Bd. 28, Teubner Verlag 1987

/2/ Heragu, Kusiak: Analysis of Expert Systems in Manufacturing Design. IEEE Transactions on Systems, Man, and Cybernetics, Vol. SMC-17, No. 6, Nov./Dec. 1987.

/3/ Spur et al.: Planning and Programming of Robot Integrated Production Cells. ESPRIT Techincal Conference, Brussels, Belgium, September 987.

/4/ Duelen, G.; Kirchhoff, U.; Bernhardt, R.; Schreck, G.: Algorithmic Representation of Workcells and Task Description for Off-line Programming. Robotics and Computer Integrated Manufacturing, Volume 3, No 2, pp. 201-208 (1987).

/5/ Denavit, I.; Hartenberg, R.S.: A Kinematic Notation for Lower Pair Mechanisms Based on Matrices, Journal of Applied Mechanics 77(1955), pp. 2154-221.

7. Acknowledgement

The following companies, research institutes and universities are working together in the CIM-PLATO (ESPRIT 2202) project:

Fraunhofer-Institut für Produktionsanlagen und Konstruktionstechnik,IPK-Berlin
KUKA Schweissanlagen + Robotertechnik GMBH, Augsburg
FIAR SPA, Divisione Robotica Industriale, Milano
Universität Karlsruhe, Lehrstuhl für Prozesstechnik
University College Galway, CIM Research Unit
Universidad Politecnica de Madrid, Madrid
Universidade Nova de Lisboa, Lisboa
Politecnico di Milano, Milano
LADSEB Consiglio Nationale delle Richerche, Padova
PSI, Gesellschaft für Prozesssteuerungs- und Informationssysteme MBH, Berlin
University of Amsterdam, Dep. Computer Science, Amsterdam
BULL, Paris, Angers
INVESTRONICA, Division Productos Industriale, Madrid

CIDAM - modules for the creation of CIM

A. Hars

Institut für Wirtschaftsinformatik, Germany

1. THE CURRENT SITUATION

Despite apparent enthusiasm the implementation of complete CIM systems is still far from being state of the art in world industry. Advances in technology have led to widespread use of information technology in manufacturing but have mainly generated islands of automation. Figure 1 illustrates the current situation in industry. The organizational and planning functions on the left hand side belonging to the area of production planning and control (PPC) are seperated from the technical functions of the CAE/CAD/CAM chain situated on the right. The distance is stronger on the upper level of planning and decreases but is still existent on the realization level on the lower half. In addition to this barrier, within both chains additional barriers exist. They correspond to the boundaries of the traditional isolated subsystems of an enterprise which are characterized by their functions, the level of EDP-support, the EDP systems themselves, the organizational structure, the allocation of personnel, the administration of their own data and the relationships with the other subsystems. Therefore CIM as integrated information processing for the technical and operational tasks of an enterprise[1] has not yet taken off in industry.

2. APPROACHES FOR CIM

As a CIM system cannot be implemented from scratch, the implementation has to start out from an analysis of the different subsystems of the enterprise. Subsystems as characterized by the Y of figure 1 cannot be reduced to EDP-technical aspects but also regard organizational aspects in an enterprise. Therefore for the implementation of a CIM system modern technology and management science have to be brought together. From such an integral system view the following approaches to CIM can be derived:

1) Increase the level of EDP support within the subsystems. This approach focuses largely on automation and does not lead by itself to integration. The focus of automation is technical. Isolated automation, however, has some inherent dangers. It should very carefully be planned so as not to interfere with integration (see below). The productivity potential of isolated automation within a subsystem of a company should not be overestimated.

But CIM is not equal to automation. For the implementation of CIM, the emphasis has to be laid upon integration. Integration is concerned with the relations between the different subsystems and can be split into two approaches:

2) Harmonize the structures within different subsystems. One of the basic approaches to integration has to be the adjustment of different systems to a common standard. As much as standardization is necessary in a global perspective, standardization is necessary for the different subsystems within the enterprise. This affects each structural element of the subsystems:[2] Data, functions, organization and EDP-systems. For the data held in the different subsystems, harmonization can be accomplished through the creation of enterprise-wide unified data structures. Harmonization of functions requires that in different areas similar functions are utilized. For the organizational structure it requires the use of consistent structures for all the subsystems and for common structures above the subsystems. Harmonization of EDP-systems starts out with the decision to enforce one type of hard- and system

286

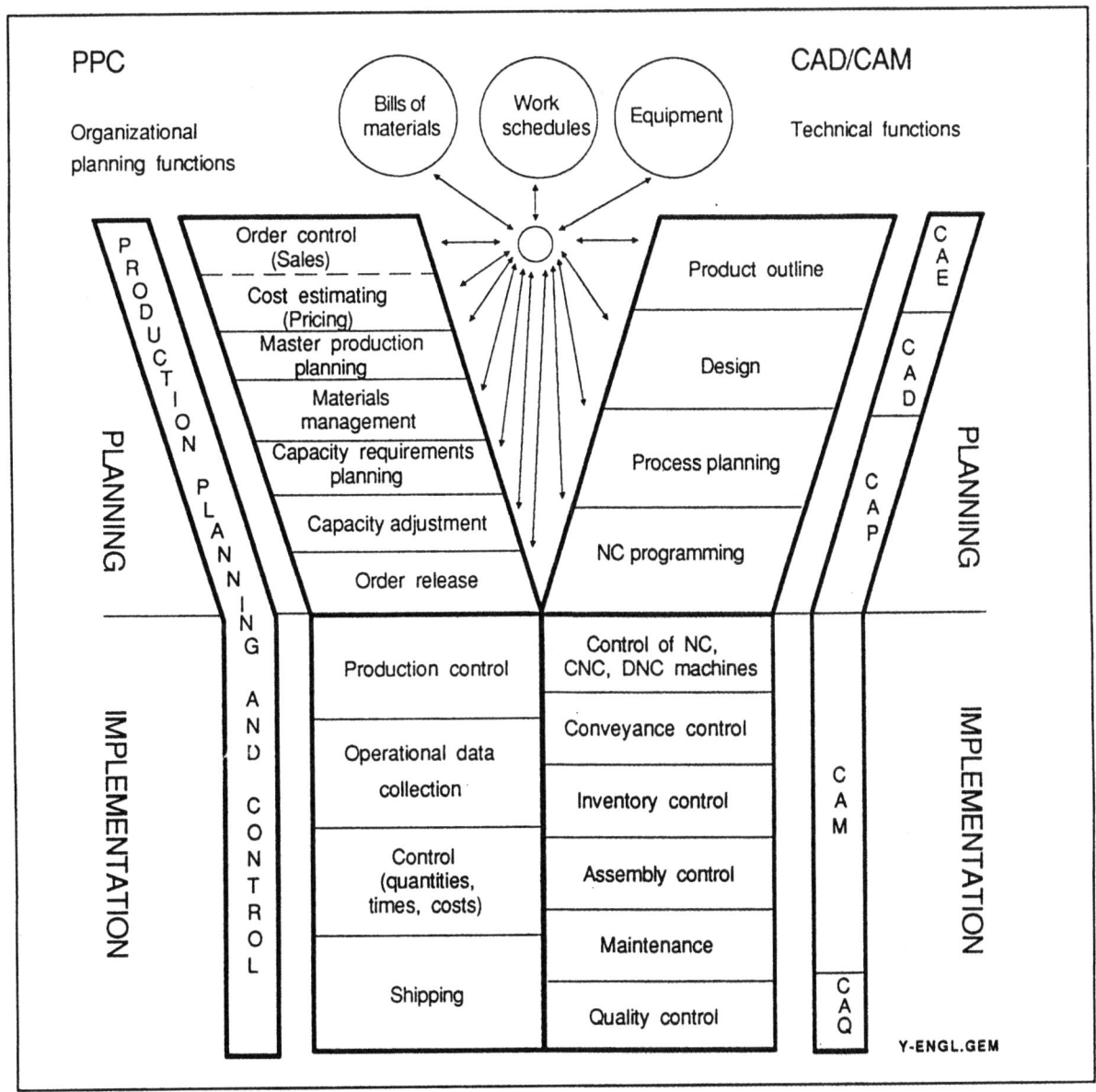

Figure 1: Information Systems in Production[3]

software. The need of harmonization is striking for the interaction between human and machine. Man machine interfaces have to present an identical "look and feel" so as to allow the same persons to be active in different application systems.[4]

3) Establish links between different isolated systems. Besides adjusting the structures of the different subsystems, integration requires the different subsystems to be linked, making the exchange of information possible. On the physical level linking requires the implementation of a network between EDP systems. On the logical level, links however depend upon the different characteristics of the systems. To link processes, one process of one system has to be made capable to trigger one or more processes within another system. To link systems for the exchange of data, a system which allows the exchange and update of data redundantly held in different databases has to be implemented or central or distributed company-wide data bases have to be established. To link organizational structures, consulting units can be formed. User interfaces can be linked for example by establishing interface windows.

Both linking and harmonizing the different systems will in many cases result in the merging of subsystems. The boundaries between the subsystems are eliminated and the original systems as such vanish. Within these new merged, integrated systems, new functionalities can be envisioned.

The goal of achieving integration by harmonizing and linking subsystems can be targeted primarily to certain of the different types of structural elements within the subsystems. Of these the most efficient structural components on which implementation acitivities should focus can be selected:

4)   Data integration realizes linking and harmonizing for data structures and databases. Prerequisite is a logical model which describes the data structures of the whole enterprise - an enterprise-wide data model.[5] This model then is implemented in a two-step process on the physical level of a database management system. The appropriate design language for logical data models is the Entity Relationship Model[6]. To link the databases which are distributed over different locations of an enterprise, a distributed database approach has to be used. Data integration is a very efficient approach to integration as it tends to increase the integration of procedural and organizational structures also.

5)   Functional integration is an approach where processes which previously had been divided according to Tayloristic principles are recombined.[7] This also is an effect of linking and synchronizing the processes in the different subsystems of an enterprise. The effect of functional integration is a considerable reduction of lead-in and transfer times for the corresponding tasks.

For an enterprise striving for CIM, an appropriate choice of implementation activities has to be made. Any path will contain risks, and requires effort. To ease this, a set of modular tools is developed, which provides assistance for the implementation approaches chosen.

3. SOLUTIONS FOR INTEGRATION

The CIDAM (CIM System with Distributed Database and Configurable Modules) project no.2527 is an ESPRIT II project which aims at facilitating the implementation of CIM-systems through modular components by
-      developing tools for integration
-      developing CIM-components
-      realizing CIM-solutions in the actual world in the testbeds.

Figure 2 depicts the integral approach of CIDAM showing the combination of software integration tools and CIM solutions offered by the project.[8][9]

CIDAM started in March 1989 and is planned for a time horizon of four years, the funding having been agreed upon for the first three years.

The CIDAM modules are developed in a common effort by the industrial and academic partners:
- Mannesmann Kienzle, Germany (project leader)
- FIAT Aviazione, Italy
- Institut für Wirtschaftsinformatik (IWi), Germany
- Syseca Temps Reel, France
- Trinity College Dublin, Irland.

In addition, five subcontractors to Mannesmann Kienzle are participating in the project:
- Mannesmann Datenverarbeitung
- Mannesmann Demag
- Mannesmann Fichtel und Sachs
- Mannesmann Hartmann und Braun
- Mannesmann Procad.

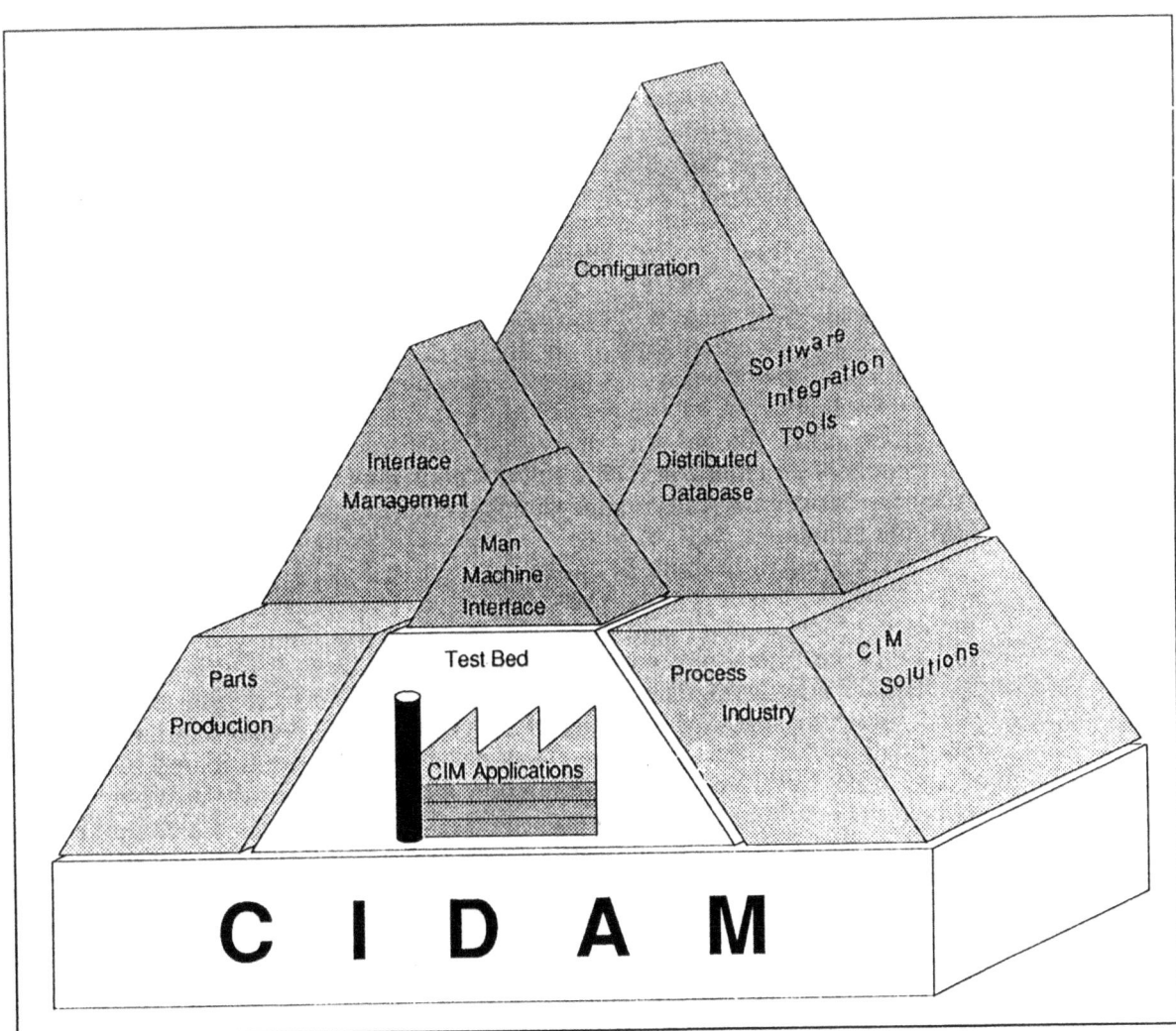

Figure 2: The CIDAM project

The CIDAM project is divided into three categories of modules built, which are vital for the design and operation of a CIM system:

3.1 Configurable components

CIDAM builds various components which support the integration already before runtime by being adaptable to the requirements of an enterprise. They will be part of the configuration carried out by COSY which is the subject of the next chapter. The components built are:

- New CIM Components for missing key functionalities: An analysis of the deficiencies of current EDP-systems in industry shows that both for the long term functions and for the short term functions situated on the shop floor, the EDP support is low. In contrast, the current EDP support in process planning focuses on the medium term range with the classic MRP functions.[10] The analysis of the situation at CIDAM's testbed partners has confirmed this analysis and has shown, that on the shop floor level functionalities like short term scheduling and monitoring in real time are insufficiently supported. Therefore multifunctional components are developed corresponding to the short-term scheduling and shop-floor monitoring requirements of the testbeds both in the case of part production and process production environments. This includes the shop-floor control and data collection functionalities. For the design of the scheduling modules, the object-oriented approach is followed. A more detailed description is given by HERTEL[11]. The development is carried out in close cooperation with the testbed partners. The new components do not only automate functions previously carried out manually but also introduce previously not existent functionalities. In addition, functionalities to link the shop floor with the higher level MRP systems are developed. Thereby the scheduler will eventually become a medium for shop-floor integration.[12]

In a second step the existing CIM-components within the testbeds and the background material of the information technology vendors are made compatible with the CIDAM environment. In particular they are adapted to the requirements of the Configuration System, the Distributed Database and the Interface Management System. The adaptation also regards the links of the process application to a conventional MRP system. These adaptations are mainly carried out in the CIMCO workpackage. The adaptation is in line of the harmonization of the different system corresponding to approach two for a CIM implementation.

Another effort of the project aims at developing a tool to build Man Machine Interfaces. This allows the design of uniform graphcial user interfaces in all subsystems of an enterprise. This activity conforms to alternative two harmonizing the structures between different subsystems, centering on the most important element of CIM: the human.

The tool to be built is called Interactive System Builder which will generate user interface source code. The Interactive System Builder consists of the User Interface Model builder and the User Interface Builder. The User Interface Model Builder is the interactive part of the Interface System Builder residing on a logical designer-friendly level. There a model of the user interface to be developed is created based upon the CIDAM user interface style which contains all the user interface specifications for CIDAM. There are three aspects to be considered to build the model of the user interface: the presentation technique, the dialogue control and the application interface where the connection between system actions and user actions and the access to application data is specified. After specifying the model of the user interface, the User Interface Model Builder generates a user interface description file which serves as input for the User Interface Builder. The user interface builder uses graphical libraries and the description file to automatically create the source code for the user interface.

The Interface System Builder relies on standards. While X-Windows is definite, additional standards as OSF Motif, PCTE and EIFFEL are currently examined for suitability. During specification time the different testbeds are utilized to supply requirements for the user interface style.

3.2 Configuration system

Maybe the most ambitious part of CIDAM is to allow the configuration of a CIM-system by tailoring the system itself, CIM-functions and data structures to the requirements of a customer. This is done by the configuration system COSY.

It consists of four components: The CIM System Configurator selects and combines the CIM components which match some customer's CIM functional and operational requirements. These CIM components are stored within a software component base consisting of three description levels. The functions-catalogue is the most abstract level containing an easily understandable functional description which is independent of the technical aspects of the realization. On the second level the data processing realization is described containing the structure of programs, screens and windows. The lowest level is the Development Dictionary which manages the data structures for all applications.

In addition the Component Data Structure Configurator allows the tailoring of the individual component to the requirements of the customer, especially regarding data necessary for creating interfaces between different functions and for the user-interface. Here a close cooperation to the MMI interface is assured. The Component Functionality Configurator selects, modifies and combines different functions from an atomic objects library according to the requirements of a customer. As the CIM systems and the requirements for the functionalities of a given component are constantly changing, a Version and Configuration Manager is needed to control the changes made to the components and the system from time to time. This Version Manager resides on the same level as the CIM System Configurator.

A typical usage of the Configuration System COSY then can be envisioned as follows: A customer who wishes to build a CIM system uses the CIM System Configurator to select the functionalities required from the library. The functionalities are taken from the library, configured to the situation of the client and an installable system in source code is generated. The Version Manager works in the background and comes in use at every change of the CIM system. If certain functionalities of the system have to be changed or enhanced or new functions have to be implemented then the Component Functionality Configurator comes into use. During the configuration the databases for the system and the data interfaces have to be generated. This is done by the Component Data Structure Configurator. The Version Manager keeps an eye to the changes on all the levels of the configuration system.

Thereby already before runtime, applications stored in a library are linked and harmonized to become an integrated system.

3.3 Runtime system

These modules operate during the runtime of a CIM system. To establish links between different islands of automation, facilities for the common management of distributed data, for the exchange of data between different systems and to ensure the consistency of data held redundantly are required.

- Therefore in the Distributed Database workpackage a distributed server based database and a distributed file system of a homogenous kind are developed. They allow to access and store data on physically distributed databases or file systems. In addition a data dictionary is designed which is to work together with both systems and which stores the location, access control and CIDAM data conversion rules. This is a first step towards common data (approach two) and advances the degree of harmonization between the different systems.

- For heterogenous database systems, functionalities are developed to link these systems. A CIDAM gateway functionality is designed to link heterogenous relational database management systems. Furthermore CIDAM bridge functionalities are developed to link different file systems and to allow a limited link between relational systems and file systems.

- An Interface Management System INMAS serves as individually configurable module for the exchange of data. This link ensures consistent data over all systems and eliminates the risks inherent in the set-up of individual links. Instead of utilizing point to point links, all systems are connected to INMAS and all data exchange is carried out over this module. INMAS contains a neutral enterprise-wide data structure to which the data structures of the existing systems are referenced. Via transformation rules it then becomes possible to transfer data from one source system to as many destination systems as required. Thereby the update of data which is redundantly held in different databases can be automated. INMAS also contains the process logic to ensure that for a certain event which affects the data in one source system the corresponding data transfer to the destination systems is carried out.[13] The working of the INMAS system is shown in figure 3. An action in the source system on the left leads to a trigger message to the Communication Control Module (CCM) of INMAS in the center. There the event is decoded and the data is passed on to the General Information Control Component which initializes an activity chain corresponding to the event. The activity chain consists of read and write operations which use the SQL-Gateway to access both source and destination database management systems or file systems. After having completed the transaction, the Communication Control Component sends confirming messages to the source system and an information message to the destination systems when required.

The Interface Management System does not only increase integration through the establishment of links between islands of automation. The creation of the neutral enterprise-wide data structure in a relational standard format is also a means to harmonize the data structure of the different systems on the long term thereby serving

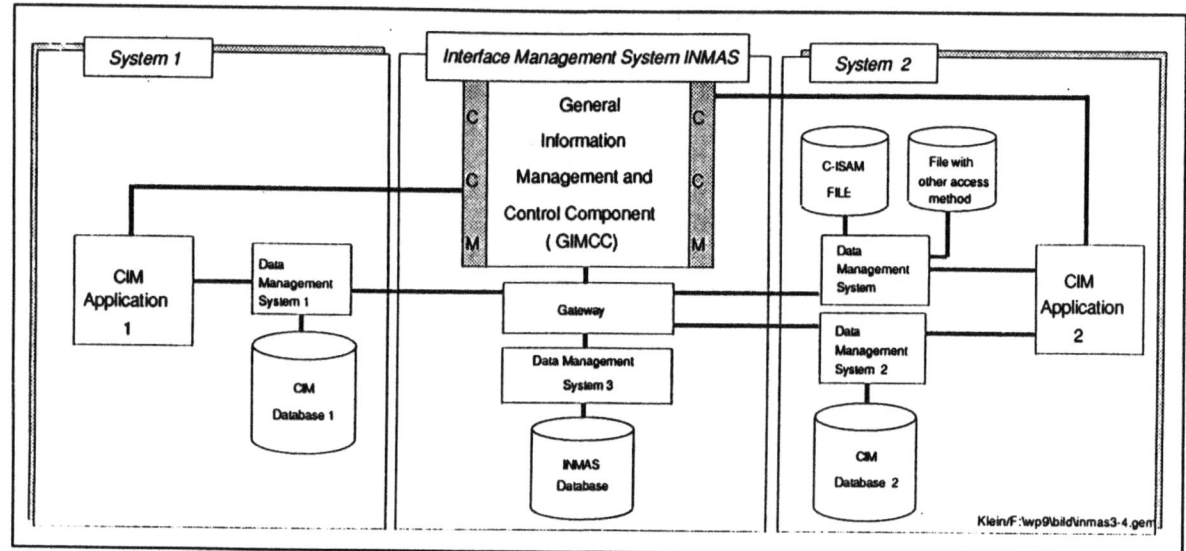

Figure 3: The Interface Management System INMAS

as a reference guide which should be followed for the implementation of new systems and for enhancements of present systems.

All modules developed within CIDAM work together to form the project's CIM systems. Figure 4 shows how the different components interact. Each block shows one of the key modules being developed by CIDAM. The creation of a CIM system is divided in three phases. In the first phase the standard configurable modules are developed. These are the configurable components created by the CIM components (CIMCO) workpackage which are the short term scheduler and shop floor monitoring system, the adapted existing EDP systems (shown as triangles) and the Man Machine Interface represented as flat rectangle (see above). These activities are necessary once to create the projects CIM system but will not have to be duplicated for other companies utilizing the CIDAM modules to install their CIM-system. The second phase comprises the interactive configuration of the CIM system for a specific enterprise. The basic module here is the Configuration System COSY. The designer of a CIM system will interactively be lead to specify the functionality of a system which consecutively will be generated by the Configuration System COSY both as functional definition (source code) and as data structure definition. The third phase shows the CIM system during operation. Here the Interface Management System and the Distributed Database play a central role for the exchange of information between the different systems and for the storage of data in distributed homogenous database management systems.

4. VALIDATING THE APPROACH

Every methodology and tool set developed to implement CIM systems has to be measured for the compliance to new scientific approaches and for practicality against the current situation within the companies.

The conformance to the scientific state of the art in all areas of application is not only ensured by the presence of two academic partners, Trinity College and Institut für Wirtschaftsinformatik (IWi) but also through the creation of a special workpackage, the Academic Scientific Foundation, which manages the information flow between CIDAM and the scientific and technological environment. The Academic Scientific Foundation monitors the developments in the relevant areas of technology and transmits it to the appropriate members of CIDAM. In addition, this group ensures the rapid dissemination of CIDAM results.

The practicality of the CIDAM solutions is ensured by a close involvement of the three testbed partners over all phases of developmentand implementation. The testbed

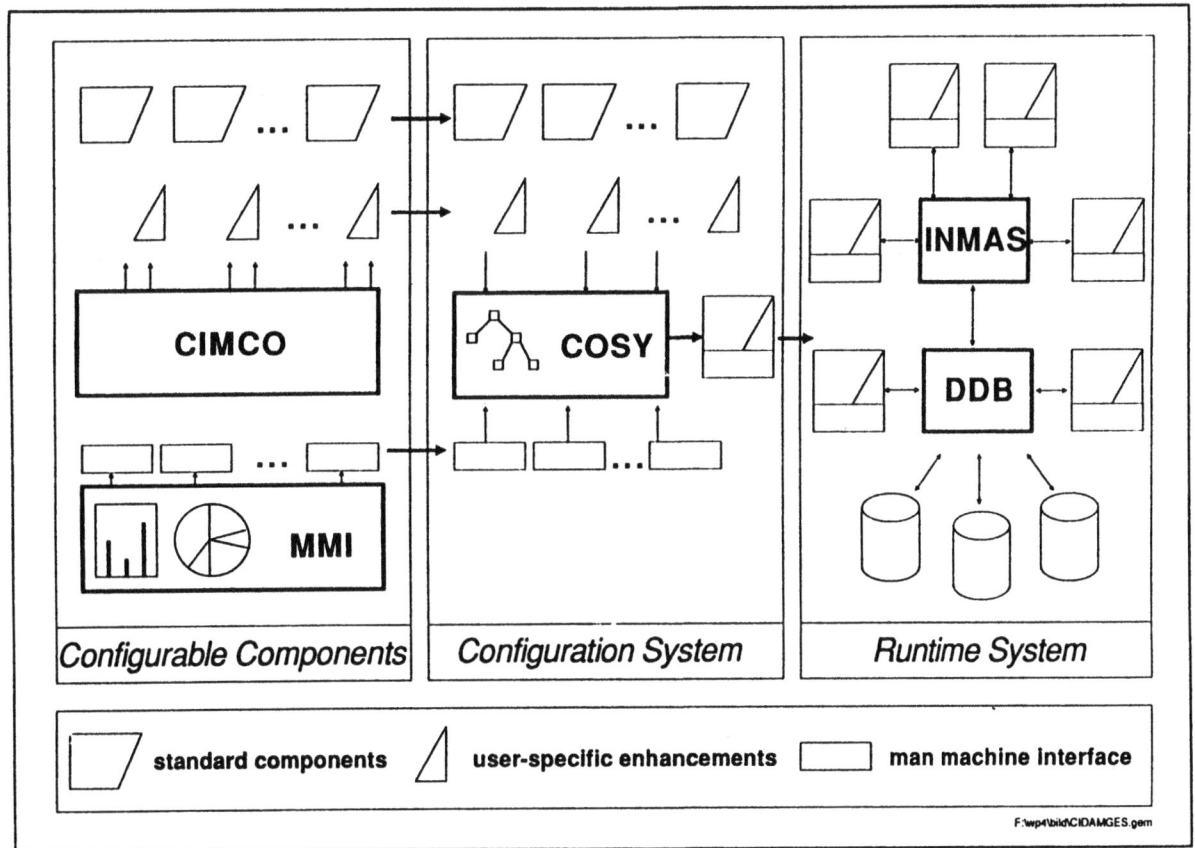

Figure 4: The working of CIDAM

partners represent the three different manufacturing environments discrete parts manufacturing and continuous and discrete process environments. The participation of testbed partners belonging to different industries ensures the independence of the project's developments from the specific circumstances of a single type of industry. In the early phases the testbed partners influence the development of the modules by defining requirements which stem from an analysis of their testbed situation. At all later stages the conceptions for the modules developed within the workpackages are presented to the testbeds for consent. This procedure ensures the practicality of the concepts developed and has given rise to many fruitful discussions between developers and testbeds. The influence of the testbeds is further increased by a conception of early prototyping with stepwise enhancements. Already after the first year of the project the first versions of prototypes are available. In CIDAM terminology they are called "mock-up". Their aim is not as much to provide first functionalities but to show the utilization and working of different functionalities increasing the capability of examining different development options. In a revolving process of subsequent stepwise enhancement then the final prototypes are designed. Thus at all times the interaction with the testbed partners who will apply the prototype is ensured.

Adherence to CIDAM key principles further ensures successful development. Modularity, conformance to standards, instant testbed validation and the use of modern engineering methodologies such as object-oriented design form the essence of the CIDAM philosophy. Conforming to standards, the CIDAM developments take into account the modular approach of CIM-OSA as far as applicable. The Man Machine Interface bases upon X-Windows and closely monitors the Motif Standard. The developments focus on the Unix operating system, the standard language for development being C.

Object-oriented design is tested in several applications for CIDAM modules. The Man Machine Interface evaluates object-oriented methods such as EIFFEL and C++.[14] Some of the CIM components developed in the CIMCO workpackage base upon the object oriented paradigm. Currently, ADA and Smalltalk is in use within the project. The conception of the

Interface Management System also takes object-oriented principles into account. However, the choice of the implementation language will be made carefully, weighing the advantages of OOD methods and conventional methods.

As CIDAM enters the second project year, emphasis is laid on the development of first prototypes focusing on the evaluation of the CIDAM concepts developed.

5. CONCLUSION

CIDAM very much benefits from the close cooperation of international partners - industrial, software-technical and academic - and in addition identifies the cooperation with other ESPRIT CIM projects as vital to the project. This has led CIDAM to initiate the organization of an ESPRIT workshop with the working title "Implementing CIM - Models, Tools and Strategies" which aims to actively involve similar projects thereby establishing an efficient information exchange platform for all participants.

The modular approach of CIDAM directed both at build-time and at run time modules is the essential factor for the implementation of enterprise-specific integrated information systems in industry. This hands-on approach focuses on direct implementation starting out from the present hard- and software environment of the companies and will efficiently support industry in migrating towards integrated systems.

6. ACKNOWLEDGMENTS

The author would like to thank all members of the CIDAM project who have contributed to this paper, in particular R. Herterich, J. Klein, B. Mathews, and D. Morin.

7. THE AUTHOR

Alexander Hars is researcher at the institute for information systems (IWi) in Saarbrücken, Germany. He joined the IWi after graduating from the Universität des Saarlandes in Business Administration and currently is responsible for the IWi activities in the Academic Scientific Foundation of the CIDAM project. His research work focuses on data modelling, scheduling and systems integration.
Alexander Hars can be contacted at the Institut für Wirtschaftsinformatik, Im Stadtwald Geb.14.1, D-6600 Saarbrücken (e-mail: hars@campus.sbuvax.uni-sb.de).

(⌖ LISSAB90.TXT)

[1] Scheer, A.-W.: Computer Integrated Manufacturing - Computer Steered Industry. Berlin-Heidelberg-New York-London-Tokyo, 1988, p.3.

[2] for the categorizing into structural elements see also: CIM-OSA ESPRIT Project AMICE Nr.688: Reference Architecture Specification. Brussels 1989.

[3] Scheer, A.-W.: Computer Integrated Manufacturing - Computer Steered Industry. Berlin-Heidelberg-New York-London-Tokyo, 1988, p.3.

[4] Shevlin, F.: Man Machine Interfaces and Systems for their Management and Development. In: Neelamkavil, F.; Buntzel, W.; Hars, A. (Eds.): CIDAM Quarterly Report August 1989, pp.12-37.

[5] Scheer, A.-W.: Enterprise-Wide Data Modelling - Information Systems in Industry. Berlin et al. 1989.

[6] Chen, P.: The Entity Relationship Model - Toward a unified view of data. In: ACM Transactions on Database Systems, Vol.1 No. 1, March 1976, pp.9-36.

[7] Scheer, A.-W.: Computer Integrated Manufacturing - Computer Steered Industry. Berlin-Heidelberg-New York-London-Tokyo, 1988.

[8] for a description of the project see also: Neelamkavil, F.: CIDAM - CIM System with Distributed Database and Configurable Modules. In: ESPRIT CIM - Results and Progress of Esprit Projects in 1989. 1989, pp.163-173.

[9] for a description of the project see also: Lutz, P.; Beyer, A.: Entwicklung von Integrationswerkzeugen im europäischen Verbund. In: Zeitschrift für wirtschaftliche Fertigung (ZwF), Vol. 84 (1989) No. 9, pp. 483-486.

[10] Scheer, A.-W.: Enterprise-Wide Data Modelling - Information Systems in Industry. Berlin et al. 1989, p.259.

[11] Hertel, U.: Object-oriented realisation of a CIM-system for shop-floor control and production monitoring and short term scheduling. In: Proceedings of the 6th CIM-EUROPE Conference, 1990.

[12] Hars, A.; Scheer, A.-W.: Entwicklungsstand von Leitständen. In: v. Behr, M.; Köhler, C. (Eds.): Werkstattoffene CIM-Konzepte - Alternativen der Werkstattsteuerung und CAD-Vernetzung. To be published by Projektträger Fertigungstechnologie, Karlsruhe 1990.

[13] Scheer, A.-W.; Herterich, R.; Klein, J.: INMAS - Eine individuell konfigurierbare Schnittstelle. In: Information Management 5 (1990) No.1, 1990.

[14] Mathews, B.: Object Oriented Programming Languages for CIDAM. In: Neelamkavil, F. (Ed.) CIDAM Workpackage 9 Quarterly Report May 1989, pp.61-70.

ENABLING TECHNOLOGIES – PART II

Object oriented realization of a CIM-system for shop floor and production monitoring and short term scheduling

Dipl.-Ing. Uwe Hertel
Mannesmann Kienzle Ratingen, West-Germany

Existing MRPII-systems (production plannng and control) support long term scheduling of production orders and theoretical availability checks of the needed resources. production data collection systems are able to collect the real production data like resource availability, setup times, working times, downtimes, etc.). Between the MRPII-systems and the production data collection-systems there exists a lack of functionality such as:

- short term scheduling of orders under the constraints of real availability of resources, based on the long term scheduling of the MRPII

- online shop floor or process monitoring with visualisation of the order flow and resource availability

- optimising of the order flow based on the experience of the past

This paper describes the approach in realizing the required CIM-application, using the object oriented language Smalltalk80.

1. Introduction

Normally in discrete part manufacturing industries or process industries the data flow in MRPII-systems starts with the long-term scheduling of production orders or recipes (figure 1). The basis for this scheduling are the due dates from the customers, the delivery dates of material and the theoretical availability of the resources. For the enforcement of the scheduled orders or resources on the actual shop floor or in the plant the calculated order data have to be changed because,for example

- resources or material are not available
- other orders or recipes have higher priority

Figuaratively speaking,all prescheduled data have to pass a filter and only a few can pass this filter without change . In the shop floor and in the plant production and process data collection systems are installed.

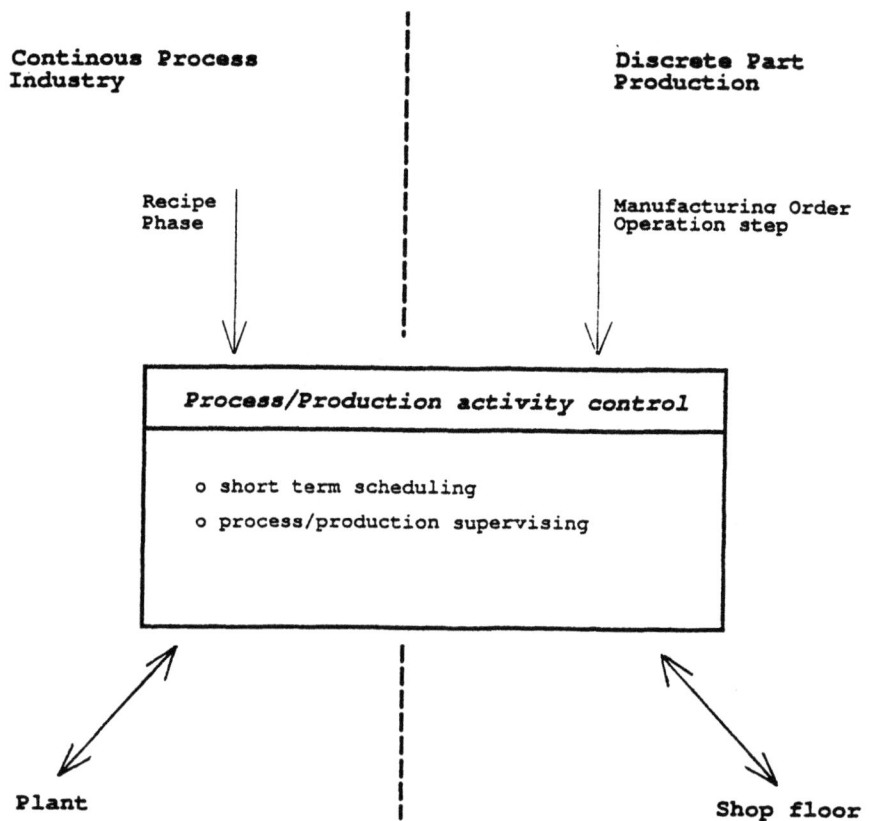

Continous Process   Discrete Part
Industry   Production

Recipe   Manufacturing Order
Phase   Operation step

Process/Production activity control

o short term scheduling

o process/production supervising

Plant   Shop floor

figure 1 : common functionalities

        Data from these systems could be input for the MRPII-
system making the best of the future scheduling. In fact,
inside the MRPII-system there aren't any mechanisms for
processing these data. Using our analogy the production data
also have to pass the filter, losing a lot of data before the
MRPII-system is able to process these data. In order to
remove this filter we need additional functionalities such as:

- short-term scheduling of orders and recipes under the
  constraints of actual availability of resources based on the
  long-term scheduling of the MRPII
- graphical visualisation of the scheduled orders and recipes
  with interactive graphics for manipulation of order flow and
  sequence
- online shop floor or process monitoring with visualisation
  of order flow and resource availability.

The main difference between the requirements of discrete part production and process application requirements are as follows:

- in process applications additional constraints for continous processing must be taken into consideration.
- operation steps in discrete part production and phases in continous process application have different relationships to each other.

Within the Esprit Project 2527 partners from discrete part manufactures and process industry decided to design a common tool that complies with the above described requiremnents. The realisation is done by each partner,based on existing software systems.This paper describes the approach of Mannesmann Kienzle. The project has reached a stage where we are looking for companies for beta - test installations.

2. Our Approach
_____

Our basic idea was to realize a computer internal model of the problem space based on an object oriented approach.Changes in the real word, the status of resources or orders in the shop floor or new orders arriving from an MRPII system, have to update the model. For presenting the changes to the user the changes in the model are updating the views via the MVC - mechanism of Smalltalk80. This approach guarantees that the actual status of the shop floor is always taken into consideration at every user interaction like an information request to the system or the scheduling of orders (figure 2).

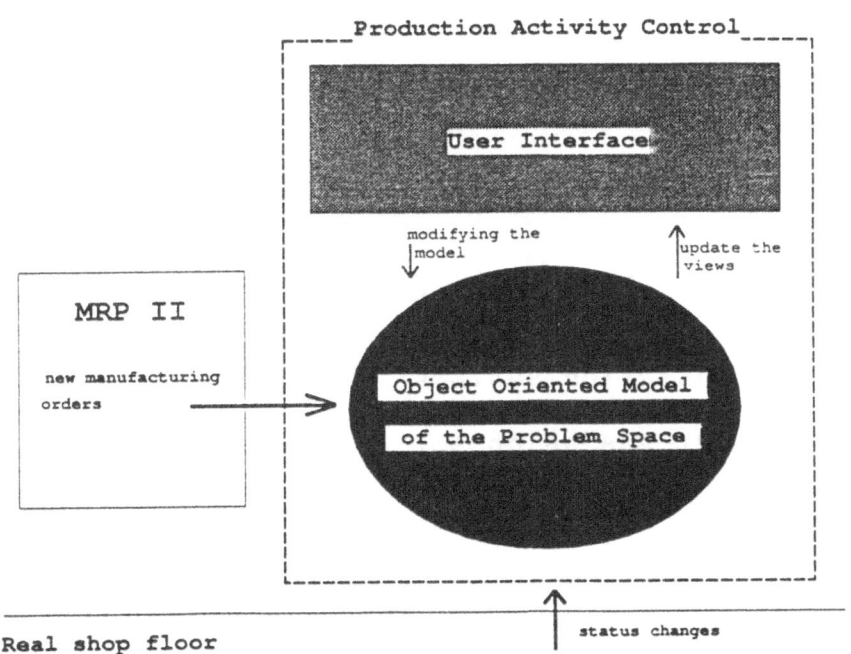

figure 2: Basic concept

2.1 General Requirements

The end user´s reason for using the described missing CIM-application is to minimize production costs. For that reason the system has to support him in optimising his shop floor according to his objectives. The best way to do this is to visualize the current status of the shop floor by an IT-system realized on IT-technologies availible today. That means showing the end user the current status of:

- the main resources
- the staff
- and the order flow.

Based on this information the end user needs tools for comparing the current shop floor situation with the ideal shop floor situation.

For the rescheduling of orders it is necessary to visualize the current schedule, to outline the status of each operation and to give the end user tools for changing the schedule regarding to his current objectives and the status of the shop floor.

It has to be possible to simulate different schedules and to compare different schedules for finding the best solution.

After finishing the optimization of the schedule it has to be possible to guarantee that all necessary resources like tool-machines, staff, fixtures, tools, NC-programms are available befor proceeding with each scheduled order.

2.2 Advantages of the choosen approach

The realization of the above described requirements contains some problems of IT - Technology :

- The system must be highly interactive and suitable for a large amount of data.
- The system needs a lot of simulation functionality.
- The system needs a graphical user interface.
- The system must have flexible configurable interfaces for connecting different MRPII systems and production data collection systems.

Our approach was to use an object oriented language and we chose Smalltalk80 for the realization. One advantage is that the object oriented language Smalltalk80 has some functionalities for solving the described problems of IT-Technology.Additionally , the software development enviroment of Smalltalk80 contains a lot of predefined objects, which are easily adabtable to our problems.

With Smalltalk80 it is possible to realize both the
application and the user interface via an object oriented
approach.The graphical user interface of the realized system
must be able to change the representation of any part of the
application after changing the application part itself.
Moreover, changes in the user interface have to change the
application itself.Both described changes must work under
different constraints.The Model-View-Controller metaphor
realized in Smalltalk80 supports the described communication
between user interface objects and application objects.

The behavior of application processes and user
interface processes requires three different mechanism :

- after starting a new process the currently active process
  has to continue its activities,
- after starting a new process the currently active process
  has to wait until the new process is finished,
- after starting a new process the currently active process
  has to stop.

The Smalltalk80 internal object Process Scheduler supports all
the above described functionalities.

Smalltalk80 allows the compilation and linking of
source code modules at runtime.Based on this functionality we
were able to formulate rules for modifying the scheduling
algorithm and generate interfaces to other CIM-applications.

Smalltalk80 contains a lot of predefined objects which
support the simulation of dynamical systems.
Last Smalltalk80 is fully object oriented.The advantages of
object oriented programming,i.e.

- reusability
- extensibility
- code sharing

are thus fully supported by Smalltalk80.

3. The Realized System
_____

The realized system consists of four major blocks which were
developed independently (figure 3).

- The model of the problem space containing a shop
  floor model and models for all products
  manufactured on the shop floor.
- The monitoring system allows the communication with
  other CIM - Systems.Based on the information flow
  from and to other CIM - System the monitoring
  system is responsible for updating the shop floor
  model and the product models.
- The scheduling system supports the user in
  scheduling the future situation of the shop floor.
- A set of views builds the graphical user interface
  to the System.

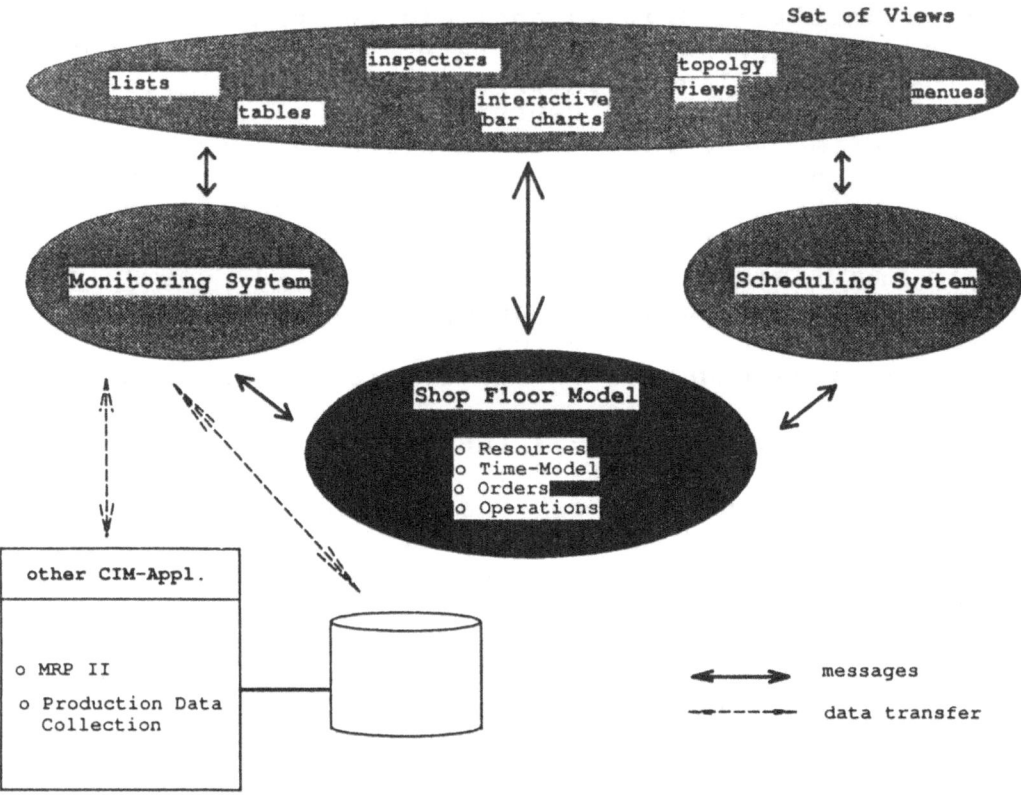

figure 3 : basic architecture

3.1 The Model

3.1.1 The shop floor model

For the shop floor model we implemented the objects´ resources and time-model as part of a class hierarchy.

The objects from class resources and their subclasses like machines, personnel,tools etc. are necessary for modelling the physical topology of the shop floor. Resources represent the master data of the system. The monitoring system is able to change the status of resources by sending messages to the instances of the different subclasses. Resources know their current shift model,a part of the later described time-model,and their current and future availibility. This information is implemented by additional objects like operating time counter,maintenance intervals,status representations (break down,running,...),etc.

A very important part of the model are the objects of the time-model.Time-model comprises all those components which are necessary to depict and represent time interdependencies in production scheduling and monitoring.The time-model is divided into low level objects like time coordinate,duration and interval and high level objects like operating calendar and shift model. The low level objects are used for modelling the basic functionalities of time.Operation steps for example need different intervals for representing the duration of the different activities.The scheduling system needs the time coordinate for fixing the due dates in the system.

The high level objects of time-model represent the available intervals for scheduling operation steps or monitoring the shop floor. An operating calendar, like a normal calendar, consists of days, weeks etc. (the smallest unit in the operating calendar is a day). However, it takes account of criteria specific to operational requirements, e.g. the operating calendar contains only those days of the calendar on which work is performed. The operating calendar forms the basis for any shift model.The shift model provides information relating to a resource's (machine, staff) anticipated availability at a specific time.

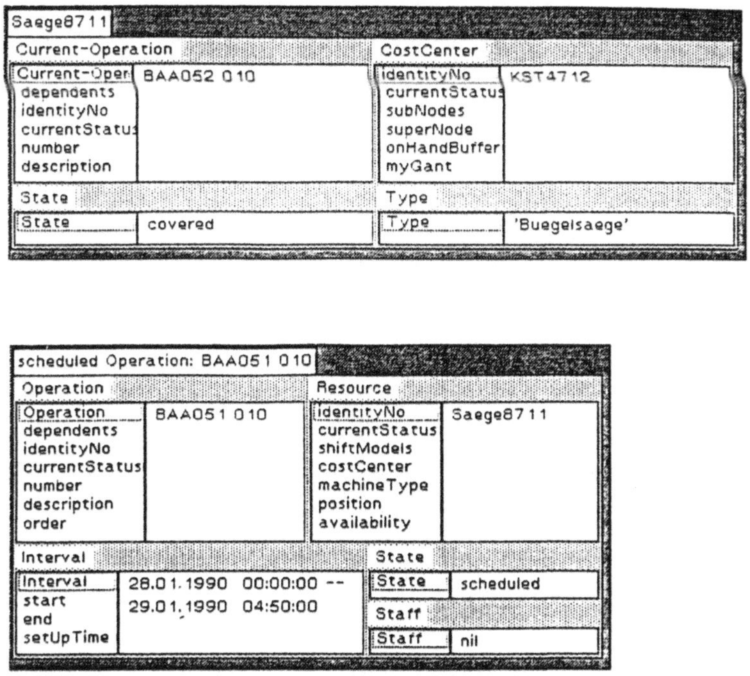

figure 4: Inspector views for model objects

3.1.2 The Product Model

The product model is modeled with the objects order,operation and NC-program. An order is a pre-scheduled sequence of operations. Order data are generated from the master data in the MRPII-system for the parts and the operating sheets. An operation is indivisible within the short term schedule and unambiguously allocated to an order. It describes a self-contained process sequence applied to a process component or, alternatively, a self-contained production sequence applied to a production component. Prior to execution each operation has to be explicitly enabled by the short term schedule. To do this the short term schedule has to check that the resources required for the next operation are available.

Instances of the classes operation and order are generated from the monitoring system based on the information flow from the MRPII-system.During the different scheduling manipulations operations and orders guarantee the consistency with the redundant information in the MRPII system.

Instances of class NC-program are necessary to check the availility and to transmit the NC-programs to the DNC system.

305

The task of the Monitoring System is to recognize changes in other CIM-systems and to update the model with regard to these changes. The Monitoring System is doing this by using a set of configurable interfaces
These interfaces supervise the sources of information for the model. The sources could be other CIM applications, databases or file-systems. For example: A production data collection system gets an event that a special operation in the shop floor has finished. This event is recognized by the interface, configured for this production data system. The interface sends this event to a queue and continues its job of supervising.This queue is, in turn, supervised by another object called "supervisor". The supervisor interprets the event, identifies the receiver in the model and generates a message to the model object. The model object performs this message in the example an operation would change its status (from running to finishing). Via the Model-View-Controller mechanism of Smalltalk80 all dependencies (that means: all views connected to this operation) also change their status (i.e. a bar representing this operation would change its colour).

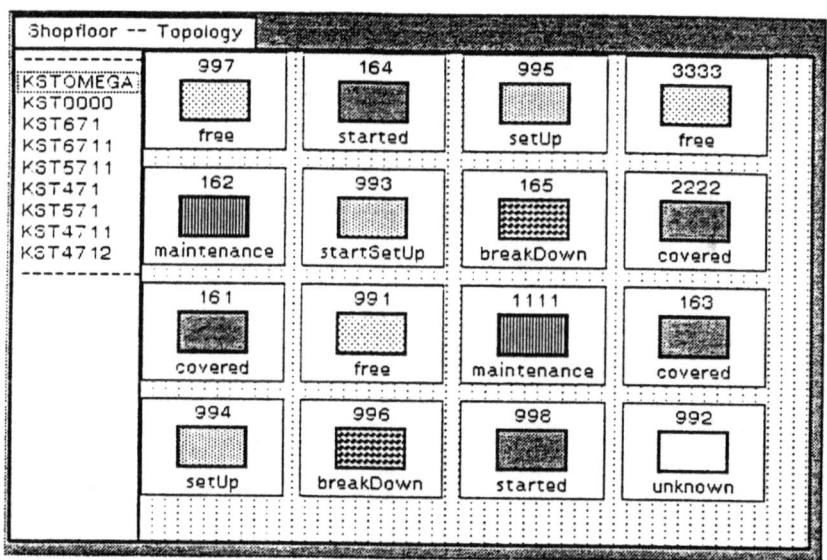

figure 5: Shop floor topology

The Monitoring System is build with the class External Interfaces with different subclasses,Supervisor and Alarm Handler.

For each source of information a subclass of External Interfaces is generated. The instances of these subclasses recognize changes in the different sources,interpret these changes,find out the receiver in the model and generates a message for updating the receiver.This message is sent to the supervisor who collects all incoming messages in a queue.

The supervisor is an asyncronously running process, which is triggered by the incoming messages. The messages are distributed and the model objects can perform different necessary actions. Additionally urgent mesages are send to the Alarm Handler who is updating a window for displaying the messages on the screen.

3.3 The Scheduling System

The task of the Scheduling System is to schedule the future order flow on the shop floor. This is done partly by using automatic scheduling algorithms and partly by user interaction using different views. During the interactive schedule the system has to check the consistency of the product model and the feasability of the schedule. Additionally the Scheduling System allows the creation of different schedules and the comparing of the results.

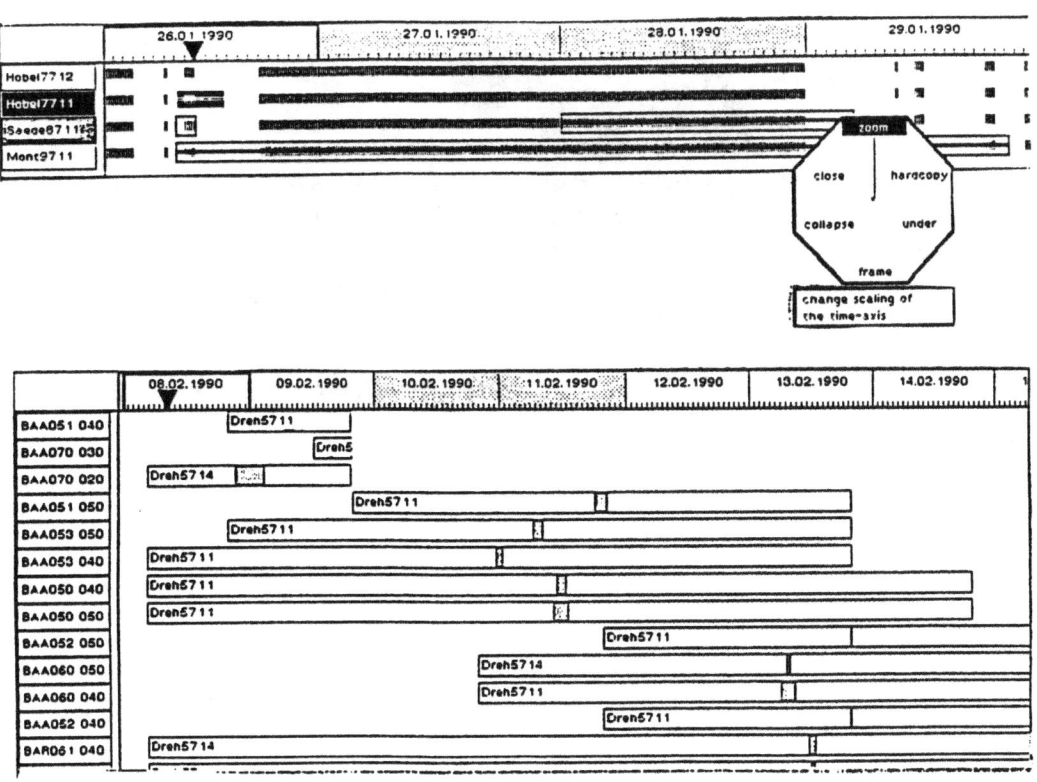

figure 6: Interactive bar chart for scheduling

It was our goal to have a mechanism for the flexible incoporation of different scheduling algorithms in our model. For that reason we first realized a framework into which we could put any scheduling algorithm. It is our intention to use knowledge based tools for scheduling in the future. Today the sytem uses a rule base without back-tracking, which can be manipulated at runtime. The framework has to be able to handle one special schedule and also different simulations for optimising the scheduling. This framework must also connect the Scheduling System with the model and the views.

In order to build this framework we implemented a class called schedule. Instances of schedule represents schedules for simulation as many as you like.Each instance of class schedule is a container for storing the load schedule of every machine on the shop floor.

Instances from class load schedule represent a list of free and covered intervals along the time axis.

To incorporate different scheduling strategies and to adapt the system to user specific requirements we implemented a rule base with rules for scheduling.The rules are formulated in the Smalltalk80 language.That means that they use the Smalltalk80 syntax and a subset of the existing Smalltalk80 vocabulary.At runtime existing rules could be modified and new rules could be added to the system.The rules mainly describe necessary actions if the user creates inconsistencies in the system,how to handle the dependencies between different operations and orders and which changes in the pre-scheduled data from the MRPII are allowed.

3.4 The Views

With the above described objects we realized a computer internal representation of the current situation on a shop floor and tools for manipulating these representations. One tool,the monitoring system, supports model manipulation based on the information flow from other CIM-application. The second tool, the scheduling system supports user manipulation.

The problem in realizing the user interface for our model is that we only have two dimensions on the screen.However, the model has more than two dimenensions and a lot of dependencies between the dimensions. Additionally, the model consists of a large amount of data (normally more than a thousand operations currently active in the system). For example, the user is interested in the currently existing network of manufacturing orders ,the relation to the necessary machines and personnel and the interdependencies between the orders. With one single view it is imossible to show all the required information.

Our solution contains a flexible configurable set of views which show different aspects of the model.The user can choose from this set the views he uses for seeing the aspect of the model and for supporting the manipulations of the model he currently wishes.

At this stage, we have realized different inspector views for model objects (figure 4) a topology view for resources or orders for the Monitoring System (figure 5) and different interactive bar charts for the Scheduling System (figure 6).

4. Conclusion

Based on an object-oriented approach we realized a highly interactive system. The use of Smalltalk80 has allowed the realization of a

- CIM-system with a graphical user interface
- CIM-system with flexible configurable modules and interfaces

which is able to be integrated in the data flow between the long term scheduling and the shop floor.

Additional experiences we gained are:

- An accelaration in realizing software

- Smalltalk80 allows the interactive design of software-systems.

  That means: The realization starts with a frame of needed functionalities. The first solution for a special component is developed after a meeting with the end user specifying his requirements. The development team realizes the requirements and presents the results to the end user in a next meeting. Now some corrections could be done. After a few meetings between the end user and the development team the final version of a special component is finished.

References

Adele Goldberg "Smalltalk-80 Programming language" Addisson Wesly Publishing Company 1984

Glenn E.Krasner,Stephan T.Pope "A Cookbook for using the Model View-Controller User Interface Paradigm in Smalltalk-80" ParcPlace Systems.Palo Alto

Mujin Kang " Entwicklung eines Werkstattsteuerungssystem mit simultaner Termin- und Kapazitätsplanung" Carl Hanser Verlag 1987

CIDAM-D1.1.1-H&B-Fure#1-Nov89 "Specification of Process Application requirements"

CIDAM-D1.1.3-H&B-FIAT-MDF-MDV-#1-Feb90 Requirements for short term scheduling and shop floor supervising useful for part production and process application"

ON-LINE HIERARCHICAL MODEL SIMULATION FOR PROCESS FAULT DETECTION AND LOCALISATION, AN OBJECT ORIENTED APPROACH.

A.J.Billington, A.R Boucher:
Control Systems Centre Sunderland Polytechnic England.

INTRODUCTION

The requirement for safe operation of complex process plant has led to the development of a wide range of monitoring hardware and software dedicated to the task. Earlier systems were implemented using limit checking sensors and were usually autonomous from control related instrumentation. The use of digital computers as control devices enabled these simple monitoring strategies to be embodied within the main controller itself using data gathered for continuous control purposes. As the power and compactness of modern computers has increased much higher levels of automatic plant supervision have become possible due to the resulting development new of fault detection methods. The methods currently in use can be broadly divided into "rule based" and "model based" techniques.

The "rule based" techniques use a data base of plant operating conditions, faults and symptoms either pre-computed or built up from historical data. The monitored plant parameters are input into the system and decisions about plant health made, based on the data contained in the rule base. The drawback with these methods lies in the likelihood that the database is incomplete or inaccurate which may lead to inaccurate or incorrect detection and diagnosis of faults.

The "model based" techniques use a dynamic model, or models, of the system being monitored which are run in real time in parallel with the actual plant and are driven from measured plant data. In some cases models of specific fault conditions are defined and statistical algorithms used to compare the measured outputs with the estimated fault output. In other cases models are used to generate unobservable plant parameters which are subsequently used in fault detection. In either case accurate plant models are required if the systems are to operate correctly. The method to be described falls into the latter "model based" category and as such requires the definition of accurate plant models.

Currently a large proportion (approximately 70 - 80%) of the effort and hence cost in the design of industrial control and monitoring systems is attributable to the modelling phase. In general two approaches to the modelling task are available, these being the physical modelling techniques and the system identification methods. In this instance a system identification technique was chosen at it provides the best route for automatic model generation.

HIERARCHICAL MODEL GENERATION.

Complex process plant can be considered to be made up of a number of sub processes which may in turn be sub-divided into simpler processes as suggested by Gawthrop and Leary(1987), Guariso(1989) and others.

Figure 1 shows a possible structure for simple two tank system.

Hierarchical model generation is concerned with the semi-automatic construction of model of the individual sub processes. This is achieved by capturing the dynamic behaviour of each sub process in response to changes in their inputs. Statistical algorithms are then used to rapidly estimate an accurate mathematical model of the process under scrutiny. The individual models can then be built into a hierarchical structure which mirrors their operation in the actual plant.

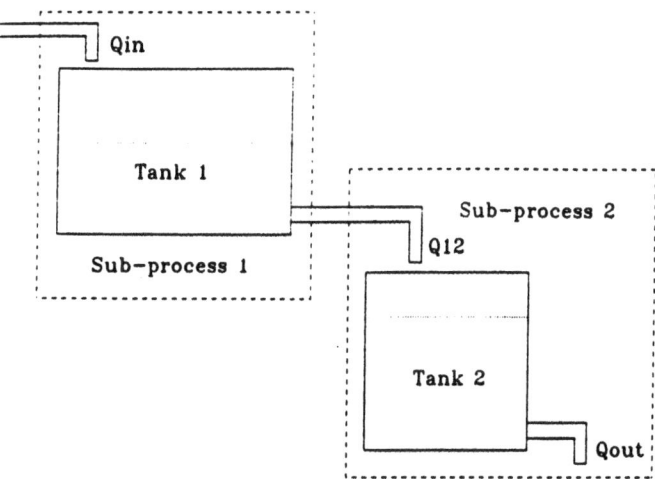

Figure 1. Two tank system

Process Models.

The sub process models generated by estimation are causal discrete models of the form shown in equation 1 where the z^{-1} operator represents a backward time shift of one sampling interval. The parameters $a_1..a_p$ and $b_d..b_q$ are scalars for single input single output systems. This form of model is capable of predicting the output of a sub process at instants in time, separated by the sampling interval from a sequence of the same form.

$$G_p(z^{-1}) = \frac{b_d z^{-d} + .. + b_q z^{-q}}{1 + a_1 z^{-1} + .. + a_p z^{-p}} \quad [1]$$

The subscript "d" represents the integer time delay of the process measured in sampling intervals, and is the prediction factor of the model.

Figure 2 shows how a process with sub processes can be hierarchically modelled and introduces two the types of model needed, the "simple process model" and the "composite process model". The simple process model has no sub processes and is used to represent the lowest level in the hierarchy.

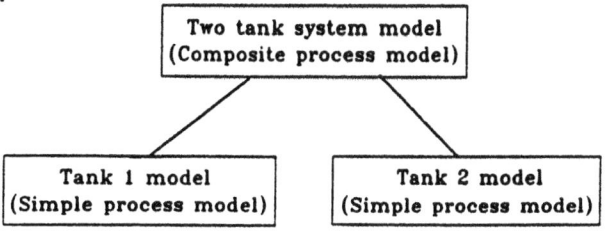

Figure 2. Hierarchical system model

The simple process model must be capable of predicting an output sequence from a sequence of inputs and also estimating the input sequence from a sequence of outputs. This latter anti-causal behaviour is derived by inversion of the causal model.

The Composite process is used to build the hierarchical structure in the system and is defined, not in terms of dynamic behaviour, but from the set of sub processes which it represents. It should be noted that a composite process may contain further composite processes in its make up, hence the depth of hierarchy is not artificially restricted.

FAULT DETECTION USING ON-LINE HIERARCHICAL SIMULATION.

Fault detection is carried out using the hierarchical model of the total process by driving it from the measured input and output of the actual plant as shown in figure 3. Measured data from the input (uk) of the plant is propagated forwards through the models while measured output plant data (yk) is propagated backwards through the models as shown in figure 4. By differencing the forward and reverse propagated data at each sub process model, after imposing a suitable time shift, a set of residuals (rk) are generated. These residuals are used to determine the presence of a fault in the following manner.

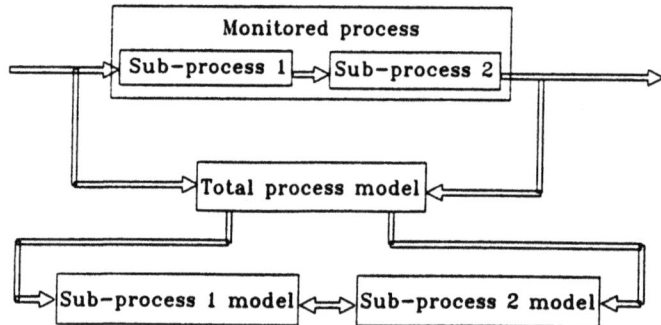

Figure 3. On-line hierarchical simulation

Under no fault conditions the process behaviour and model behaviour will be nearly identical and hence the residuals generated will be near to zero. If however a fault occurs at some point within the process the output will be altered in some way. Since the measured input to the plant will not be affected by the fault all predictions generated by propagating the measure input in the forward direction will

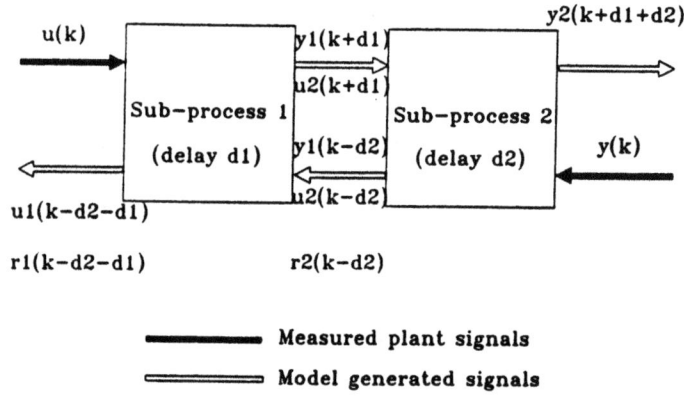

Figure 4. Signal propagation through model

remain unchanged and the final model output will be a prediction of the expected behaviour with no fault. However when the measured output data which contains the modified behaviour is propagated backwards through the plant model the estimated inputs will differ from those predicted by forward propagation and the inter-model residual signals will become non-zero indicating the presence of a fault.

It is clear that for a model with many sub-models the computational effort in propagating the forward and reverse data through the hierarchy will be large. This problem may be overcome by summarising the behaviour of the substructure at each Composite process model. Hence when a Composite process model is defined an overall dynamic model of its substructure must be automatically generated. A Composite process model can therefore propagate signals either using the overall dynamic model of its substructure or by using the substructure itself.

During initial fault detection it is intended that the overall plant model is used generating a single residual sequence. If a fault is detected by the presence of a non-zero residual the fault may be localised in the following manner. The immediate substructure of the total process model is driven from a sequence of historical plant data up to the time the fault was detected. This will generate a number of inter-model residual signals which are used to isolate the fault to a particular model, how this is achieved will be demonstrated in the subsequent sections. If the fault is isolated to a Composite process model then the process is repeated moving down through the hierarchy until the fault has been localised to a Simple process model. Hence the maximum accuracy with which a fault can be localised is determined by how the process has been subdivided.

OBJECT ORIENTED SOFTWARE DEVELOPMENT.

In this section some of the important implementation aspects of the fault detection method described will be highlighted, and the advantages of the Object Oriented approach will be shown. The Language chosen for the development was a PC implementation of Smalltalk. The language is described in detail by Goldberg(1983). Software development within an Object Oriented environment is principally concerned with the definition of object classes and their behaviour. The standard Smalltalk system contains a large number of useful object classes, with predefined behaviour, arranged into a class hierarchy. At the top of the hierarchy is the class "Object" which provides the behaviour required for objects to be created, respond to messages, trap errors and provide information about themselves. Classes can be thought of as factories for the production of usable objects in the system. The class contains the "blueprint" of the required object and specifies the variables contained by it, and the procedures or "methods" necessary for the performance of its required behaviour. All variables possessed by an object are private and hence data encapsulation is a natural feature of the system. It is worth noting that the variables possessed by an object are themselves objects as is everything in the Smalltalk environment.

The definition of new object classes involves the creation of sub classes in the system. The software developer must decide which of the classes already in existence is nearest to the required one, a sub class is then created and given a name. The new class will now inherit all the internal variables and methods of its super class. Further internal variables can now be added and the available methods may be extended to either include new methods or to re-implement some of the super class methods.

The Smalltalk system provides extensible windowing objects and applications are usually designed to make use of the built in features for application control via a flexible menuing system. This enables the rapid development of a highly graphical user interface for applications programs.

Development of the System Building Blocks.

In this section the development of some the objects needed in the application will be shown. Reading through the description of the proposed system it is clear that the following types of object were needed: -

(i)   Interface objects for the output and input of analogue data via suitable hardware.
(ii)   Data processing objects for the generation of mathematical process models
(iii) Data display objects for the display of both stored and real

time data.

(iv) Process model objects to allow Simple and Composite models to be defined and operated.

) Controlling application windows for System Identification, and Online Hierarchical Simulation.

Some of the less obvious objects required were those for implementing matrices, polynomials and their associated operations.

Interfacing with the Real World. The standard Smalltalk system contains a group of classes known as "streams" which can be thought of as very flexible files, though only one type of "stream" actually involves the use of secondary storage media. The stream classes use a common protocol for the storage and retrieval of object instances. In a similar manner the output of analogue signals can be considered to be like sending numbers representing voltages to a write only stream while retrieval of analogue signals can be thought of as a read only stream of numbers. Hence to provide the ability to use analogue input and output with the stream protocol two new sub-classes of the class "Object" were defined, these were named "ADCStream" and "DACStream".

The interface hardware used provided 16 channels at 12 bit resolution for analogue input, 2 channels at 12 bit and 2 channels at 8 bit resolution for analogue output. To illustrate how a a new object is defined the definition of the "ADCStream" class is explained.

The "ADCStream" is to be created for use with a particular analogue channel. Channels are selected with the use of an analogue multiplexer on the interface card and hence an instance variable was needed to hold the channel selection address for multiplexer control, this instance variable was named "channelValue". Armed with only this information a sub class of Object was defined as shown in listing 1.

In order to provide the required behaviour for the "ADCStream" class, procedures or methods were defined for the new subclass. Methods fall into two categories "class methods" and "instance methods".

```
Object subclass: #ADCStream
    instanceVariableNames: 'channelValue '
    classVariableNames: ''
    poolDictionaries: ''
```

Listing 1. ADCStream class definition

A class method defines behaviour related to the object class itself and is usually concerned with the creation of instances of the class while instance methods define the behaviour of object instances   It is only necessary to define class methods for a new class if initialization of the internal instance variables is required during object creation. If this is not necessary then the inherited class methods may be used.

Methods are provided with a "selector" rather like a Function or Procedure name in a procedural language. The method selector may optionally have arguments in much the same way as procedures have parameters. In this case a single class method was defined for the creation of an ADCStream for a selected channel. Methods are invoked in Smalltalk by sending messages to objects rather than by procedure calls, this provides very elegant source code and influences the choice of method selector; listing 2 shows the class method defined for the creation and initialization of an ADCStream instance for a particular channel.

```
onChannel: anInteger
  "answer an ADCStream for channel anInteger or report an error"
  (anInteger class == SmallInteger and: [ anInteger between: 0 and: 15])
  ifFalse: [ ^self error: 'only integer channels 0-15 avialable' ]
  ifTrue: [ ^(super new) initialize: anInteger ]
```

Listing 2. ADCStream class method

The first line of the method is the selector which has one argument
"anInteger" inferring that the argument should be an integer, however
this is not checked during compilation, indeed their is no concept of
type in the Smalltalk language. The second line enclosed in double
quotation marks is a comment. The third line checks at run time if the
argument belongs to the class SmallInteger and also that its range
lies between 0 and 15, if this test fails the inherited super class
method "error:" is invoked with a suitable error message as its
argument. This will cause a "walkback" window to be displayed and
commence a debugging session. If the argument "anInteger" meets the
run time requirements then the super class method "new" is invoked to
create an instance of class ADCStream. The new ADCStream instance
must then be initialized and this is to be done by the instance method
"initialize:" with the argument "anInteger". Note this code can be
compiled into the system without the initialization method having been
defined.

The "initialize:" instance method must carry out two tasks, firstly
the interface card hardware must be prepared for use. This is done by
programming its on board PIA, and secondly the instance variable
"channelValue" must be initialized to ensure that subsequent A to D
conversions are carried out from the required analogue channel. One
more instance method "next" is required to allow the channel voltage
to be returned. As stated above the choice of "next" as the message
selector allows an ADCStream object to be used in place of a stream.
Listing 3 shows the two instance methods required to complete the
ADCStream definition.

```
initialize: aChannel
    "PIA 1 ports A & B inputs port C all output"
    (Dos new) outByte: 2r10010010 toPort: 16r703.
    "initialize channelValue"
    channelValue := (aChannel bitShift: 4) + 2!


next
    "Read the analogue voltage and return
    voltage from -10 to 10 volts"
    ¦ dos ¦
    "create a Dos object for sebsequent I/O operations"
    dos := Dos new.
    "select analogue channel and start conversion"
    dos outByte: channelValue toPort: 16r702;
        outByte: (channelValue + 1) toPort: 16r702.
    "read ADC result and convert to a %"
    ^((((dos inByteFrom: 16r701)
                bitAnd: 16r000F)
                bitShift: 8)
            + (dos inByteFrom: 16r700)) - 2048 / 204.8
```

Listing 3. ADCStream instance methods

In a similar manner the DACStream class was defined with

315
```
```

appropriate methods to allow the output of voltages on a selected channel. Listing 4 shows how an ADCStream and DACStream may be created and subsequently used.

```
| adc dac voltage | "define object identifiers"
adc := ADCStream onChannel:1. "create an ADCStream for channel 1"
dac := DACStream onChannel:3. "create a DACStream for channel 3"
voltage := adc next. "read voltage from channel 1"
dac nextPut:voltage / 2. "output half measured voltage to channel 3"
```

Listing 4.   Creation and use of ADC and DAC streams

Data Processing Objects.   The rapid generation of accurate dynamic models of process behaviour is essential to allow a model based fault detection scheme to be easy to configure and reliable in use.   This requirement was fulfilled with the use of least squares recursive estimation techniques.   However these algorithms require matrix operations to be carried out and hence the definition of a new class "Matrix" was undertaken to allow matrix operation to be implicitly performed.   The important matrix operations required were creation, definition, addition, subtraction, multiplication, inversion and the calculation of norms.   The standard Smalltalk environment contains a set of number classes with methods to allow scalar operations, these were also modified to allow valid mixed scalar and matrix operations to be performed.

Recursive estimation techniques based on least squares cost functions provide a simple yet powerful tool for the identification of system parameters. The general form of the recursive estimation alogrithm using a modified notation due to Young(1984) is shown in eqns. 2.

$$\underline{k}_k^T = (\underline{x}_k^T \cdot P_{k-1} \cdot \underline{z}_k + \alpha)^{-1} \cdot \underline{x}_k^T \cdot P_k$$

$$P_k]S (P_{k-1} - (P_{k-1} \cdot \underline{z}_k \cdot k_k^T)) / \alpha$$

$$\underline{a}_k^T = \underline{a}_{k-1}^T - ((\underline{a}_k^T \cdot \underline{z}_k - y_k) \cdot \underline{k}_k^T)$$

[2]

where $\underline{z}_k$ is the vector of measured states.
$\underline{x}_k$ is the vector of estimated states used in instrumental variable estimation. *Note if $z_k = x_k$ the estimation algorithm degrades to the basic recursive least squares form.*
P_k is the covariance matrix.
$\underline{k}_k$ is the Kalman gain vector.
α_k is the forgetting factor.
y_k is the measure system output.
$\underline{a}_k$ is the vector of parameter estimates.

With the capability for implicit matrix mathematics it becomes a trivial problem to implement recursive estimation algorithms.   Though there are a wide range of model types the same basic algorithm can be applied to all.   For this reason the common features of all least squares estimation algorithms were built into an abstract super class RecursiveEstimator.   Abstract classes are a useful way of defining the common structure and behaviour of a family of object classes.   The abstract class itself is not intended for the production of objects as in themselves they will serve no useful purpose.   Rather the intention is that further sub classes will inherit the features and add functionality to provide useful services.   In this manner the general form of the estimation algorithm was implemented in this super class using the Smalltalk code shown in listing 5, which clearly demonstrates the elegance of implicit matrix mathematics. The variable "initCount" is used to unsure that the data vector is filled before the estimation is allowed to commence.

```
iterate
    "Compute the next values of the parameter estimates.
     Note initCount is to ensure the date vector is filled
     before the first estimate is computed"

    | kkT pkzk |
    initCount == 0
    ifTrue: [
        pkzk := pk * zk.
        kkT := (xkT * pkzk + alpha) inverse * xkT * pk.
        pk := (pk - (pkzk * kkT)) / alpha.
        ek := akT * zk - yk.
        akT := akT - (ek * kkT)
    ]
    ifFalse: [ initCount := initCount - 1 ]
```

Listing 5. General recursive estimation step

The basic recursive estimation algorithm shown can be used to carry out parameter estimation of a wide range of model types the only difference between the various methods lies in the control of the data vectors x and z and the forgetting factor α. Specifically sub classes of the x_k RecursiveEstimator were defined to provide the following types of estimator:

Auto Regressive	AR
Auto Regressive Moving Average	ARMA
Integrated Auto Regressive	IAR
Auto Regressive eXogenous	ARX
Auto Regressive Moving Average eXogenous	ARMAX
eXogenous	X
Moving Average eXogenous	MAX
Integrated eXogenous	IX
Bilinear Auto Regressive eXogenous	BARX
Instrumental Variable ARX	IVARX
Instrumental Variable BARX	IVBARX
On line instrumental variable	OnLineIVARX

The recursive estimation algorithm shown will produce a vector of parameters estimates which contains no information about the model structure; the number of numerator terms q, denominator terms p, or noise terms. To enable model specifications to be efficiently stored and passed around the system the new class ModelSpec was defined to contain the parameters of the model and information about its structure. Additional methods were defined to provide stability information, steady state gain and dc offset which must be included to completely characterise the systems static and dynamic characteristics.

Graphical Data Display. In order to provide easy to use facilities for graphical data display two types of display object were defined. PenRecorder objects provide real time chart recorder type scrolling displays while Graph objects produce more detailed displays of multiple data sets in a number of plot styles with automatic scaling . Examples of the displays produced by these objects appear in screen dumps from the model identification, fault detection applications.

Simple and Composite Process Models. An object sub class ProcessModel was defined to implement the behaviour of the Simple process described above. These were made up of a number of simpler objects as shown in figure 5 and provided the following behaviour.

(i) Forward and reverse propagation of signals.

(ii) Generation of residuals.

(iii) Maintaining a history of past signal propagation.

(iv) Displaying real time chart recorder type displays of forward, reverse and residual signals.

(v) Replaying history data.

(vi) Automatical alarm indication on forward and residual data.

To provide the behaviour of the Composite process a sub class (CompositeProcess) of ProcessModel was defined with additional instance

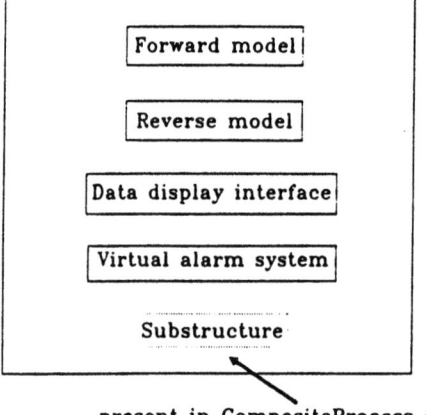

present in CompositeProcess only

Figure 5. The ProcessModel object

variables to provide a substructure and the ability to operate from either the overall model or the substructure itself. By incorporating the above behaviour in the process model implementations the complexity of the controlling applications could be reduced to a minimum.

THE MODEL IDENTIFICATION APPLICATION WINDOW.

Using the recursive estimator classes and the built in windowing objects an application was developed to provide on line dynamic test facilities for step, PRBS (Pseudo Random Binary Sequence) and random multi-level excitation sequences. Once the dynamic behaviour of a process has been captured a discrete time model can be automatically generated by selecting the "identify model" menu option. When initiated the application first determines the best delay d for the model and then searches for the best structure to ensure that the model produced will have as few parameters as possible while maintaining an accurate model of the dynamic characteristics. Figure 6 shows the search pattern used by the identification algorithm and

```
Searching for best delay
with p=2 q=2 d=1 YIC=-10.381508
with p=2 q=3 d=2 YIC=80.104075
Searching for best structure.
with p=1 q=1 d=1 YIC=-15.008091
with p=1 q=2 d=1 YIC=-10.199993
with p=2 q=1 d=1 YIC=-10.087577
with p=2 q=3 d=1 YIC=-8.3408153
correcting model dc offset.
best model is:-
            +0.1751z^-1
Gp(z^-1) = ---------------
            1-0.8138z^-1
T = 1
```

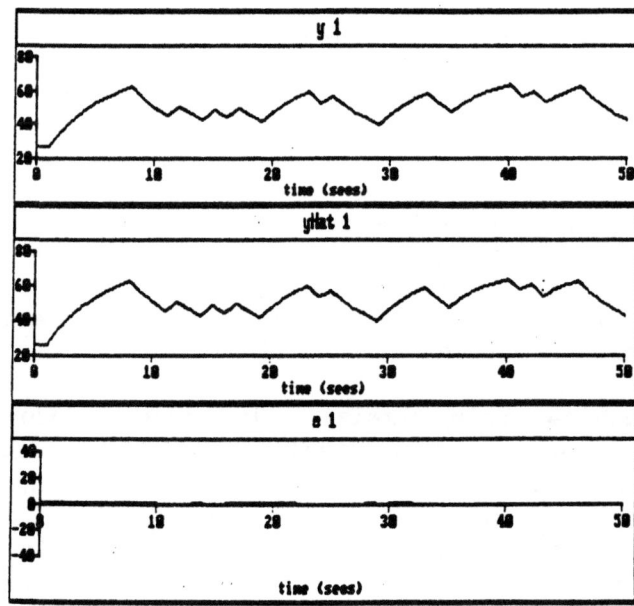

Figure 6.   Model Identification.

the final model produced from a data set captured during a PRBS test on tank 1. The three plots show the measured (y1) and model generated (yHatk) output and the model error (e1).

The procedure of dynamic testing and identification was carried out on tank 1 and tank 2 and the models stored in the model dictionary. Finally a Composite model of the total two tank system was defined using the "definemodel" option by simply naming the processes of which it is made, this composite process was stored as "twotank".

THE PROCESS MONITOR APPLICATION WINDOW.

A ProcessMonitor window is opened for the "twotank" process with the message "ProcessMonitor openOn:"twotank"". The user is then be prompted for the analogue channels to be used for measurement of the process input and output, after which the window will be displayed on the screen. The window consists of three sub panes, the top left hand pane is a ListPane which will display the process "twotank" and all subprocesses in its hierarchical substructure, in this case only two sub-processes exist, "tank1" and "tank2". Clicking the left mouse button on a process name will cause it to print information about itself in the top right hand TextPaneand additionally cause chart recorders to be displayed in the lower graphical display area.

Pop up menus provide options to set the alarm levels for predicted outputs and residual signals in the individual model virtual alarm systems. Monitoring of the process "twotank" is begun by selecting the "monitor process" option. When initiated five chart recorders will be displayed in the graph pane. The left hand recorder displays the measured input signal to the plant, the right hand recorder displays the measured output data from the plant while in the center of the display three recorders arranged vertically display the following signals for the selected process. The top recorder displays the model output due to forward propagation, then centre recorder displays the estimated input due the reverse propagation while the bottom recorder displays the residual generated by the model. If the selected process is "twotank" then by default it will use its overall model of the process to generate the signals, however if "tank1" or "tank2" are selected then "twotank" will automatically begin to use its substructure and the display will change to show the signals from the new selected process.

Figure 7 shows the displays generated over a period with the process "twotank" selected. Towards the end of this period a leak was induced in the inlet to tank 1. It can be seen that during the period when there is no fault the residual signal is near to zero despite changes in the operating levels in the tanks, however when the fault occurs a fall in the flow from the second tank causes the estimate of the input flow to drop suddenly, reproducing the fault condition, this in turn causes to residual to jump to a no-zero value which is automatically detected by the virtual alarm system of the "twotank" model and reported as an error as shown in figure 8.

Once a fault has been detected the next operation is to isolate the fault to a particular sub process. This is done in the following manner.

(i) Monitoring is stopped.

(ii) The process "twotank" is forced to use its substructure by selecting the "view/hide substructure" option in the menu.

(iii)The stored history maintained by "twotank" is replayed through the substructure using the replay option. This causes each model in the "twotank" substructure to generate their inter process signals.

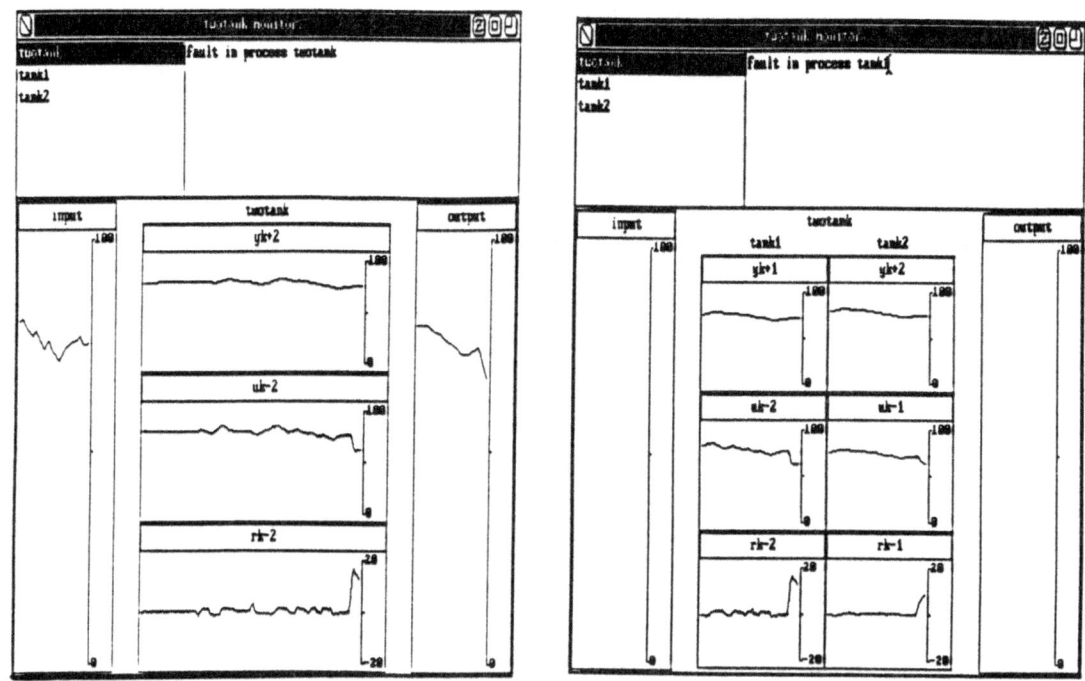

Figure 7. Tank 1 Inlet leak detected   Figure 8. Localising to tank 1

Figure 9 shows the signals generated by replaying the history through the substructure of the model. The residual generated by model "tank2" is an exponential rise brought about by an exponential fall in the estimated input to tank 2. However the residual generated at model "tank1" undergoes a step increase. Hence the residual alarm for tank 1 is triggered before that of tank two confirming that the fault is located at tank 1.

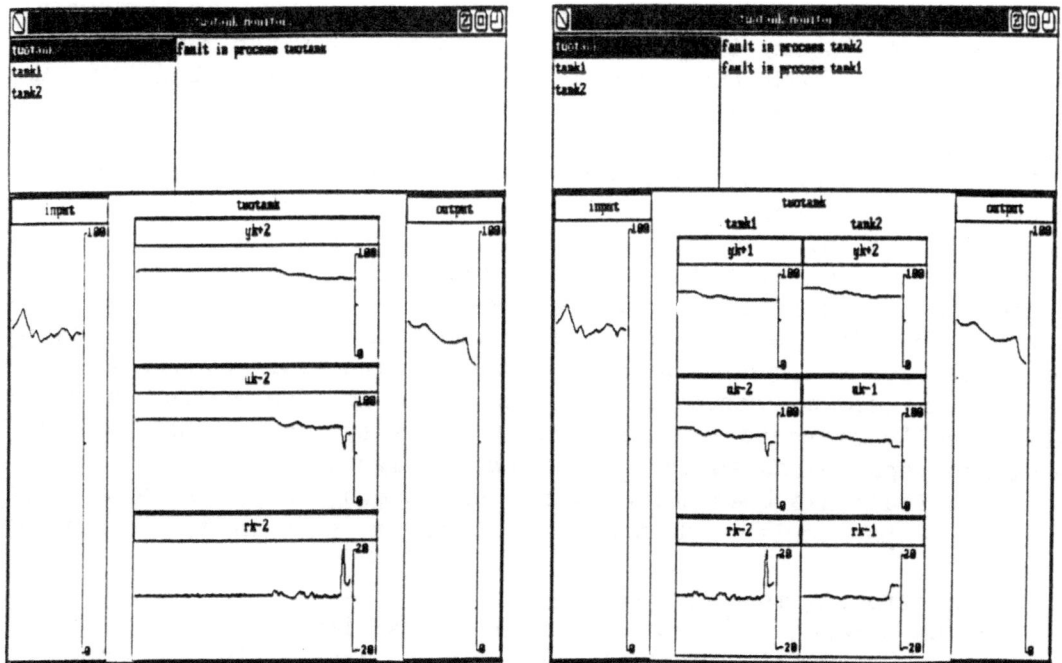

Figure 9. Tank 2 leak detected.   Figure 10. Localising to tank 2.

A leak was then induced in the inlet pipe to tank 2 with the effect shown in figure 9. Isolating the fault in the manner described above yields the signals shown in figure 10. In this case the residual generated at model "tank2" increases suddenly due to a sudden drop in the estimated inlet flow. When this estimated input is propagated further back through the model high frequency spikes are generated which do not represent legitimate process behaviour and will also trigger alarms. However since backward propagation takes place for output to input the error at tank 2 is detected first indicating that the fault is located in tank 2.

CONCLUSIONS.

The use of Online Hierarchical Simulation provides a method for the detection and localisation of process faults using only measured input and output data. This may have particular benefits in processes in which intermediated process states can not be reliably measured in normal operation. The system described is in prototype form and in particular the virtual alarm system is very crude. Further research is in progress to provide automatic localisation and diagnosis of faults once detected. Afurther problem to be tackled is the isolation of gradually occurring faults in which analysis of the residuals will be carried out by a higher level of expert system pattern recognition software.

The use of the Object Oriented Paradigm is a natural choice for this type of application and allows rapid development and experimentation to be carried out.

REFERENCES

Gawthrop P.J. & Leary J.J "Process fault detection using constrain suspension". IEE Proceedings, vol 134, Pt. D, No. 4, July 1987.

Goldberg A. & Robson D. "Smalltalk-80, The Language and its Implementation". Addison Wesley Publishing Co. 1983, ISBN 0-201-11371-6.

Guariso G. "An intelligient simulation model generator" Simulation, August 1989.

Young P.C. "Recursive estimation and time-series analysis" Springer-Verlag, Berlin, 1984, ISBN 3-540-13677-0.

ACHIEVING FLEXIBILITY IN ROBOT PROGRAMMING

BY MEANS OF EXPERT SYSTEMS TECHNOLOGY

Roberto Gallerini, Antonio Pezzinga
FIAR Spa, Milano, Italy

1. INTRODUCTION

The main difference between programming a computer and programming a robot for an assembly task is that, in the first case, mainly symbolic reasoning is required to the programmer; in the second case, three-dimensional geometric reasoning is essential.

Because the human geometric intuition does not work well without a visual support, the only way to program a robot is to have it in front of the programmer. This requires to have the physical cell available, and leads as a consequence to the impossibility of using it for production during the programming phase.

This of course is a rather expensive and not flexible way of programming. Each time the production changes it is necessary to stop the production line in which the robotic cell is placed for programming, debugging and testing. For this reason a great interest exists in different approaches to robot programming, not requiring the use of the real robot: these approaches are commonly grouped under the heading 'off-line programming'.

The basic idea is to have sophisticated simulation tools and graphic environments, on powerful workstations placed elsewhere from the production line to develop and test the programs. Once this has been done the programs can be downloaded to the robots for actual execution. The advantages, apart from avoiding the use of the real cell include the possibility to simulate an assembly immediately after the CAD definition of the workpieces, allowing the 'design for assembly' concept, and the possibility to easily try and evaluate different layouts for robots, pallets, etc.

Two main approaches exist in off-line programming: the first one, often called explicit programming, relies on the sophistication of the simulation tool and the programming environment provided to the user, to ease programming. Within this approach some products, as IGRIP (Deneb Technology Inc.), have been realized and are commercially available. The second one, called task-level or implicit programming, is based on automatic tools able to obtain the program from a very high level description of the robots' task, using appropriate models of the robot and of the cell [Spur87]. Within this approach only partially developed prototypes have been designed, mainly the classical LAMA [Loz77] and AUTOPASS [Lie77].

In the ESPRIT Project 623 these two approaches have been studied, and systems based on these concept developed. The team of partners was organized into the subgroups Explicit (realization of an explicit, or motion oriented programming and simulation system), Implicit (realization of an implicit, or task oriented programming system) and Planning (realization of a planning system for robotized cells).

The main drawback of the explicit programming approach is the difficulty to obtain an effective 3D representation of the environment: even using quite sophisticated graphic techniques (e.g. solid rendering with shadowing) it is always difficult to move the simulated robot, for instance to insert a peg into an hole. The drawback of the implicit programming is the difficulty to gather the technological experience to deal with the variety of problem that a robot program can pose. Moreover the implicit programming poses major technical problems, requiring a quite intelligent behavior of the system. As a matter of fact implicit programming involves many classical Artificial Intelligence areas (planning, geometric reasoning, etc).

In the following we describe the experiences we have gathered in participating in the ESPRIT Project 623, "Operational Control for Robot Systems Integration into CIM", designing and developing the expert system IMPRES for implicit programming of assembly robots at cell level.

2. WHY DEVELOP AN EXPERT SYSTEM

The two questions which have to be answered first of all are: first why we have chosen the implicit approach, and second, why the decided to develop an expert system and not a conventional program.

The answer to both question can be summarized in one word: flexibility. The implicit approach insures in fact, in our opinion, a greater speed of execution. This means both producing programs faster, and adapting to changes in the workpieces to be produced (to the limit of completely changing production) in a shorter time. This happens because a large part of the analysis of the task to be executed by the robots, which has normally to be performed by the human operator, is done by the system itself. In other words, the border between the automatic system and the human is moved closer to the latter, automatizing many tasks (as for instance determining the order in which the parts have to be assembled, or scheduling the assemblies to the robots, if the cell is a multi-robots one) which have normally to be done 'by hand'.

The price to be paid for this is clearly that a certain degree of 'intelligence' is required by the system. And this leads to the second question. The traditional software approach (data plus fixed control flow) is totally inadequate to take into account the variability in data, execution conditions and rules to be followed in the assembly problem. To be able to successfully attack this problem, only a 'non-conventional', or knowledge based, or expert system, approach, whatever you want to call it, can be used.

Note that the difference between these two software solutions is a difference in the philosophy of the approach, rather than a difference in the specific language used. One could in principle implement an expert system in COBOL, apart from the effort needed, as he could as well use LISP to imitate conventional programming.

The point is to identify and make explicit in the system the knowledge used to solve the problem, rather than using application specific data. In fact, to be correct, one should use the term expert system when in the system are codified strategies, methods, rules used by a domain expert which relies on his/her personal expertise for solving problems. In our case, however, the system uses search methods and scheduling algorithms which are general, and the knowledge on the assembly problem and related issues to apply them in a 'smart' way. The correct name of our system should therefore be Knowledge Based System rather than Expert System.

3. IMPRES

As it has been anticipated in the previous section, the definition of the task to be performed is given by the operator to the system only in an implicit way.

This task definition is composed by two parts: the first one is the description of the workpieces to be assembled (also called "Problem Description"). The second is the description of the workcell that has to carry out the task ("Cell Description").

The Problem Description is made by the geometric description of each workpiece, the initial position (before the assembly), final position (after assembly) of all the workpieces and by the definition of the type of contacts between objects in the assembled configuration (e.g. screwing, insertion, ...). The initial and final positions plus the geometry of the parts can be easily obtained using an interactive CAD-like system, able to define relative positions of objects.

The Cell Description is the geometric description of all the elements in the cell (robots, pallets, etc.) except the workpieces, plus the model of the robots (kinematic parameters, definition of the robot language) and the cell layout (position of the robots and fixtures).

IMPRES builds from these elements the Knowledge Base and determines from it the executable program for each robot which, when executed, carries out the assembly task.

As it can be seen, the definition of the task is only implicitly given to the system. The user does not define in what order the parts have to be assembled, or which robot has to execute which operation. He does not synchronize the robots to avoid collisions, nor determine collision free trajectories. All these functions, including the generation of programs for each robot in the cell, possibly in different languages, are carried out by the system.

It is clear that to be able to perform this task, a number of sophisticated capabilities are required by the expert system: planning, scheduling, motion planning, program generation and translation are the most important. IMPRES is composed by four modules (see fig. 1), which are described in the following, and which realize these functions.

ACTION PLANNING: in this phase the geometry of the parts, their positions and type of contacts are used to determine the assembly graph. The assembly graph is a partial ordering of the operations, of the type:
(step 1: part2; step 2: part4, part5; step 3: part1, part3)
The assemblies within one step can be performed in parallel. Each step has to be completed before the next one can be started. If this is verified no conflicts (i.e. collisions between parts) will be generated during execution. To determine the assembly graph, the Action Planner uses geometrical considerations (e.g. the assembly of a workpiece must not obstruct the following assemblies) and stability considerations (a workpiece must be stable in the final position).

This phase is carried out independently from the actual number and positions of the robots in the cell. Only the geometric models of the parts and the grippers are used, that is, only the task constraints are exploited. This means that whether the assemblies within one step will be actually executable in parallel depends on the Cell Description, which will be considered by next module of IMPRES.

The Action Planner determines also the parameters of each assembly action (assembly direction, departure,...).

To determine the assembly graph the Action Planner carries out a simulation of the assembly, according to the following control cycle:

1. the parts candidate to be assembled are the parts which are in contact in the goal configuration to be achieved with parts that have already been assembled at the present stage. At the beginning all the parts are not assembled, and the ones in contact with the assembly fixture are chosen.

2. for each part in the candidate set, the Action Planner checks if the final position is stable. If a workpiece is not stable in the final position, it is eliminated from the candidate set: it is likely that this workpiece will be made stable by some of the not already assembled parts: therefore its assembly is postponed.

3. For each workpiece in the candidate set, the system finds the allowable assembly directions. Then these directions are checked for collision: if during the approach to the assembled position a collision with an already assembled workpiece occurs, that direction for that assembly is discarded. Also the collisions due to the gripper are taken into account.

If an assembly is not possible in any direction, it means that one or more already assembled objects block that assembly. At this point the system tries to solve the problem using a set of strategies to reorder the assembly graph. Strategies try for instance to disassemble objects (remember that since the assembly is only simulated this corresponds to altering the order in which the assemblies are performed) or to build separate subassemblies in different areas of the workplane.

All the remaining operations within one step are then checked to verify that they can be executed in parallel (the execution of assembly-i does not collide with the execution of assembly-j).

4. All the remaining objects in the set are collected into a step of the assembly graph, as they can be assembled in parallel -without taking into account additional constraints due to the robots-. The assembly operations are performed and the cycle is repeated, until all the workpieces are assembled.

The output of the Action Planner is an assembly graph which contains also all the parameters needed to execute the operations, except the trajectories to be followed, because this will depend on the specific robot which will execute the assembly.

SCHEDULING: purpose of the Scheduler is to order the operations of the same step of the assembly graph and to assign them to the robots. Usually the number A of operation in the same step is quite small (5-6 as a maximum) and the number R of robots is also small (2-3). Because each operation of a step must be performed before each operation of the following step, each step can be considered independently from the others. This is actually a simplification which reduces the complexity of the problem, reducing also the optimality of the solutions found.

A tree is built generating the possible assignments of operations to the robots; if a robot cannot perform an operation, because either the initial position or the final position of the workpiece is out of its working area, the assignment is discarded.

A search of a pseudo-optimum path is performed on this tree, using the A* algorithm. The Motion Planner is called by the Scheduler to find collision-free paths for the robot movements.

It can happen that the Motion Planner is not able to find a trajectory of this sequence, either because this trajectory does not exist or because it is too complex and a computational time limit is reached. In this case the Scheduler backtracks, chosing a different path on the tree. This choice is done trying to save as many trajectories as possible, because the motion planning is a quite time consuming job. Of course, the method presented works only when the number of operations that can be performed in parallel is small: if this is not true, the system asks to the operator to divide the operations into subsets in order to keep the size of the search space small.

For what concerns the synchronization between the robots in a multi-robots cell, the user defines regions reserved to one robot (for instance, each robot has its own pallet from which to grasp the workpieces) and shared regions (for instance, the assembly area). The Scheduler assures that only one robot at a time is within a shared region, by means of explicit synchronization (i.e. semaphore variables).

As it is clear, there are a number of approximations of the scheduling found with respect to the optimal one.

This happens because finding optimal scheduling requires a complex interaction between scheduler and motion planner, with many backtrackings from the scheduling level and the motion planner level. Because both motion planning and optimal scheduling are time consuming activities, an exact approach is not applicable or a too high computational time will be obtained.

The result of the scheduling activity is a high-level program, which contains all the details necessary to execute the task.

MOTION PLANNING: the Motion Planner has the task of determining collision free paths. This is done discretizing the space into n-dimensional cells (for a robot with n degrees of freedom) with a given discretization factor. This space is not explicitly represented, but one can think of it as an n-dimensional grid, in which each cell represents a robot configuration in joint coordinates; each cell has a side length depending on the discretization factor.

The cell corresponds to a node of a graph, in which the arcs represent adjacency relations. The starting position and the final position are two of the nodes: the system generates a search tree in graph, with the A* algorithm, using as cost function the distance in joint space. With this approach the system finds the minimum length path in joint space. The complexity of the search depends of course on the number of cells: this number depends on the number of degrees of freedom of the robot and on the selected discretization.

It can be noted that this method is similar to the widely used configuration space method, in which the free regions where the robots can move are pre-computed, with two major differences:
- not all the configuration space is generated, but only the part required in the search;
- the forbidden space representation is simply the list of the cells in which a collision occurs.

We have chosen this approach because it allows a very easy reconfiguration of the system with respect to the robot kinematics and geometry.

Moreover, since in an assembly task each time a workpiece is assembled the situation changes, the configuration space method would require the re-computation of the free space after each assembly. This is clearly not applicable because it would lead to an overall computation time in the range of hours.

The system works at two levels: the first level finds a preliminary path with a rough discretization; the second one refines this path, finding -with the same method- the subpaths between the intermediate positions of the first level path. At both levels the search of the path can fail, either because it does not exist a collision free path or it is too complex to be found. A fail at the second level is reported to the first one, allowing backtracking between the levels. A fail at the first level is reported to the Scheduler, that must try a different scheduling.

The Motion Planner plans the trajectories for one robot at a time, taking into account the current positions of the other robots, because the synchronization between the robots has already been handled by the Scheduler.

PROGRAM GENERATION: at the end of the scheduling and motion planning activity the system has produced an high level program for each robot in the cell in which all the details necessary for execution are defined. The task is now to translate this program into the specific language of each robot controller.

To allow an easy reconfigurability for different robot languages, without having to insert all their definitions in the system beforehand, the following approach has been chosen.

The user is required to define, for each language needed, its syntactic structure, defining how an instruction is defined in terms of more elementary ones. For instance one can define that the instruction 'assembly' is composed by the instructions (approach, grasp, transfer, assemble, release). Each sub-action is then again detailed in terms of more elementary instructions, until the actual language of the robot controller is reached.

The top language is the one in which the high-level program is generated by the Scheduler, the lowest level is the target language. In defining this decomposition the user 'teaches' to the system the rules by which the translation process has to be carried out. This allows to generate different robot languages with a minimal effort: all that is needed is to know the internal language and the target one. It is not necessary to develop each time a different software compiler.

The semantic network built by the Program Generator can be completed by the user with the definition of the inheritance mechanisms which allow the system to derive the parameters of a low level operation from an high level one.

4. WHAT THIS EXPERIENCE TEACHES

The size of the IMPRES is 33,300 lines of source code; 15,500 lines are Pascal code (for geometric computation, CAD interface and motion planning) and 17,800 lines are OPS83 code, a rule based language descendent from OPS5 (for the inferential parts). A large part of the Pascal code -collision detection, geometric model management- is used by both Action Planner and Motion Planner. Two versions of IMPRES exist: one for the VAX Station family and one for the Hewlett-Packard 9000 serie.

A first point which emerged is the fact that, due to the complexity of the problem, it is not possible to solve all the problems at the same level of abstraction. In order to obtain a solution, it was necessary to split the activity of the system in phases, namely planning (determination of the assembly graph), scheduling, motion planning and program generation.

If we define the optimal solution as the one which produces the minimal overall assembly time when executed, to determine it, it would be necessary to allow complex interactions and backtracking between levels. For instance, if the number of the assemblies within the same step of the assembly graph is large, this could be reduced by means of a cooperation between the Action Planner and the Scheduler. Another approximation is to consider the steps of the assembly graph to be independent, and therefore perform the search of the 'best' assignment robot-assembly only within one step at a time. To be correct the search would have to be performed on the overall graph, and not only on a part of it.

However it is literally impossible to allow this interaction, because the computing time would grow to unacceptable values: hours, or even days. Moreover, the solution found by IMPRES, is always 'good', meaning that it is obtained with computing times ranging from minutes to less than an hour, it is executable by the robots, and generates assembly times and usage of the robots which are fully acceptable. The cost of going from this 'good' solution to the 'best' one is not bearable, at the present state of the art.

The second point which has to be stressed is that the performances of IMPRES depend on the complexity the assembly task. Difficult cases have one or more of the following characteristics:
- complex logical structure: the structure to be assembled looks like a 3-D puzzle. This solution is however not very common in the industrial field.
- assemblies requiring special techniques, for instance the use of ad hoc tools or of intermediate fixtures.
- cluttered environments which require to plan complex trajectories.

The solution to these problems, and to others of this kind, rely on the degree of interactivity of the system. In other words, by letting the user cooperate with the system, guiding it in difficult cases, a solution can be determined. Therefore, it can be stated that to build a really helpful tool for assembly programming, it is necessary to allow a high interactivity between this tool and the operator.

The last point which emerged from this development is the following. IMPRES generates programs which are executable and carry out the task in nominal conditions. Because the program is generated off-line, it is not possible to react to unexpected events, like defective parts, or errors during execution (for instance the loss of a part from the gripper during a transfer movement). In these cases the workcell stops, and a human operator has to determine what is the cause of the malfunction, and remove it.

This problem, the error recovery problem, is in fact a crucial point in robot use, whether the robot has been programmed on-line, off-line explicitly or off-line implicitly. At the present state of the art no solution has been implemented and industrialized. Many studies are being carried out in research centers and universities, but none is used in a real factory. Solving this problem, by giving to the robot some capabilities for autonomous error recovery is the key to widen the use of robots in industry.

5. CONCLUSIONS

The system has been tested on a set of examples. Among them one is based on the "Cranfield Benchmark" assembly. This benchmark has been defined mainly as a test for robot accuracy; it has been chosen because it is representative of a quite large class of assembly tasks. Our example task requires to assemble 14 workpieces with two 6 degrees of freedom PUMA260 robots, using a fixture for the assembly and two pallets for the workpieces.

Within this example the geometry is quite complex (about 2000 faces for describing workpieces and fixtures and 1000 faces for describing each robot). Conversely, the logical structure of the assembly is quite simple -in the Action Planner no backtracking is required. The CPU time required to obtain the program -on a DEC VAXStation- is less than 17 minutes, excluding time for graphic presentation and files reading and writing.

The complete execution, including graphical simulation of the various planning phases, requires about 75 minutes. As reported by the authors of the benchmark, the time required when the programmer is a human operator -with only one PUMA robot- is about 8 hours.

This study and development was greatly benefited by the ESPRIT frame and in particular by the specific competences of the various partners involved in the project. The consortium included various industrial partners producers of robots and utilizators of them, and important universities and research institutions across Europe with specific expertise in robot programming, planning and information management. The prime contractor has been IPK - Fraunhofer Institut fur Produktionsanlagen und Konstruktionstechnik, Berlin. It can be stated that none of the partners alone could have realized the systems, among which IMPRES which has been described by this paper. The complementary expertise of the partners, both in theory and industrial utilization has allowed great sinergy and an excellent level of cooperation, which has been proved by the creation of the three subgroups Explicit, Implicit and Planning. The ESPRIT project 623 can be regarded, in our opinion, as a very successful project, which has fully reached its initial goals, and caused important fallouts in different fields and in education.

The work on IMPRES is now continuing in the aim of the Italian Program for Robotics (PFR), in which the current prototype is being extended and tested on industrial assembly cases.

Another important follow-on of our work is in the space robotics field. For a space robot, on-field programming is almost impossible; therefore off-line programming systems will play in the next future an important role.

The Autonomous Robot Control and Simulation project (with contract issued by the European Space Agency) has the aim to study and identify an architecture of an Autonomous Robot for future space applications.

In the definition of this architecture, the experiences gathered in the development of IMPRES will play an important role.

IMPLICIT PROGRAMMING EXPERT SYSTEM

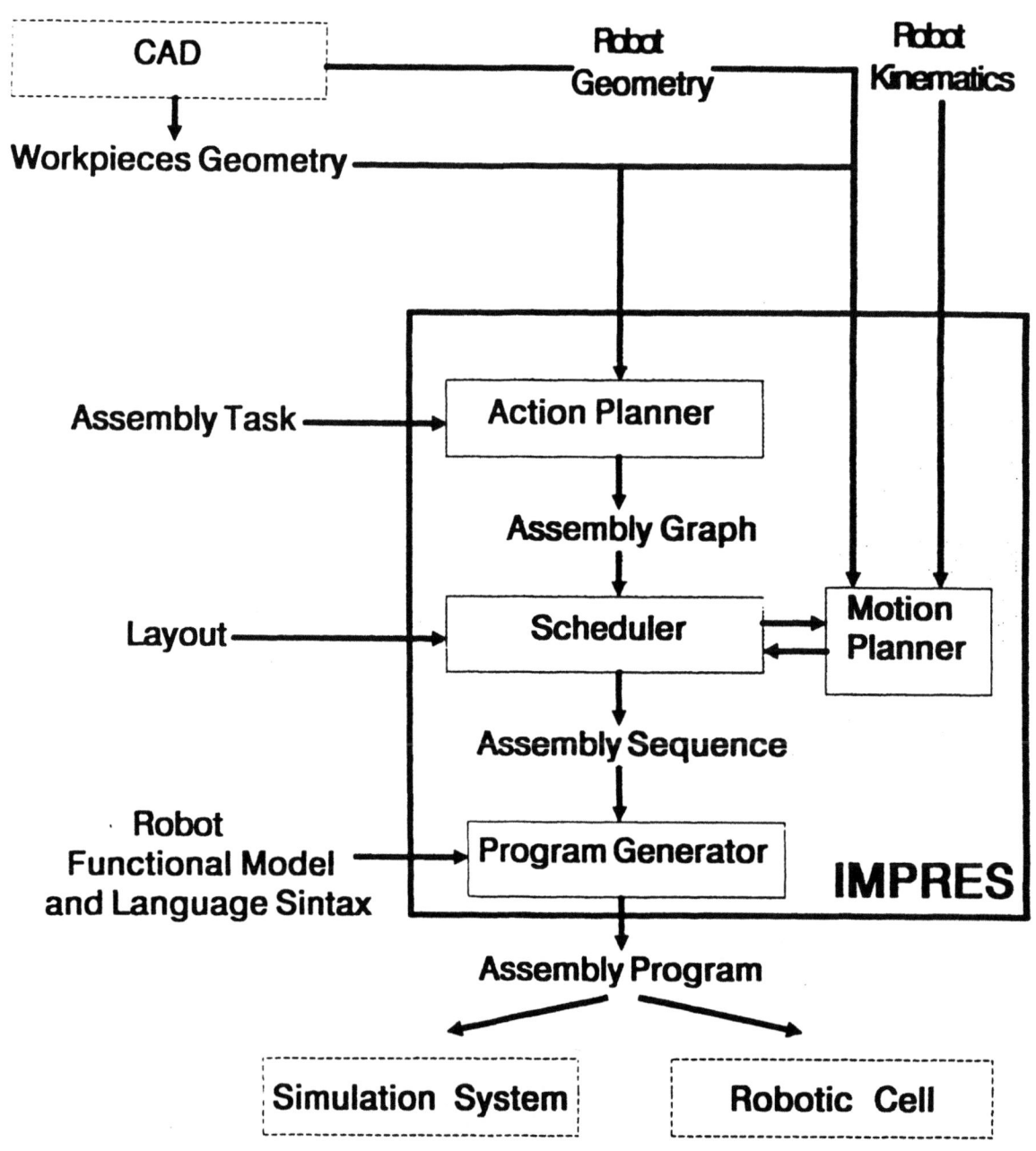

REFERENCES

[Fav84] Faverjon, B.: "Obstacle avoidance using an octree in the configuration space of a manipulator". IEEE Int. Conf. on Rob. and Autom, Atlanta (1984).

[Hom86] Homem de Mello, L.S. and Sanderson, A.C.: "AND/OR graph representation of assembly plans". AAAI-86, Philadelphia (1986).

[Kan87] J.J.Kanet and H.H.Adelsberger "Expert systems in production scheduling" Europ. Journ. Oper. Res., vol 29, pp. 51-57, 1987

[Kha86] Khatib, O.: "Real Time Obstacle Avoidance for Manipulators and Mobile Robots". Int. Journal of Robotics Research, vol. 5 pp. 90-98, 1986

[Lie77] Liebermann, L.I. and Wesley, M.A.: "AUTOPASS: An automatic programming system for computer controlled mechanical assembly". IBM J. Res. Dev. vol 21, no. 4, pp. 321-333 (1977).

[Loz77] Lozano-Perez, T. and Winston, P.H.: "LAMA: a language for automatic mechanical assembly". Proc. 5th IJCAI, Cambridge, MA (1977).

[Loz82] Lozano-Perez, T.: "Task Planning" and "Automatic Planning of Manipulators Transfer Movements", Robot Motion, Planning and Control, MIT Press (1982).

[Loz86] Lozano-Perez, T.: "A simple motion planning algorithm for general robot manipulators". AAAI-86, Philadelphia (1986).

[Sha89] Sharir M. "Algorithmic Motion Planning in Robotics" Computer, March 1989

[Smi86] S.F.Smith, M.S.Fox and P.S. Ow "Constructing and maintaining production plans: investigations into the development of knowledge-based factory scheduling systems", AI mag. vol 7, no 4 pp. 45-61, Fall 1986

[Spu87] Spur, G. et al.: "Planning and Programming of Robot Integrated Production Cells". Proc. 4th ESPRIT Conference, pp. 1716-1743.

[Ver83] S.A. Vere "Planning in time: windows and durations for activities and goals" IEEE Trans. Pattern Anal. Machine Intell. vol PAMI-5, no 3, pp 246-267, may 1983

METHODS AND TOOLS FOR INTEGRATING CIM ELEMENTS – PART II

A KNOWLEDGE-BASED APPROACH FOR DESIGNING INDUSTRY-SPECIFIC PRODUCTION MANAGEMENT SOFTWARE

I.P. Tatsiopoulos and I.A. Pappas

National Technical University of Athens
Dept. of Mechanical Engineering
Sector of Industrial Management & O.R.
28is Octovriou 42 - 106 82 Athens - Greece

A methodology and an expert system are presented to help solving the problem of determining software requirements for production planning and control systems. A dual taxonomy of production systems is proposed based both on functional types and industrial sectors. The main theme of the research is to resolve the implementation dilemma 'change the standard software or the organisation of the firm?'. The methodology is implemented with the help of software engineering tools ranging from structured systems analysis to an object-oriented expert system shell with simultaneous support of the hypercard technique.

INTRODUCTION

The large volume and complexity of the computer programs needed to support the production planning and control function (PPC) in manufacturing firms, and the large variety of different types of manufacturing settings, puts a particularly heavy load on the task of determining software requirements for PPC systems. Those requirements will serve the purpose of selection, customization and implementation of a suitable commercial software package or, alternatively, they will form the basic guide for an in-house development project.

This problem is crucial particularly for small and medium sized firms (SMEs), due to their lack of internal expertise and resources for in-house software development and maintenance.

The usual PPC software package, which is offered by hardware or software vendors, has shortcomings in its restriction to specific manufacturing settings, usually those of machine construction and assembly of complex final products. Here, the management of numerous production data in the form of bills of materials and routings, and consequently the release of production orders via MRP procedures are the overriding considerations.

However, serious deviations from the standard offered PPC model can be seen in other industrial sectors. For example, much simpler systems are needed in manufacturing firms, where a multiplicity of final products are made using a small set of raw materials. This is typical of the ceramic, paper, chemical, food and plastics industries, where final products are often only distinguishable through packaging or size differences. In this type of production primary considerations are the smoothness of material flows, the balancing of assembly lines and the optimization of the sequence of production, given set-up costs. This type of manufacturing structure is typical for the majority of industries in a threshold country, like Greece, and a suitable PPC software package should not include unnecessary complexity aiming at the heavy engineering sector.

It is suggested here that an industry-specific PPC software design procedure should be developed, based on the knowledge domain of production management theory about the needs of the utilizing production system, according to its characteristic structures and processes (e.g. industrial sector, type of production flow, product variety, production volume, factory layout, etc). This knowledge domain should be incorporated in the form of a generic software package with specific modules and features for a series of industrial sectors (e.g. clothing industry, foods industry, plastics industry, etc). The PPC design procedure should try to customize the generic software package according to the above needs.

The reported efforts in the literature for the determination of requirements and the design of manufacturing software have their origin in different science fields using different approaches. Three main research streams can be identified. The information systems, the industrial engineering and the artificial intelligence approach.

The information systems approach is represented by the CASE (Computer Aided Software Engineering) tools supporting a variety of structured systems analysis and design methodologies (Yourdon, De Marco, Jackson, see Leslie, 1986).

An example of the industrial engineering approach is the well accepted PPC software evaluation system BAPSY of the TH Aachen (Speith and Brief, 1982, Hoff and Virnich, 1986) which uses multi-criteria and value analysis methods to compare different PPC commercial software packages and select the most appropriate for a specific manufacturing firm.

Representatives of the expert systems technology are: The system TWAICE of Nixdorf Computer (Krallmann 1986, Mensel and Michel 1985) which is mainly a software configuration system to support the Nixdorf COMET integrated manufacturing package and the needs of its customers for customization and consultation. The GRAI system (Doumeingts, 1989) which is a computerisation of the GRAI method for the design of production management systems, and the CASE-I system (Tatsiopoulos, 1989) for the selection and implementation of materials management software.

The next section of this paper uses an extension of the Data Flow Diagrams technique for the design of expert systems (Kellner, 1987) in order to address the overall problem of determining PPC software requirements with the help of a knowledge-based system, which is broken down into smaller subsystems.

The proposed methodology is implemented using the 'KnowledgePro' expert system shell which supports object-oriented programming and the hypercard technique.

THE OVERALL KNOWLEDGE-BASED SYSTEM

The proposed knowledge-based system (Figure 1) aims at judging the degree of "suitability" of every particular software requirement. Any one of the main factors that go into the final decision may represent a mini expert system which can be further broken down into many hierarchically structured factors.

There is a broad spectrum of "Suitability" of a particular software module from very good to very bad. For this reason there is often a temptation to use quantitative methods like value analysis which specify the decision on a numerical scale, say from 1 to 100. However, the factors talked about in this study are general guidelines, often not very well specified, and in a real-world situation they may not be known with great accuracy. This suggests that a 0 to 100 decision may be a meaningless breakdown given the fuzziness of the factors that go into making the decision. A better choice would be to limit ourselves to a three-way decision: CRITICAL, DESIRED, NOT NECESSARY.

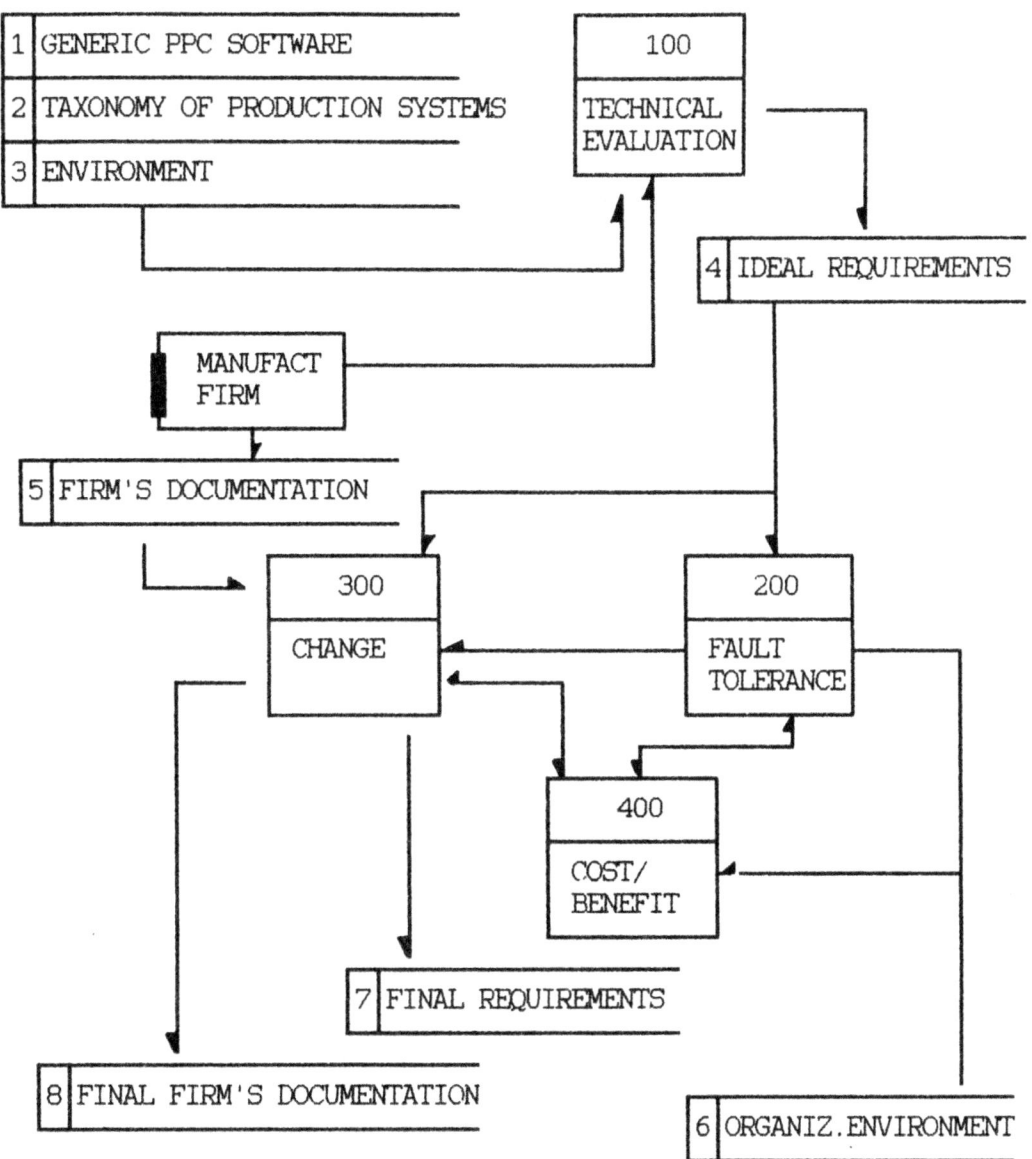

Figure 1. The overall knowledge-based system.

Subsystem TECHNICAL EVALUATION (100).
It addresses the problem of evaluating PPC software packages on the basis of their pure technical characteristics, as compared with each other and with the production management theory for the various types of production systems. A knowledge-based model is proposed whose objects and relations describe the type of production system, the generic PPC software package and the environment.

Subsystem FAULT TOLERANCE (200).
Its objective is to judge the sensitivity of any particular PPC software module feature and its parameters, as a result of specific database and/or transaction faults due to a loose organizational environment.

Subsystem CHANGE (300).
The factor CHANGE is considered in two directions. The vertical direction concerns the adoption of existing vs. "ideal" procedures. The horizontal direction is concerned with changing the package to come closer to the

339

specifics of the organization, against changing the organization to adopt the design philosophy of the package. A synthesis of directions is also possible.

Subsystem COST/BENEFIT (400).
It judjes the value of information, provided by PPC software modules, for attaining production management and overall business objectives, against the cost of hardware/system software requirements, software maintenance, and operational and organizational costs.

The present work concentrates on the technical evaluation subsystem (100) and that part of the change subsystem (300) which concerns the determination of inconsistencies between the "ideal" requirements and the documentation of the existing firm's information flow and organisation. The other subsystems which are prerequisites for the succesfull implementation of a PPC system are currently under research and development.

TECHNICAL EVALUATION

It is suggested here that the software technical evaluation procedure should produce a set of "ideal" PPC requirements at three levels:

   - The functional production type (e.g. job-shop)
   - The industrial sector (e.g. clothing industry)
   - The specific manufacturing firm

The requirements for all functional production types will be based on a taxonomy according to functional characteristics (e.g. product standardisation, factory layout, etc) and the knowledge domain of production management theory about the needs of a particular production type. The industry sector—specific requirements are imposed by technological processes, special problem areas, objectives, measures of performance and environmental factors, even aliases of production management terms which are characteristic of each separate industrial sector.
For the determination of PPC requirements of a specific manufacturing firm, a "normative" approach (Davis and Olson, 1985), using an "ideal" generic PPC software package will be followed at this stage for three main reasons:
a) It is quite common that the proposed information system embodied in the production management generic software package may be fundamentally different from existing patterns (in its content, form, complexity, etc.), so that anchoring on an existing information system or existing observations of information needs will not yield a complete and correct set of "ideal" requirements.
b) Usually in practice there is not enough time at this stage to proceed to a detailed analysis of the company's current production information processing and organisational procedures.
c) A database of existing commercial packages would not be practical to maintain, given the continuous release of updated versions and new products in this field.
The systematization of expert knowledge needed at this stage includes:

- Description of the Generic PPC Software Package using hybrid techniques of Data Dictionaries (Gane and Sarson, 1978) and object-oriented

programming (OOP).
- Description of the firm's production structure and classification of the firm's production system according to a knowledge base of functional production types and industrial sectors.
- Comparison with the methods included in the generic software package, which are formulated in classes of technical features.
- Selection of technical features suitable to be included in the "ideal" software package.

There follows a more detailed discussion of the main knowldge bases.

Production Systems Taxonomy

Many authors have proposed classifications of production systems, e.g. Schmenner (1981), Schmitt et al (1985) and Kettner et al. (1984). Their common characteristic is that they adopt a narrow view of the physical production system dealing only with functional characteristics like shop layout and process flow classifications, which are not enough to determine production management requirements. Factors like management objectives and measures of performance, main problem areas, special technological processes and specific environmental constraints which are connected to the various industrial sectors and subsectors are equally important for the determination of PPC requirements.

The proposed taxonomy in this study is of a twofold nature, a functional taxonomy and a taxonomy based on industrial sectors. For the first one, a system of classes of parameters and values has its origin in the Schomburg (1980) classification model which was built specifically for engineering firms. However, the taxonomy proposed here is extended and built in a software tool using object-oriented programming concepts. This software tool facilitates the connection of the functional taxonomy to the industrial sectors taxonomy and all relative classes, i.e. problem areas, objectives, environmental factors and aliases (terminology of production management across industrial sectors) as seen in Figure 2.

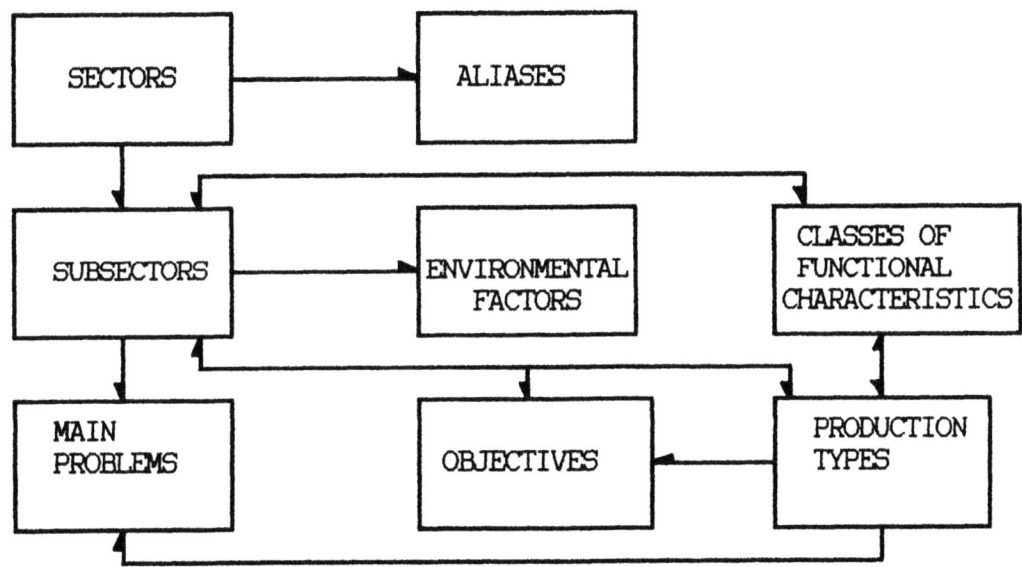

Figure 2. Knowledge Base of Production Systems Taxonomy

As far as industrial sectors and subsectors are concerned, the formal classification of the National Greek Statistical Service has been used.

For the classes of functional production characteristics, the classification model of Schomburg (1980) enlarged by Tatsiopoulos (1988) is used. This model uses 9 classes, each one of them having several instances, as shown in the example of Figure 3.

00		– Product Standardisation
	00.1	Non–catalogued customized products
	00.2	Catalogued products with non–standard, customized options
	00.3	Catalogued products with standard options
	00.4	Standard products without options
01		– Product structure . .
02		– Form of independent Demand (products)
03		– Form of Dependent Demand (components, raw materials)
04		– Form of Purchasing
05		– Process form (time dimension)
06		– Shop–layout (space dimension)
07		– Existence of Hybrid Production Structures
08		– Process complexity
09		– Warehouse structure

Figure 3. Classes of functional production characteristics
and an example of instances

The Generic PPC Software Package

Most of the available commercial software modules for production management belong to large integrated packages of manufacturing management software which follow in general the MRP II design philosophy. Even though they use slightly different names for their modules, it can be said that the terminology and content of the most functions offered is fairly standard as in the following example:

PDM = Production Data Management   CRP = Capacity Requirements Planning
MPS = Master Production Scheduling   SFC = Shop Floor Control
MRP = Material Requirements Planning PUR = Purchasing

An excellent review of most of the available packages and their characteristics both from large hardware manufacturers and independent software houses in Central Europe can be found in Foerster et al. (1986), while a review of american packages can be found in Sepehri (1985).

The purpose of this study is to build an Extended Data Dictionary using object-oriented methodologies in order to describe and document a parametrised generic PPC software package which contains, in terms of parametric "modules" and "features", all the Knowledge domain of PPC information technology and production management theory. The Knowledge Base for this Generic PPC Software Package is depicted in Figure 4.

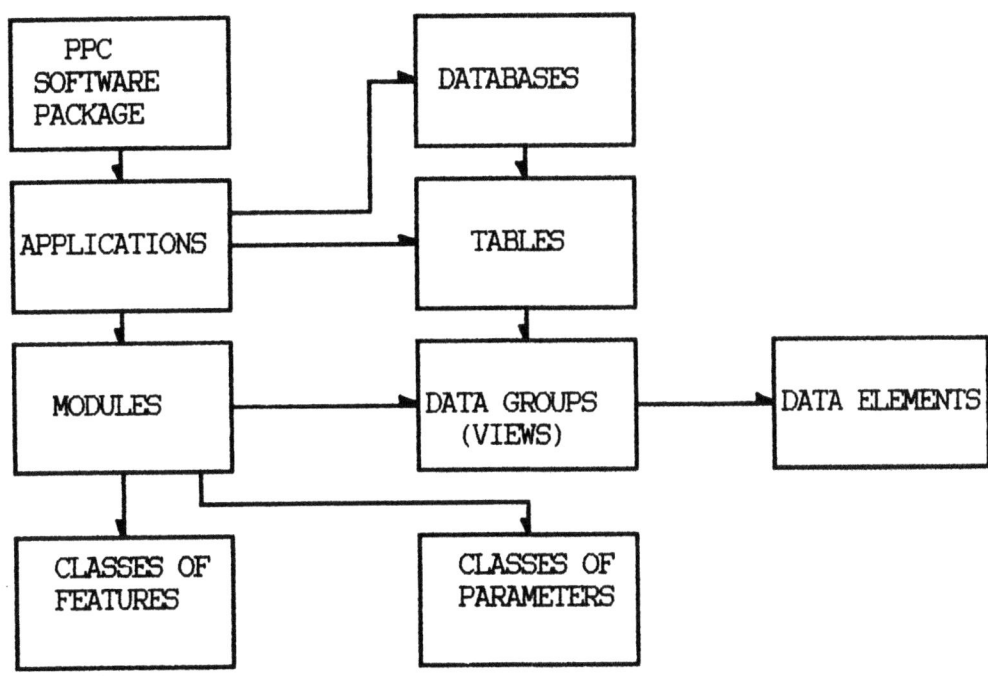

Figure 4. Knowledge Base of Generic PPC Software Package

The typical Data Dictionary for documenting software applications (Gane and Sarson, 1978) contains the entities Application (Program), Module, Database, File (Table), Data Group (User View), and Data Element. The proposed in this study Extended Data Dictionary additionally provides for the classes "Features" and "Parameters" with their corresponding instances.

For the module parameters, the terms used in this study correspond to the following definitions which come close to the definitions given by Burbidge (1984).

A DESIGN PARAMETER is a variable which can take one or a combination of a limited number of alternative values, defined in advance. Examples of design parameters are the 'planning frequency', the 'planning horizon' and the 'length of planning periods'.

A REGULATORY PARAMETER is a variable to which a range of different values, not defined in advance, can be assigned.

An OUTPUT VARIABLE is a variable which can only indirectly be altered by changing parameter values (usually measures of performance).

An extensive set of parameters and values is given by Burbidge (1984) which has been adopted in this study to be included in the module structures and to establish relevance relations with the classification parameters and coverage relations with the module features given below.

Examples of Regulatory Parameters and Output variables are: Run Quantity, Transfer Quantity, Buffer stock (Regulatory), Throughput time, Storage time (Output). What interests in terms of the production management requirements is the capability of the PPC software package to provide decision-making for the Regulatory Parameters and to measure the Output Variables.

An ENVIRONMENTAL (or Input) VARIABLE is a variable whose value is imposed by the Environment in which the system exists and over which it has little or no control. Examples of this kind of parameters are given below:

- Traceability regulations (LOT TRACKING, CONFIGURATION CONTROL, SERIAL NUMBERS)
- Stock recording by government contract (YES,NO)
- State tax regulations, e.g. obligation for tracking of tax-free materials (YES,NO)
- State import-export regulations (e.g. tax-free raw materials if included in export products (YES,NO)
- Foreign currency transactions (YES,NO)
- Number of possible material suppliers (ONE,FEW,MANY)
- Reliability of suppliers (GOOD,FAIR,POOR)
- Number of competitors (NONE,FEW,MANY)
- Strike potential (RARE,POSSIBLE,FREQUENT)
- State regulations regarding hiring, firing and overtime (NEGATIVE,RESTRICTIVE,POSITIVE)

Module Features (the example of the MRP module)

The concept of module feature is characteristic in the world of commercial software being of a rather verbal nature. Usually it represents some sort of functionality and how it is performed. Its analytical description and documentation would require extensive data flow diagrams and flowcharts which is not practical for our purposes. Therefore, the "feature" is described by a short phrase and a set of values corresponding to alternative methods of performing its function. In this study an attempt is undertaken to codify, standardize and store in a knowledge base all known PPC software features and their alternative values. In terms of the requirements of the production management system, first we are interested in the very existence of a specific feature and then in its value, i.e. the method used. Next is given an example of codified features for the MRP module.

01. MRP operation (BATCH, INTERACTIVE, BOTH)
02. Form of MRP explosion (REGENERATIVE, NET-CHANGE, BOTH)
03. Requirements presentation (BY PERIOD, BY DATE, BY JOB)
04. Calculation of lead times (BY ITEM, BY PARENT-SON RELATION, BY BATCH PROCESS TIME)
05. Form of orders issued (WORK ORDER, FLOW ORDER, BOTH)
06. Pegging of Requirements (SINGLE LEVEL, FULL LEVEL EXPLOSIVE, FULL LEVEL IMPLOSIVE, MRP BY CONTRACT)
07. Selection of depth of requirements explosion (YES,NO)
08. Simulation net-change capabilities (YES,NO)
09. Lot sizing techniques (USER INPUT, PROGRAMMED RULES OF THUMB, PROGRAMMED OPTIMIZING)
10. Phantom items (YES,NO)
11. Blanket material acquisitions for large orders (YES,NO)
12. Allocations (BY PRODUCTION ORDER, BY CONTRACT, TIME-PHASED, AUTOMATIC MULTI LEVEL, ALL, NO)
13. Allocation of safety stocks (USER INPUT, PROGR.RULES OF THUMB, MATHEMATICAL)
14. Backflushing (YES,NO)
15. Government contract support, i.e. purchase and production orders by contract (YES,NO)
16. Lot tracking (YES,NO)
17. Variable Yield (YES,NO)
18. Configuration control (YES,NO)
19. Support of co-product and by-product creation (YES,NO)

In order to decide on the selection problem of module features and parameters, the production systems taxonomy and the generic PPC software knowledge bases have to be combined and inferenced under the logical schema of Figure 5.

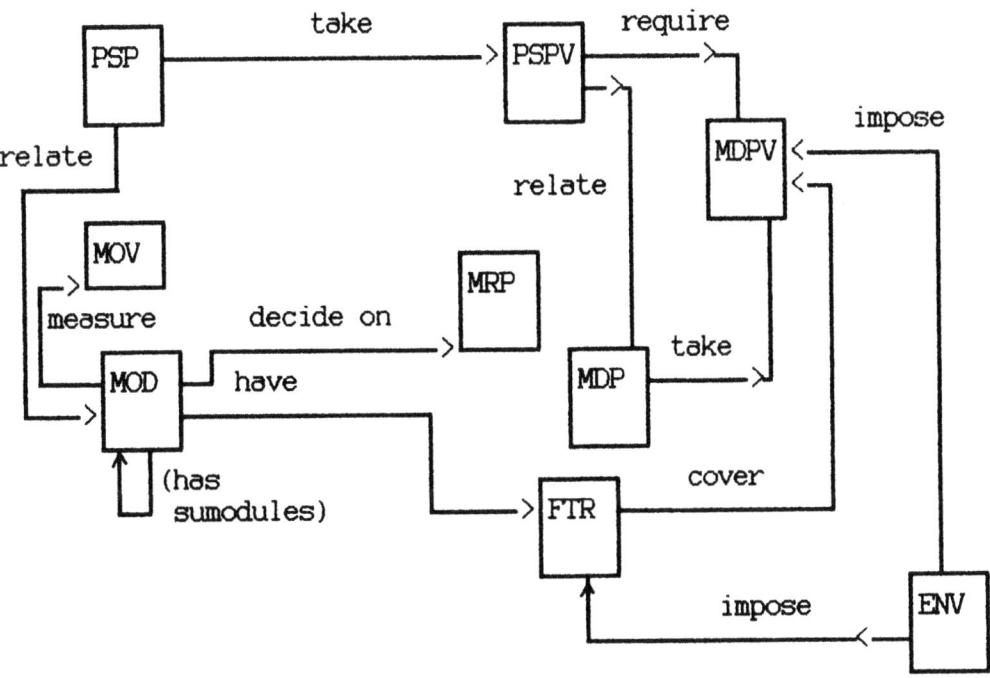

MDP: Module Design Parameters
MDPV: Module Design Parameter Values
MOV: Module Output Variables
MRP: Module Regulatory Parameters
EVN: Environmental Variables
FTR: Features

FTR: Features
MOD: Modules
PSP: Production System Parameters
PSPV: Production System Parameter
       Values

Figure 5. Logical Schema of expert system "technical evaluation"

A PPC software structure derived from a certain production system type can be covered by the existence of a particular software module (MOD) and/or the value of a module feature (FTR). For example the module MRP has as a submodule the "KNOW ABOUT THE CONSEQUENCES OF A LATE PURCHASE ORDER ON THE PROGRESS OF A CUSTOMER ORDER" which is required by the PSP parameter value "Demand Dependent on the Customer Order" and is covered by the software module feature MRP-06 (FULL LEVEL IMPLOSIVE).

SOFTWARE CUSTOMIZATION

The suitability of a packaged software module for a particular manufacturing firm cannot be judged only on the basis of its coverage of "ideal" requirements based on a technical evaluation model such as the one discussed in the previous sections of this paper. Every specific factory has some special requirements and an existing organisational and information system.

  Suitability means above all high chances for successful implementation depending on a plethora of basically qualitative factors. The problem in our opinion is mainly centered on the factor "change" . The change may affect either the software package or the manufacturing organization and its direction may be either towards adopting the "ideal" requirements or towards better organizing the existing methods.

The two extreme decisions are that either the procedures of the selected software package should be accepted exactly as they are, which should lead to extensive reorganization within the firm, or the software package should be modified to the extent that it should completely reflect the specific needs of the firm. Both solutions are impracticable in real production management environments. What is really done is to find a solution in between, i.e. to modify both the software package and the organization of the firm to a certain limited extent which represents an accepted total of tangible and intangible costs.

The first step towards the above direction is to find out the inconsistencies between the "ideal" PPC requirements proposed by the 'technical evaluation system' (100) and the current existing system in the manufacturing firm. Figure 6 represents a further brake-down of subsystem 'Change' (300), as defined in Figure 1, in more detailed mini expert systems, which serves this purpose.

The subsystems are the 'Determination of incosistencies' (310), which compares the PPC software profile to the organisational profile of the existing system, the 'Software customization' (320) which identifies the truly needed modifications of PPC software for the specific firm, and the 'Organisational change' (330) which determines the unavoidable organisational measures in order to implement the PPC software.

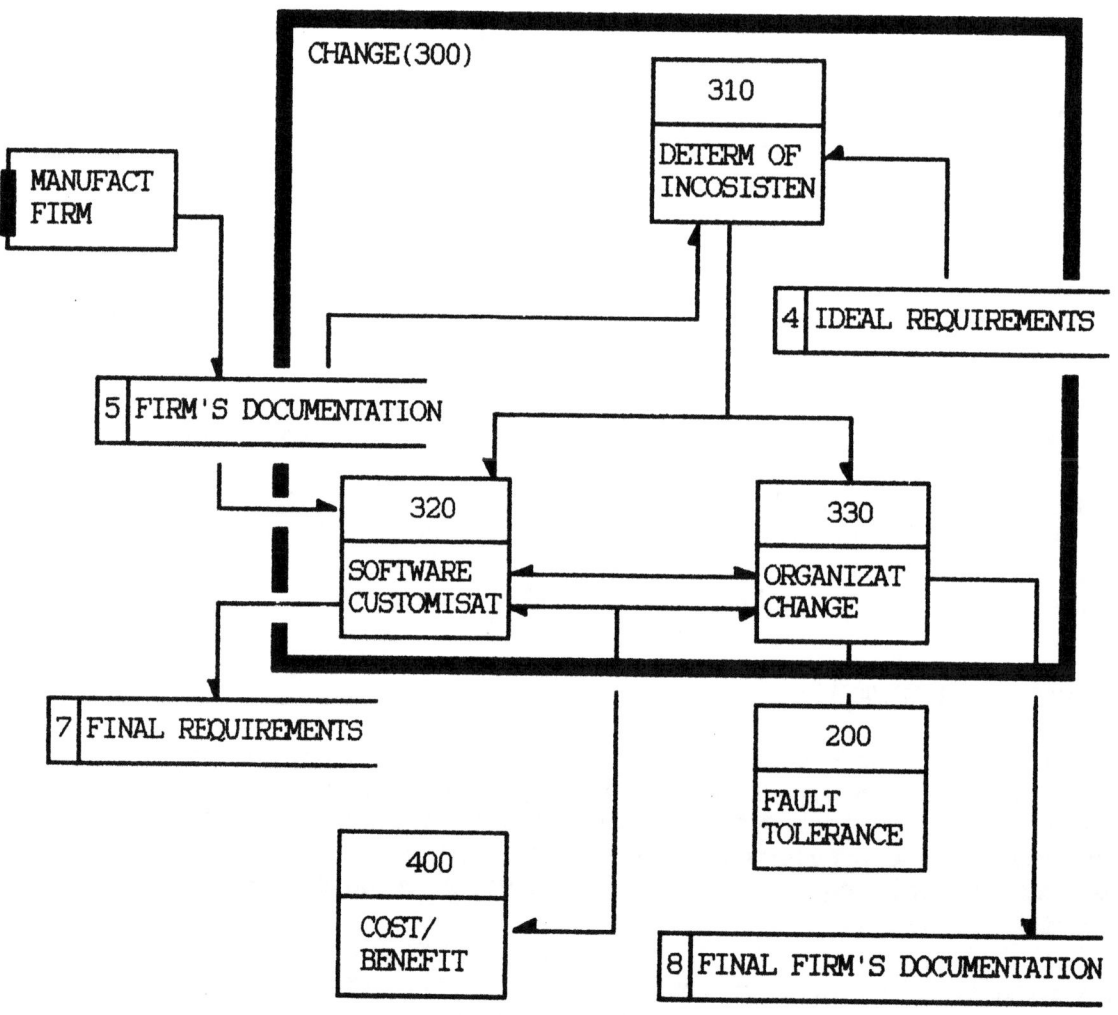

Figure 6. The subsystem Change (300)

The necessary documentation of the firm's current existing system
(Knowledge base 5, Fig. 6) follows the Extended Data Dictionary method
(Leslie, 1986), implemented with object-oriented programming techniques.
The classes of knowledge used are depicted in Figure 7.

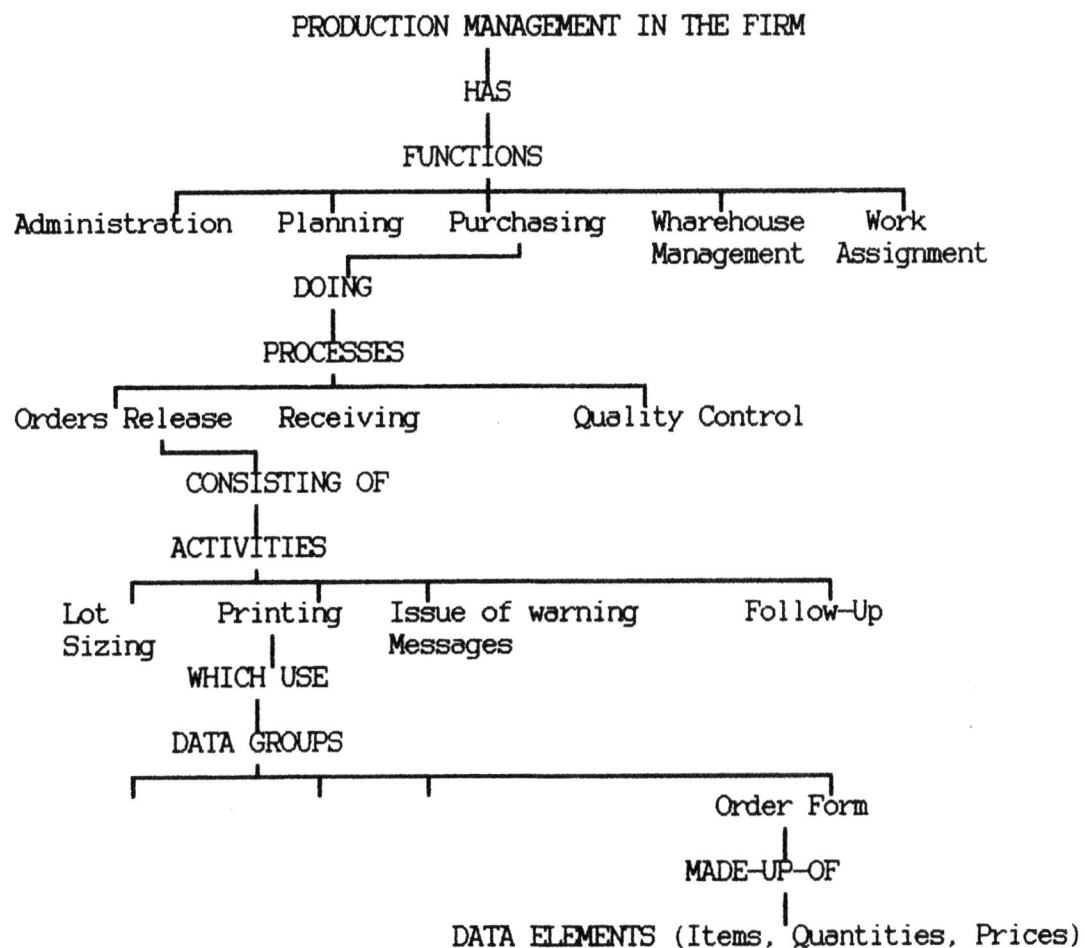

Figure 7. Firm's Documentation.

CONCLUSIONS

The proposed methodology, as far as the technical evaluation phase
described is concerned, can be applied during projects in manufacturing
firms seeking to buy/develop and implement integrated production management
software. In all these cases the main problem is to define suitable
software requirements, which will form the basis for either selecting
commercial software or starting an in-house development project.

Determination of PPC software requirements concerns both its pure
technical features, as compared with the characteristics of a certain type
of manufacturing setting, and its chances for successful implementation in
a particular environment of a specific manufacturing firm characterized by
a plethora of human, cultural, financial, educational, technological,
marketing, strategic, and other possible factors, mainly of a qualitative
nature.

Such evaluation and selection decisions are far too complex to be faced
either by quantitative multi-criteria methods or by simple guessing. The
systematic description of knowledge that AI technology requires is a
valuable distillation of a vast amount of experience (Sheil,1987) which
only can lead to a rational decision. This gave the initiative for

starting an AI project in order to codify the required knowledge across the above described guidelines in an actual expert system. A prototype system is now implemented using the 'KnowledgePro' expert system shell.

It is expected that the above expert system will be of great help to both potential buyers of production management software facing the selection and implementation problem and to software houses facing the software configuration problem for each particular customer. It also is expected to be a significant educational tool that will help those trying to implement advanced production management methods and software in uninformed production environments. The hypercard technique, which is part of the expert system, is particularly helpful for training and communication of knowledge as opposed to simply inferencing logical rules.

REFERENCES

Burbidge,J.L., A Production System Variable Connectance Model, Cranfield Institue of Technology, Cranfield, 1984.

Davis, G.B. and Olson, M.H., Management Information Systems, McGraw Hill,1985.

Doumeingts,G. et al, Knowledge-based System for the Design of Production Management Systems, in: J. Browne (Ed), Knowledge Based production Management Systems, Elsevier (North-Holland), IFIP, 1989.

Foerster,H., H.Hoff and Miessen,E., Marktspiegel PPS-Systeme auf dem Pruefstand, Verlag TUV Rheinland Gmbh, Koeln, 1986.

Gane,C. and Sarson,T., Structured Systems Analysis, Prentice-Hall, 1979.

Hoff,H. und M.Virnich, Wie man ein PPS-System richtig auswaelt und einfuert, io Management-Zeitshrift 55 (1986) 9 (1.Teil), 55 (1986) 10 (2.Teil).

Kellner,R., Expert System Technology, Prentice-Hall, 1987.

Kettner,H., J.Schmidt and H.R.Greim, Leitfaden der systematischen Fabrikplanung, Carl Hanser Verlag, Muenchen, 1984.

Krallmann,H., Expertensysteme fuer die computerintegrierte Fertigung, FB/IE 35 (1986) 3, pp. 100-106.

Leslie, Systems Analysis and Design, Prentice-Hall, 1986.

Mensel,G.und J.Michel, Moeglichkeiten des Einsatzes wissenbasierter Systeme in der Fertigung, ZwF 80 11, pp. 495-500, 1985.

Sepehri,M., Newest Manufacturing Software Packages Offer Modules Which Meet Specialized Needs, Industrial Engineering, 28,October, pp. 32-43, 1985.

Schmenner,R.G., Production/Operations Management, SRA, Chicago, 1981.

Schmitt,T.G., T.Klastorin and A.Shtub, Production classification system: concepts, models and strategies, Int.J.Prod.Res., 23 (1985) 3, pp. 563-578.

Schomburg,E., Entwicklung eines betriebstypologischen Instrumentariums zur systematischen Ermittlung der Anforderungen an EDV-gestuetzte Produktionsplanungs-und-steuerungssysteme im Maschinenbau, Dissertation, TH Aachen, 1980.

Sheil,B., Thinking about artificial intelligence, Harvard Business Review (1987) July-August.

Speith,G. and U.Brief, BAPSY - ein Instrumentarium zur Beurteilung und Auswahl von Produktionsplanung - und steuerungs - Systemen, FIR - mitteilungen (1982) , TH Aachen, Nr. 43 (Juni).

Tatsiopoulos, I.P., A systematization of knowledge for the selection and implementation of materials management software, J.Browne (Ed.), Elsevier (North-Holland), IFIP, 1989.

COMPUTER ASSISTED TECHNIQUES FOR THE PRESENTATION OF PRODUCT DESIGN INFORMATION TO THE DESIGNER

Luis M. T. Cortez Pereira, (1) Francisco J. L. Figueiras, (2)

Paulo J. S. P. Leal, (2) Luciano L. O. Faria, (2)

(1) SOREFAME, Amadora, P

(2) IST/CEMUL, Lisboa, P

1 - INTRODUCTION

The development of Information Technologies - IT leads to the use of CAD in design and project bureaux. However these IT applied to automated manufacturing and to automated or robotic assembly created the needs of revising the rules used by the designers in order to take into consideration the problems raised by these new technologies.

For this purpose some studies have been done by different projects in the CIM-ESPRIT area [1] [2]. These projects analysed the product design procedure and developed an overall strategy and functional design procedures; they arrived at the conclusion that:

  - an overall strategy is required for design with respect to automated manufacture;

  - the product range should be rationalized;

  - the product structure should be determined;

  - the design of components should consider the general strategy.

The overall strategy should include not only the conventional criteria - such as: performance, reliability, maintainability, safety, aesthetics, etc..., - but also the aspects concerning the robotic or automated manufacture and assembly - such as: chain tolerance, reduced level of positional uncertainty, etc....

The rationalisation of the product range should lead to reduce the possible variants, ensuring the use of similar methods of assembly, minimising the capital investments in hardware, maximising the utilisation of robots, attaining and maintaining the pre-defined product quality.

The product structure should be analysed in order to define the function and form of the components, minimising their numbers and not increasing their cost. So, before commiting to manufacture, the design should be changed as a result of an iterative procedure.

The design of components should consider the capabilities of the available automated manufacturing processes and the simplification of all assembly operations, such as: presentation, orientation, relative positioning, parts mating, joining. At this late stage the iterative design procedure can lead to change the product structure.

In order to achieve this iterative procedure all the departments involved in production should participate in it throughout the different stages.

Having in mind the possibilities offered by the application of these concepts, but knowing that putting CIM into operation should consider the real problems of the industry and the specific problems of each company, an entreprise - SOREFAME - has been chosen for a practical application, in order to know how the presentation of the product design information to the designer should be done using the computer assisted techniques. This choice took into consideration that SOREFAME has both an important design bureau and workshop specialised in designing and manufacturing mechanical equipment for hydro-electric power stations and rolling stock.

2 - IDENTIFICATION OF THE PROBLEM

One of the problems that affects big, medium and small size companies is the compatibility between the design specifications and the cost of the product, forseeable or pre-determined. And obviously this problem concerns not only the design bureau, but also the manufacturing sectors (methods, commissioning, workshop, quality control, etc...).

The designer is responsible for the definition of the product, should define the corresponding specifications, having in mind its performance and quality level; and each component should be detailed, considering its function in the final product.

Unhappily, in many cases these specifications are prepared not having into consideration the manufacturing cost; for instance, tolerances or dimensions (e.g. hole diameter) are chosen not in accordance with the available facilities at the workshop: machine-tools or tools able to perform the work as specified.

In many design bureaux the information available, concerning machines and tools, is rather unsuficient to the designers; as a matter of fact an information such as the maximum diameter that could be drilled by a machine tool, or the standadized drilling range, does not give enough guidance.

What the designer should know, in this specific aspect of holes drilling, for a certain drill range, for a material to be drilled with a defined accuracy given by a machine tool (drilling machine, milling machine, boring machine, etc...), is the existing preferential tool set that assures a definite tolerance and an acceptable manufacturing cost. Not having this information, the designer could be led to impose chimeric specifications or feasible specifications but at a very high cost.

The workshop tries, as a rule, to comply with these specifications coming from the design bureau. If the workshop is not in a close touch with the design bureau, as it happens often, the people directly in charge of manufacturing does not know the role of each component, the required tolerance level, or the performance of the product.

When the workshop faces a procedure that cannot be performed, or considers this procedure very costly, decides:

- either to change the procedure or the tolerance, without informing the design bureau, leading probably to a lower quality level or bad performance of the product;

- or to inform the other sectors of the factory of their inability to comply with the specification; this leads to interrupt or modify the manufacturing plan, as the workshop should await for the required alterations: meetings with the other departments to discuss the alternatives to be adopted, new approval from the client, etc... .

In any case, this is time consumming and could be very costly.

3- THE PROPOSED SOLUTION

In order to settle the difficulties above referred the designer could ask for a computer software giving the following information for each manufacturing operation:

- cost,

- time performing,

- machine (s) where the operation is achieved,

- tools required for the operation,

- tolerance (dimensional, form, position, rugosity).

This requires for each company the settlement of a data base, allowing the designer to make the best choice. The software should give to the user the different possible solutions by increasing cost degree.

As this data base should be used in a personnal computer, the data included should be limited to the computer capacity; so, the choice of data should consider the real needs of each company and in our approach the data should include only:

- materials,

- dimensions range,

- machine tools,

- operations or procedures,

- parts finishing quality.

Other factors influence the quality level of the operation procedure or process. But, if we supply to the designer an excessive information, this could lead to :

- a great increase of the data base,

- difficulties in the data base structure,

- system with a long delay response,

- a not friendly user's software.

And, in consequence, the designer could be tempted to ignore the existence of this powerful "tool".

These considerations led the authors in this study applied to SOREFAME, and after discussed the problem with different departments of this company, to limit the data bank to the 5 groups of information, considering the types of equipment they manufacture. So:

a) The materials should include only:

- carbon steel, as rolled,

- cast steel,

- stainless steel.

b) The dimensions range will include:

- small dimensions (up to 12 mm)

- medium dimensions (between 12 and 500 mm)

- high dimensions (larger than 500 mm)

c) The machine tools will be grouped in the following families:

- portable drilling machines

- fixed drilling machines

- milling machines (NC or conventional)

- boring machines: floor type

                         table type (NC)

- planning machines

- vertical lathes: big size

                    medium size (NC)

- sliding lathes: big size (NC)

                  medium size

                  small size

d) The machining operations to be considered are:

- planning   - screwing   - slotting   - milling

- boring   - grinding   - drilling   - honning

- threading   - sawing   -turning

e) Parts finishing quality:

- very high

- good

- fair

Some of the other factors not included in this software could be related to:

- machine tools: in a certain machine tool with a certain tool, variations in the accuracy
of positioning and finishing could arise, due to the age of the
machine, lack of its rigidity, geometrical errors in its initial
manufacturing, etc... ; these errors are more relevant in the final
finishing operations.

- parts to be machined: their physical proprieties can vary, as well as the resistence of
defects (pores, inclusions, cracks) influence the quality of
finishing.

- tools: lack of rigidity of the mounting devices, defects in the tool material,
tool duration not allowing a complete machining operation,
influencing the quality of finishing.

- operations: their experience and skilfulness influence the operation mainly in
conventional machine tools.

These factors have more or less influence in the machining operation depending from
the required accuracy.

In the present phase of this study the authors consider that the group technology concepts
could not yet be applied as their introduction requires a manufacturing organization the
SOREFAME has not yet attained. And this is due to the great variety of one of a kind

manufacturing parts belonging to different equipments produced; gates, valves, hydraulic turbines, lifting and handling equipment, passenger cars, locomotives, etc; so, in the case of SOREFAME the adoption of the group technology is not for the moment, an easy task and it has been considered preferable to limit the study to the 5 groups of information above mentioned.

4 - SOFTWARE DEVELOPED

The software developed was written in DBASE III Plus language and manages the data base. The data base is fulfilled by the user and could be subjected to alterations or additions.

The first group of data includes the finishing accuracy, the dimensions range, materials, operations and equipment (machines) to be used.

The first routine of this software comprises the different menus allowing the user to make the choice of the data. The data are stored for each application and in consequence, the user can alter one or more data and get the corresponding information in other applications. The menus give access to other sub-menus introducing new options if required.

After chosen the data the software accesses to a "key" file that provides a value necessary to access in turn to the results file.

This structure is very worthwhile as reduces considerably the dimension of the results file, considering that different data combination can lead to the same results.

At the end, the results are presented in the screen to the users: equipment, tools, tolerances, cost and time length of operation.

In fig. 1 are presented the data base structure adopted, as it is detailed in section 3 of this paper.

In fig. 2 are presented the five results for each particular choice.

TYPE OF MATERIAL
Carbon Steel
Cast Steel
Stainless Steel

DIMENSIONS RANGE
Small (up to 12 mm)
Medium (between 12 and 500mm)
Large (greater then 500mm)

MACHINE TOOLS
Portable Drilling Machines
Fixed Drilling Machines
Milling Machines
Boring Machines
Planning Machines
Vertical Lathes :- Big Size   - Medium Size (NC)

Sliding Lathes :- Big Size (NC)
               - Medium Size
               - Small Size

MACHINING OPERATIONS
Planing
Screwing
Sloting
Milling
Boring
Grinding
Drilling
Honning
Threading
Sawing
Turning

FINISHING QUALITY
Very High
Good
Fair

fig. 1

RESULTS
Cost
Time   Performing
Possible   Machine(s)
Tools   Required
Tolerance

fig. 2

5 - SOME FINAL REMARKS

The practical application of different studies, mainly [2], described in this paper, leads to the conclusion that the concepts adopted are quite adequate to the use of designers working in design bureaux of factories which manufactures are related to machining operations.

The time and cost of creating the required data base is largely compensated by the advantages of the user of such software. As a matter of fact the methdology developed lead to the creation of a specific data base, as the data base is intended to settle specific problems of an entreprise. With the introduction of data in the software we hope to reduce at SOREFAME 70% to 80% the number of corrections that in the past occured during the working preparation and the manufacturing itself. Besides this, the data base presents other advantages for the sectors of the factories related to production; as a matter of fact, the information included in the base will help considerably other aspects of internal organisation of the workshop, easing the subsequent introduction of CAM and CIM.

Although it is not yet developed, it seems possible to extend this kind of software to other types of manufacturings, for instance, forging,stamping,deep drawing, etc... . And for certain types of equipment this also could apply to automated or robotic assembly; in this case the study should take into consideration the tolerances allowed by the joining processes (welding, bonding, fastening, riveting, etc...), and the adoption of the group technology concepts.

But it belongs to each company to create its own data bases avoiding the inconvenients above mentioned, as the software developed is very effective and its use is simple and does not require an imposing hardware.

REFERENCES

[1] "Design Rules for the integration of Industrial Robots into CIM Centres,
Esprit Project 75

[2] "Product Design for Automated Manufacture and Assembly",
Esprit Project 338

INTEGRATED MODELLING OF PRODUCTS AND PROCESSES USING ADVANCED COMPUTER TECHNOLOGIES - IMPPACT (ESPRIT PROJECT NO. 2165)

Andreas Meier

1. OBJECTIVES OF IMPPACT

The aim of IMPPACT is to promote CIM technologies by developing and demonstrating prototype software for integrated product and production process modelling systems. The desired information flow between design, process planning and production requires a conceptual approach that is based on an integrated view of design, planning and production activities.

To reach that goal, a flexible, adaptable system architecture is necessary with the capability of handling various software packages with individual structures and environments. To enhance the performance and functions of different systems, their existing structures have to be combined in order to compare the results with an overall product and process model. Additional information on machining facilities, standards, etc. as well as knowledge, such as production experience and rules, must be provided.

The architecture must be open and has to permit the integration of existing autonomous systems with future software.

The objective is, on the one hand, to develop tools which support the integration of various software modules and, on the other, to build up and maintain company-specific product model structures, while providing knowledge and information support functions, as well as simulation modules for evaluation and optimization of the product through simulation of machining processes at an early stage.

The integrated product and production process model for IMPPACT will be adjusted to design tasks and should be able to involve all information about standards, supplier parts and processes.

The capability of generating intermediate workpiece stages is regarded as an essential requirement in providing planning systems for successive manufacturing processes. It is also very important for the project not only to look at the integration of design and planning production in the sense of a directed information chain, but as an information loop providing the upward feedback of planning and manufacturing experiences (e.g. to the design process).

The modelling tools given to the user provide knowledge of materials, the production processes and further production possibilities as an aid in the modelling process.

Knowledge-based information for both product and process modelling will in principle be made available on the basis of production experience, material properties, machine and tool facilities. This is to bring production knowledge back to the designer and process planner, guiding him to do the manufacturing "right - first time". The system is to include rules both for design and production problems.

2. PRODUCT AND PRODUCTION PROCESS MODELLING: STATE OF THE ART

System Architecture

The goal of Computer Integrated Manufacturing (CIM) draws upon many of the traditional areas of manufacturing automation including CAD, CAM, MRP, etc. It is not the sum or totality of these components but the linking of them into a complete system that will satisfy the enterprise´s business strategy and objectives.

Concerning the architecture of the system, IMPPACT will take into account the results of the AMICE project (CIM-OSA, Esprit Project 688) (1). This architecture is developed to be appropriate to a wide range of users, who would then select or reject components according to their needs. CIM-OSA shows how current systems can evolve towards possible future systems, in discrete, manageable steps.

Important international standardisation activities for open system architectures take place in ISO TC184 (Industrial Automation Systems) SC5 (System Integration and Communication). IMPPACT influences the development of STEP (Standard for the Exchange of Product Model Data) within ISO TC184 SC4 (Industrial Data and Global Manufacturing Programming Languages).

Reference Models for Product and Production Process Modelling

The state of the art in the area of product and production process modelling for CIM results from various research and standardisation activities. Integrated applications which are in use within companies are in general based on older (CIM) technologies. Only a few advanced CIM applications which are in use now are based on dedicated software.

The current developments of a reference model for data exchange in an open integrated environment are strongly influenced by US research programs. The most important programs are

ICAM	(Integrated Computer Aided Manufacturing),
PDDI	(Product Definition Data Interface),
GMAP	(Feature Modelling) and
CALS	(Computer Aided Logistic Support),

sponsored by the DOD (Department of Defense) (2, 3, 4, 5). Most of the results are not publicly available, but they eyert a strong influence on standardisation efforts such as PDES (Product Data Exchange Standard) and therefore on STEP (6).

Two ongoing ESPRIT projects have some influence on the development of a reference model for product and production process modelling, the AMICE-project (7), leading to an Open System Architecture, and the CAD*I-project (8), aiming at the development of standard interfaces between CAD systems.

Data Handling

CIM components may store, access and exchange data in various ways. Each application may use its own database management system, but it is also possible that a database is shared by several applications. When systems use their own local database, they may exchange data through neutral files such as IGES, VDA-FS, SET or STEP. A general concept of data-handling that supports both aspects has been developed within the US Air Force PDDI project.

An important reference model for the development of DBM Systems is the ANSI/X3/SPARC reference model.

Within the ESPRIT CAD*I project (9), a concept for a common database has been developed for analytical and experimental models and results, and aims to decrease the amount of redundant information.

For the exchange of neutral files, preprocessors and postprocessors are required. It is necessary to make processors independent of the conceptual scheme and therefore the concept of metafiles, containing formal specifications of the conceptual scheme of the files to be exchanged, is used in the PDDI project. Processors are then written on the basis of Data Definition Language (DDL). STEP applies the same concept, based on the object oriented specification language for information modelling EXPRESS (10, 11).

Interfaces for several CAD systems are beeing developed for the CAD*I neutral format (which is close to STEP) within ESPRIT Project 322.

Application Interface

To improve the system modularity, to enable independent development of application modules and to ensure software exchangeability, neutral programming interfaces are required. CAM-I´s Geometric Modelling Program (GMP) has developed the Applications Interface Specification (AIS) (12). This neutral programming interface is unfortunately restricted to mainly geometrical functions. CAM-I´s AIS has also a poor separation of the logical content and its physical, language-oriented representation. Standardisations are also discussed in German organisations such as DIN (Deutsches Institut für Normung) and DFN (Deutsches Forschungsnetz) (13).

Developments are needed to extend neutral programming interfaces to the reference data model that will be the basis for IMPPACT.

Part Modelling

The geometric modellers with respect to solid modelling in today´s CAD/CAM systems are mostly based on two approaches. The first is called CSG (Constructive Solids Geometry), where geometric volume primitives are combined with special volume operators. The second is a boundary structure oriented approach using analytical defined geometry.

New directions in the development of modellers are using the Non Uniform Rational B Splines (NURBS) representation, or combine NURBS and analytical representation. No system has so far managed to fully integrate NURBS and a boundary structure oriented modeller based on analytical representation. A system integrated along these lines will offer the user high functionality and will have a strong competitive position in the market (14, 15, 16, 17, 18, 19).

Another trend in the development of geometric modellers is to use object-oriented techniques. At least one system on the market claims that the whole system was developed using an object-oriented language. Experience has shown that object-oriented techniques can contribute significantly to the productivity when developing geometric modellers.

It is known that the conventional ways of modelling with primitive basic elements and purely geometric data are not sufficient to achieve the goals set for product modelling and CIM. Worldwide, most of the approaches in integrating the computer further into the design and production process are using feature-based and constraint modelling techniques to relate most of the product related information to the geometric model.

Some experiments have shown the usefulness of form feature data, particular in the interface between solid modelling and process planning (20, 21, 22, 23, 24).

Process Modelling for Parts

Process planning, being the link between design and production, is affected to a high degree by company internal and external developments and requirements. This fact is expressed in an increase in planning capabilities and new additional planning tasks (16).

As an important aspect of integration, it becomes evident that - in spite of many partially integrated systems - an integrated overall solution is still missing.

Basically, one has to distinguish between simple and complex planning tasks in this context. Complete vertically integrated solutions, that is to say working connections between CAD, CAP and NC programming, have only been realised in some instances e.g. for rotational parts (14).

Looking at company internal planning processes and information in the case of complex tasks, we see the information exchange between design and process planning is nowadays still being done conventionally by means of technical drawings and parts lists. The drawing is set up manually by the designer or with the help of CAD systems. Today´s existing feasibilities in creating information as well as the interfaces of existing CAD systems (e. g. IGES) do not sufficiently consider the requirements of process planning.

New methods for information processing, such as expert systems and modelling techniques for generating intermediate workpiece stages, offer new possibilities of developing CAPP (Computer Aided Process Planning) systems which support or even perform heuristic activities of the process planner. Process planning faces a number of influences which will change the contents of the planning process (25, 26).

For the development, new software development techniques are necessary (e. g. artificial intelligence and method banks). Special importance is attached to the idea of integration by product data models.

3. THE IMPPACT CONCEPTUAL FRAMEWORK

The goal of IMPPACT is the development of a new generation of integrated modelling systems for product design, process planning, and operation planning including machine control data generation.

Based on a product and production process model, highly sophisticated systems for part and process modelling for sheet metal and complex shaped-parts applications will be developed. Fig. 1 shows the corresponding project structure.

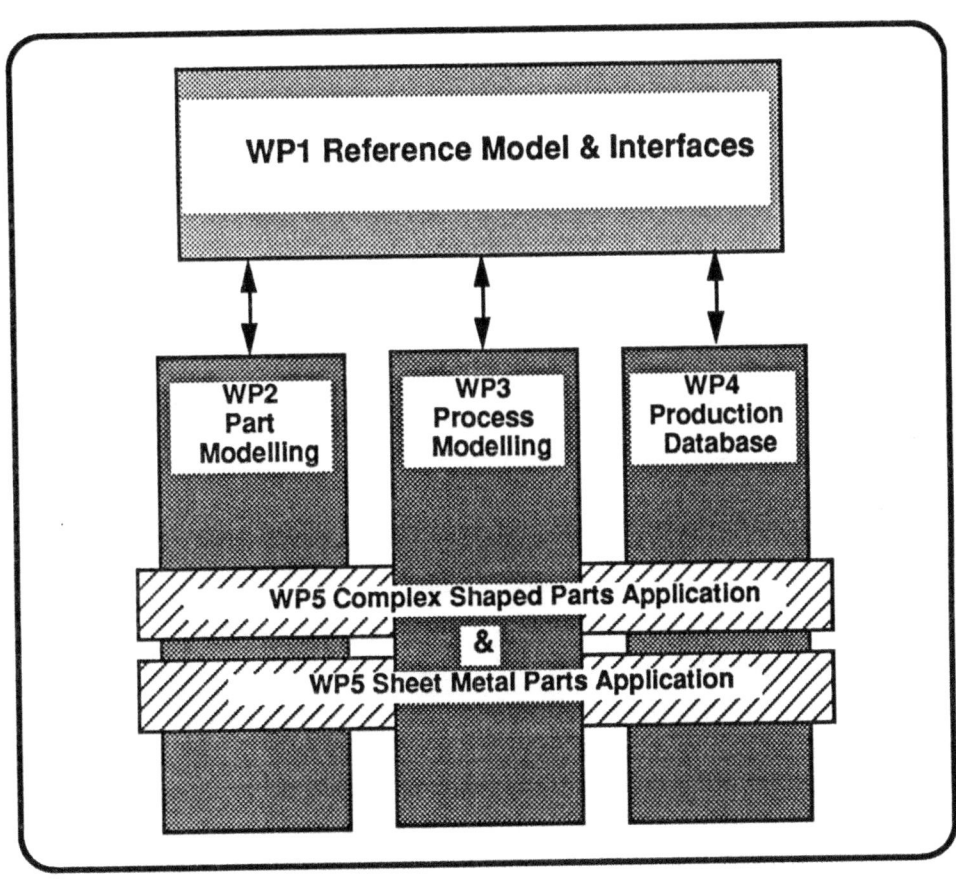

Fig. 1: IMPPACT project structure

Reference Model and Interfaces

The goal of the reference models is to provide a general specification and concept for an open system architecture describing an integrated product and process modelling system. The reference models have to consider detailed requirements from engineering, design, process planning and shop floor production. From these, more general reference models for functions and data structures, which form the common basis to the project, will be derived. This will permit work to be done especially for part (or product) and process modelling and for the design of a production database.

The development of the reference model comprises a function model and a data reference model (Fig. 2). Starting with the definition of the requirements for each application area, followed by the development of reference models for parts design and their production processes, a generic reference model will be derived.

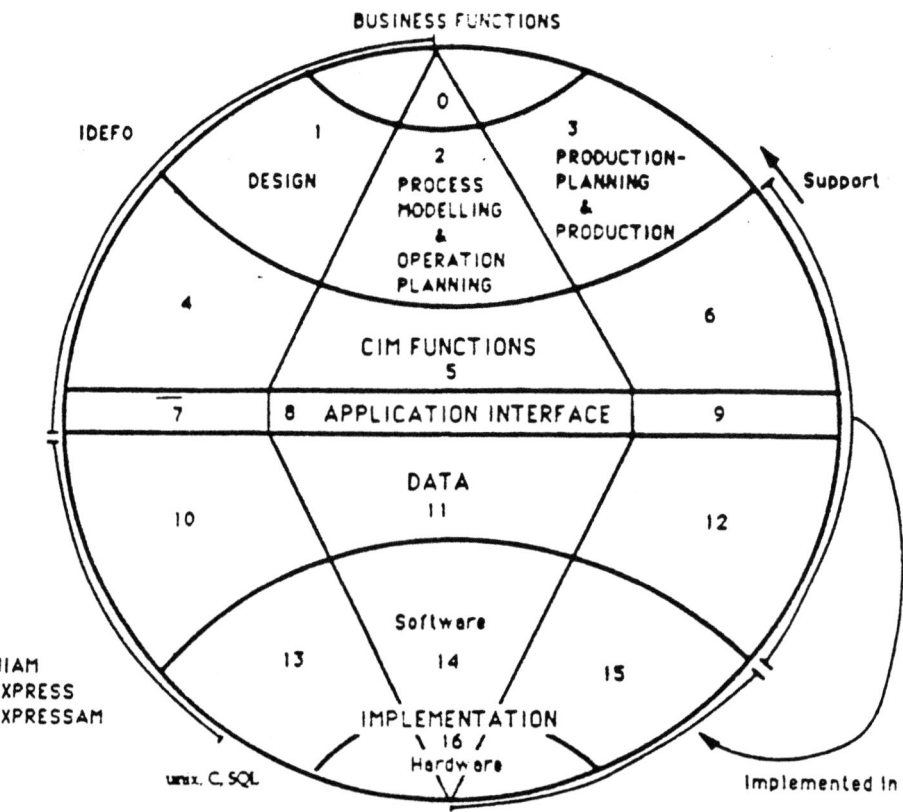

Fig 2: **Concept for a reference model for an open system architecture**

The global data reference model for CIM applications in the IMPPACT project will be developed in a sequence of logical steps. From the discussions within IMPPACT about areas of common interest, such as the part model, the feature model, the production process model and others, a first proposal for the reference model already indicated how these topics are all interrelated. Work will proceed with the development of two detailed reference models, one for part modelling and one for process modelling. Once agreement exists about the global architecture of these models, the detailed definition in EXPRESS will follow. Then both models will go through the final integration stage. The interface boundaries between the CIM applications have to be indicated in the reference model, and application interface functions for selected entities will then be defined. During the project, the model will be refined and improved on the basis of experiences with implementation.

The reference model will then consist of general terms and more specific extensions for complex shaped part modelling and sheet metal part modelling, or even very specific terms for the demonstrators. Attempts will be made to keep the reference model as general as possible, so that the common basis for the project will be provided.

Part Modelling

The main goal for IMPPACT is to integrate product or part modelling with production process modelling. Integration, however, only makes sense, if the parts to be integrated already have a high standard before their functionality is considered. Therefore, the main issue of part modelling must be to overcome the functional limitations in the design of parts by making several views available to the user through the combination of solid and surface techniques.

The second issue is of course the integration aspect itself. If it is assumed that parts to be manufactured with the integrated system are unique in the sense that no part is entirely the same as the previous one, the handicap for the upstream of data can be solved by using either

- a parametric definition of parts or
- a decomposition of the parts into regions (features) which are used or applied more frequently or
- both, by parametric definition of features.

Based on the general goals for part modelling and the selected applications, there are three activity areas in Workpackage 2 on which IMPPACT will focus:

- Complex shaped parts
- Sheet metal parts
- Feature modelling

In part modelling for complex shaped parts and sheet metal parts, the specific needs for the design of this part spectrum will be addressed; the functionality will be improved to fulfil advanced user requirements (Fig. 3 and Fig. 4).

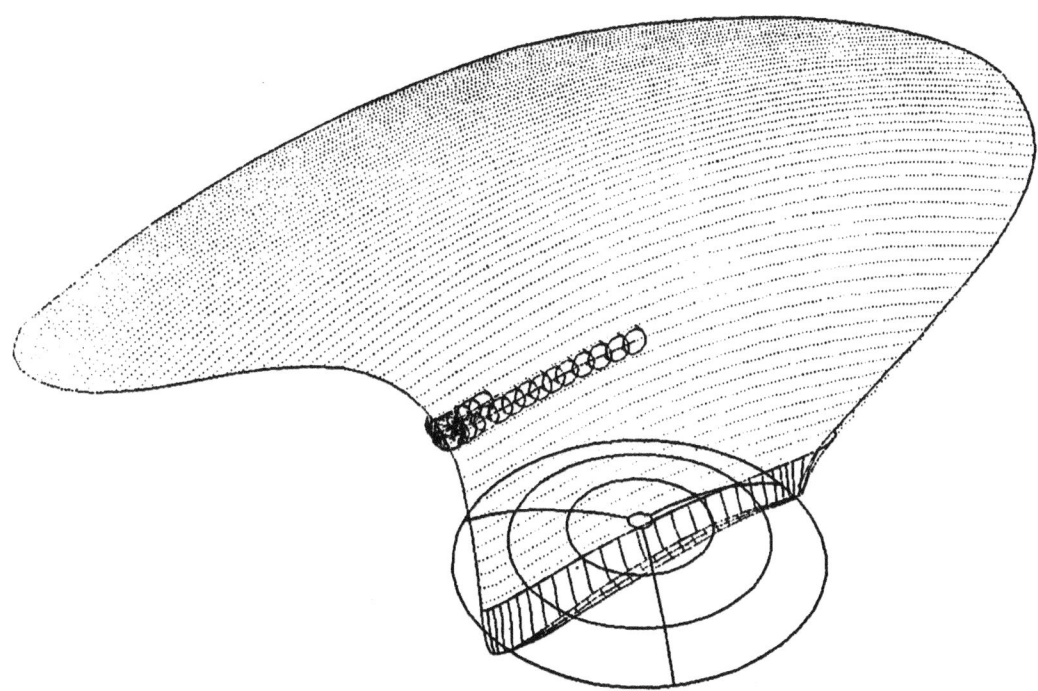

Fig. 3:   Smooth transition between a combined solid/surface model

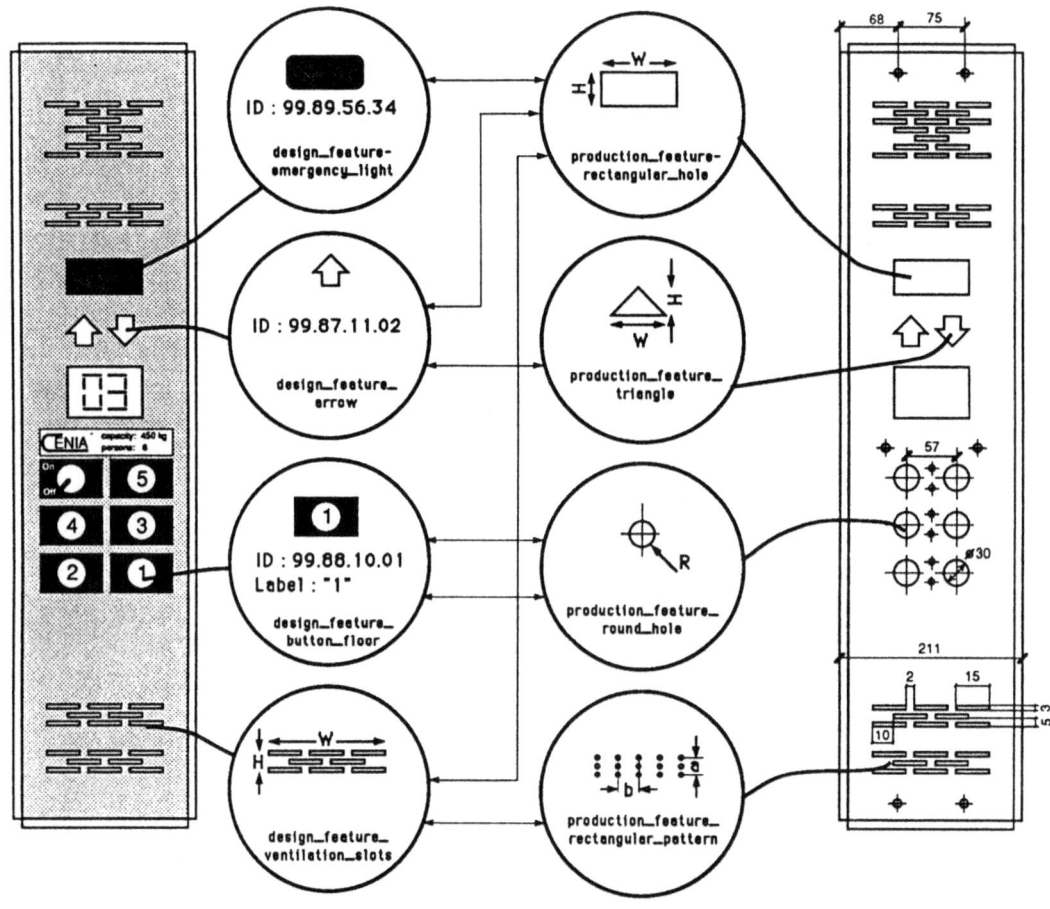

design view   production oriented view

Fig. 4:   **Different views of design and production**

Different experts in the manufacturing process have different views of the product. The designer must first satisfy the functional requirements, whereas the production planner looks at the produceability. Both views lead to different descriptions of the same product. If a part is described in terms of features, the analysis at the demonstrator sites leads to the definition of four types:

a. Design features or functional features. These are defined from the viewpoint of the functional behaviour of the product and are oriented towards their primary function.

b. Analysis features play a role in all stages of the manufacturing process, from design to production. They differ from design features in the sense that they may not refer to the primary function of the feature. For example a feature whose primary function is to reduce stresses may have important characteristics for thermal analysis (the casting/cooling process for example).

c. Production features. For these, one or more production processes exist. Although they can be produced, no direct selection of a specific production process is made.

d. Production activity features, as a minimum requirement, relate one-to-one to a specific production activity. Up to now, their need is not yet clear.

The example in Fig. 5 shows the possible use of a feature.

The feature modelling approach will apply to both complex shaped and sheet metal parts. A common concept will be developed, to be implemented in various specific software systems.

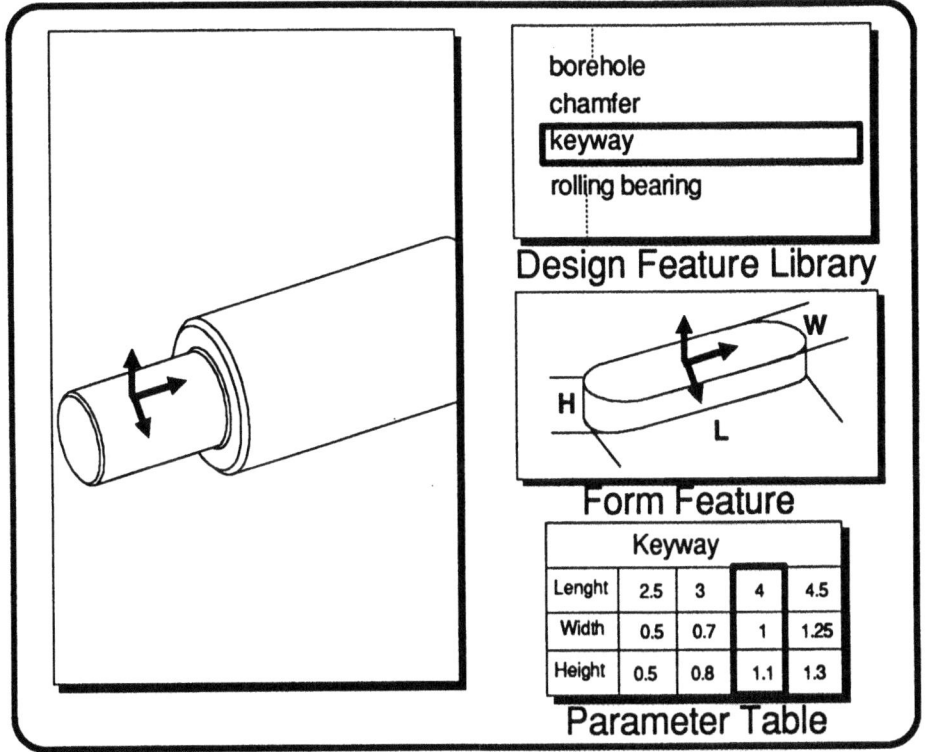

Fig. 5: **Selection and parameterisation of a feature for a shaft and collar connection**

Process Modelling

Process modelling is the processing, manipulation, extension and representation of general as well as application-oriented process knowledge and information referring to the product spectrum and available production facilities. The general process modelling concept is based on the factory model oriented part containing available equipment such as machines, tools and fixtures.

In particular, the access to all information and the extension of company-specific process knowledge must be guaranteed. This will be carried out on the basis of planning-oriented workpiece models containing all geometric and technological information needed for the planning task. According to each machining operation, all intermediate stages will be generated and represented.

The process model represents a special connection between the "product model" and the "factory model". It will provide information which is necessary to fulfil the planning tasks in an optimised manner and can roughly be divided into the parts described in the following.

One central part contains the general technological knowledge about available processes in the company and possibly at their suppliers. It contains, for example, methods for the determination of cutting data, times and costs, machining strategies, production features and standard process plans as well as company-specific heuristic knowledge about machining processes that is based on experience from the production process.

The product-model-oriented component of the process model contains a detailed description of the manufacturing process for every specific workpiece, including all necessary process, operation and NC information. Alternative process and operation sequences attached to intermediate workpiece stages are also provided.

The process model is a complete description of how to make the part. It includes all the necessary instructions to human operators and machine controllers for direct manufacture. There should also be the commercial data (costs, times) and management information for satisfactory control of the factory. The precise content and format of the information will vary from company to company.

The process modelling activity is concerned with the development of prototype systems for process planning and operation planning for sheet metal parts (e.g. punch/nibbling, contour routing and bending) as well as complex shaped parts for the machining processes of drilling and milling.

A Process Planning Supervisor will evaluate the product design (i. e. analyse design features) and administrate submodules (both knowledge-based and generative) to make complete process plans for all types of products and processes (Fig. 6). A challenge will be to prepare the system to take full advantage of the new design techniques using features. Such features will be preplanned in the process planning system, and selected and entered into the total process plan where necessary.

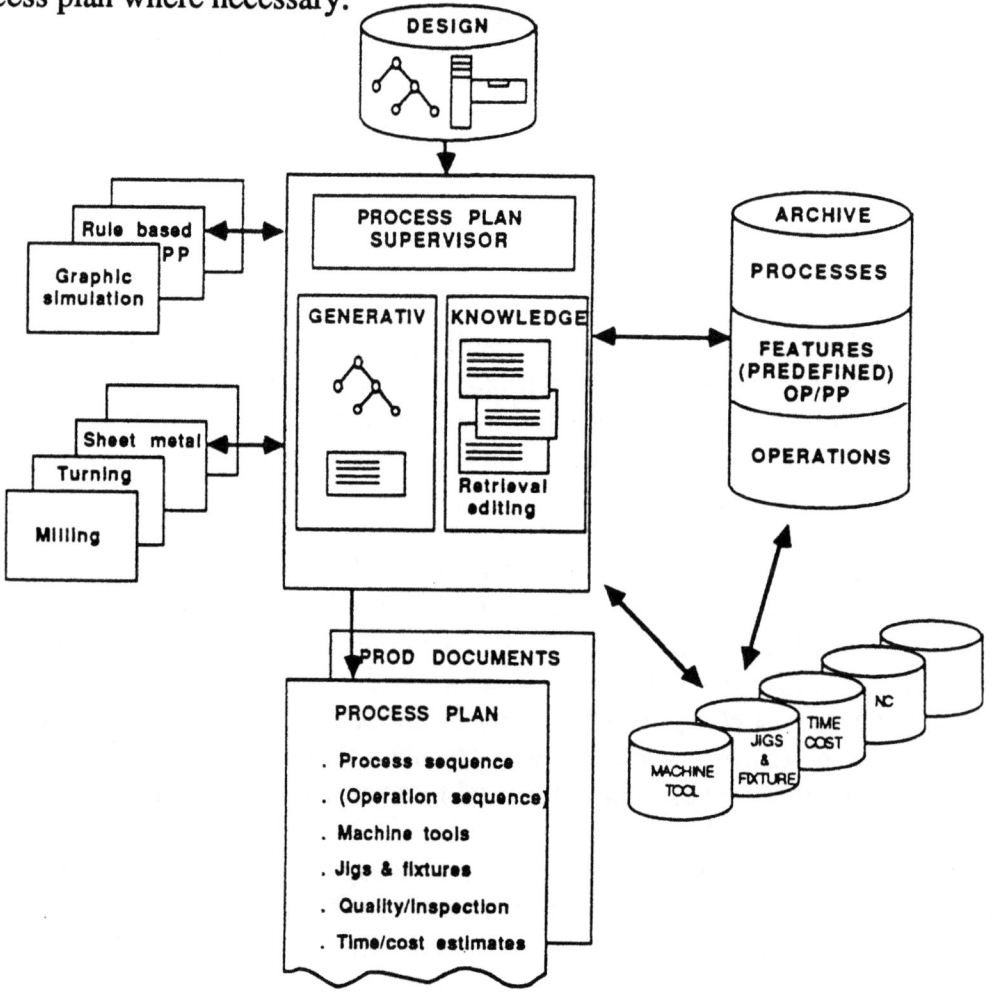

Fig. 6: Architecture of the Process Planning Supervisor

368

For operation planning, the basis will be the feature oriented workpiece description. (Fig. 7). The operation planning system as developed caters for two sorts of operation planning. It will either request an external system to generate the operation details, or it can make them internally using ist own knowledge base. The system will try to run automatically, but it has the ability to ask for user interaction where the manufacturing logic is incomplete, or the part parameters are outside the limits of validity for the available logic. This is important, as it allows the user to progress from an interactive to a more automated system, yielding steadily increasing benefits.

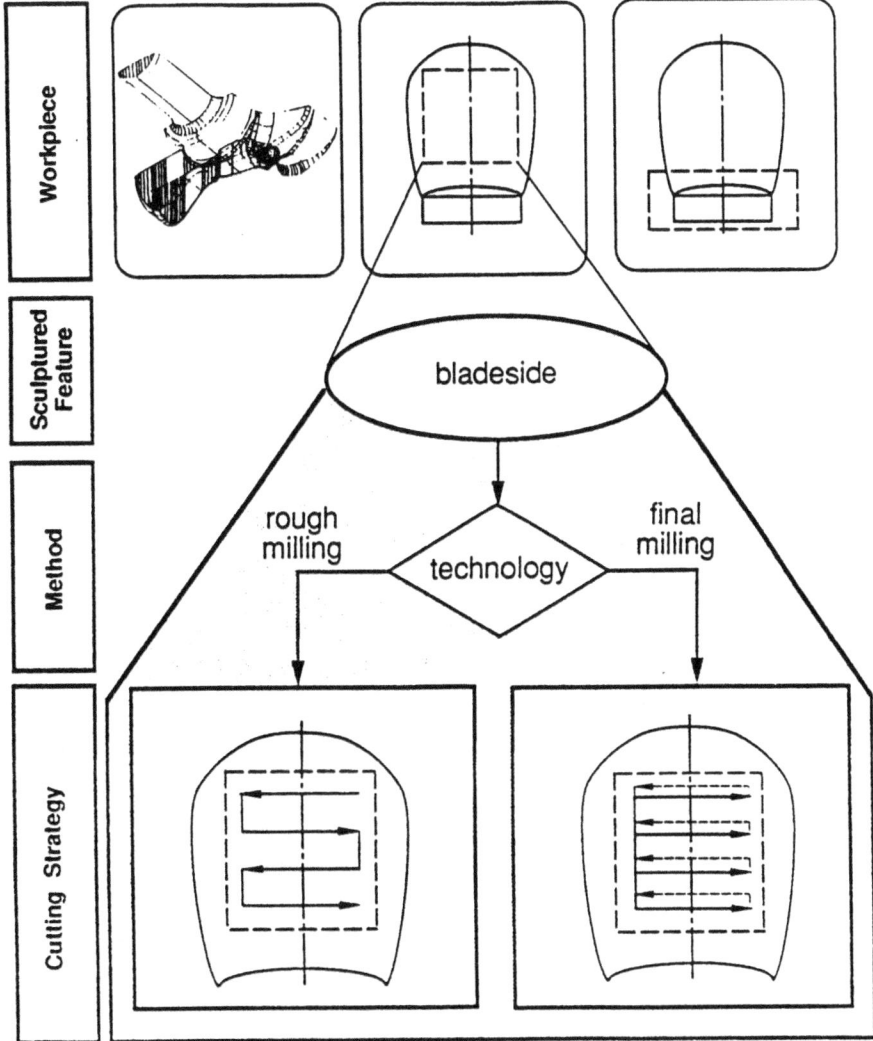

Fig. 7: **Automated operation planning on the basis of production features**

Production Knowledge Database

For both part modelling and process modelling, it is necessary to have access to relevant information whenever needed. Of course IMPPACT will provide a product database and knowledge database system for the integrated product and process modelling environment. This system is requried for accessing all the information relevant to integrated product and production process modelling in a manner independent of the physical storage. This information may comprise design requirements, geometric models at various intermediate stages, process plans, NC programs, analysis results, production and quality reports. This product database serves to make available the experience gained with the design and manufacturing of the product to a similar, later case.

The approach is to implement a corresponding database and an interface, and, on top of this, specify and implement comfortable end-user functions to operate on the product database.

4. CONCLUSION

The production process itself consists mainly of integration. Each action, whether performed by man or machine, is a combination of objects and information. Which items of information and how many of them are needed depends on the smoothness with which this process takes place. It is not enough to have a common database into which all subsystems input what they know, trusting that whoever has a need (and perhaps requires help) will search in it to find the information he needs. The quality of the items of information available must be structured so that they will satisfy an event-controlled need, i.e. so that they will then be present at the required time and place. Information should not have to be requested and searched for, but should be active and make itself available. Integration means the readiness of all subsystem departments to take action themselves.

Without this, computers and manufacturing are not combined in a way that really increases efficiency. Total monitoring from receipt of order up to delivery is then certainly possible, but the aim of being able to react more flexibly and more quickly to the requirements of the market is not achieved. The amount of work needed for this task will be enormous. With IMPPACT, the partners will all contribute a step towards this aim, and will set an important milestone.

IMPPACT was initiated in 1987 and started in 1989 to develop a new generation of integrated systems for product design, process and operation planning in order to "do it right the first time". The goal of reducing the time from market requirements to the availability of a product without increasing the cost of manufacturing but, on the contrary, reducing, leads to this functional investigation of the manufacturing areas of design, process planning and production.

(1)   CIM-OSA: Reference Architecture Specification, ESPRIT-Project Nr. 688-
        Report, 1988
(2)   Product Definition Data Interface - System Requirement Document
        Dept. of the Air Force document SRD5601130000, July 1984
(3)   Product Definition Data Interface - Needs Analysis Document
        Dept. of the Air Force document NAD560130000, July 1984
(4)   Product Definition Data Interface - System Specification Document, Draft
        Standard; Dept. of the Air Force document SS560130000, July 1984
(5)   Product Definition Data Interface - System Specification Document
        Dept. of the Air Force document SDS560130000, July 1984
(6)   Weiss, J. STEP & Project Plan - Document o. O, ISO TC184/SC4/WG1
        Document Nr. 297, Aug. 1988
(7)   Trippner, D. CAD*I Interface development - Experience Gained for the
        Benefit of Improvement of CAD/CAM Data Exchange; Presentation CAD*I
        Workshop, Karlsruhe, December 1986
(8)   Schlechtendahl, E. (ed.); Specification of a CAD*I Neutral File for CAD
        Geometry,Version 3.3 Springer 1988
(9)   ANSI/X3/SPARC Database System Study Group. Reference Model for
        DBMS Standardization, U. S. Dept. of Commerce, National Bureau of
        Standards, Report NBSIR-85/3173 (1985)
(10)   Altemueller, J. The STEP File structure, ISO TC184/SC4/.WG1
        Document N279, 1988
(11)   Altemueller, J. Mapping from EXPRESS to Physical File structure,
        ISO TC184/SC4/.WG1 Document N280, 1988
(12)   Pratt, M. J. Solid Modelling and the interface between design and
        manufacturing, IEEE Computer Graphics and Applications 4, 7,
        page 52 - 59 (1984)
(13)   Normung von Schnittstellen für die rechnerintegrierte Produktion (CIM);
        DIN-Fachbericht 15, July 1987
(14)   Bjorke, O. (ed), APS - Advanced Production System
        Tapir-Verlag, Trondheim, 1987
(15)   Grabowski, H., Anderl, R., Paetzold, B., ESPRIT, Technical Week 1987
(16)   Spur, G., Krause, F.-L. CAD-Technik, Carl Hanser Verlag, München, 1984
(17)   A. G. Requicha, Reproduction for Rigid Solids, Theory, Methods and
        Systems, Computing Surveys, Vol. 12, No. 4, Dec. 1980
(18)   A. G. Requicha, Solid Modelling: Current Status and Research Directions,
        IEEE C G & A, Oct. 1983
(19)   A. G. Requicha; Growing and Shrinking Solids blending,tolerancing and
        other application (CAD-Kolloquium) Nov. 1986, Berlin, SFB 203
(20)   Bunce, P. G., M. J. Pratt, S. Pavey, J. Pinte; Features Extraction and Process
        Planning; CAM-I report R-86-GM/PP-01, Arlington 1986.
(21)   Falcidieno, B. and F. Giannini. Extraction and Organisation of Form
        Features into a Structured Boundary Model, Proc. Eurographics 87   Conf.,
        North Holland (1987)
(22)   Henderson, M. R., and D. C. Anderson, Computer Recognition and
        Extraction of Form Features: a CAD/CAM link, Computers in industry 5,
        page 315-325 (1984)
(23)   Krause, F.-L.; Vosgerau, F. H.; Yaramanoglu, N.; Using Technical Rules and
        Features in Product Modelling;Proc. IFIP Working Group 5.2 Workshop on
        Intelligent CAD, Boston, Oct. 1987, (24) Pavey, S. G.; Hailstone, S. R.;
        Pratt, M. J. An automated interface between CAD and Process Planning;
        Proc. International CAPE conf., Edinburgh, April 1986, Mech. Engineering,
        Publications Ltd., Bury St. Edmunds, England
(25)   Major, F., Grottke, W. Knowledge Engineering within Integrated Process
        Planning Systems, Proceedings of the Int. Conference on Intelligent
        Manufacturing Systems,Budapest, 1986
(26)   Mäntylä, M. Feature-Based Product Modelling for Process Planning

MANUFACTURING COMMAND, COMMUNICATIONS, CONTROL (MC^3) -
THE INTELLIGENT LINK

Dave Brankley
SAC HITEC Ltd.

1   INTRODUCTION

MC^3 bridges the gap between a production plan and the
shop-floor, by translating a production requirement into a
production schedule, maximising the use of available
resources. This schedule is then applied to the
manufacturing facility and closely monitored to ensure
that its progress is within user defined limits of the
required schedule. Should the shop-floor situation change
during a production run, MC^3 is capable of re-routing and
re-scheduling on a real-time basis. It therefore
optimises the use of shop-floor resources and ensures that
production targets are met.

MC^3 also brings together existing systems already in
place, such as mainframe Manufacturing Resource Planning
(MRP II), Computer Aided Design (CAD) and Computer Aided
Process Planning (CAPP), stock control, robots,
Programmable Logic Controllers (PLCs) and Automatic Test
Equipment (ATE), to form an integrated flexible system,
which can be easily adapted to reflect the changing needs
of a modern manufacturing or process plant.

MC^3 will help to maximise the use of existing systems,
but it does not depend on them for its operation. It can
operate simply as a stand-alone system, requiring nothing
more than a standard terminal for data input and output.

Within MC^3, extensive use is made of an Artificial
Intelligence (AI) kernel in order to define and implement
complex decision strategies. User definable rules,
together with the use of an Oracle relational database
provide the key to MC^3's flexibility, allowing it to be
employed in a wide range of manufacturing and process
industries.

In this way the characteristics of the production plant
can be modelled and an operating strategy defined to
provide efficient plant operation based on
up-to-the-minute shop floor information.

The rules-based model may be used in an off-line mode
enabling experimentation with different production set-ups
or requirements in order to "fine tune" the production
system. A schedule will only be implemented on the shop
floor when it is committed to the schedule queue.

2 INNOVATION

In order to produce a portable and robust product with a
high level of integrity, and also because of the short
timescales imposed on the development of MC^3 by market
forces, extensive use is made of third party products,
which are already developed and well proven. The main
ones are:-

* an Oracle database.
* an AI kernel, Leonardo.
* graphics utilities, BGUL and S-GKS.

Much of the innovative nature of MC^3 lies, not
particularly in the major elements that are used, but in
the way they are combined, and the targeted application
area. Specifically, the strength of MC^3 lies in its
ability to use existing systems and plant to create an
integrated and efficient production facility, whilst
providing total real-time control and monitoring, 100%
product traceability and quality management, stock control
functions, and off-line modelling exercises, all within a
single product.
The Scheduler, however, has been developed in-house,
and this offers a very high level of flexibility, and
because of its heuristic nature, it can benefit from the
acquisition of knowledge over a period of time.

3 THE BENEFITS TO INDUSTRY

Up-to-date information about the shop-floor, together with
intelligent scheduling, combine to increase production
efficiency and quality, whilst decreasing down-time and
plant maintenance problems. MC^3 will interface with
existing plant and computers to provide:-

* Real-time shop-floor information in an easily
  understandable form.
* Real-time re-routing to avoid bottlenecks and maintain
  maximum use of equipment.
* Automatic monitoring and re-scheduling of the
  shop-floor to keep the production run on target.
* Production schedules generated by Artificial
  Intelligence.
* The ability to perform off-line simulations of the
  production line to examine the effect of possible
  changes.
* 100% traceability of manufactured goods for improved
  quality.
* Mixed batches, and the ability to economically produce
  very small batch quantities.
* The ability to increase plant efficiency whilst
  minimising the effects of down-time caused by
  maintenance.

4   COMPUTER ARCHITECTURE

The architecture of the computer system is influenced by the requirement to cater for small-batch manufacturing, and a hierarchical and modular approach is indicated [1], making use of "standard" interfaces for easy expansion. This approach leads to a general concept of four major software component technologies:-

1. Manufacturing Systems Control.
2. Distributed Data Administration.
3. Communications Systems.
4. User Interfaces.

The following sections describe how this concept has been implemented in the design of MC^3.

5   DETAILED DESCRIPTION

MC^3 has been designed with modularity as a major consideration. This means that it can be run on a single computer on a small installation, or it can be distributed over several computers networked together for larger installations. Modularity also ensures that the system is easily expandable, so that it can grow to keep pace with increasing demands.
There are five major modules which make up an MC^3 system, and they are described more fully in the following sections. They are:-

1. The Database.
2. The Overall System Manager.
3. The Scheduler.
4. The Real-time Shop-floor Interface.
5. The House-keeping Manager.

A further two modules are required to complete the functionality, and these are:-

1. The Communications.
2. The Error Handler.

These are implemented as low level routines and drivers, whose functions are self-evident, and as such do not warrant detailed descriptions.

5.1   THE DATABASE

The Oracle database holds all the historical, stock level, maintenance, target and production rules data on which the line operation is modelled. All of the information is readily accessible through pre-defined screens and printed reports.

By using an industry standard database in this way, considerable flexibilty is built into the MC^3 product. Interfaces to spreadsheet packages are readily available and the Structured Query Language (SQL) used is an accepted standard in the data processing world.

A maximum of two databases may be in use at any one time, and each one contains approximately forty different data tables. The first contains all the tables necessary to run MC^3 in the on-line mode, whilst the second is used when performing a "what-if" exercise, and it enables the user to extensively vary parameters, without affecting the on-line data.

5.2 OVERALL SYSTEM MANAGER

The Overall System Manager, controls and regulates access to all the information and data that the system needs. In practice, most of the functions of this module are performed by the Oracle database, which is ideal for this type of work.

This module can be considered to be the central "core" of the whole system, since all the other modules transfer data to and from it, and it is used as a central data store for the system. Consequently, this module can be performing many different and varied functions at any one time.

5.3 THE SCHEDULER

The Production Scheduler, on the other hand, has a very specific task to perform, and that is to generate a schedule which will allow the shop-floor to produce the desired quantities of goods, at the required times. This seemingly simple task is, in fact, one of the most complex functions of the system, since a great many factors must be considered when generating a schedule, and it is the ability to handle all these factors which is one of the notable features of MC^3. Some of the factors to be considered include:-

* Quantities and timescales.
* Mixed batches.
* Stock levels of raw materials.
* Plant maintenance.
* Shop-floor performance.
* Production line rules.

The Production Scheduler uses an AI kernel, Leonardo, in order to perform this difficult task, and from any given set of criteria, it can produce an optimized schedule which best meets all the requirements. It also has the ability to base its results on past performances of the production line under similar circumstances by using the historical data held in the database, and it can therefore "learn" from previous experience to produce more accurate results.

5.3.1 The Rules

The AI kernel uses rules which consist of English clauses connected by conjunctions "if", "then", "and" and "or". They have the following general format:-

if <antecedent> then <consequent>

where <antecedent> and <consequent> have the form "object is value".

The Scheduler operates by means of structured rule-sets, which generate decisions and possibly quantitive data also. A collection of decisions must be satisfied to complete each stage in the generation of the schedule. For example, rules of the following type will govern the overall flow of the Scheduler.

if the day is not saturday
    and the day is not Sunday
then the work_schedule is normal

if the resource is a bottleneck
    and the shift_capacity is exhausted
    and the optimisation is done
    and the capacity_check is not good
then split_jobs_across_shifts is required

If the work is identified
    and the bottlenecks are found
    and the bottleneck work is sequenced
    and the non-bottleneck work is sequenced
    and the schedule reports are generated
then the schedule is generated

The Scheduler makes use of four different types of rules. They are:-

1. Production Rules.

2. Core Rules.

3. Constraint and Batch Rules.

4. Application Specific Rules.

Production Rules are declared in product Bill of Materials, process plans, and so on, whilst the Core Rules are those which govern the Scheduler's operation and are therefore fixed.
Constraint and Batch Rules, however, are user-imposed constraints and cater for:-

* Preferred product routes or machines for specific products.
* The defininition of the breakdown and scrap and rework statistics.
* The mutually exclusive production of certain products.
* The phasing of the use (or mutually exclusive use) of machines when they are:-
  - producing specific products.
  - operating irrespective of product type.

* The re-ordering of product priorities or ranking interpreted by the core rules due to:-
  - 'customer satisfaction' issues (as determined by the user).
  - material shortages.
  - shortfalls in the capacity of resources.

The priority of usage of machine and materials by products is initially made on a due-date basis. In other words, where an alternative exists, the product with the earliest due-date takes the major share of material or resource capacity. However, a set of constraints and qualifiers within the Scheduler lay down the framework for its operation [2].

The fourth type of rules are the Application Specific Rules. As part of the MC^3 installation procedure, one activity will be the acquisition of knowledge about the application. This will include the specific requirements of the user, and proven heuristic scheduling methods currently implemented by him. This knowledge must be mated with the core rules of the Scheduler, in order to produce schedules which satisfy each individual user's requirements.

The Scheduler uses rules which prove suitable for meeting the criteria selected by the user, in order of importance, and when producing a schedule it can cater for the following criteria:-

1. Maximise a resource utilisation.
2. Maximise the attainment of product due-dates.
3. Maximise the product throughput.
4. Maximise the profit and minimise the cost.

Certain combinations of criteria produce conflicting requirements, and this leads to a three level hierarchical approach to produce a solution [3].

5.3.2 The Input Data

The following data, which is extracted from the Oracle database tables, is required to generate a schedule:-

* The Manufacturing Requirements.

* Inventory Levels, including those of tooling.
* Resource Availability.
* Shift information, including Schedule Period start and
  finish dates and times, and the allocation of
  personnel.
* Work In Process.
* The Process Plan data which gives preferred routings,
  standard times, tool usages, setup requirements, and
  resource alternatives for performing the processes.
* The Bill of Materials.
* Quality data, giving anticipated scrap rates and
  resource failure rates.
* User priorities.
* Historical data concerning the performance of
  resources and personnel.

5.3.3   The Output Data

Having successfully created a schedule which best meets
all the requirements and criteria, the Scheduler outputs
the following data:-

* An itemised schedule table for transfer to the
  Production Controller. This table will be used to
  generate a number of reports.
* Inventory pick lists. A list of material and tools,
  itemised by time and identity, is produced for the
  schedule period.
* Recommendations for achieving the desired production.
* Optionally, a log of all major scheduling decisions
  can be provided.

5.4   THE REAL-TIME INTERFACE

The Real-time Shop-floor Interface, has the task of
implementing the production schedules produced by the
Scheduler, and of communicating with all the shop-floor
devices and plant. Because of the real-time nature of the
task, this module normally runs on its own dedicated
computer, on all but the smallest of installations. It
constantly monitors the flow of goods around the
production area, and it is capable of re-routing materials
to avoid bottlenecks developing, whilst reporting the
current situation to the Overall System Manager.
    Specifically, the Real-time Interface provides the
following facilities:-

* Intercell communication.
* The transfer of data between individual cells and the
  "main" MC^3 computer.
* The tracking of products around the shop floor.

* Configuring alterable cells to handle different
  product types.
* Shop floor device handling (Terminals, displays,
  bar-code readers etc).
* The logging and reporting of product "in cell" times,
  and other historical information.
* Limited local re-routing between similarly configured
  cells.
* Loading application programs to cells from local or
  remote storage.

5.5 THE HOUSEKEEPING MANAGER

The last of the major modules, the House-keeping Manager,
is responsible for all the other sundry tasks, such as
general communications, terminal displays, reports,
printouts, and so on. In other words, it handles all
those tasks not directly concerned with the immediate
running of the production line, including all
communications with management and supervisory personnel.
    The most important aspect of any interactive software
package, as far as the user is concerned, is the user
screens. Within MC^3, the screens are arranged in a
hierarchical manner, and are nested several layers deep.
All user screens are of the following type:-

* Menu.
* Tabular.
* Graphical.
* Text.

    Menu screens present to the user a number of lower
level screens for selection, and on-line help is available
for each of the selections. The number of selections
available on any menu screen is restricted according to
the security access level of the user.
    Tabular screens present to the user alphanumeric data
to be displayed or modified, and they consist of
pre-defined fields where data may be displayed. On-line
help is available.
    Graphical screens present to the user graphical data to
be displayed. They consist of barcharts, histograms,
Gantt charts and mimic diagrams.
    Text screens present to the user text data which may
not be modified by the user. Help screens are an example
of text screens.

6 AN EXAMPLE INSTALLATION

In order to put the previous sections into context, an
real example installation is described, which utilises
many of the capabilities of MC^3.

Standard Electrica S.A. (SESA) are a subsidiary of the Alcatel group, and have a site in Malaga, Spain, which produces a wide range of telephones for the consumer and public service markets. MC^3 is responsible for controlling a flexible assembly line, containing a mixture of robot cells, manual workstations, ATEs and PLCs, and the line is supplied by Automatic Guided Vehicles (AGVs).

The manufacturing requirement is supplied from an MRP II system in Madrid, and the two sites are connected by a network link. This installation of MC^3 is implemented on DEC equipment, with the main MC^3 software running on a microVAX 3400 under VMS, and the Real-time interface running on an rt-VAX under VAXELN, a real-time kernel.

The MRP II requirement is sent from Madrid weekly, and MC^3 uses this to schedule work for the following week. It controls the stock of components held in the local stores, and stock entered or removed is recorded by means of bar-code readers at the stores terminals.

The benefits to SESA are:-

*   The ability to produce small batches economically.
*   Increased throughput and reduced costs.
*   The ability to automatically implement the remote MRP II output on the production line.
*   Full shop floor monitoring and trend analysis, down to the wear of individual tools.

The hardware block diagram illustrates the hardware configuration used at this site.

7   DESIGN STANDARDS AND ACKNOWLEDGEMENTS.

The following section defines the terms used in this document.

*   "Oracle" refers to Oracle V6.0, supplied by Oracle (UK) Ltd.
*   "Leonardo" refers to Leonardo V3.18, supplied by Creative Logic Ltd.
*   "BGUL" refers to BGUL V1.0, supplied by Scientific Software Ltd.
*   "S-GKS" refers to S-GKS V2.1, supplied by Xelion.

S-GKS is a full implementation of the Graphical Kernel System defined by the International Organisation for Standards, ISO 7942-1985.

References to "C" refer to the language whose features and functions are detailed in "The C Programming Language" [4].

DEC, VAX, microVAX, rt-VAX, VMS, VAXELN are trademarks of Digital Equipment Corporation.

8 CONCLUSION

A software product has been described which is flexible enough to be applied to a wide range of industries, and which offers a great many benefits to the user. Its design has ensured that it reflects the real needs of industry today, and will continue to do so in the future.

SAC Hitec's many years' experience of designing and installing factory automation equipment and production lines were a major consideration when deciding to develop a package such as MC^3, and this was no doubt one of the factors which the Department of Trade and Industry took into account before deciding to give the project its backing. In addition, our involvement with the Eureka initiative highlights our commitment to European technology, and this is reflected in MC^3 being granted FAMOS backing (Project Number EU196).

9 REFERENCES

[1] McLean, C., M. Mitchell, and E. Barkmeyer "A Computer Architecture for Small Batch Manufacturing" IEEE Spectrum (May 1983), pp. 59-64.

[2] Fox, M.S., B.P. Allen and G.A. Strohm "Jobshop Scheduling: An Investigation in Constraint-Directed Reasoning" In Proc. of the 2nd National Conference on Artificial Intelligence" AAAI (August 1982), pp.155-158.

[3] Emmons, H. "A Note on a Scheduling Problem with Dual Criteria" Naval Research Logistics Quarterly, 22, 615-616 (1975).

[4] Kernighan, B.W., D.M. Ritchie "The C Programming Language", Second edition.

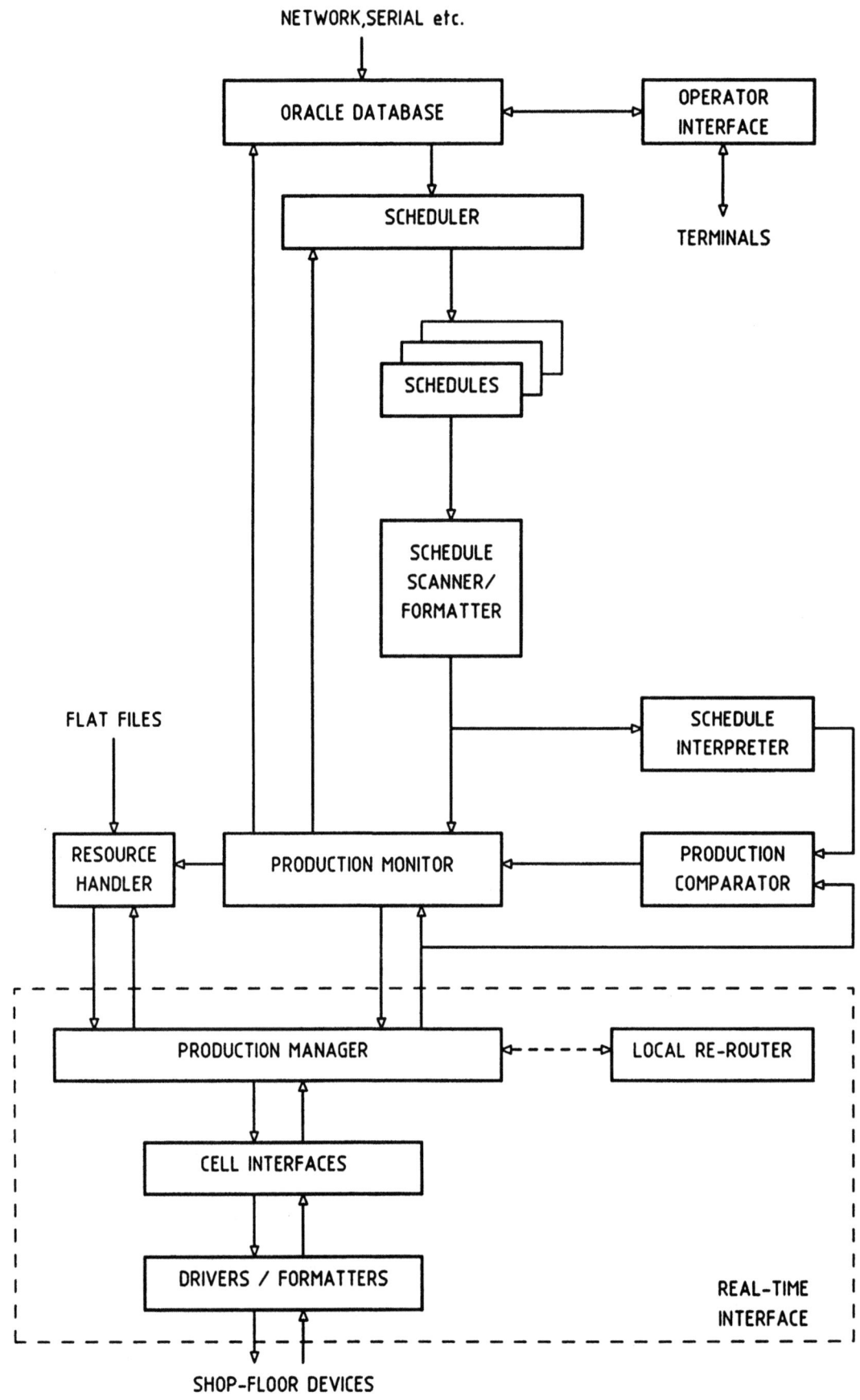

NETWORK,SERIAL etc.

ORACLE DATABASE

OPERATOR INTERFACE

TERMINALS

SCHEDULER

SCHEDULES

SCHEDULE SCANNER/ FORMATTER

FLAT FILES

SCHEDULE INTERPRETER

RESOURCE HANDLER

PRODUCTION MONITOR

PRODUCTION COMPARATOR

PRODUCTION MANAGER

LOCAL RE-ROUTER

CELL INTERFACES

DRIVERS / FORMATTERS

REAL-TIME INTERFACE

SHOP-FLOOR DEVICES

SYSTEM BLOCK DIAGRAM

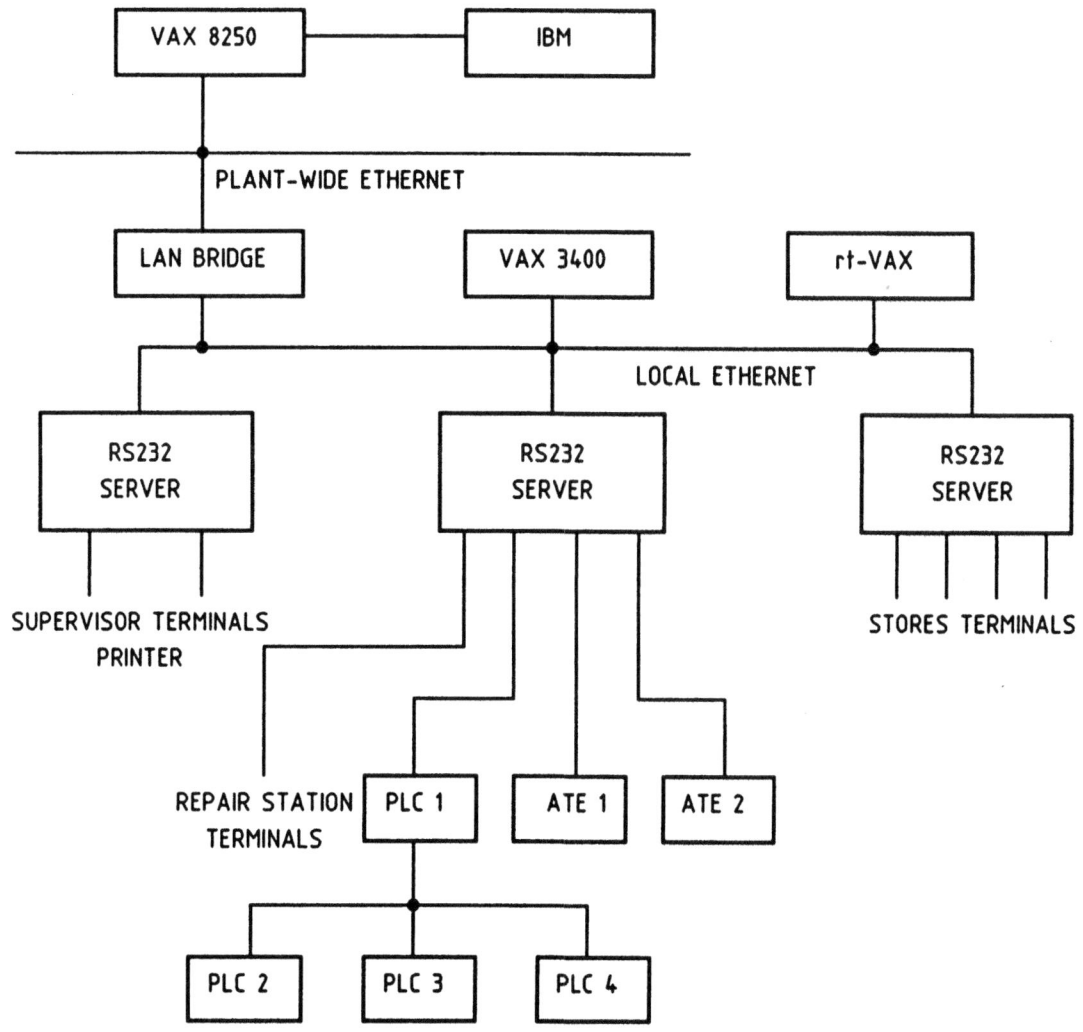

HARDWARE BLOCK DIAGRAM

HUMAN-ORGANISATIONAL AND MANAGERIAL ASPECTS OF CIM

ORGANISATION, PEOPLE AND TECHNOLOGY: TOWARDS CONTINUING
IMPROVEMENT IN MANUFACTURING.

P T Kidd
Cheshire Henbury Research & Consultancy, Macclesfield, UK.

INTRODUCTION

To compete in the world markets of the 1990's it will not
be sufficient simply to seek cost efficiency as
manufacturing companies have done in the past. Success in
the customer driven global markets will depend upon
achieving, without tradeoff, three potentially conflicting
goals: cost efficiency, quality and flexibility. But this
is only part of the story. Improved competitiveness, once
achieved, has to be retained. This will be achieved by
companies transforming themselves into learning
organisations seeking continuing improvement in every
aspect of their activities (Hayes, Wheelwright and Clark
1988).

Before any of the above can be achieved however,
companies will have to learn that there is no such thing as
a quick technological fix to the many problems faced by
manufacturing industry. New technology is important,
but is not the answer. In the majority of cases, it is
people who transform raw materials into profitable high
quality products. New technology and organisational change
must be shaped by this fact.

People and the question of how best to help and support
them, will be one of the central manufacturing issues of
the 1990's. Technology will still be important, but
companies can no longer afford to use technology to
minimise the role of shopfloor people. It is these
employees who are, in many cases, the key to business
success.

Organisation, people and technology and the pursuit of
continuing improvement in manufacturing are therefore
strategic issues. Companies need new organisations and work
practices. They need different attitudes towards shopfloor
people, and they need new ideas about the role of
technology in manufacturing.

Companies will have to make a strategic decision to put
people first, to build organisations that support these
people, and to use technology as a means of making more

effective use of peoples' skills and abilities. The keywords are empowering, team working, reduced hierarchies, decentralised control and decision making, and networked organisational structures.

The paper explores the requirements that arise from markets that are becoming more dynamic and customer driven. Problems with traditional manufacturing practices are described. A technique, still under development, called Organisation, People and Technology is outlined. This is both a manufacturing strategy and a method for achieving continuing improvement in manufacturing. The technological implications are explored.

A NEW ERA IN MANUFACTURING

Requirements

It is evident that market conditions have become more dynamic, more global, and more customer driven. Product price is also no longer the main factor affecting business performance. Other non-price competitive factors such as quality, design, delivery, and customer service have become equally, if not more important.

In the 1990's there will be an increasing trend towards tailoring of products to meet customer needs. This will lead to a further reduction in production volumes, increased product variety, shorter product life cycle, and a reduced number of repeat orders. Opportunities for improved profits will come from economies of scope rather than economies of scale. Competitive advantage will be derived from flexibility, rather than from low costs (Williamson 1989). The key business issues are therefore to reduce manufacturing lead time, improve customer service, reduce inventory and increase responsiveness, against a background of highly variable customer demand patterns (Hamlin 1989).

European companies will need to achieve cost efficiency, quality and flexibility without tradeoff, and this objective is now posing a major challenge to manufacturing industry. This however is only part of the story.

It is often not fully understood that a company's competitive position at any point in time is less important than the rate of improvement as compared with its competitors (Hayes, Wheelwright and Clark 1988). Manufacturing competitiveness is not static. There is no project, no investment, that will provide the magic solution to lack of competitiveness. Manufacturing competitiveness is a dynamic process, a never ending rising and falling of performance in relation to that of competitors. To remain in the game it is therefore necessary for companies to transform themselves into learning organisations seeking continuing improvement in all aspects of their activities.

Organisational and Management Problems

The advent of dynamic customer driven markets has not yet

had a significant impact upon manufacturing strategy. There is a need to question the taken for granted assumptions that underlie manufacturing strategy. Research and development also needs to be re-evaluated.

Manufacturing industry still uses Victorian models of organisation and control. Technology is still designed and used on the basis of replacing or reducing the role of shopfloor people. Manufacturing industry typically uses the pyramid shaped organisational structure shown in Fig. 1. Information flows upwards, and control is exercised from the top downwards. Five or six levels of management are the norm in many companies.

Manufacturing facilities are also commonly arranged on a functional basis, with turning machines located in one part of the factory, milling machines in another, assembly work facilities in another, etc. Moreover, control of these functions is often based on the hub and spoke model shown in Fig. 1, with control and decision making centralised in the planning office.

These organisational and control models are based upon ideas developed during the late 19th century. Frederick Taylor, a 19th century mechanical engineer, for example, advocated a system of factory management based upon centralisation of all decision making. Workers were to be told exactly what to do. Working methods were to be standardised. There was only one best way of doing work, and this could be scientifically established. Functional specialisation and work simplification were to be widely adopted (Taylor 1902, 1907).

Max Weber, another man of the 19th century, argued that the most efficient form of organisation was one in which (i) the duties and responsibilities of all members of the organisation are clearly defined; (ii) the various positions in the organisation are arranged hierarchically with each person responsible to a superior and responsible for subordinates (with the exception of course of the extreme top and bottom of the pyramid); (iii) an elaborate system of rules governs the manner in which each person carries out his duties (see Kempner 1976).

In the 1990's manufacturing companies should be using the organisational and control concepts of the 1990's. They are still using the ideas developed in the 1890's. Taylor and Weber's insights, however, are based upon the markets of 1890's. They had no knowledge or understanding of flexibility and customer driven markets. Manufacturing industry however still continues to use these outdated ideas.

Technological Problems

Price, for many decades, was the most important competitive factor. In this economic regime the dominance of cost efficiency in manufacturing can be understood. To remain competitive and profitable, costs must be reduced. This was achieved through mass production, product rationalisation, division of labour, and automation.

Technology has often been seen as a means of reducing costs. It has been used to reduce the size of the workforce and to reduce skill levels so that cheaper labour can be used. In both cases the emphasis has been solely upon improving labour productivity.

Technology has also been seen as a solution to organisational problems. An example is the application of Artificial Intelligence (AI) in computer-aided design (CAD) to achieve design for manufacture. There is a potential for AI to contribute towards improving design, but it should not be used as a solution to problems created by functional specialisation (i.e. division of labour).

In the past the application of technology has resulted in improved competitive performance. It is natural therefore, but not necessarily correct, to continue to regard technology as the prime source of improved competitiveness. Before developing and applying more technology, it is important to gain a better understanding of what needs to be done in order to improve competitiveness, and not just to assume that technology is the answer. This does not mean that investment in new technology is no longer required. It is a statement that accepted thinking about the role of technology needs to be updated and adapted to changed circumstances.

Analysis of the Situation

The new markets lead to opportunities for improvements in competitiveness and profit growth through the ability to be highly responsive to market changes. There is therefore a need for flexible manufacturing.

Flexible manufacturing should allow manufacturing companies to increase their market share and reduce their costs. The former can be achieved by: shorter lead times, better due date performance, improved product quality, and increased customer responsiveness. The latter can be attained through: lower inventory levels, more effective management and decision making, reduced rework and scrap, lower management and indirect labour costs, and less capital expenditure on complex computer systems.

The key to achieving these benefits lies in achieving cost efficiency, quality and flexibility without tradeoff. This requires a fundamental change in thinking.

Quality control and rectification of defects is an obvious way of achieving improved quality. This approach however, creates a necessity for a tradeoff between cost efficiency and quality because quality becomes a cost. A less obvious approach is to build quality into the product, procedures and systems, and to make quality an issue for all employees (Bolwijn et al. 1986).

This is the strategy that the Japanese have adopted. It has proved to be very successful. Doing things right the first time is more cost efficient than creating a functionally separate quality control section and rectifying defects after manufacture. Tradeoff between cost efficiency and quality can therefore be avoided. This needs major changes to established organisational practices.

Combined use of organisation, people and technology is one of the factors that has made Japanese manufacturing industry so successful. Comparisons between Japanese and UK manufacturing industries typically report that the Japanese regard people as assets to be developed, while many UK companies see people as a cost to be minimised. The organisational structures of many Japanese enterprises are also generally regarded as being more flexible and responsive than those of UK companies, with fewer hierarchical levels, fewer specialists, fewer but more integrated departments, and greater decentralisation of decision making (Burnes and Weekes 1989).

The Japanese have shown that achieving cost efficiency and quality without tradeoff is primarily a matter of organisation and people. It is not a purely technological problem, although technology does contribute towards improved quality. Likewise, achieving cost efficiency, quality and flexibility without tradeoff is not a purely technological problem. These three potentially conflicting goals can actually be achieved without technology, by making organisational changes and by making better use of people (Caulkin 1988). The addition of appropriate technology can further improve performance.

Moreover, to achieve continuing improvement in manufacturing it is necessary to tap into the intelligence and creativity of all people. It is not just a question for management and other professional groups. Well motivated, highly trained shopfloor people are also a source of knowledge, creativity and innovation, and can help bring about continuing improvement in products and processes. Without continuing improvement in manufacturing it will be difficult if not impossible to remain competitive, and without people there can be no process of continuing improvement.

Well trained, highly motivated people at all levels are the key to survival and improved competitiveness. Manufacturing strategies need, therefore, to be built on the idea of using people at all levels. Technology should be used to support people, rather than to replace them.

ORGANISATION, PEOPLE AND TECHNOLOGY

Organisation, People and Technology is a strategy for achieving cost efficiency, quality and flexibility without tradeoff, and a method for implementing continuing improvement in manufacturing. It is based on: flattening of hierarchies, pushing decision making and control towards the bottom of the organisational pyramid, increasing the competence of shopfloor employees, and product-based organisation (see Fig. 2).

Organisation, People and Technology is therefore about people working in appropriate organisational structures supported by technology, transforming raw materials into profitable products. In short, the philosophy underlying Organisation, People and Technology is that people make it happen.

The first step is to consider organisation. An organisational structure needs to be designed that will allow shopfloor employees to take controlling decisions and actions; it should allow them to determine how to meet agreed manufacturing targets. This is rather different from the usual approach which begins and ends with technology; organisation and people are normally seen as secondary issues.

The factory organisation can be based on production islands, which are an application of group technology. Production islands usually comprise clearly identified areas within a factory where resources, both machines and people, work in a highly co-ordinated fashion to produce a range of products. This is in contrast to the more conventional method of organising production by grouping machines according to process (as shown in Fig. 1). Production islands are not the same as islands of automation, as this latter term usually refers to an isolated highly automated process somewhere within a manufacturing organisation.

Production islands (see Fig. 3) are characterised by (Brodner 1984):

. Grouping of parts, i.e. parts with similar manufacturing requirements are grouped together.
. Grouping of machinery, i.e. equipment needed for complete manufacture of a part family is placed in the island.
. Grouping personnel, i.e. appropriately skilled workers cooperate to completely manufacture a part family using appropriate equipment.
. Organisational grouping, i.e. integration of design, planning and control tasks for complete production of a part family.

Clearly there are a number of possible variations on this theme, made necessary, for example, by the need to share expensive resources.

The second step is to consider people. Organisation, People and Technology is concerned with all employees in the enterprise, but most especially with shopfloor workers. There are two reasons for this. First, most shopfloor employees in manufacturing companies are intelligent, flexible, enthusiastic, well motivated, creative, and loyal, but many companies have made sure that these people will never get the chance to demonstrate these characteristics.

Second, shopfloor employees are far more flexible and creative than any of the complex machines and computer systems devised by technologists. No machine can be creative enough to recognise when and where improvements can be made to products and processes. People can. No machine can use past experiences and knowledge to foresee new problems and then devise a strategy to avoid the problems. People can. Organisation, People and Technology is about improving company performance by fully exploiting this potential.

The people aspects are therefore concerned with: empowering shopfloor employees, increasing their responsibility and competence, and involving them in activities such as planning and design.

The final step is to address the question of technology.

This is largely a matter of providing the tools and machines that will speed up tasks, provide user support, and facilitate problem solving and bottom up product and process innovation.

Much of the technology currently used in manufacturing takes little account of organisation and people. Typically it is expected that the organisation will have to be adapted to take account of the new technology. Jobs have to be redesigned to fit the changed technology (Bolk and van Manen 1989), regardless of whether the outcome of this is effective or productive. By way of contrast, a strategy based Organisation, People and Technology, must be concerned with developing technology that takes account of the desired organisation and the jobs and skills of the people working in the organisation. Technology is therefore simply a tool to support organisation and people, and is not therefore the prime determinant of organisational change.

Analysis of Benefits

The benefits of a strategy based on Organisation, People and Technology include: shorter throughput times, reduced inventory, improved product quality, more economic operating conditions, and improved responsiveness. These benefits can be achieved because this strategy can lead to:

. Better management of manufacturing operations and improved decision making on the shopfloor.
. A reduction in avoidable problems.
. Maximisation of opportunities to control product quality.
. Improvements in product design and the design process.
. Increased opportunities for preventive maintenance.
. Faster recovery from operational failures.
. Improved efficiency.
. Better industrial relations and improved motivation.

Product-based organisation leads to:

. Shorter throughput times (a 60 to 80% reduction in time and a 44 to 60% reduction in in-process inventory can be achieved).
. Simplified material flows.
. Easier production planning and control.
. A high degree of compatibility with JIT manufacturing techniques.
. Greater job satisfaction leading to more highly motivated people.

There have been some notable successes arising from the reorganisation of conventional process oriented production systems into production islands. In one particular German example, the claimed improvements were (Klingenberg and Kranzle 1987):

. A 25% increase in turnover per capita.
. A 10% reduction in cost of plant.

. A 40% reduction in production space.
. A 50% reduction in total space.
. A 15% reduction in energy costs.
. A 30% reduction in costs.
. A 30% reduction in tied up circulating capital.
. A 60% reduction in throughput time.
. A 40% increase in management certainty.
. A 28% reduction in indirect labour.
. A 71% reduction in waste rate.

Direct labour increased by 10% because of organisational changes. However, the company has achieved: improved due date performance, a high speed of response to quantitative and qualitative changes of demand, improved quality, reduced expenditure on computer equipment, and a reduction in breakdowns.

Costs can be reduced because less indirect labour is needed and because less capital is needed to fund the business. Quality is also improved because operators are made responsible for quality. Defects are therefore detected at an earlier stage or are avoided altogether. This also brings cost savings in reduced rework and scrap. Flexibility is also achieved.

Flexibility is the ability to respond quickly to changing situations. Flexibility is largely determined by throughput time. This is not just determined by the speed that design drawings are produced, or by machine set-up times, etc., but also by organisational characteristics and the time taken to respond to signals arising from various points within the organisation.

Throughput time determines the company's ability to introduce new products and to make product changes, and to manufacture products in a way that enables rapid delivery against a background of customisation and ever changing specifications. Unless throughput times are short, the reaction time of design and manufacturing will be slow and the company will not have the flexibility to respond to dynamic markets.

The organisational characteristics that effect flexibility are: the complexity of procedures, the number of levels in the hierarchy, and the degree to which the organisation is based upon functional specialisation. The greater the complexity of procedures, the more levels in the hierarchy, and the larger the number of specialised functions in the organisation, the longer it takes to get things done.

The time taken to respond to signals arising from various points in the organisation is dependent upon the time taken for these signals to be fed back and then acted upon. Response time is increased by the use of direct feedback loops and prompt control actions. The lack of direct feedback and prompt control actions slows down the response time which lengthens throughput times.

Reduction in throughput times can be achieved using production islands because response time is shortened by more direct feedback and prompt control action. Simplified procedures, a reduced hierarchy, and the removal of functional specialisation also contribute to the reduction in throughput time.

The factory planning and control functions also become much simpler as a result of production island organisation. Systems are no longer required which attempt to organise the correct schedule of work for every single operation on every single resource at every moment in time. Instead the system only needs to provide the factory planner with information allowing each island to be loaded with a sensible volume of work over a period of time (Hamlin 1989). The detailed planning and scheduling can be done by the operators within the islands. This results in effective, but simpler and cheaper computer systems.

Production island based organisation also provides the basis for achieving continuing improvement in manufacturing. More traditional employee participation schemes such as quality circles and suggestion boxes are limited and not always effective. There can be lack of feedback to employees, ideas can get lost in the organisation, shopfloor people lack the power to pursue and develop their ideas, and recognition is not always given where it is due. These things can discourage participation. Production islands, with there emphasis on functional integration, make better use of people. Competence is increased at the lower levels, and decision making is shifted towards the shopfloor. This can create ideal conditions in which to implement a policy of continuing improvement in manufacturing.

For example, a rejected part, a half finished item standing next to a machine awaiting new tooling, a CNC programme that can be improved, a unnecessary design feature that leads to machining difficulties, all represent time and money that is being wasted and competitive opportunities missed. Acting on and remedying these types of problem is what continuing improvement in manufacturing is about. This however, is not just an issue for management. It affects all employees. Because production islands are based upon the concept of empowering shopfloor personnel to take decisions and actions, these types of problem can be more easily identified and remedied.

CONCLUSIONS - MAKING THE CHANGES

The theme of the paper is that there is no quick technological fix to the many problems faced by manufacturing companies. Technology is only part of the solution, which must be based upon a strategy of a balanced approach using Organisation, People and Technology.

Organisational changes must come first. These must be supportive of the people who will have to work in the system and who will provide the basis for a policy of continuing improvement in manufacturing. Technology is important but must not come first. It must be shaped by the organisational requirements and by the need to achieve continuing improvement.

Using such a strategy of course brings new difficulties. In addition to new learning, both shopfloor employees and managers must unlearn years of ingrained philosophy and practices. The strategy also tends to bring into sharper focus problems such as outdated attitudes, resistance to change, over specialised narrowly based roles, vested interests, power relationships, conflicts, company politics and cultures, etc. All the things in fact which need to be exposed, addressed and changed if companies are to become more competitive.

Organisation, People and Technology forces management to face up to the long-standing problems that hold back the development of the company and keep it from achieving its full potential. The strategy can unlock unused resources and smooth the way for organisational and technical changes. It can also change the outlook of company employees from that of departmental turf defenders to that of networked individuals engaged on a common purpose of improving company competitiveness, profitability and business performance.

There is little doubt however, that a three dimensional strategy based upon Organisation, People and Technology will pose some severe problems for industry. It will need a major transformation of attitudes and practices. It will need an interdisciplinary manufacturing systems design approach that will involve the application of knowledge that lies beyond the range of experience of many engineers and managers. It will also generate a need to manage major changes in organisation, conventions and company culture.

Bringing about these changes is no easy matter. There is very little experience to call upon, and few tried and tested tools. The application of a manufacturing strategy based upon Organisation, People and Technology, and the transformation of a company into a learning organisation committed to continuing improvement in manufacturing, is no easy road to follow. It is essential however, that manufacturing companies make the journey if they wish to improve their competitive position in world markets.

REFERENCES

Bolk, H. and van Manen, M., 1989, Technological Developments and Educational Demands: The Dutch Situation. In H. Bolk, H-U. Forster and B. Haywood (eds.), Implementing Flexible Manufacturing. Challenges for Organisation and Education in a Changing Europe, Eburon Publishers, Delf.

Bolwijn, P.T., Boorsma, J., van Breukelen, Q.H., Brinkman, S. and Kumpe, T., 1986, Flexible Manufacturing. Integrating Technological and Social Innovation, Elsevier Science Publishers, Amsterdam.

Brodner, P., 1984, Group Technology - A Strategy Towards Higher Quality of Working Life. In T. Martin (ed.), Design of Work in Automated Manufacturing Systems, Pergamon Press, Oxford.

Burnes, B. and Weekes, B., 1989, Success and Failure with Advanced Manufacturing Technology: The Need for a Broader

Perspective, Advanced Manufacturing Engineering, 1, pp 88-94.

Caulkin, S., 1988, Britain's Best Factories, Management Today, September, pp 58-80.

Hamlin, M., 1989, Human Centred CIM, Professional Engineer, 2(4), pp 34-36.

Hayes, R.H., Wheelwright, S.C. and Clark, K.B., 1988, Dynamic Manufacturing. Creating the Learning Organisation, Free Press, New York.

Kempner, T. (ed.), 1976, A Handbook of Management, Penguin Books, Harmondsworth.

Klingenberg, H. and Kranzle, H., 1987, Humanisation Pays - Practical Models - Volume 2 Production and Production Control, RKW, Eschborn.

Taylor, F.W., 1902, Shop Management, Transactions American Society Mechanical Engineers, 24, pp 1337-1480.

Taylor, F.W., 1907, On the Art of Cutting Metals, Transactions American Society of Mechanical Engineers, 28, pp 31-350.

Williamson, I.P., 1989, Integrated Manufacturing: Developing Your Strategy. In C Halatsis and J. Torres (eds.), Computer Integrated Manufacturing, Proc. 5th CIM-Europe Conf., IFS Publications/ Springer-Verlag, Kempston.

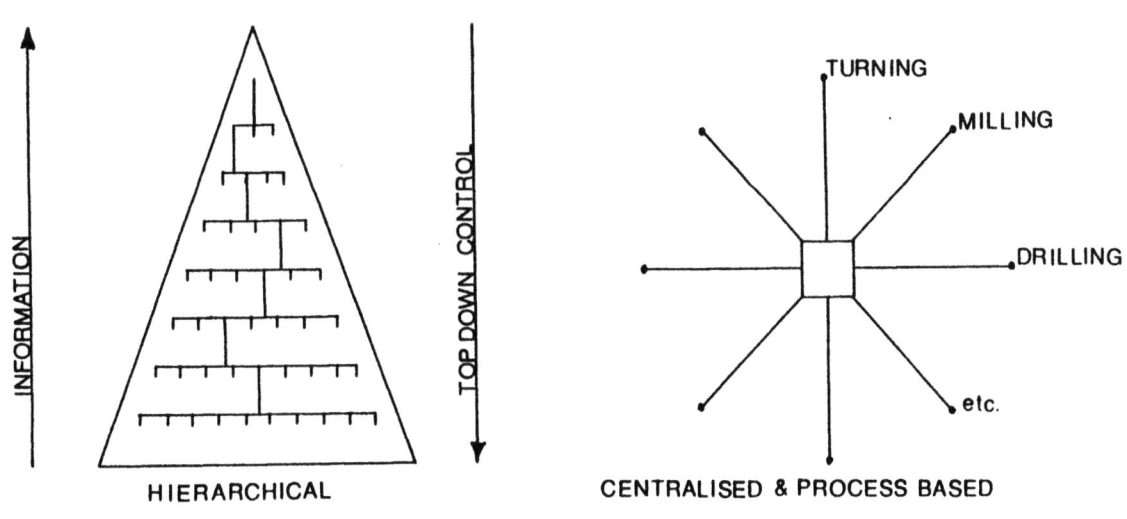

Fig. 1. 1890's Practice in the Factories of the 1990's

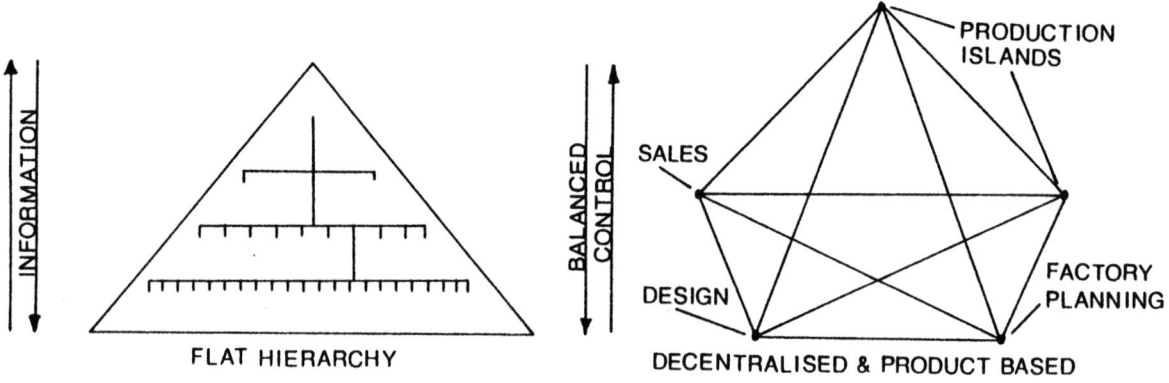

Fig. 2.   1990's Practice for the Factories of the 1990's

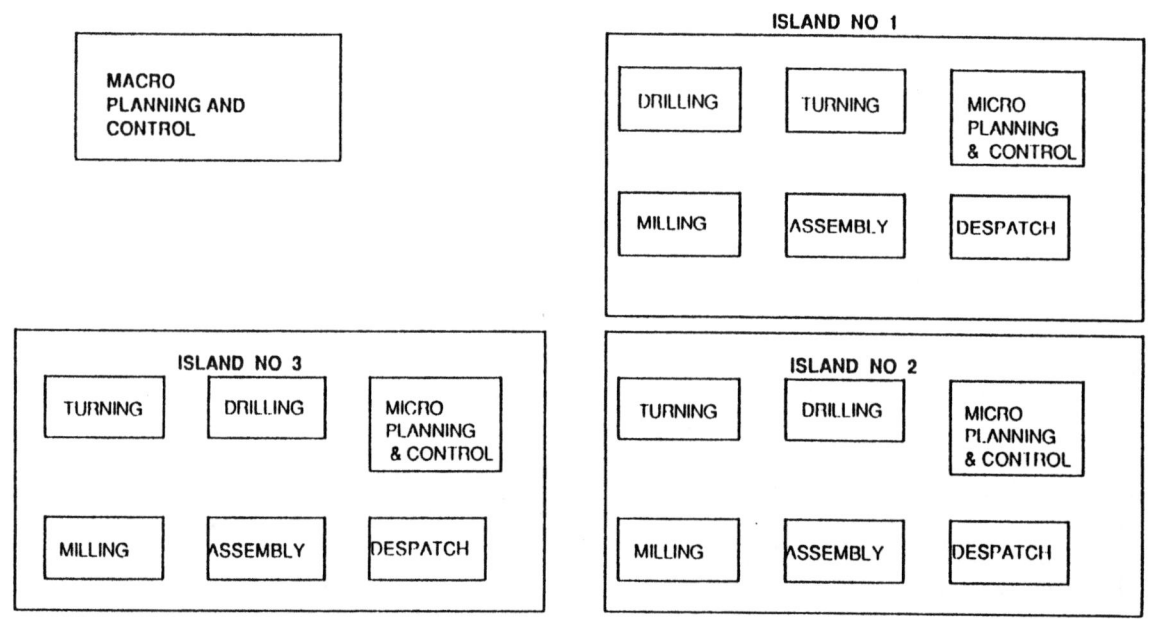

Fig. 3. Production Island Factory Organisation

HUMAN, ORGANISATIONAL AND MANAGERIAL ASPECT OF CIM

dr.ing. Roberto VIO - FIAT Corporate

0. SUMMARY

The aim of this document is to illustrate the changes to professional skills and labour organisation brought about by the incessant diffusion of automation in our production systems over the last ten years. It is structured as follows:

(1) The introductory section contains a brief presentation of the Fiat Group as a whole and its main operating division, Fiat Auto.

(2) The second part features the Cassino plant, where the Tipo and Tempra models are manufactured, and where there is one of the world's greatest concentrations of automation. A brief video will be shown.

(3) The central section describes the new professional categories that are necessary to deal with the requirements of such a highly automated plant and, where possible, variations with respect to the current situation.

The basic principles of the new organisation of labour are also described, specifically those applying in production plants.

(4) The final section describes the problems involved in training these new professional categories.

(5) The conclusion is that - even in high-tech factories - -organisation and the human factor are critical success factors, and are today those with the widest margin for improvement, and, by their very nature, unique in every situation.

1. PRESENTATION OF THE FIAT GROUP AND FIAT AUTO

Fiat's present situation is reflected in its commercial success on markets all over the world, in its technology standards, highly diversified product ranges and its balance sheet figures. Let us start with the latter.

TABLE-01 sums up the most important figures and the way they have changed over the last few years. In short, the Fiat Group is now a major industrial group on both the domestic and international scene, with 15 operating sectors, more than 450 subsidiaries and 160 affiliated companies in about 50 countries.

TABLE-02 gives a schematic idea of the way the Group is divided into major product lines. As may be seen from this slide, the most important operating sector is the Fiat Auto division, whose production system is centred around about 20 main plants. TABLE-03 illustrates the main characteristics of the division.

2. PRESENTATION OF THE CASSINO PLANT

Situated at the bottom of the valley at the foot of the famous Benedictine abbey between Rome and Naples, the Cassino plant started production as far back as 1972.

Following a series of extensions (1976-78) the plant reached maximum potential in 1979 with the RITMO and 131 models and subsequently (1983) with the REGATA,.

The "Cassino Anni Novanta", or "Cassino in the Nineties" reorganisation scheme was launched in 1985, but the real turnaround came in 1988, when the TIPO entered into production. Since 1989 the TEMPRA has replaced the REGATA alongside the TIPO.

The brief video I am about to show you reveals better than anything else the plant's high degree of automation and integration.

TABLES-04-05 present the principal characteristics of the plant: production volume, flexibility and, especially, the high concentration of automation systems.

3. THE EVOLUTION OF PROFESIONAL SKILLS AND ORGANISATION

As the previous chapter has shown, the Cassino plant is a highly complex system with a high degree of integration.

It seems to us that the most immediate consequence of this is that the philosophy must change. We cannot look at a limited part of the process and say that "the rest does not concern us".

What is necessary is a more global vision of labour. We must move on from automation-oriented reference models (typical of the early eighties) to integration-oriented reference models, whether they are automatic or not (a reference model is a specific framework which guides the activities of defining, designing and realising the complete industrial system).

The overall objective of these models is to generate an industrial system that functions by optimising the productivity of the whole system, and not just one factor at a time, safeguarding the development and promotion of human resources through the quality of life at work.

The objective of maximum productivity of the whole system combined with maximum quality may be achieved through four functions, which have the following aims:

- to define an industrial system in which all the principal factors are interdependent, and in which applied solutions constitute the lowest item of total costs in terms of external constraints
- to define and apply the operative tools that ensure that real responses are consistent with project requisites
- to transform the objectives of every single process into operative programmes, and to guarantee their implementation in conformity with commitments to product quality and the performance of the production equipments
- to ensure the efficiency of the production equipments and the quality of operations through constant improvements in service and prevention

The job of achieving these four aims is entrusted to the following four skills, which constitute the keystone of the high-automation factory:
- SK1: automated system operator
- SK2: integrated system operator
- SK3: process integrator

- SK4: system integrator
  (where SK stays for skill)

These four new skills, which will be described below, are the so-called "integrating skills". In the factory, of course, we continue to have the traditional specialist skills which are not considered here because they are not pertinent to the document.

3.1 Automatic System Operator (SK1)

SK1's job is to totally control segments of the production cycle from the point of view of the efficiency of the production equipments and the quality of the cycle's output.

What is new about SK1's job is that he does not manage real events directly, but perceives and controls them through the abstract representation thereof provided by computerised procedures (symbols).

In terms of increased know-how, therefore, SK1 must possess:
- knowledge of the operating systems (models) of machines and their interdependences
- general knowledge of physical flows and technical constraints
- knowledge of technical and management information flows

In terms of increased ability, SK1 must possess:
- logical capacities of abstraction and symbolisation
- memory capacity, inasmuch as procedures are complex and must be grasped
  by the worker without any uncertainty
- capacity to perform pre-diagnosis of stoppages and simple repair work

In terms of changes in behaviour, SK1 must:
- be more concerned with the final result
- be more involved in the process of constant improvement
- be strongly oriented to collaboration with contiguous skills.

In short, the systemic integration of the production equipments requires:
- on the one hand, greater technical-scientific skills in relation to the technological evolution of each of the various functions the worker has to control
- on the other hand, a new non-specialist, managerial capacity, which gives rise to a service that is not only more complex but also more discretional and creative, that demands a more direct involvement of the worker in the achievement of better productivity and quality.

3.2 Integrated System Operator (SK2)

SK2 operates at workshop level, and possesses all the responsibilities and duties required of SK1 above. In addition he is responsible for the operative management of production flows.

This responsibility involves him effectively in production programming, but, most of all, in tough operative decision-making mechanisms upon which depend not only local performance but also the transmission of disturbances to other processes.

In terms of increased know-how, SK2 must thus possess:
- knowledge of new managerial and organisational logics
- knowledge of basic technological components
- basic knowledge of management control and the information system.

In terms of increased ability, SK2 must possess:
- the capacity to manage and integrate more outstanding and diversified
  professional skills

401

- the capacity to exploit the potential of the new professional categories
under him, and plan their vocational progress.
In terms of general behaviour, SK2 must:
- share responsibility with contiguous skills
- be oriented towards the ongoing growth of the system as a whole.

3.3 Process Integrator (SK3)

Whereas SK1 and SK2 operate exclusively on the shop floor, SK3 operates on the shop floor during the production phase, and in the technical office during the design phase.

The process is an entity which completes a total function: in terms of production that means a single component or a subassembly or a finished assembly. Its success depends upon the operating quality of the single components and their mutual relations.

SK3's job is to control these mutual relations, and the influence which local results exert upon the final result. He must achieve a high degree of consistency between strategic requirements and operative results.

In terms of know-how, SK3 is not required to have detailed knowledge of the technical aspects of single machines and single production equipments. He sees them as black boxes whose behaviour he analyses in relation to parameters of reliability and maintainability, both of single elements and of the system as a whole.

In terms of ability, SK3 must have an outstanding aptitude for modelling and abstraction. Most of all, he must be able to identify and describe cause/effect trees in the quality route throughout the processes which he manages.

The very nature of the job requires a culture centred around reliability and maintainability engineering, and around problem-solving techniques.

3.4 System Integrator (SK4)

SK4's job is to define the operative means of levelling the system to the minimum point of total cost.

In other words, this means that he must prescribe consistent functional specifications in order to ensure the constant levelling of production in every component of the system. More precisely,
- production equipments
- production material flow
- information systems
- production engineering services
- plant and machinery layouts
In terms of know how, SK3 must have detailed knowledge of:
- the theory of models
- project management tools
- information technology
- maintainability
- reliability
In terms of ability, SK4 must be able to manage highly complex projects, coordinating human and technical resources of great value.

In terms of culture, this is a highly skilled professional category with many years of corporate experience.

3.5 The New Factory and Labour Organisation

TABLE-06 shows how the four integrating skills interact one with another, and how their action intersects with that of specialists at different levels, orienting labour organisation towards teamwork and continuous improvement at all levels.

The above is the essential basis for the process of the pursuit of total quality, and for addressing the great challenges of the 90s.

Culture, seen as the set of small-scale decisions made by single individuals and spreading to all, has to change. As we have already said, we can no longer look at part of the process and say that "the rest does not concern us". What is needed is a global vision of labour. Otherwise, we run the risk of achieving poor results from the point of view of quality and fulfilment of customer requirements.

In meeting the necessities listed above, the guiding criteria for the redesigning of organisation have been:
- delegation of decision-making responsibilities and attribution of managerial levers wherever knowledge of problems and problem-solving capacity are strongest
- attribution of the responsibility for the final result to the greatest number of skills possible
- maintenance of the logic of specialisation solely for the skills of true specialists
- for non-specialists, strong orientation towards management, with multidisciplinary training and definition of joint responsibilities for the global objectives of the production system.

The guiding criteria listed above are implemented in practice in two typical examples:
- the plant's organisation chart. In the traditional chart, the three functions of Manufacturing, Production Engineering and Logistics were jointly controlled by the Plant Director, who could be too high up to be in constant touch with each and every problem. So response times could be too high and the control of production flows is fragmented over a number of different departments.

In the new chart the tendency is towards unitary management of the technical and managerial aspects of every type of product. Hence Operative Units have been set up. These combine the logistic, technical and production functions according to finished product families: engine, gearbox etc.
- the technological team, which focuses on one machining line, and must ensure that it functions perfectly. The leader of the technological team is the process integrator (SK3). He may count upon the help of specialist technologists, experts in the various disciplines, maintenance operatives and integrated system (SK2) and automatic system operators (SK1).

4. THE NEW PROFESSIONAL SKILLS AND PROBLEMS OF TRAINING

The evolution of professional skills and of organisation described in the previous chapter have demanded a notable commitment to training. The latter has been planned and organised in close collaboration with the structures of ISVOR, the Fiat company whose mission it is to provide specialist and managerial training.

ISVOR provides an annual total of about 22,000 days of training to about 30,000 course members, and boasts a syllabus of almost 600 programmes. These figures put Fiat in the forefront in terms of update and development initiatives, not only at a domestic level, but also in relation to the highest standards of some of its international competitors.

As to the retraining of the skills described above, the biggest commitment in terms of quantity has been directed towards the more operative categories, whether traditional, such as maintenance operatives, or innovative, such as the system operators described above.

TABLE-07 illustrates the commitment made towards these two professional categories.

5. CONCLUSIONS

The organisational transformation of the factory has kept in step with technological modernisation. In this context, organisation and the human factor have emerged as critical success factors.

This assessment results mainly from the following considerations:
- as a general rule, new technologies are equally available to all: moreover, anyone can capitalise upon the experience of those before him
- the new technologies are, by definition, non-mature: hence the advantage goes to those who manage to complete the learning curve most swiftly
- the new technologies have great potential, but also constraints and demands in terms of service, which must be understood, and around which it is necessary to develop superior and distinctive know-how
- the new technologies reshape organisation and skills to provide the opportunity to point up the human contribution in terms of creativity, know-how and direction of effort. This is now regarded as the competitive factor with the widest margin for improvement, and is, by its very nature, unique in every situation.
In a recent interview to a house review, one of Fiat's top production managers declared: "automation has changed organisation, which is now based more on collaboration than on a tayloristic division of labour. It has been like passing from the crew of a boat to a football team. In the first it is the cox who coordinates the boat's progress: the other members simply row. In the second, on the other hand, everyone has a wider responsibility: teamwork is the secret behind good results".

FIAT GROUP EVOLUTION
(BILLION LIRA)

TABLE 01

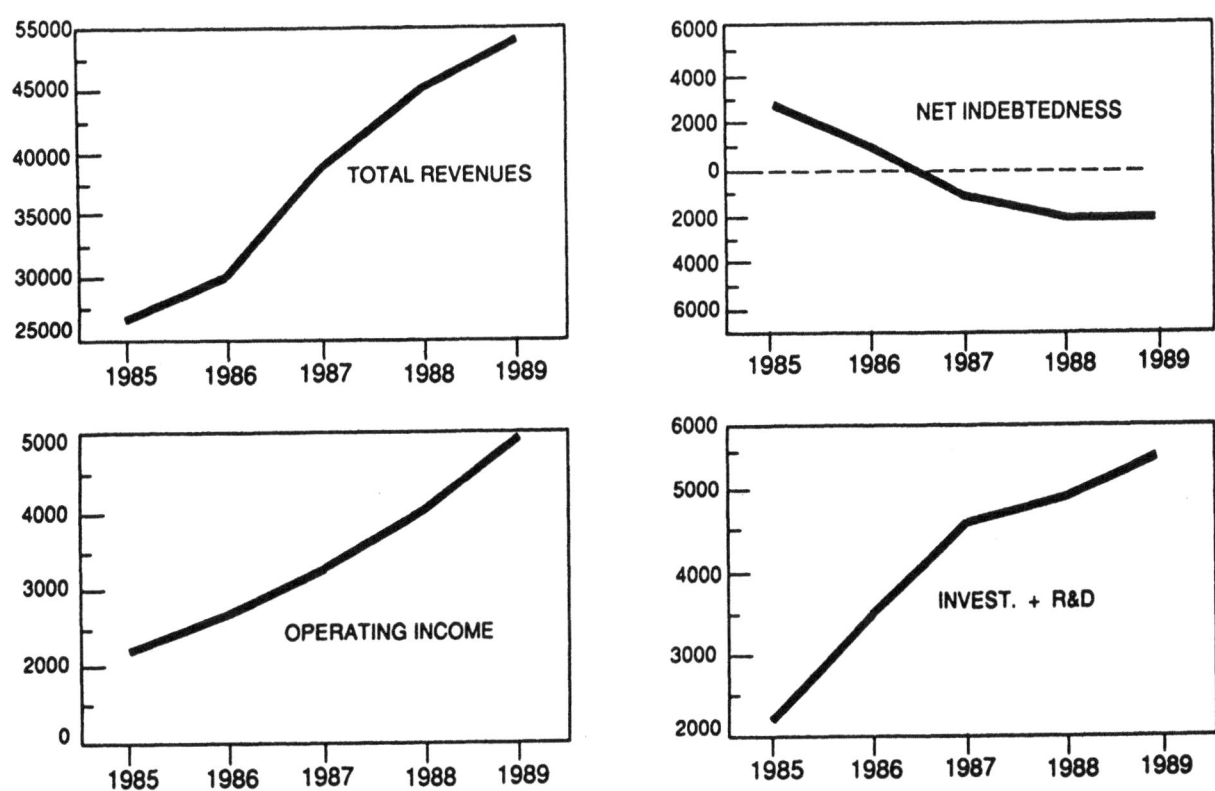

TABLE 02

FIAT SALES BREAKDOWN
1989

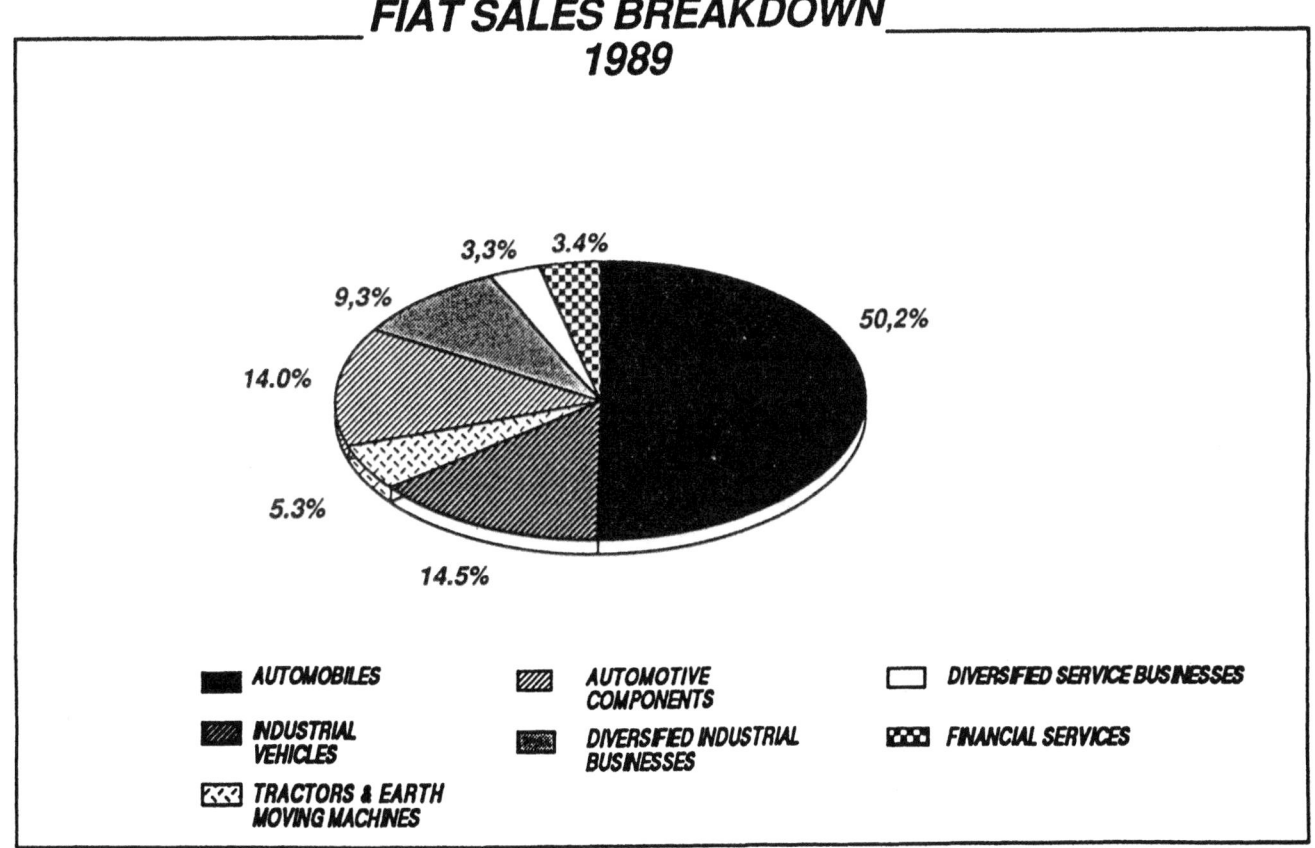

TABLE - 03

FIAT AUTOMOBILE DIVISION

4 TRADE MARKS

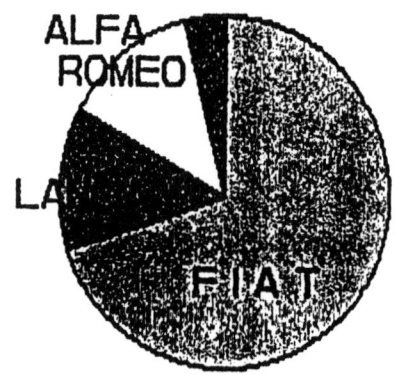

FERRARI
ALFA ROMEO
LA...
FIAT

TOTAL 1988 PRODUCTION:
2.141.000 UNITS

TOTAL 1988 REVENUES:
25.500 BILLIONS LIRAS

LEADER IN ITALY:
59% OF MARKET SHARE

LEADER IN EUROPE :
14.9% OF MARKET SHARE

14 COMMERCIAL MODELS

100 COMMERCIAL VERSIONS

FIAT spa

TABLE - 04

CASSINO PLANT AUTOMATION IN FIGURES

- 450 ROBOTS

- 61 LASER EQUIPMENTS FOR WELDING AND QUALITY INSPECTION

- 50 VISION SYSTEMS

- 570 A. G. V.

- 110 COMPUTERS

FIAT spa

CASSINO PLANT FLEXIBILITY

TABLE - 05

☐ **PRODUCT AND PROCESS FLEXIBILITY**

▷ ABILITY TO LAUNCH A NEW MODEL
WITHOUT STOPPING PRODUCTION

☐ **MIX FLEXIBILITY**

▷ TOTAL DAILY PRODUCTION : 1800 UNITS

▷ NUMBER OF MODELS : 3 (A,B,C)

▷ FLEXIBILITY CONSTRAINTS : 1800 A or
600 A +
1200 B,C

▷ FACTORY TECHNOLOGICAL : PRATICALLY
REACTION TIME INSTANTANEOUS

FIAT spa

TABLE · 06

WORK STRUCTURE IN AUTOMATED FACTORY

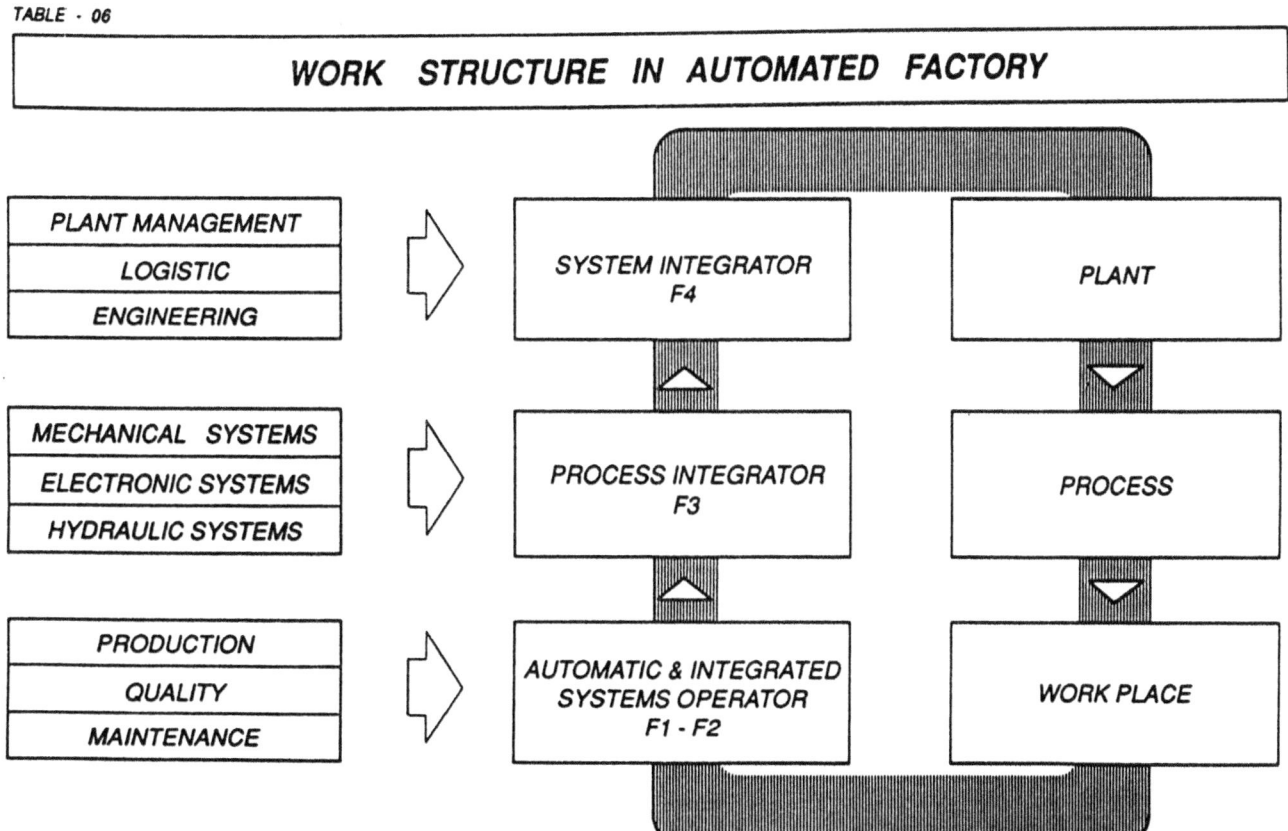

TABLE - 07

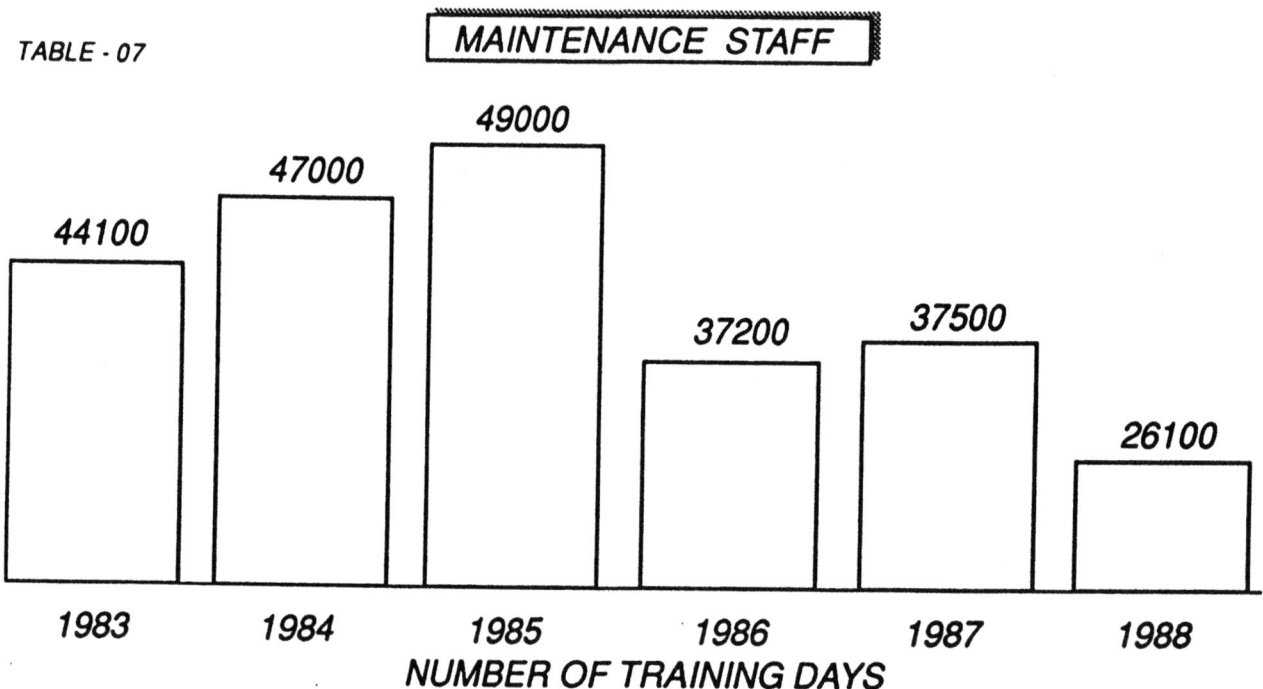

MAINTENANCE STAFF

44100 — 1983
47000 — 1984
49000 — 1985
37200 — 1986
37500 — 1987
26100 — 1988

NUMBER OF TRAINING DAYS

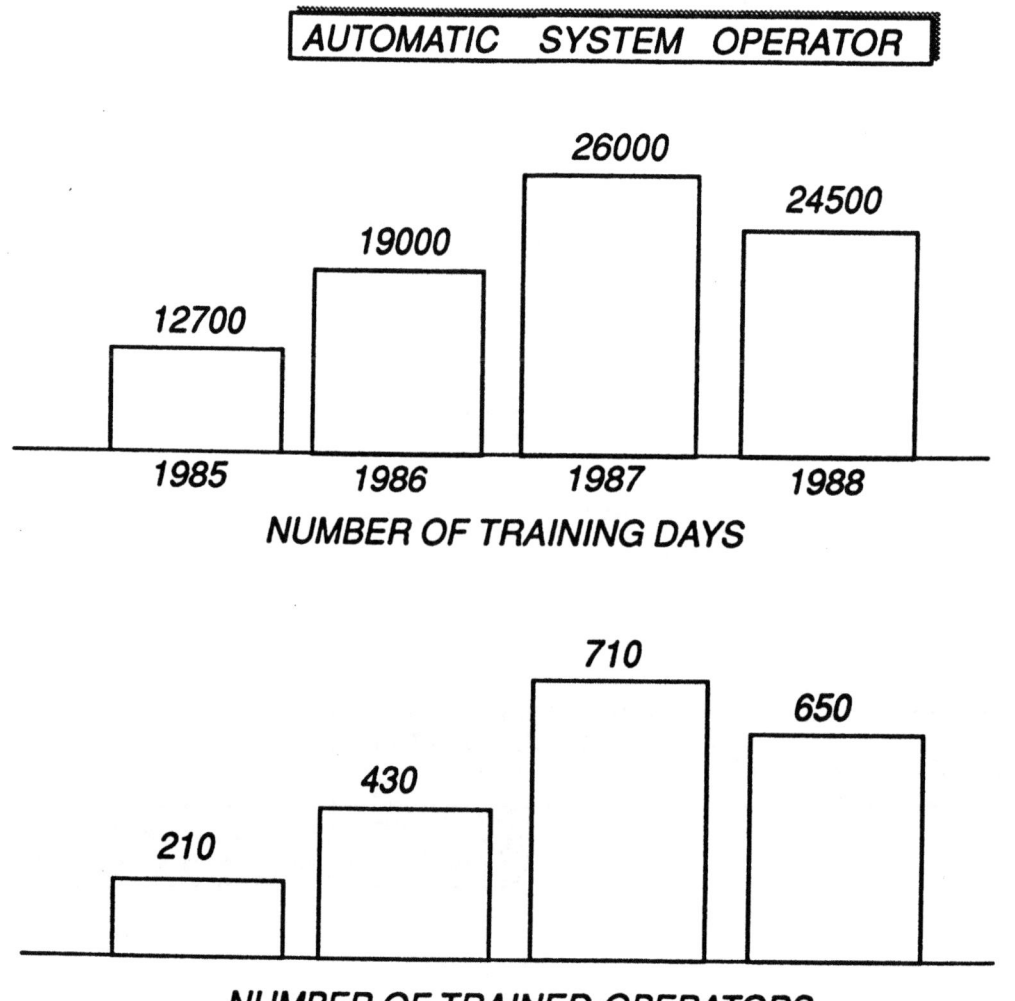

AUTOMATIC SYSTEM OPERATOR

12700 — 1985
19000 — 1986
26000 — 1987
24500 — 1988

NUMBER OF TRAINING DAYS

210
430
710
650

NUMBER OF TRAINED OPERATORS

Collaborative Research - Issues, Benefits and a Rationale for CIM Implementation

(Dr V.B., Prabhu and D.W., Wainwright)
Newcastle Business School,
Newcastle upon Tyne, England, U.K.

This paper is based upon the experiences of formulating and running three major collaborative (CIM) projects. In each case the same academic partner was involved but with different organisations. Even though there were differences in the outcomes of each project, it is contention of this paper that many common threads could be drawn together from these experiences.

The paper reviews the authors' experiences in the initiation and execution of such projects and the organisational issues involved in implementing the CIM systems using the collaborative research approach. Some of the benefits to be gained from this approach are also highlighted. Finally the paper puts forward a rationale for the use of collaborative research as a vehicle for implementing CIM systems.

1. Introduction and Aims

The Newcastle Business School (NBS) has been actively engaged in the formulation and running of three major CIM Implementation collaborative research projects concurrently from 1985.

Collaborative research projects at the NBS currently take two forms. The first is a more academic form such as provided by the SERC ACME Directorate where researchers are expected to formulate new ideas and usable frameworks which they disseminate to industry in general. As such they are required to produce generic benefits as well as perform specific in company project tasks. The second are projects funded by the Teaching Company Directorate and are of a more applied form where research associates perform a much greater proportion of the work for the host industrial organization. This has the aims of providing a catalyst for and enhancing the companies CIM implementation progress.

Even though each company has had different benefits and experiences it is the contention of this paper that there are numerous examples of generic benefits and learning which has been derived from our (NBS) experience.

There are two main aims of this paper. Firstly it is to share our experiences in operating collaborative research projects (CRP's) with other researchers in the field. There are several issues worth noting in connection with the setting up and execution of such projects and the benefits to be gained from them. Secondly it is to put forward a rationale for adopting such an approach to introduce change within a CIM environment. On the one hand companies feel unsure of what is possible from such initiatives and are naturally sceptical of the benefits to be gained from these activities. On the other hand, there is a danger that academically driven projects could result in theoretical ideas looking for a solution which is of little benefit to the business needs of the industrial partner. Our experiences on this matter will be put forward.

The paper is divided into three sections. Section 1 provides a summary of the three CRP's in question and in particular the nature of the tasks undertaken in each case. Section 2 highlights some of the problems and issues associated with collaborative research. In particular attention is focussed on the setting up of CRP's and the management of both the CIM implementation and collaborative processes. The benefits to be gained from such an approach are referred to in Section 3. In this section a rationale is proposed for adopting CRP's as a way forward for introducing change in a CIM environment. In the final section some of our conclusions are presented.

Section 1
The Three Collaborative Research Projects

Company A: (Work undertaken from 1985 to 1989)
This company is a division of the world's leading manufacturer of pressure sensitive materials, operating on a 24hr, 7 day a week basis.

A large amount of capital investment was made in new technology. This covered the areas of statistical process control, automated materials handling, stock control systems, coater planning, slitter scheduling, bar code reel tracking and time and materials recording with shop floor data capture facilitated by the integration of distributed database technology with various communication packages and networking.

Company B: (Work undertaken from 1986 to 1990)
Company B is part of a multi-national corporation involved in the mining, smelting, working and selling of aluminium.

In recent years the company has made small improvements to the smelting and casting processes. It has updated its process control capability through the use of PLC's in order to improve the quality of the product, reduce scrap (which is sold for remelt) and create a safer working environment for the work force.

More recently the firm became aware of the potential of the application of IT to industrial problems. A steering group co-ordinated by the financial director and with inputs from the NBS began to investigate the potential value of I.T. to the firm. After an "awareness raising" training programme a number of personnel identified potentials applications in a number of areas from the power station to the sales department.

Company C: (Work undertaken from March '87; to be completed by November 1990)
Company C is the U.K. leader and one of Europe's largest manufacturers of PVC compounds, and a leading manufacturer of PVC resins. Just prior to the CRP the company had made a major investment in replacing much of its outmoded equipment in the compounding plant. The aim of the CRP was to develop an integrated manufacturing planning and control system to both understand and manage the complex PVC compounding business.

Several major projects have been completed within this overall scheme or are in the process of being completed. These include the development manufacturing strategies (via the use of computerised simulation methods) to be adopted under varying market and environmental conditions; development of forecasting models for product demand, raw material requirements, plant capacity etc; the development and implementation of plant loading and sequencing systems which interface with the company's underutilised MRPII system.

For further background information and analysis of these cases see AMBIS (1988).

Section 2

2. Problems and Issues Associated with Collaborative Research

All our collaborative projects have a fairly rigourous form of management and control as they are externally funded. In each case the appropriate number of graduates with the specific qualifications and experience were recruited on fixed term contracts, usually for two years, to undertake particular parts of the research. Strict guidelines were laid for monitoring the projects. Short term control was through a project management group (PMG) consisting of the industrial and academic supervisors and the graduate researchers. This group met regularly to review progress and set future targets. Overall project guidance and control was provided by a Local Management Committee (LMC) which consisted of all the members of the PMG plus directors of the Industrial and Academic partners and the funding body. The LMC controlled and approved any expenditure within the set budgets for the project and monitored the PMG's performance against overall goals and targets set on a quarterly basis. It is against this background that our comments on the collaborative process have been made.

2.1 The Setting up of CRP's

In our experience this has been one of the most crucial aspects of the collaborative process. It is important that both partners establish a close relationship prior to embarking on any major project, even though this process could easily take a few months to achieve. For example with Company A contact had already been established via undergraduate student placement work and incompany personnel attending part-time management courses at the Polytechnic. In case of Company B, the process started with a training initiative on Information Technology which was run by the Polytechnic for their senior managers. With Company C we were introduced to them via our Engineering faculty colleagues and over several visits to them the present project was developed.

In this initial 'courtship' the industrial partner needs to know precisely what the academic partner has to offer them. The academic partner on the other hand needs to understand reasonably well the specific issues and concerns of the industrial partner and be able to focus on them. In our particular case the average time was between six to nine months. It is worth stressing that all our research projects had been proposed initially by the industrial partner and after some debate and consultation the expertise for them provided by the academic partner.

Meticulous planning is essential for a CRP. This normally took the form of several meetings especially with the funding body. As such it could be a major resource constraint especially for the academic partner. Finally the content of the proposal has to be agreed and written down in great detail with clear indications given of how the researcher's time will be spent and to show the range of benefits to all partners of such an research exercise.

2.2 The Management of the Collaborative Process

2.2.1 Project Control Mechanisms

Collaborative research projects are complex to initiate and manage primarily due to the plurality of views that exist between the project team members. The project team typically has a diverse membership consisting of key staff from both an industrial organization and an academic educational institution. Members from both have different backgrounds, objectives and skills. Typical membership of such CRP would be:

- senior managers from the industrial collaborator
- middle managers with project responsibilities
- consultants external to the organization
- applied academic researchers
- research grant co-ordinators and academic advisers

Such a diverse membership can be responsible for the initiation of actions that result in rapid organizational change, project success or failure, conflict, increased competitive advantage and contribution to knowledge. Or, if proper care is not exercised, the disillusionment.that comes from the rejection of perceived 'good ideas'.

At the CRP initiation stage the key instigators are responsible for setting project aims, objectives and plans. The academics and grant co-ordinators will be primarily concerned with those elements concerning contribution to knowledge. In this situation of applied industrial research it will take the form of construction of a framework, techniques, methods and/or methodology for encapsulating the research findings into a usable and highly pragmatic management tool. The generic extent of the tool would also have to be addressed in terms of applicability to different industrial contexts.

Members of the industrial collaborative organization will also have a significant degree of ownership and therefore desirable outcomes from the project. These objectives are usually very practical and shorter term than the academic membership. Typically these objectives will centre around the acquisition of more in depth skills and knowledge to increase managerial confidence and awareness for controlling complex project initiations, justification and implementation. The rationale being that this injection of state of the art ideas and thinking will increase the chances of project success and assist the achievement of business objectives.

Differences become immediately apparent between the collaborators. Time is of critical importance from an industrialists viewpoint in terms of implementation of new systems to achieve greater efficiency, effectiveness and competitiveness. The same time pressures are not readily apparent for the academics who prefer to distil learning experience into disseminable knowledge. Research method may not be easily reconcilable with the industrialists desire for quick action and fixes.

At the commencement of a CRP the objectives are versed in such a way as to be of equal validity to either collaborative party. When the pressure begins to unfold both from an industrial competitive dimension and an academic research perspective, these objectives if not well understood and unambiguously defined can be fragmented. The result is that the industrialists do not perceive that they have benefited significantly, and the academics feel that adequate contribution to knowledge has not been achieved.

2.2.2 Misalignment of Power

A major theme emerging from the three NBS CRP's concern the nature of power within the dynamics of research projects of this kind. Power can usually be associated with responsibility of ownership. In the case of a collaborative research initiative there is usually a question mark over who actually owns the project or exercises dominant control. The ethos of the type of practical CRP embarked upon by the NBS requires that applied researchers become an integral part of the decision making managerial team encompassing both middle and senior managerial levels at the host industrial organisation. These research personnel are not however employed by the company, but by the academic organisation. The research personnel consist of research assistants or associates and full-time lecturing staff. Research associates are also only employed usually for a maximum of 3 years

The result of this structure is that researchers never really control the company projects they are associated with. They can however be the designers and chief architects of systems with a lot of vested responsibility to initiate change and implement their innovations. The fact that they are not permanent company employees however means that their legitimate basis of power does not fully match the responsibility and requirements for successful implementation of those innovations. Slow progress and unnecessary obstruction is the practical project result from this.

2.2.3 The Changing Project Environment

Action research in an organisation inevitably contributes to change within the organisation, influences individual project outcomes and also has consequences for the research process itself. The complexities of action research as a method of inquiry is already well documented Checkland (1981). CRP's at the NBS are based on the action research model as it is considered that this method will deliver the rich contextual material considered critical to understanding manufacturing systems implementation Bowker and Wainwright (1989).

Research projects of this sort can quickly become a victim of their own success. In company A the CRP injected 8 man years of effort into the planning and control of over £6m of CIM related project activities. This resulted in the company having the confidence and requisite skills and knowledge to invest and progress far more rapidly than previously envisaged. In terms of information systems development alone, close to £1m of development took place from 1985 to 1988 and much of it was prototyped and instigated through members of the CRP team. This rapid and spectacular success prompted an over reliance on these new systems to control the business. Therefore increasing importance was placed on the early delivery, operation and maintenance of the systems. These activities were to be performed with only a finite set of resources. Hence, research personnel were distracted from innovative developmental tasks to become more operationally involved with the systems. As a result a stage was reached where little new progress was eventually made.

The generic extent of this observation was confirmed by similar experiences occurring in company C. There, as and when prototypes of parts of systems were being developed and tested, the managers concerned immediately wanted to make use of them for their day to day activities. They insisted on having a copy of that software on their own PC and then immediately sought all sorts of modifications to it that would allow them to obtain more control information. Thus the original task of the development work was often delayed and strict discipline had to be applied to redirect attention on the original objectives. A natural conclusion is that this rapid success should be anticipated and managed more effectively.

2.3 The Management Process of CIM Implementation

2.3.1 Increased Awareness Alters Project Structures

All of our research projects contributed to the individual company CIM systems strategy. This took the form of both practical assistance in managerial planning, decision making and control activities together with the development of ideas of how CIM should grow within the overall business objectives of each company.

There were spin-offs resulting from the 3 research projects running concurrently. Numerous inter company visits and seminars were arranged by the NBS for the purposes of propagation and dissemination of this learning. The effects of this knowledge on speeding up various CIM implementation projects cannot be underestimated. A pilot systems and prototyping implementation framework is currently being disseminated amongst the NBS portfolio of companies. This framework was derived from the experiences arising out of the CRP at company A.

2.3.2 Increased Awareness Spawns a Number of Different Reactions

An early consequence of CRP's is the increased awareness of the potential of information technology. As prototypes are built and demonstrated to the users and to senior management as part of the monitoring process, there is a sudden realisation that major operational changes appear to be feasible only if the technology could be applied fast enough. The developments in Company C are a case in point. Successful demonstrations of prototype PC based systems to the board of directors of the company resulted in the employment of a firm of specialist consultants to look at the whole question of new technology being applied across all divisions. Even though the CRP has a few months more to run, the specialist consultants' report on their study has been accepted by the Board of Directors and their recommendations will be introduced over the next two years. Our belief is that the general interest in IT applications demonstrated by the CRP had a very significant bearing on the developments now taking place in CIM in the company. Similar consequences were observed in Companies B and A; in the former case collaborative work was actually dropped in favour of vendor led solutions.

2.3.3 Prototype System Responsibility is Contracted out to an External System Supporter.

This particular type of problem cannot be anticipated at the start of a CRP, though it is quite likely in organisations where the entire system support is provided by an external software house. Such a situation has arisen with Company C. Unknown to the CRP team the software house which maintains the company's partially installed MRPII system, had been working independently on writing a separate sequencing and capacity management module as an add-on to their software. Their developments were still at an early stage though similar in principle to that being developed by the CRP team. Company C has decided recently that at the appropriate time the more advanced CRP software will be handed over to their software maintenance people so that they could build a system using the CRP's prototype at no cost to the company, and in turn they (the software house) have the right to market it to other organisations with similar problems.

Two implications arise from such a decision. The company has lost the learning curve in this development as once the graduate associates leave the business at the end of their two year work, there is no one in the company who will have the confidence to maintain it. The second issue arising from this situation is that the urgency to provide user training for as many employees as necessary, in operating and maintaining the systems, is no longer there.

2.3.4 Beneficiaries of the Learning Curve

Our experience at the NBS indicates that the knowledge and skills gained from CRP's is totally underestimated, seriously misunderstood and that the potential benefits are rarely capitilized upon.

8 man years of research effort was injected into a CIM implementation strategy at company A. The company questions what it gained from playing host to that effort. One member of that research team is now European information systems manager for that company. More people in the company use the fourth generation developed systems created by the research team members than any of the other plant systems. All senior and middle managers now have on-line workstations and ad-hoc decision support applications transparently linked to any of the plant databases regardless of computer architecture. Large increases in plant volume output and decreases in scrap percentages coincide and can be attributed to the implementation of these applications. Yet people question what benefit did they derive from the research project.

It is argued here that a mere fraction of that progress would have been achieved without the injection of the requisite skills, innovation and state of the art knowledge from the research team. The injection of knowledge is totally under-estimated, undervalued and quickly forgotten by company fire-fighters in the quest for quick superficial technological as opposed to systemic or organizational and cultural changes.

2.3.5 Importance of Prototypes and Incremental Delivery

The results from Company A indicate the benefits of adopting a pilot study and prototyping approach to both the management of a CRP and the implemented innovative system itself in terms of incremental or phased delivery. The benefits of this in research terms are that increased flexibility is built into the research process to enable response to changes in both the company project needs, governmental and academic requirements and the needs of the individual researchers themselves. This can be extremely important where the natural disparity in project members views previously discussed will tend towards dysfunctionalism. More tangible shorter prototype deliveries together with feedback of knowledge and experiences so gained enable greater control and performance measurement against the overall and much broader based objectives of the CRP.

In company A information systems prototypes were rapidly constructed, implemented and evaluated in time periods of typically one to six months. This enabled early feedback of performance measures for the CRP itself, which in this case concerned the planning and implementation of a totally new manufacturing strategy in the form of a pilot system. This early feedback was very favourable and enabled the justification of rapid scaling up of the pilot system and much earlier implementation. The confidence for this was provided by early incremental delivery of tangible innovations prototyped by the CRP team.

In pure research terms, this early success also provided the material upon which to base further lines of enquiry and disseminate the knowledge so gained.

Section 3

3. Some Key Benefits of CRP's

3.1 Stimulate and Act as Catalysts for Organizational Change

In all three CRP's initiated and successfully run jointly by the companies and the NBS significant if not spectacular change has occurred. Company A has undergone a major re-organization and totally restructured the information systems function. An additional senior management position was created together with two other high middle management posts. The company now has integrated information systems with managerial decision support work stations. Production has increased and the company has become more competitive. This change has all taken place in the period involving the planning and implementation of the CRP from 1984-1989 and is continuing into the present time. This rapid change could not have been initiated or effectively managed over such a relatively short time period without the catalyst provided by the CRP and the underlying thinking that preceded it.

Company B has undergone relatively less change than company A primarily due to the loss of several key 'champions' within the organisation. The CRP has therefore had to operate with far more caution due to a more risk averse culture embedded within the company. Resources were therefore hard to find.

Company C has just completed a major IS review. This has led to the creation of a number of middle management positions in this area together with a re-arrangement of the IT reporting responsibilities. Two of the researchers have moved on to the company payroll.

3.2 Translating Theory to Practice

Company C achieved rapid progress from the CRP with early delivery of prototype production scheduling systems, simulation modelling, and customer product specification database. These prototypes have gone into early operation providing the business with a far more innovative production and marketing potential. This innovation would not have been achieved without outside injection of ideas, knowledge and skills provided by the synergy of a CRP. This synergy results from the attempt to translate theory into practice. Industrialists benefit from the injection of new ideas (new to them but not always new in themselves) and academics gain from the learning process of testing those ideas in practical situations. Company A have had similar experiences with rapid implementation of customer service and stock allocation systems together with materials tracking and shop floor data collection systems.

3.3 CRP as a Cost Effective Method of New Developments

CRP's can be viewed as a low cost addition to a manufacturing companies portfolio of resources especially regarding the skills base. These state of the art skills are commonly regarded as critical for implementation success. Companies who do not have the expertise in house or the culture for frequent management re-training have previously relied on outside vendors or consultancy houses to compensate. CRP's do not replace these existing resources but become complementary to them.

CRP's can provide a large proportion of the necessary skills and knowledge injection badly needed by companies embarking on ambitious CIM strategies. As such they are a very valuable and presently under-rated mechanism for aiding any CIM strategic planning through to implementation process. Also, companies can tap academic partners for advice and consultancy on an informal basis in a variety of different areas. The academic partner can therefore act as a sounding board for related projects.

3.4 CRP as a vehicle for Management Training and Education

An opportunity exists to foster long term links between academic and industrial establishments. This opportunity is due mainly to the nature of a CRP being at least 3 years in duration. Prior to this there are the conceptualization, planning and justification phases which can take many months or years to deliver. Post project there is always scope to further the work or to set up some ongoing consultancy operation. This works both ways between the collaborators with practical anecdotes and skills provided by industrialists for an educational audience and theoretical or applied research knowledge provided by academics for the companies concerned.

The longer timescales involved reveal a growing relationship between the collaborators which fosters more mutual trust, a process of continuous improvement and a rare insight into the increased performance of both institutions over a measurable time scale. This releases pressure on managers to learn at unrealistic rates and provides time for academics to test theories and reflect on the knowledge gained together with the needs of industry. Academics therefore become far more relevant in their teaching and their research activities.

4. Towards a framework for CRP's as a Way Forward for Introducing Change in a CIM Environment

4.1 CRP's as a Complementary CIM Implementation Process

At the present time our growing experience from participating in CRP's at the NBS indicates an urgent need for a rationale or framework for both the project management process itself together with a methodology for CIM implementation using the CRP as a complementary mechanism to existing company CIM strategies.

The methodology and body of knowledge concerning implementation issues and change initiation is evolving out of the research projects. This progress towards a coherent methodology can be traced through a series of papers disseminated by active NBS researchers at the NBS (AMBIS (1988), Bowker & Wainwright (1988), ACME (198)9, Wainwright and

Thomas (due 1989), Kimble (1989), Prabhu & Kimble (1988) & (1989)). To date the most successful and promising implementation strategy has been found to be a pilot system and prototyping framework. Research is still being planned and undertaken regarding the management and control of this process (Wainwright (due 1990) and indicates that power/political factors are of greater importance to the process than previously envisaged. Technological issues have given way as a cause of concern to 'softer' organizational and business validity issues. These findings compare favourably with those of the latest Price Waterhouse 89/90 IT review which finds that for the first time concern for matching IT and IS strategy to business strategy is seen as of prime importance in relation to other issues such as project deadlines and staff recruitment e.t.c.

4.2 A Framework for CRP Initiation, Planning and Management

A three dimensional framework is proposed that encompasses the critical directions concerning the initiation, planning and management of CRP's, fig.1.

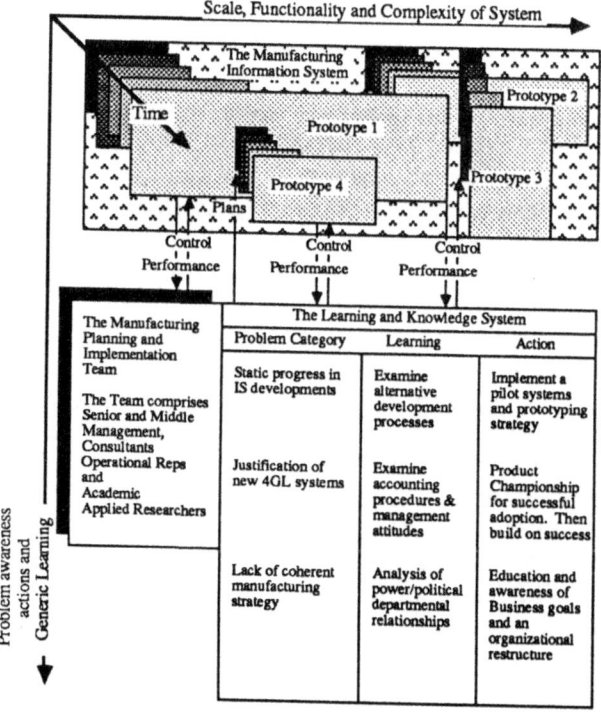

Fig.1 A Three Dimensional View of the Collaborative Research Process
(Wainwright 1990)

(a) Scale, Functionality and Complexity of Systems
Three related components have been identified which critically affect project success. The Scale of a project is an issue where there exists much cause for concern. This can be alleviated to an extent by breaking projects down into manageable 'chunks' where a defined set of success criteria can be defined. These small projects are co-ordinated against the background of an overall CIM and CRP implementation strategy. Incremental delivery occurs with tangible progress visible towards more global business objectives Tyson and Cassells (1986).

In smaller scale project modules there must however be sufficient functionality to provide a meaningful contribution to an overall implementation plan. This is where the current research performed on pilot systems and prototyping provides the requisite framework, confidence and flexibility to implement functional and fully operational prototypes.

Prototypes are either incrementally extended over the project duration or evolve and gain enhanced functionality in response to environmental changes and the learning assimilated. Prototypes form part of a pilot system which is a fully working business system only scaled down to reduce risk and gain understanding of both the business problems and the implementation process itself.

The pilot system can then be scaled up on the basis of early smaller project success and the increasing complexity more rationally managed due to the increased knowledge, learning and performance measurement of the implementation system.

(b) Problem Awareness and Generic Learning

As the CRP progresses suitable research methodology should be applied to capture and process the essential generic learning that is taking place. Standard techniques can be used such as longitudinal case study recording, action research methodologies such as soft systems methodology Checkland (1981), and empirical evidence wherever relevant. Project milestones should be carefully examined in order to determine significant events, sequence of activities and the actors involved leading to successful or failed outcomes.

(c) Problem Cycles over Time

The problem awareness, categorization, learning and action cycle Wainwright(due 1990) is an iterative process with generic learning the final product. It is important that this iterative process is recognized from the start of the CRP in order to understand the complex interactions that tend to occur over time. This is due to the evolution of projects, growing awareness amongst CRP team members of implementation issues, power political implications, environmental change and changing business strategies.

5. Results and Conclusions

Our experiences so far lead us to believe that collaborative research projects have a role to play in the implementation of CIM systems. It is an approach which meets a lot of the requirements of both partners involved in it. From the industrial partner's point of view, the biggest advantage is that it, as a user, is involved in every stage of the development. There is a strong involvement of all the key players and a feeling of ownership right from the beginning. The users can gradually build their confidence in using the new technologies with the assistance of the academic partner and develop systems that really meet their requirements. The learning curve in this type of collaborative process is very steep indeed and results in a very positive approach to the use of in formation technology.

As illustrated, our experiences bear this out very strongly. Indeed the chances of having CIM systems that do operate effectively and meet the user's needs are considerably enhanced by this learning and development experience.

From the academic partner's point of view there are equally many benefits. The most significant one is the practical experience gained from such implementation projects enrich considerably the teaching environment within the institution. Current live issues related to the development of CIM systems can be demonstrated and explored very effectively.

Academic credibility and competence is enhanced in the eyes of other industrial contacts and partners.

Undoubtedly there many pitfalls in the design and operation of such collaborative projects. However, if they are planned carefully and close relationships developed between the two partners, then our experience shows that this is a very powerful means of introducing new (particularly information) technology into an organisation.

6. References

AMBIS., (1988), The Management of Advanced Manufacturing Technology, A Collection of Papers produced by the AMBIS (Advanced Manufacturing and Business Systems Research Group), Department of Business Analysis, Newcastle upon Tyne Polytechnic, Unpublished Internal Document

Cassells D., Tyson M.J., (1986), The Introduction of New Technology to Develop a Manufacturing System based upon Part Processed Stock", Proceedings of the ACME Grantholders Conference, Salford University

Checkland P., (1981), Systems Thinking Systems Practice, Wiley

Kimble C., (1989), An Investigation into the Nature of Computer Integrated Manufacturing (CIM) in the North East of England, M.Phil. Thesis.,Newcastle Polytechnic

Kimble C., Prabhu V.B.,(1988), CIM and Manufacturing Industry in the North East of England: A Survey of Some Current Issues.,in Ergonomics of Advanced Manufacturing Systems., Ed. H.R. Parsaei and W. Karowski, Pub. Elsevier

Prabhu V.B., Kimble C.,(1989), Managing Change in CIM Applications in the North East of England, in Integration and Technology for Manufacturing., ed. Dr. E.H. Robson, Pub. Peter Peregrinus

Thomas P., (due 1990), The Management of Advanced Manufacturing Technology - An Integrative Framework, unpublished PhD thesis, Newcastle upon Tyne Polytechnic, Newcastle Business School

Wainwright D., Cassells D., Bowker P., (1989), The Introduction of New Technology to Develop a Manufacturing System Based upon Part Processed Stock, SERC/ ACME final report no GR/D 40739

Wainwright D., Bowker P., (1988), Towards an Holistic View of Manufacturing Information Systems Implementation, Factory 2000 Integrating Information and Material Flow Conference, Aug 1988, IERE Conference Publication

Wainwright D., (due 1990), A Framework for Prototyping Manufacturing Information Systems, unpublished PhD thesis, Newcastle upon Tyne Polytechnic, Newcastle Business School

CIM TECHNOLOGY IN VARIOUS INDUSTRIAL SECTORS

KB MUSICA : COMMERCIAL OPPORTUNITIES VIA PRACTICAL DEMONSTRATION
D Boland, A Campbell, A N Levin
(KB MUSICA Project Management)

1. INTRODUCTION

The objective of KB MUSICA is to develop an integrated framework
for process control and other industrial CIM applications bringing
together new generations of fieldbus technology, knowledge-based
systems, control techniques and intelligent multi-sensor systems.

The work programme addresses a number of practical measurement
and control problems covering a range of variables from standard
equipment such as pressure measuring devices to new systems
coping, for example, with object recognition of tactile information.

The aim is to develop validated prototype systems based on the
specific technical requirements of four demonstrator plants in
widely differing industrial sectors (chemical production, metal
deburring, glass manufacturing, plastic processing).

A considerable number of hardware and software products are
expected from KB MUSICA. Plans for their exploitation are given
a high priority by the Project Partners.

2. WHAT IS KB MUSICA?

KB MUSICA is an acronym describing the Project scope:

K	=	Knowledge
B	=	Based
MU	=	Multiple
S	=	Sensor systems
I	=	In
C	=	Computer integrated manufacturing
A	=	Applications

An alternative title might be 'Low Cost Intelligent Sensing and
Control Systems'.

A consortium, made up of five contractors and nine associate
contractors from seven countries, has been established to execute
the project on a 3-year time scale (Fig 1).

3. TECHNOLOGICAL HORIZONS

The objective of KB MUSICA is to develop and integrate a range of
technologies in order to enhance industrial control systems. The
total scope includes technological developments in the areas of
sensors, fieldbus, knowledge-based systems and multi-variable
control schemes.

Sensors

The Project Plan includes the development of enhanced sensors
which will possess a capability for self-validation of their
operation. These sensors are 'enhanced' by their ability to
self-compare actual operating parameters with reference data;
when these comparisons indicate that degradation or a
malfunction may be occurring, further routines are invoked to
confirm that an error condition is present and to determine
the cause or type of failure.

Exemplification of the technology is provided by its application to
four specific Demonstrators, selected to represent a wide range of
potential users. It is intended, for example, to extend existing
work on toxic gas monitoring to provide faster and more complete
transmission of data coupled with a more informative presentation
of results at both the operators' terminal and within a global
database. Data leaving the enhanced sensors is to a much higher
quality and reliability than previously available from conventional
sensors; these data, presented through a Man Machine Interface
(MMI) to the operator, will reduce the incidence of spurious plant
shutdowns and lead to better economic and safety performance.
At a higher level, Multi-sensor Systems (MSS) are being devised
for more demanding applications aimed at providing decision
support systems for use in the automated supervisory control of
robotics and industrial production processes. Sensor data fusion
techniques are also being developed, to overcome partial or
incomplete information supplied from imperfect sensors.

Fieldbus Technology

It is already clear that, if European Industry is to enjoy a
dominant position in the immediate future, there must be an
emerging role for the use of Knowledge Based (KB) Intelligent
Measurement Systems (IMS). One of the constraints to
widespread deployment of such a system has been the high cost
of data transfer using conventional cabling systems. This
problem can now be solved by utilising a fieldbus system to
convey the data between actuators and sensors. The main
requirements for a successful fieldbus system are high speed
and low cost, the specific goal being to achieve a chip cost
of less than 10 ECU with the acquisition period for at least
32 simple variables guaranteed to be less than 10
milliseconds. KB MUSICA will demonstrate that digital bus
technology is the best way to perform monitoring and controlled
calibration of sensors.

Multi-variable Control

Multi-variable Control (MVC) systems provide maximum interactive
control and, consequently achieve a much faster overall response
to a complex processing problem than can be achieved by
conventional techniques.

MVC systems, although model based, are able to handle model uncertainty. This is achieved, for example, by re-creating design requirements and accommodating the effects of noise, perturbations and non-linearities in actual plant behaviour. A generalised frequency response approach is being used for the control system analysis and design. This approach will give the required physically-based intuitive grasp of the complex systems behaviour and is flexible in that links with alternative approaches can be easily assessed.

Knowledge-based Systems

Using knowledge-based systems (KBS) to supervise processes can avoid damage to plant personnel and environment, as well as reducing plant down-time. KB MUSICA is investigating situations which require responses in timescales of a few seconds.

The project will produce KBS for online fault detection and diagnosis on process equipment and sensors; for example the Fault Detection and Identification (FDI) system on the Chemical Plant Demonstrator will provide continuous monitoring of environmental and plant operational parameters in order to detect faults and advise operating personnel in real time. Also, the project will produce KBS for online quality analysis eg systems for controlling wall thickness, bottle shape and transparency within a specified quality band for the Plastic Processing Demonstrator.

These systems will be developed using a combination of qualitative and numeric techniques, cooperating in the same environment. Artificial Intelligence techniques such as blackboard architecture, qualitative modelling and multi-agent systems will be used to produce these KBS. To enhance operator perception of the situation, advanced man-machine interfaces (MMIs) will be developed permitting strategies for handling identified breakdowns to be appropriately varied.

4.   DEMONSTRATOR DRIVEN PROJECT

KB MUSICA centres on the development of modules of technology in the areas of Sensors, KBS, MVC and Fieldbus. A strong concept of the Project is that these modules are able to 'stand alone' so that they may ultimately be used in multiple diverse applications, possibly integrated with other new technology outside the scope of KB MUSICA. Clearly, this is consistent with the ESPRIT spirit of cross-fertilisation.

However, in terms of demonstrating the viability of the technology modules, it has been necessary for the Project to focus on four specific Demonstration units, selected across widely differing industrial sectors.

5. THE DEMONSTRATORS

The technologies developed in KB MUSICA will to be exemplified by their demonstrated application in the chemical industry, the plastics industry, deburring processes and (glass) furnace temperature control. Fig 2 describes the concept of these four Demonstrators controlled from a central Project Management while Fig 3 indicates the roles of the four Technical Areas with respect to each Demonstrator. The Demonstrators are outlined below:

Chemical Plant (Fig 4)

This Demonstrator, based on a membrane chlorine production unit, will use the following areas of technology:- atmospheric sensors, enhanced pressure transmitters, fieldbus, fault detection and diagnosis systems, simulation and multi-variable control.

Enhanced sensors will provide a quality index for the integrity of individual sensors and a status of the local process and associated plant equipment. Developments proposed for monitoring of atmospheric conditions will result in intelligent sensor heads with self-checking functions. The data generated from the sensors will integrate with available system data via a rapid fieldbus communication system and thus provide a highly effective interface with the process operator (via the MMIs).

The objective of a fault detection and identification system is to provide high level inferences from plant data which can then be used with a high degree of confidence for plant control.

Multi-variable control techniques are to be applied as an alternative to conventional single variable control. This will utilise the benefits achieved by the improvements in the quality of sensor data and will make more effective use of this data.

Plastics Processing (Fig 5)

This Demonstrator is a stretch-blow moulding machine for production of hollow plastic bodies eg polyethylene terepthalate (PET) Coca-Cola bottles. A hierarchical product quality control system automatically compensates for the influence of time-varying ambient conditions.   The blow moulding machine, when equipped with the new control system, will produce hollow bodies of precise specifications at maximum machine throughput.

The areas of technology used on this Demonstrator will be:- advanced sensors, KBS, simulation and multi-variable control. The generic technology will be applicable to a wide range of plastic and rubber forming processes.

Glass Manufacture (Fig 6)

High quality lead crystal glass, suitable for blowing by craftsmen and forming into cut-glass artefacts is produced on a semi-continuous basis. A premix of lead oxide, silica sand and potash is fed forward into a continuous furnace where the mix is melted and held for approximately 3 days. The product emerging from the furnace is blown by traditional methods and eventually fed forward for cutting and finishing.

A major inefficiency in the process is the recycle of rejected blown items which can be as high as 90%. By introducing accurate multi-variable control of the furnace system, backed by a knowledge-based supervisory system, it is expected that significant reductions in recycle rates will be achieved. The implied saving in manpower and energy should lower production costs by up to 10%.

Simulation, MVC and sensor technology will be demonstrated on this unit.

Deburring (Fig 7)

This demonstrator is designed to automate the current labour intensive and dangerous task of deburring. A robotic system based on advanced robot control with multi-sensor integration is being developed. MVC, KBS, simulation and sensor technology will all be utilised in order to achieve the goal of automated deburring.

The KBS will take into consideration the required deburring technique, workpiece geometry and other nominal information about the general manufacturing task eg the precise orientation of the workpiece with respect to the robot.

A multi-sensor system will be integrated for assessment of the layer and course of the work contour, the size and shape of the burrs and the actual process and reaction forces. This information provides the robot controller with necessary process parameters for motion planning and control.

Multi-variable hybrid position force control schemes will be introduced into an advanced robot control system to adapt the robot's behaviour to rapidly changing process parameters and to maintain the required adherence to the nominal specification of the user.

The system aims to solve the problems of automated deburring with high flexibility and performance in order to satisfy the user's needs related to quality, safety and costs.

6.   CROSS-FERTILISATION

An important concept of ESPRIT is that the Partners collaborating in a Project all contribute their individual expertise so that the value of the overall collaborative Project will be greater than the sum of the individual Partners' independent capability. This high level of cross-fertilisation is achieved by incorporation of appropriate work tasks into the KB MUSICA schedule.

Cross-fertilisation within KB MUSICA can relate to transference of technical developments within one section (ie one Demonstrator) of the Project to application in another sector (ie another Demonstrator). For example:

> Multi-variable control is being used on the Plastics Processing Demonstrator and may be used in a similar mode on the compressors on the Chemical Plant Demonstrator.

> Real time process simulators are being developed using the same principles for the Deburring and the Glass Manufacturing Demonstrator.

Further cross-fertilisation within the Project occurs at the interface between two technology modules, where ideas developed in one area may be used in conjunction with a second technology area, for example:

> The development of enhanced atmospheric sensors is interlinked with the development of an intrinsically safe fieldbus system.

Cross-fertilisation also exists at the level of technical exchange between parallel ESPRIT Projects. Here, KB MUSICA has established a liaison with other projects which are active in the same general technical area eg links with DIAS Project (Project No 2172) have been established in order to exchange technical information on sensors, FDI and control technology. In some cases specific cross-fertilisation tasks have also been established. For example, close synergy has been identified with the EPIC (2090) Project and it has been agreed that the plant simulation model developed for the Chemical Plant Demonstrator will be made available to EPIC so that further assessments of the controllability of the chemical plant can be made.

Other contacts have been made with QUIC (820), AIMBURN (2192) etc, and the question of cross-fertilisation is given a high priority in KB MUSICA since it is recognised as an essential ingredient if the KB MUSICA business objective of a successful European technology base is to be achieved.

7. COMMERCIAL EVOLUTION

Impact on Marketing

A considerable number of new hardware and software products are expected from KB MUSICA; these arise in the technical areas of sensing systems, multi-variable control and a standardised fieldbus system. The products which will be developed fall into three categories:

(a) Software Systems for simulation packages, fault detection, data processing etc.

(b) Intelligent Measurement and Data Transmission for a standardised data acquisition system and transmission of validated data.

(c) Automation of Manually Intensive Tasks for reduction of production costs and increased personnel safety.

In order to develop market areas to their full potential, it will be necessary for Companies to mount a marketing and sales campaign. Some aspects of KB MUSICA, such as the development of intelligent supervision systems will lead directly to commercial products while other areas such as the application of a standardised fieldbus will require further research and development work. Links are currently being established with end users and vendors in the areas of robotics, enhanced sensors and plastic moulding machines.

Marketing Strategy

It is proposed that the overall strategy used to achieve commercial exploitation of KB MUSICA technology should be based on the following sequence of events:

YEAR 1 (1989) PROJECT EQUILIBRIUM

Establish the Project on an equilibrium course and confirm that the technological goals are achievable within the stated work period.

YEAR 2 (1990) AWARENESS

Create awareness of KB MUSICA within the technological community and the anticipated market place.

Establish a centralised coordination function to ensure that the emerging exploitation activities of the Partners are synchronised and harmonised.

YEAR 3 (1991) PROJECT FOCUS

Focus on the most promising technical developments and identify
the most likely commercial outlets.

Evolve specific marketing strategies, sales plan for the individual
products.

YEAR 4 (1992) APPLICATION

Apply and execute the specific marketing and sales plans.

8.   CONCLUSION

KB MUSICA encompasses four technologies, seven countries and
fourteen Partners. The technologies will be developed across four
(technically) widely spread Demonstrators using stand-alone
modular concepts. This approach will offer the maximum
opportunity for subsequent exploitation into diverse industrial
application by both individual Partners and by consortia. By this
means the gearing on both money deployed in the developmental
phase and the intellectual contribution of individual Partners will
be maximised. The results of KB MUSICA will directly influence
commercial products for at least 5 years after Project completion.

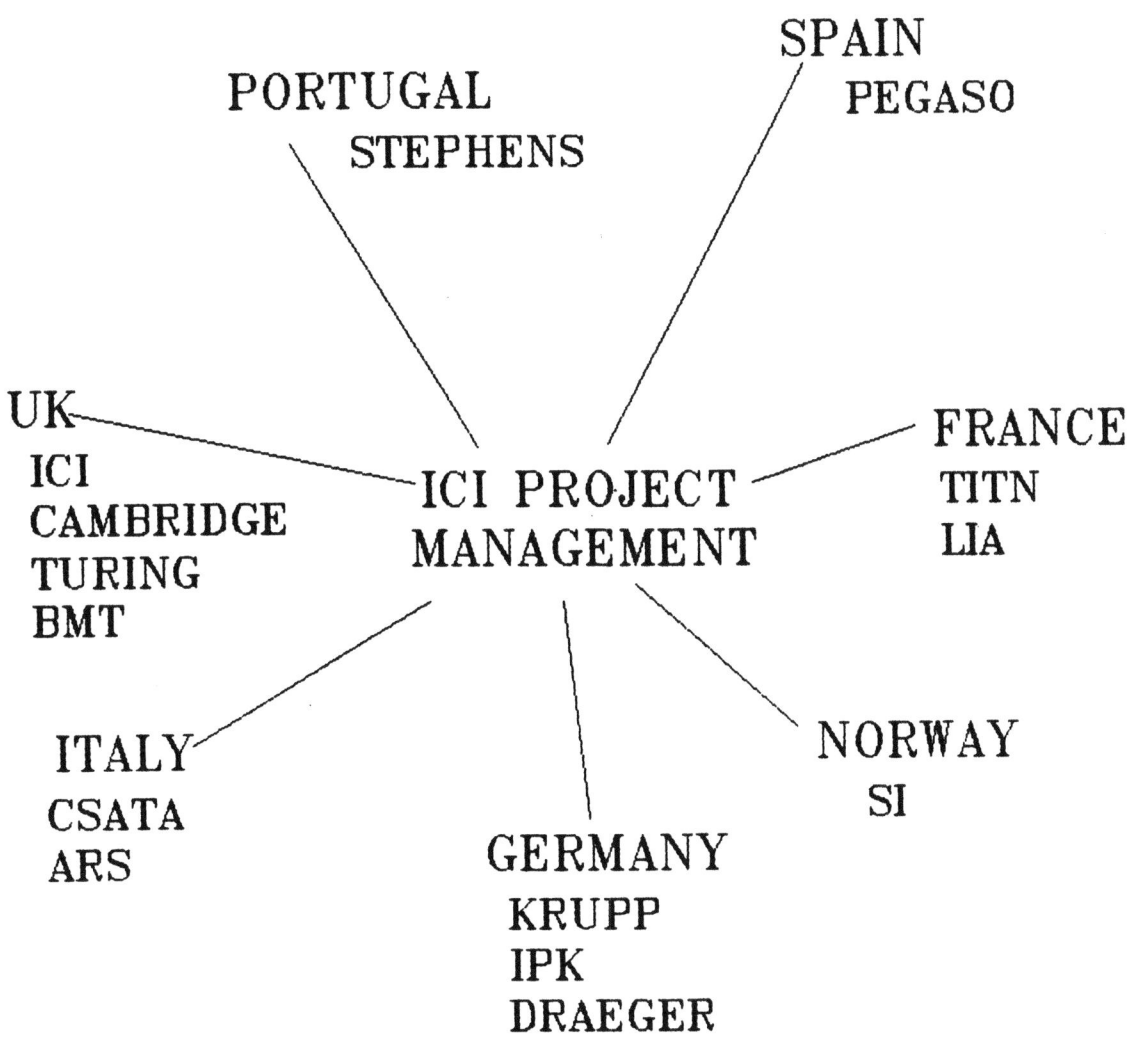

KB – MUSICA PARTNERS

Figure 1

KB MUSICA
DEMONSTRATOR DRIVEN

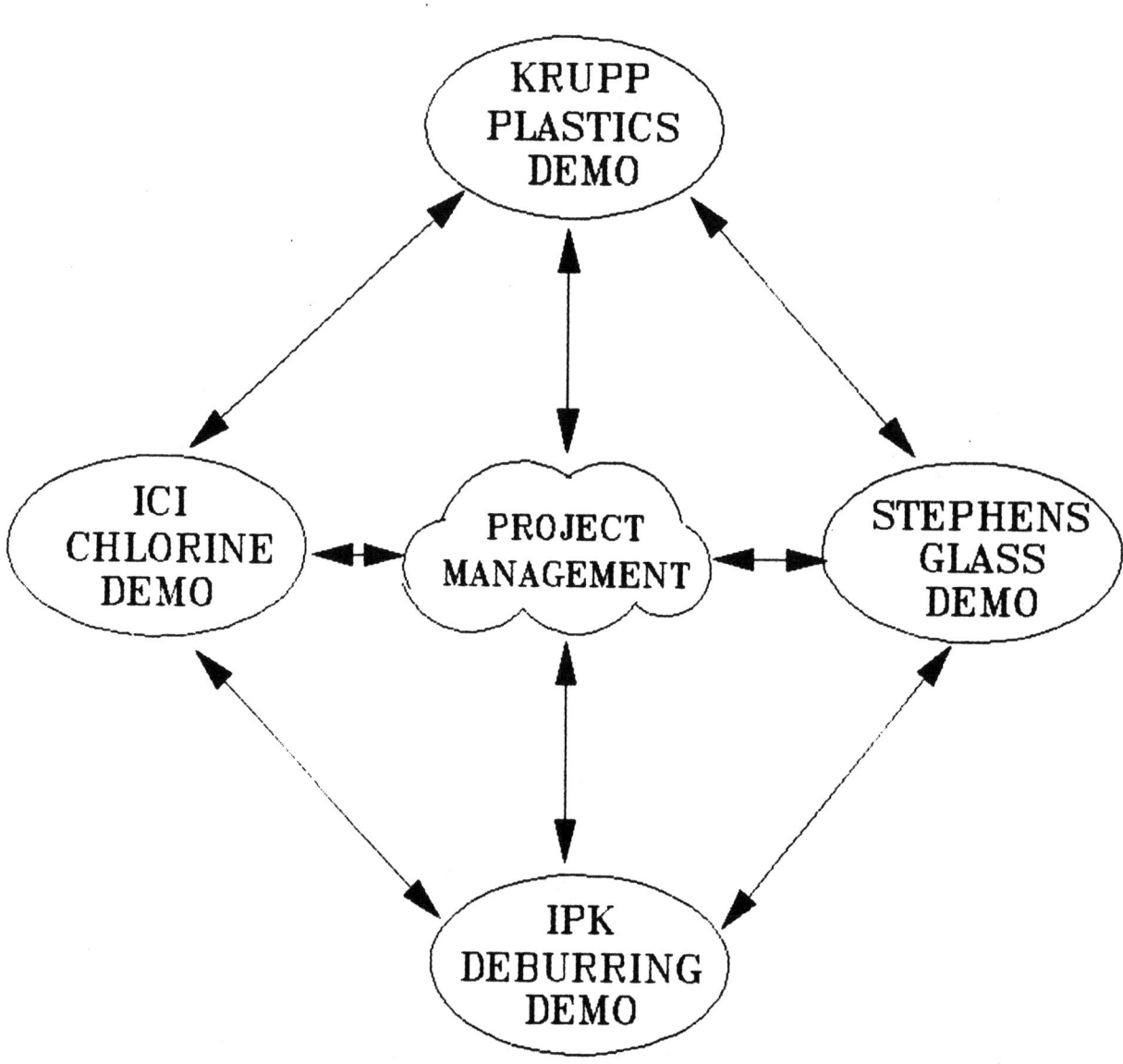

Figure 2

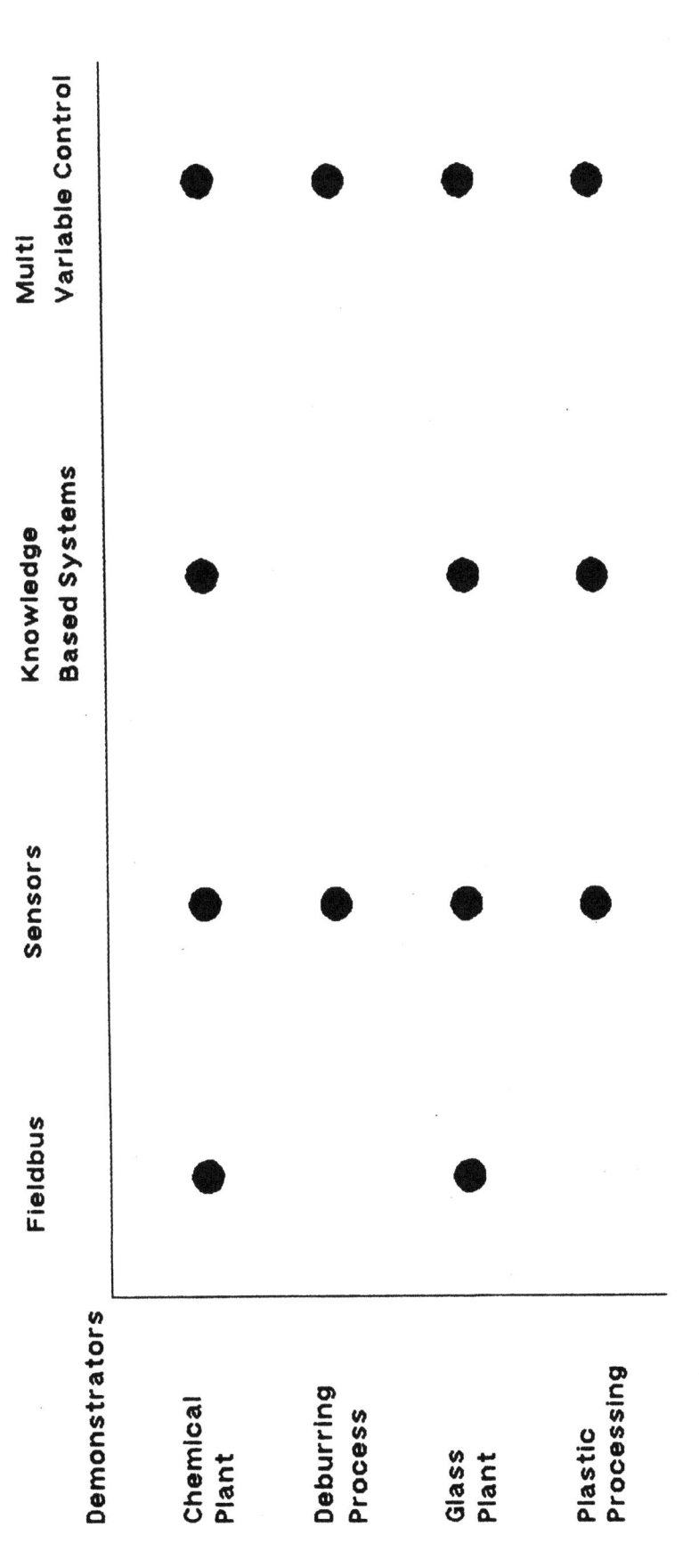

Figure 3

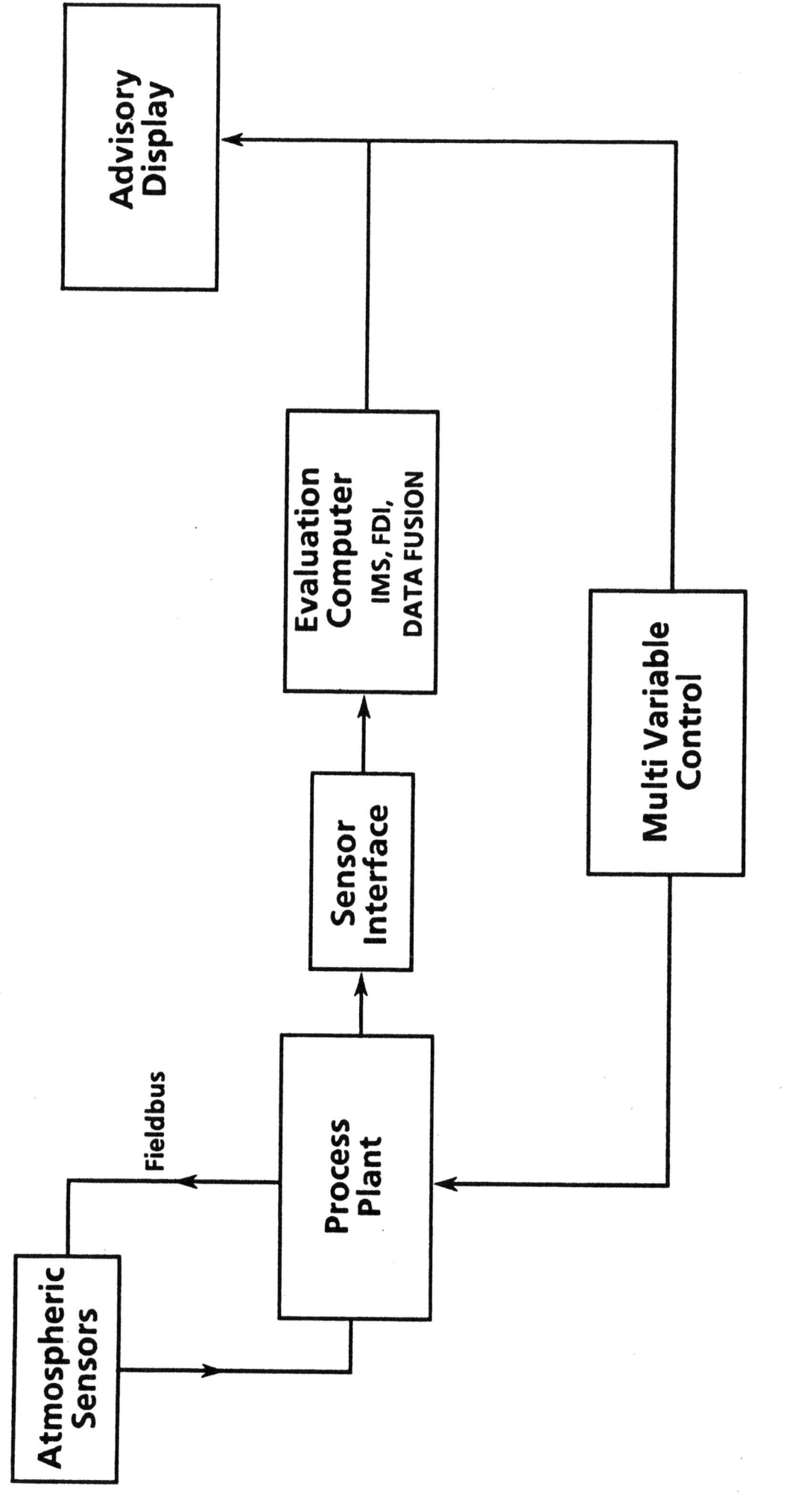

CHEMICAL PLANT DEMONSTRATOR

Figure 4

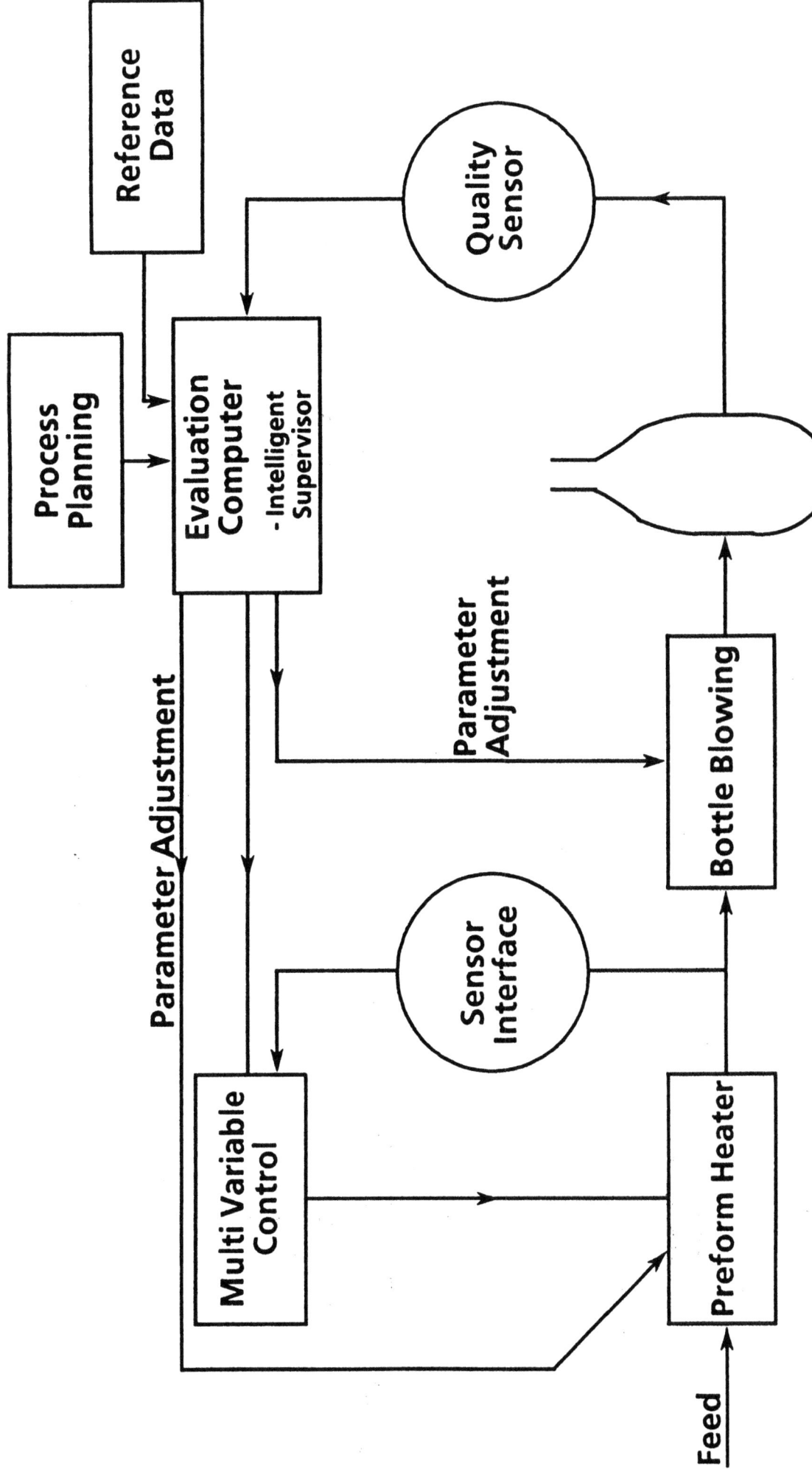

PLASTICS PROCESSING DEMONSTRATOR

Figure 5

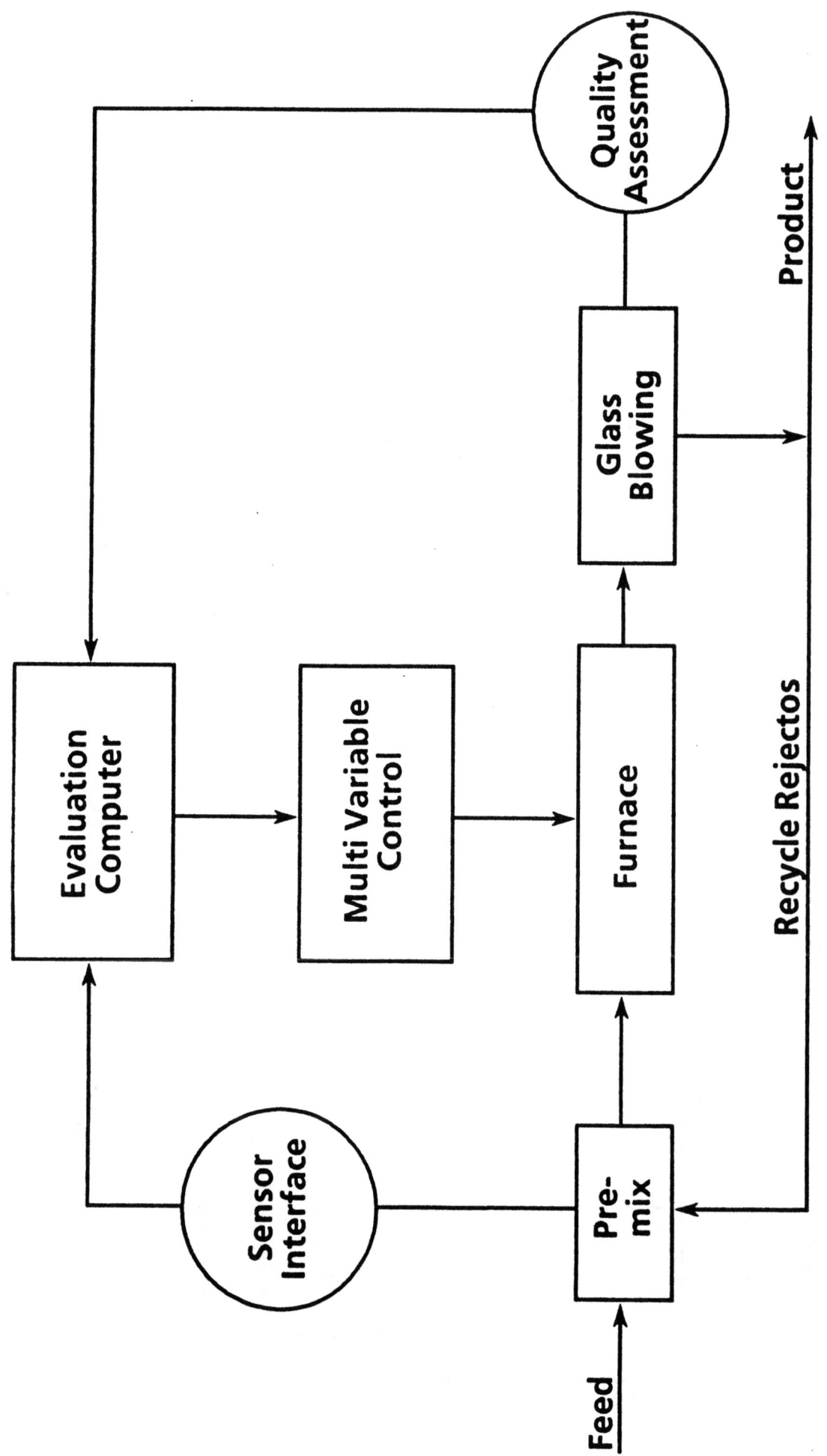

GLASS MANUFACTURE

Figure 6

436

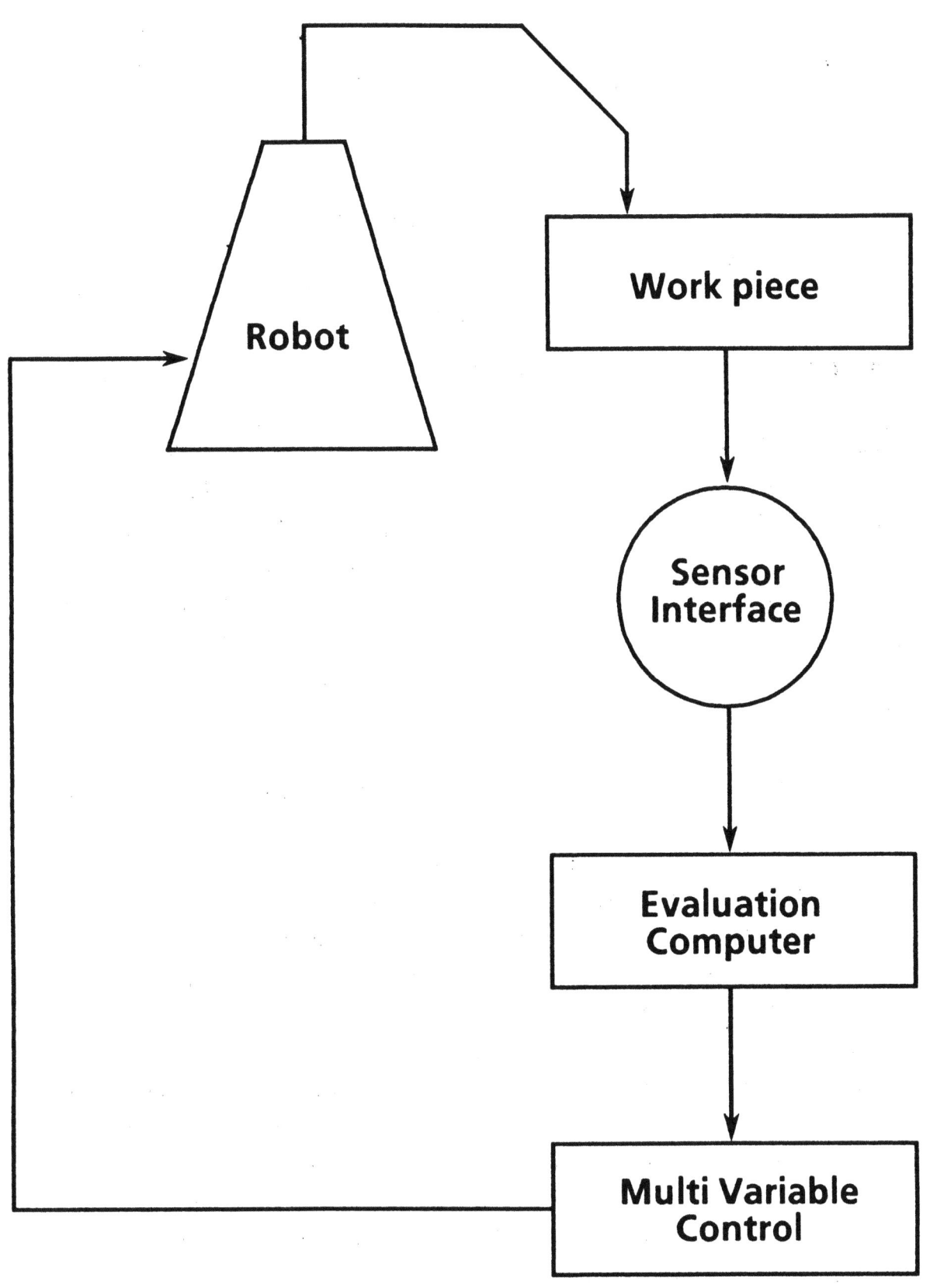

DE - BURRING DEMONSTATOR

Figure 7

IMPROVEMENTS ON 3D CAD SYSTEMS FOR SHOE DESIGN AND MODELLING

V. A. Branco, DI.ISEP/INESC.Norte
C. V. Carvalho, INESC.Norte
F.N. Ferreira, DEEC.FEUP/INESC.Norte

INTRODUCTION

The design of new products is in the start of the industrial production cycle. The use of 3D CAD systems since the initial steps of a project results in significative economy either due to the reduction of product samples (prototypes) or for the creation of information available for other equipment (CAM/CIM systems).

One of the disadvantages of a 3D CAD system is the design module, because usually the designers are unable to create a new product on the screen. Designers related to industries depending on fashion, somehow closer to artistic creation, tend to refuse the aid of 3D CAD systems in the early stages of design. This attitude is mainly due to inadaptation, not prejudice. Many designers accustomed to draw details in 3D CAD systems share the opinion that these systems do not offer an intuitive approach to design activity [ORR89].

In this paper we present an alternative, based in computer vision techniques, to help the designer's relation with 3D CAD systems. The first application of this system was related to the shoe industry but the results of our work can be easily transferred to other industries where products can be developed (at least partially) over a predefined volume (ceramics, glass, etc).

Our approach is based upon a strategy that improves the phases of conception and modelling of a shoe, allowing the designer to use the sketches made on paper, editing style lines on the screen, in a friendly way (not starting from zero), and experimenting colors and materials through texture visualization. This type of shoe conception becomes more efficient and pleasant and the designer is able to try a larger number of ideas, reducing the effort of the modelist. Furthermore, the number of samples produced is reduced to a minimum.

To achieve these objectives we intend to integrate our design module in commercial CAD systems. This module will result from the combination of a sketching system with a 3D system, joining the facility of use of the first with the descriptive power (in geometric terms) of the second.

The techniques used in computer vision studies, which fundamental basis are already implemented in the DIPS system described in this paper, allowed us to think on a set of tools more versatile and adapted to shoe 3D design. Our design module, named 'Intuitive Design System (IDES)' will be later described, focusing its utilization and architecture.

Finally we will present some future developments resulting from ideas that came up during the implementation of this system.

CAD TECHNOLOGY IN THE SHOE INDUSTRY

The production cycle in the shoe industry may include, in its several phases [FERREIRA89] the intervention of CAD/CAM technologies. The main characteristics and functions of 2D CAD systems for the shoe industry, at the present time, are the manual digitizing of the shoe shells (represent the flattened surface of the last and includes the style lines), the scaling of the patterns extracted from the shells and the manual placement of these patterns. The only CAM link furnished by these systems is for cutting tables (waterjet, knife or laser) which is extremely limited for an integrating strategy for CAD/CAM/CIM.

With a 3D CAD system (few available, yet) it is easier to integrate a broader range of manufacturing equipment. The main functions of a 3D CAD system are 3D scaling of the patterns which provides information for several machines like last manufacturing, shoe grasping and bonding, and 3D-2D flattening that furnishes information for sewing and cutting machines. In spite of these advantages, 3D systems didn't caught yet the interest of the industry [CUNDILL88].

The sketches of the new models are made on paper by the designers, including notes and referencing colors and materials to use. This drawing can also be made on a vacuum formed shell of the last.

The modelist is charged of the model pattern engineering, from the flattening of the sketch (when done over the last) till the definition and scaling of the patterns. Presently, this is the beginning point for CAD systems. The modelist converts the sketch into style lines and furnishes this information to the CAD system through manual digitizing. The process ends with the production of a sample of the model which is an expensive and time consuming operation, but also, in this scenery of usage of CAD systems, the only way to evaluate the results.

Shoe Industry and System Vendors

Looking to the products offered by system vendors we can group them into three types: Sketching systems, 2D CAD systems and 3D CAD systems.

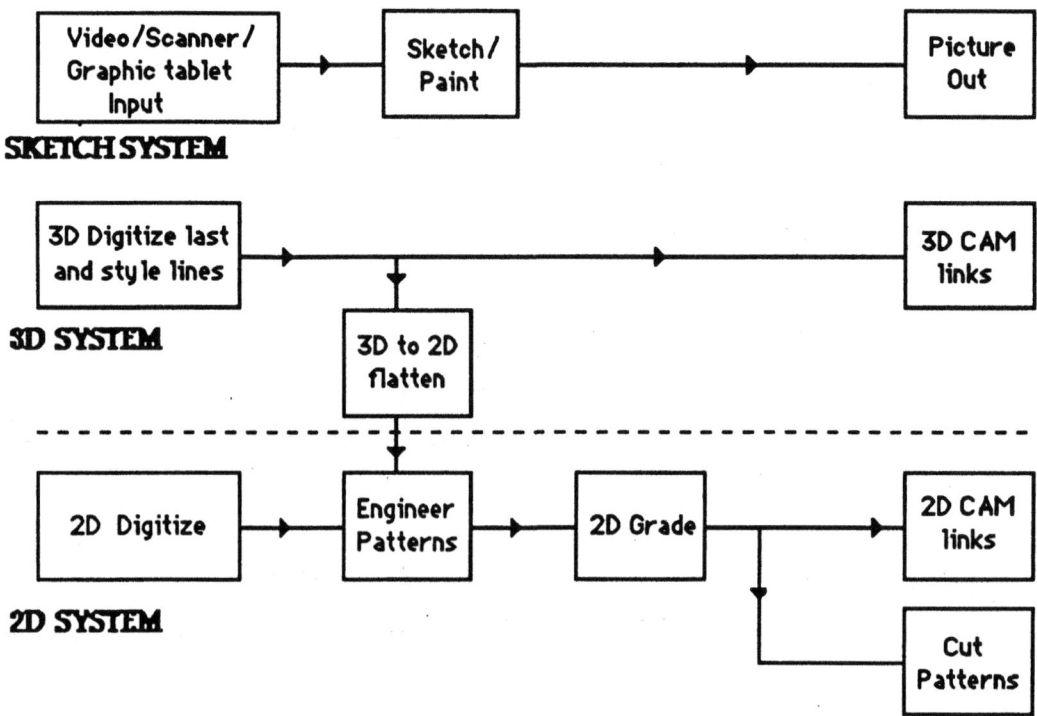

FIG. 1 - Shoe Industry and System Vendors

Sketch systems are gaining great popularity amongst designers as they offer a quick way to try alternatives in colors, textures, shapes, etc. Furthermore, they fairly emulate the preferred tools of designers: paper and pencil.

From the scheme presented above (Fig. 1), that reflects most of the situations, we observe that they are independent products and the only foreseen connection is unidirectional, between the 3D CAD system and the 2D CAD system.

All 3D systems start from a manual or automatic digitizing of the last and also from the drawing of style lines over that last. Some systems include a module for the edition of this lines. For instance, the system CIMTECH (Microdynamics) allows the entry of style lines supported on a TV camera. The operator then copies point by point the style lines from the image on screen [PERKINS86].

The flattening (3D-2D) operation is common in these systems. Once again CIMTECH system includes 3D grading and the possibility of changing style lines after the flattening, with these changes being reflected in the 3D model.

Another important feature still not included in all the actual 3D systems is the generation of high realism images. This possibility could be an interesting facility not only for design but also for marketing departments.

DIGITAL IMAGE PROCESSING SYSTEM - DIPS

The process of manual digitizing the shell, point by point, is tedious, time consuming and subject to errors due to manual imprecisions.

This was the first problem that lead us to pay special attention to the acquisition part of CAD systems. In that sense, a generic image processing package (DIPS) was developed in order to obtain knowledge on acquisition procedures. The automatic acquisition and vectorization of shoe shells is a subproduct resulting from that package, and we are now starting to develop a module to make leather measurements and to detect leather faults (through pre-established color marks), which will also be based on DIPS.

System Characteristics

The Image Processing System developed at INESC.Norte treats images with 256 gray-levels and a variable dimension between 128X128 and 512X512 points. The system is highly interactive, allowing the user to rapidly change several parameters (like threshold levels, for instance) thus providing a powerful tool for image analysis.

Beside that, the system conception is based on a modular scheme, providing flexibility to changes and improvements. The main module is the one that contains the image treatment algorithms, existing also a module for image files handling and another for the control of the graphic board.

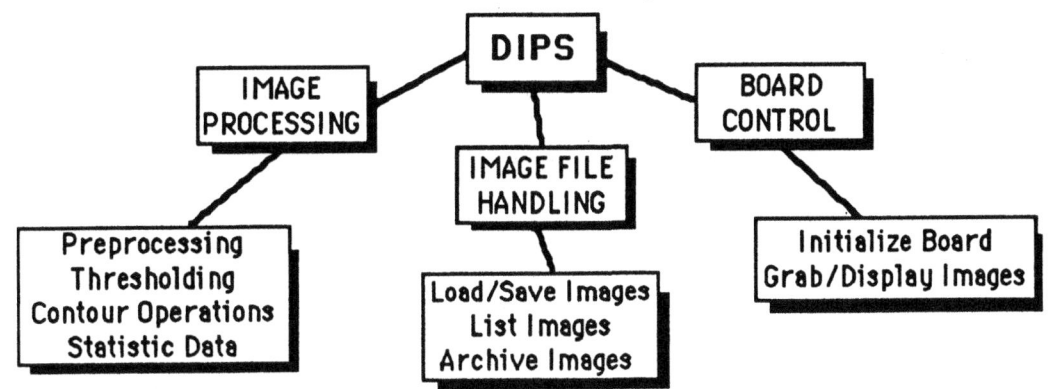

Fig. 2 - DIPS Organization

The IMAGE TREATMENT module comprehends the algorithms:
-Image preprocessing, thresholding, contour detection, contour enhancement, contour vectorizing and filtering and other algorithms like image inversion and calculation of statistical image (or window) data as histogram, mean and standard deviation.

441

The IMAGE FILES HANDLING module ensembles the listing, loading, saving, deleting image and compressing/uncompressing files.

The GRAPHIC BOARD CONTROL is responsible by the board initialization, image acquisition and display.

Acquisition and Vectorization of Shoe Shells

The process of automatic acquisition of shoe shells is the following: One digitized image of the shell is obtained (with a camera or scanner). This image is pre-processed (with a median filter and a gray-level expansion) in order to reduce luminosity deviations and to enhance the lines increasing their contrast with the background. Then a suitable threshold (based on the image mean and deviation) is chosen in order to binarize the image. By this time all the points in the image are classified as ink or paper, which simplifies the task of contour selection. The enhanced contours are later vectorized, converting the image coordinates of contour points in real world coordinates. The result has redundant information (too many points), so it is filtered in order to maintain only the points necessary to keep the contours shape (more points in high curvature and less points in low curvature). The visualization of the contours obtained is done with splines that pass through the points specified earlier.

In order to compare the traditional process and the automatic process, and considering the possible interest this module could have to CAD vendors, a connection to commercial CAD systems was established through the IGES standard. This has by now allowed the connection to several systems, either generic (AutoCAD) or shoe specific (PADSY from ATOM+VICAM).

From the first tests executed (Table 1), we observed a better time performance from the automatic system, (at least half of the time) although the precision remains identical to the manual system (This result is satisfactory in view that normally a good operator barely errors for more than a millimetre. The manual precision only starts to decay when there are many lines to digitize and the operator gets tired). In both systems there is the need of a pre-edition phase in order to eliminate acquisition errors (in the automatic system due specially to bad interpretations, for instance, of gaps) which represents normally (in both cases) 10% of the total editing time.

TABLE 1- Time comparison between acquisition methods

	Acquisition time (simple models)	Acquisition time (large models)	Pre-edition time
Automatic Acquisition	2-3 m	3-4 m	2-3 m
Manual Acquisition	4-5 m	9-10 m	2-3 m

Although the automatic system has revealed itself as better performing than the traditional one, there are yet improvements to be made. The paths to achieve these goals may be either through software optimization (faster and more accurate algorithms) or through hardware schemes(use of parallel architectures).

INTUITIVE DESIGN SYSTEM (IDES)

The links between 2D, 3D and sketch systems are, nowadays, practically inexistent like the scheme (Fig. 3.a) below shows. We want to explore the possibility of establishing the communications shown in Fig. 3b.

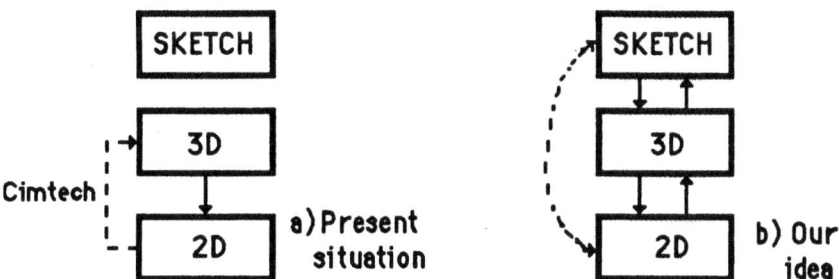

Fig. 3 - The links between 2D, 3D and Sketch Systems

The IDES system for shoe design is viewed by the user as a painter with all that common tools that he can pick to draw over an image of a last (Fig.4). The user must have previously chosen the last on a 3D CAD system database.

The image of the last will appear shadowed in an overlay (3D overlay) different from the one used for drawing (painter overlay).

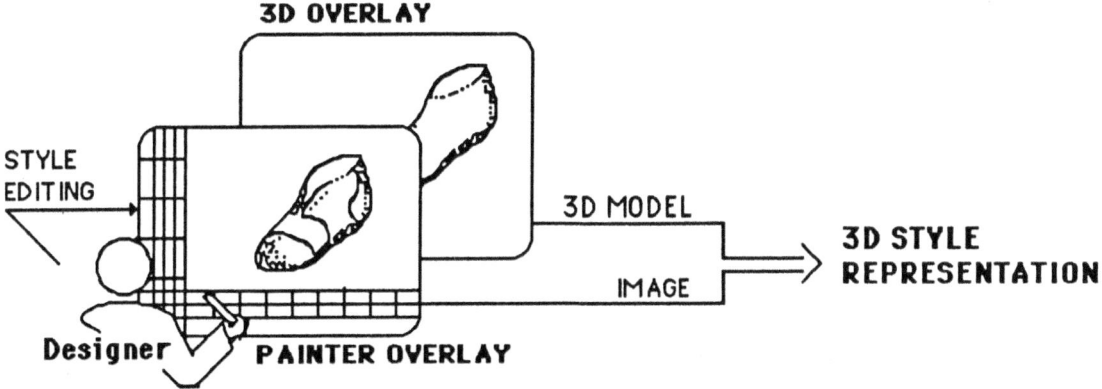

Fig. 4 - Intuitive Design System. User Point of View

The designer will also be able to rotate and position the 3D model of the last in such a way that its image adjusts to a digitized (through a scanner or TV camera) sketch or photo. Once ended the edition, the mapping of the style lines on the last surface will be made in an automatic way using DIPS capabilities.

This system will allow (thanks to the existence of a truly 3D model) the designer to rotate the last to continue the drawing or obtain automatically a mapping on the 3D model of the textures he wants to try.

System Architecture

This system will be composed by four main modules linked as shown in the scheme below (Fig. 5) where are also presented its relations to the 3D shoe CAD systems.

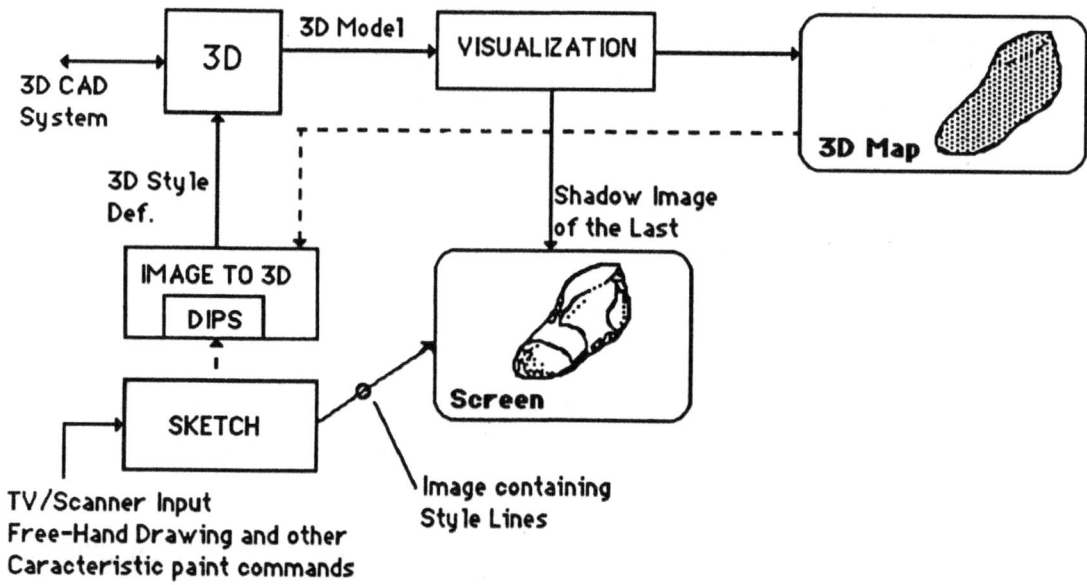

Fig. 5 - System Architecture

The 3D MODULE is responsible for the construction and maintenance of the 3D representations of the lasts and the style lines already converted to 3D.

This module communicates with the 3D CAD systems through files with an IGES (Initial Graphics Exchange Specification) format assuring this way its independence from the CAD systems. This module does not include any kind of 3D model editor.

The possibility of operating geometric transformations on the models foresees the inclusion in the scene of 3D representations of decorative accessories (buckles, lacing-holes, heels, etc) not necessarily adjusted in dimension and position to the last in use.

The VISUALIZATION MODULE does two fundamental tasks. The first one is related with the simulation of a virtual camera and the positioning functions of an observer. The second task consists in the production of shadowed images once this kind of images will ease the drawing of style lines.

A renderer based on an adaptative Ray-tracing algorithm with an increasing realism feature [SOUSA90] will be the main part of this module.

In this module a range image will be simultaneously synthesized (3D Map) [BESL85], allowing the association between each point {P3} in the last image and a group of points on the last surface {P1,P2} as shown below (Fig. 6).

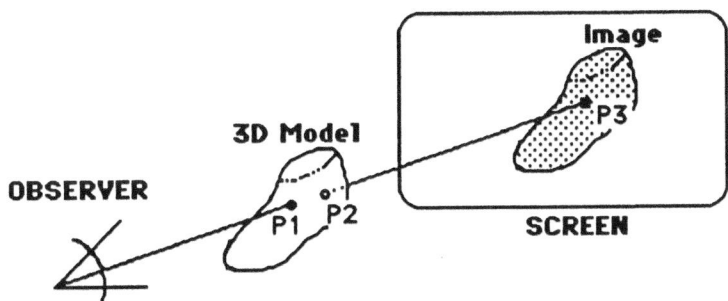

Fig. 6 - 3D Map Creation

The orientation of the surface normals on those points and the distance to the observation point permits to reduce all the hypothesis to an unique point {P1}.

The IMAGE TO 3D CONVERTER MODULE receives as input the raster images proceeding from the paint module and the 3D Maps constructed on the visualization module.

The DIPS system extracts from the raster image a set of contours that will be placed over the last surface. Departing from the 3D map, the 3D curves and surfaces corresponding to the contours and regions obtained in the raster image will then be defined.

The SKETCHING MODULE will maintain all the functionalities of a paint system. Extended facilities derive from the fact that this module will also be an interface to a geometric 3D nucleus. Another function is the retouching of images that may had lead to wrong interpretations of the contours.

CONCLUSION

This work is in an initial state of development, but its philosophy has been discussed with Portuguese industrials, which revealed great interest.

As mentioned before, the system DIPS is in the end of the development phase, and originated two important subproducts: Automatic acquisition of digitized shoe shells and leather measurement and defect capture.

New algorithms allowing increasing realism levels for renderers based in raytracing strategies are being developed, exploring transputer based strategies [SOUSA90].

The definition of this system and its development have been sponsored by the SFS-NATO programme ('PO-ShoeCAD' project).

Future Developments

In order to guarantee the success of this system, it must be fast enough so that the interaction with the user don't impose wait states that become easily annoying. The use of an architecture with parallel machines to accomplish the tasks of each of the modules presented, is in the horizon of this project's development.

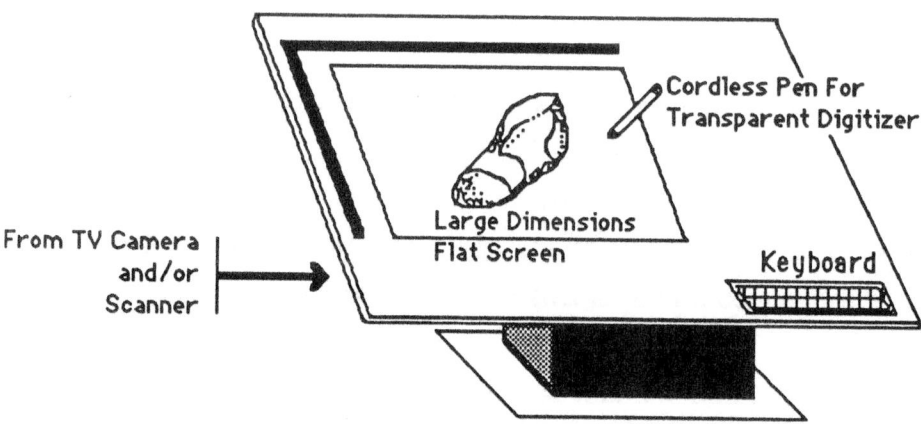

Fig. 7 - A Design Workstation

Another important topic will be the development of a new design workstation (fig. 7), for which the "Showcase" device [ORR89] seems to be a good approach, viewing that this device maintains with the user a relation identical to paper and pencil.

The inclusion of devices like TV cameras and/or scanners will allow the acquisition of information in any kind of support like paper, photos, leather samples, shoes, etc.

We also expect to increase the use of computer vision techniques to expand the range and power of our system. In [BHANU87] the author expresses the link between CAD systems and vision providing a way to obtain representations of the objects to recognize and manipulate.

Although the objectives of CAD and computer vision are in opposite sides, we defend that for CAD systems can be very important (fig. 8) the hability to incorporate easily the information that flows around it, for instance notes, sketches, photos, etc. so that he can become a true member of the project team.

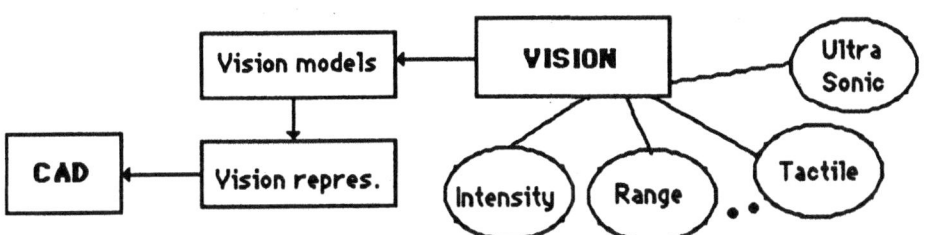

Fig. 8 - Computer Vision Based Cad System

Studies about the influence of this technique over surface and solid modelling are the first steps done within a strategy to generalize this approach to other industries.

BIBLIOGRAPHY

[BHANU87] Bir Bhanu, Chich-Cheng Ho (1987) CAD-Based 3D Object Representation for Robot Vision. IEEE Computer; Vol. 20, Nº 8

[BESL85] Paul J. Besl, Ramesh C. Jain (1985) Three Dimensional Object Recognition. ACM Computing Surveys, Vol. 17, Nº 1

[ORR89] Joel N. Orr (1989) Exotic CAD. Computer Graphics World, Vol. 12, Nº 7, July 89

[PERKINS86] Peter Perkins, Bob Hackney (1986) Computer Aided Design. SATRA Modern Shoemaking, Nº 20

[CUNDILL88] Jenny Cundill, Bob Hackney (1988) Computer Aided Design Update. SATRA Modern Shoemaking, Nº 26

[FERREIRA89] F.N. Ferreira, M.R. Gomes, A.A. Sousa, V. Branco, J. Bernardo (1989) Cad Systems for the Shoe Industry. Primeiras Jornadas PPPAC para as Indústrias Tradicionais, (In Portuguese)

[SOUSA90] A.A. Sousa, A.C. Costa, F.N. Ferreira (1990) Interactive Ray Tracing for Image Production with Increasing Realism. (accepted to be published in EUROGRAPHICS'90)

A Vision System for Analysis and Classification of Industrial Flames [1]

J.P. Costeira
J.P. Fernandes
CAPS/LRPI, Instituto Superior Técnico, Lisbon, PORTUGAL
M.V. Heitor
CTAMFUL, Instituto Superior Técnico, Lisbon, PORTUGAL
J. Sentieiro
CAPS/LRPI, Instituto Superior Técnico, Lisbon, PORTUGAL
J.P. Simões
CTAMFUL, Instituto Superior Técnico, Lisbon, PORTUGAL
J.A. Victor
CAPS/LRPI, Instituto Superior Técnico, Lisbon, PORTUGAL

1 Introduction

Optimized furnace operating conditions together with advanced furnace control systems are required to reduce energy consumption and decrease pollutant levels in industrial process furnaces, such as those used in glass, cement and steel assemblies manufacture and in power plants. Although many different types of furnaces exist, in all cases flame control is especially important in providing specified distributions of radiant and convective heat transfer together with maximum efficiency and minimum pollutant levels, complete combustion, freedom from noise and oscillation, and insensitivity to fuel changes [3,9]. Important end-products of the process are temperature, heat flux distribution to the walls and combustion efficiency, and these are intimately connected with production of pollutants, like soot and oxides of nitrogen.

A range of instrumentation, with emphasis on physical and electrochemical sensors, has been developed and used extensively in large industrial furnaces and boilers. In some cases, the information derived from those sensors has been used to control process variables, [6]. In general, the information is limited to overall characteristics in zones of easy access and, therefore, do not allow the complete characterization of the combustion processes within the combustion chambers and the identification of non-conventional perturbations.

With the advent of digital image analysis systems, the use of visualization methods may allow the implementation of novel control strategies for industrial burning systems, including the assistance of maintenance and the provision of fault diagnosis. The implementation of advanced sensing devices results in improved monitoring and control of furnaces. In particular, vision systems promote improved control of flame geometry, together with flame classification. Also, improvements in the overall furnace control can be achieved by correlating vision information with operational and performance factors. Apart from processing flame data, the vision

---

[1]Work performed under the research ESPRIT Project No.2192 - AIMBURN

system can also allow the monitoring of, for example, the state of the combustion chamber walls, the performance of bubblers or the presence of non-fused materials in the melting tank of glass furnaces.

Artificial vision systems have received special attention in the past few years in many engineering fields, namely in Robotics [5,12]. Some early attempts to use vision techniques in industrial process applications have been described [4,8,7,13]. However, they were usually restricted to the provision of information describing a furnace state. On the other hand the work of [10] is devoted to the determination of structural characteristics of the turbulent/non-turbulent interface of flames and follows the study of [2], in which digital image analysis is used to characterize intermittently turbulent flames. One important objective of the ESPRIT project AIMBURN is the development and test of a vision system that implements a pattern classifier methodology for industrial flames. In a first stage, the vision system will process flame data acquired in a laboratory environment and extracts relevant features, namely shape, separation from the burners and brightness profiles. These features will be used for flame classification according to previously learned standard patterns. In a second stage, the system will be applied to a glass furnace in order to monitor and detect small perturbations leading to malfunction states as well as to feedback information to the control and supervisor systems. Finally, the system will be optimized in laboratory environments, so that a refined artificial vision system is available at the end of the project. This paper deals with the first stage of the work and have been performed under cooperative action of the group of ADIST- 1 Instituto Superior Técnico involved in AIMBURN.

The following section describes the experimental method and gives details of the laboratory apparatus and of the instrumentation used throughout the work. Section 3 presents and discusses preliminary flame results, which are analysed on the basis of detailed mean temperature measurements in the flame. In section 4 a method to evaluate important flame image geometric features is introduced and results are presented. Section 5 outlines future developments and section 6 presents the main conclusions of the present work.

2 Experimental Method

A purpose-built experimental apparatus was designed to represent important features of industrial flames in order to test digital imaging techniques prior to their application at full scale. In industrial-scale burners required to handlelarge volume rates of fuel, the inlet stream in usually divided into a number of single jets. The size, shape and combustion characteristics of such multiple flames depend upon their mutual interaction and the effects of surrounding air streams. The interaction process of the flames, in turn, is influenced by the geometrical arrangement, the spacing of the number of burners and, in general, large flames characterized by low-frequency motions are formed within the combustion chambers. The aim of the experimental apparatus used through out this work is to represent these features in a simplified laboratory arrangement, so that an image processing system of industrial flames can be developed and tested. Jet propane flames formed on round nozzles with an exit diameter of D=5 mm were used throughout this

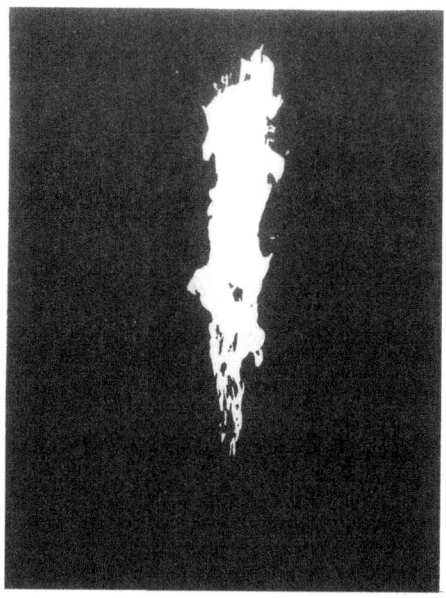

Figure 1: Typical propane flame

work. The apparatus allows the analysis of multiple jet flames by discharging gas propane through various nozzles (1 to 5) set at different distances and arranged in line for 2 and 3 flames and in a cross-shaped with a diameter of 5 mm configuration for 5 jet flames. The nozzles were set vertically and upward over a flat table, downstream of a long straight pipe of 17 mm of diameter and details of the flow downstream of the nozzles is published elsewhere.

The instrumentation used throughout this work comprised butt-welded thermocomples of platinium/platinium-rhodium for point measurement of mean gas temperature and a CCD camera with 320 TV sean lines of resolution for flames analysis. The vision system included a standard acquisition board providing a 512x512 digitized image.

3   The Experimental Characterization of the Flames Studied

Figure 1 shows a photograph of a typical propane flame and identifies an instantaneous flame/air interface, which is a stationary random variable and, therefore, can be subjected to statistical analysis. Here, attention is focused on the evaluation of the size and shape of single and multiple flames based on spatially integrated values of brightness and on related temperature measurements.

Figure 2 shows the time average values of the length of the flames as a function of their number, n, and relative distance, a. The results concern only with lifted flames and the measurements were taken as average distances between the burner exit and the end of a contiguous portion of the flame, either visually observed or visible on a photograph exposed for one second. The results are used to validate values derived from digital analysis of flame images based on the techniques described below and have allowed to quantify the accuracy of our method for defining flame lengths.

Figure 3 shows contours of mean temperature for a single flame along a vertical

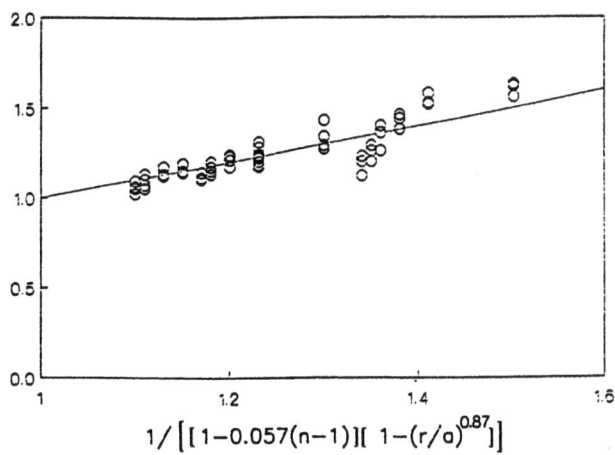

$$1/\left[\,[\,1-0.057(n-1)\,][\,\,1-(r/o)^{0.87}\,]\,\right]$$

Figure 2: Ratio between the length of multiple and single flames

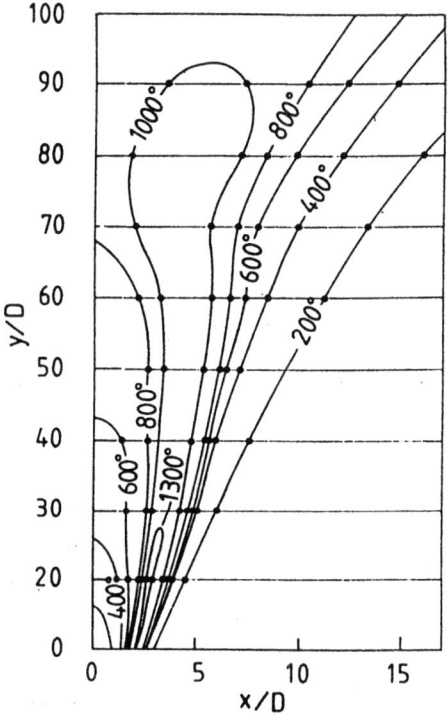

Figure 3: Contours of mean temperature along a vertical plane of symmetry for a single flame

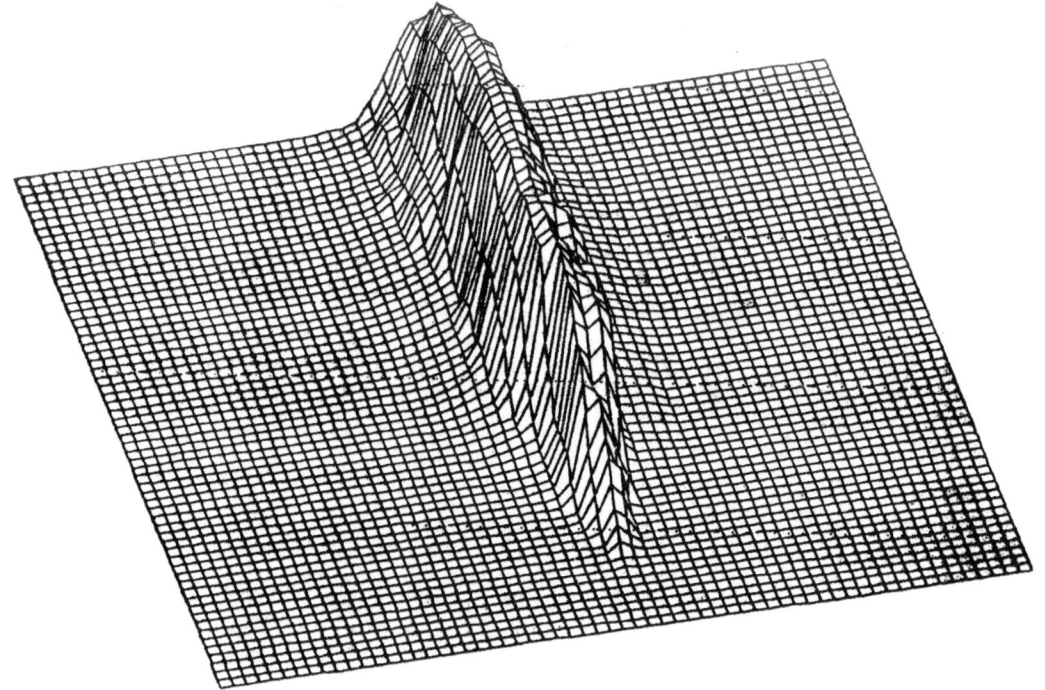

Figure 4: Mass/Brightness Distribution

plane of symetry and quantify the extent of the zones of high temperature around the core jet of each individual propane flame. A vision system integrates in space the radiation emitted by different chemical components in different wavelengths and, therefore, comparison with temperature levels is highly complex. However, the detailed knowledge of the scalar distributions embodied in the flame is important to improve understanding of the images, as refered in the paragraphs below.

4   Geometric Analysis of Flame Images

The vision system extracts from the flame digitized images four main features which are known to be strongly related with the performance of the burning process: the flame shape and boundaries, the flame separation from the burner , the brightness profiles and the spectral distribution (colour analysis) which may be related to pollutant formation.

The flame length is a constraint in the optimal operation point, being determined by the furnace geometry. The vision system evaluates this length and provides feedback to the low-level control system in order to keep the flame size within the optimal range. The flame size together with its separation from the burner leads to an important information to control both the air and the fuel rates.

The approach taken here to extract the geometric properties of the flame images consists on looking at the digitized image as being defined by set of elementary components (pixels) each one of them having associated a mass value proportional to its correspondent brightness value. This approach is illustrated in Figure 4.

4.1   Flame Shape

In order to define the flame shape the following features have been chosen and are computed by the Vision System:

452

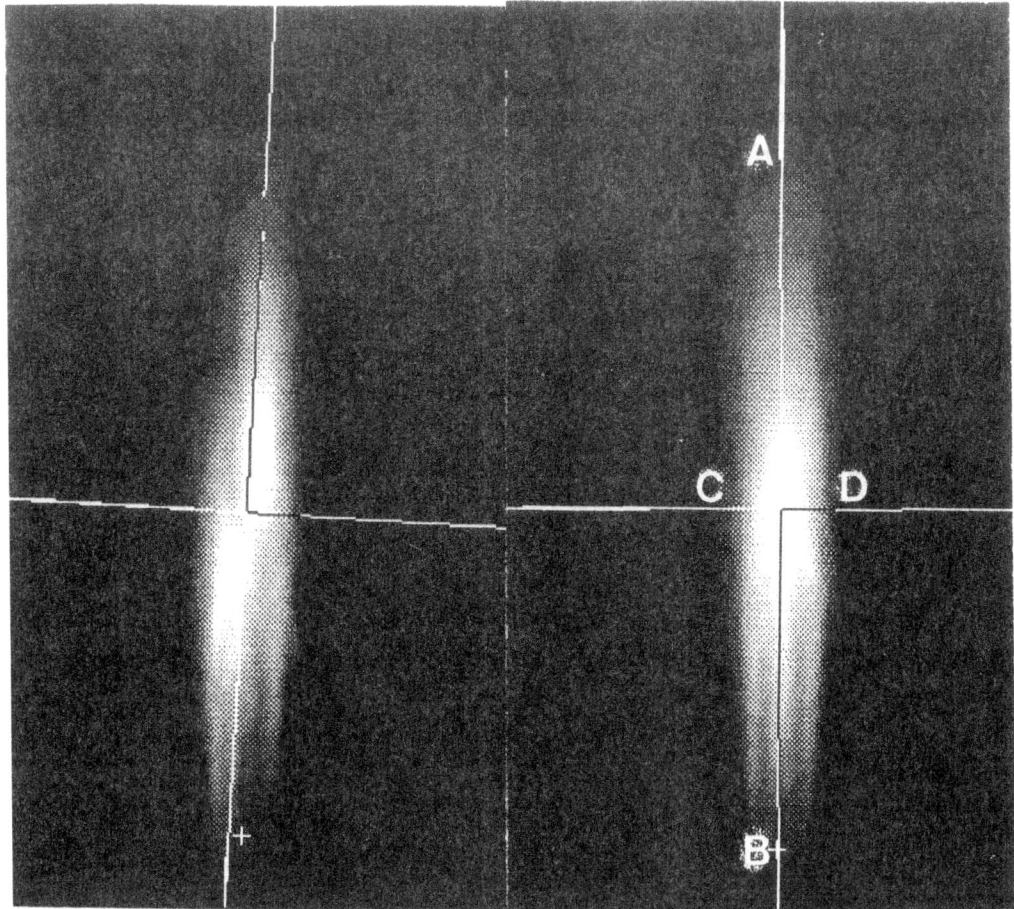

Figure 5: Simmetry axis for two different flames

- Symmetry Axis - The flame symmetry is calculated taking into account both position and brightness of each pixel in the digitized flame image.

  The simmetry axis are evaluated by minimization (maximization) of the inertia moments. Figure 5 displays these axis for two different flames. The axis orientation is related to the flame motion and its variation in time may give valuable information about flame stability.

- Maximum and Minimum Moments of Inertia I_{max}, I_{min} - The Maximum and Minimum moments of inertia are important measures of the spatial distribution of flame brightness . The ratio between their values gives an indication of how round the flame is and can be used to distinguish flames generated by one, two or more burners.

- Length and Width - These parameters are calculated by searching the flame boundaries along the simmetry axis. In figure 5 the distance between points A and B defines the maximum length and the distance between points C and D defines the maximum flame width.

- Flame Area - This parameter is calculated by summing the pixel values of the flame image. This simple calculation provides usefull information about the overall size of the flame, which can be used to discriminate flames generated by different numbers of burners.

4.2   Separation from the Burner

This parameter is determined by searching along the minimal inertia axis and detecting its intersection with the convex hull of the digitized image. The burner's position is assumed to be known.

4.3   Test Results

Once defined the set of geometrically motivated parameters and the computer algorithms to evaluate them it is important, in the present context, to test different cases (different flames) and conclude about the descriminating capabilities of those parameters.

In the graphs to follow each numbered image displayed corresponds to a time average of 20 "instantaneous" images.

The tests were performed considering the following six different burning conditions:

- **C1** - Gas flow set to 0.65 gr/s (Re=20400), only one burner activated (Images numbered from 1 to 9).

- **C2** - Gas flow set to 0.50 gr/s (Re=15700), only one burner activated (Images numbered from 10 to 16).

- **C3** - Gas flow set to 0.65 gr/s (Re=20400), two burners activated (Images numbered from 17 to 24).

- **C4** - Gas flow set to 0.50 gr/s (Re=15700), two burners activated (Images numbered from 25 to 32).

- **C5** - Gas flow set to 0.65 gr/s (Re=20400), three burners activated (Images numbered from 33 to 39).

- **C6** - Gas flow set to 0.50 gr/s (Re=15700), three burners activated (Images numbered from 40 to 58).

Figure 6 displays the value of the ratio between the minimum and the maximum inertia moments (eccentricity of the flame) for the six different burning conditions defined above. The computed values represented in the figure clearly cluster in three regions corresponding to the three different numbers of burners.

Figure 7 shows values of flame area and reveals the clustering tendency of Figure 6, even though only five regions are sharply defined. The ambiguity between conditions **C1** and **C2** can be solved by looking at the distance to the burner, as displayed in Figure 8.

In general the results obtained validate the approach taken since they confirm the chosen parameters as potentially good discriminators of the different burning conditions considered.

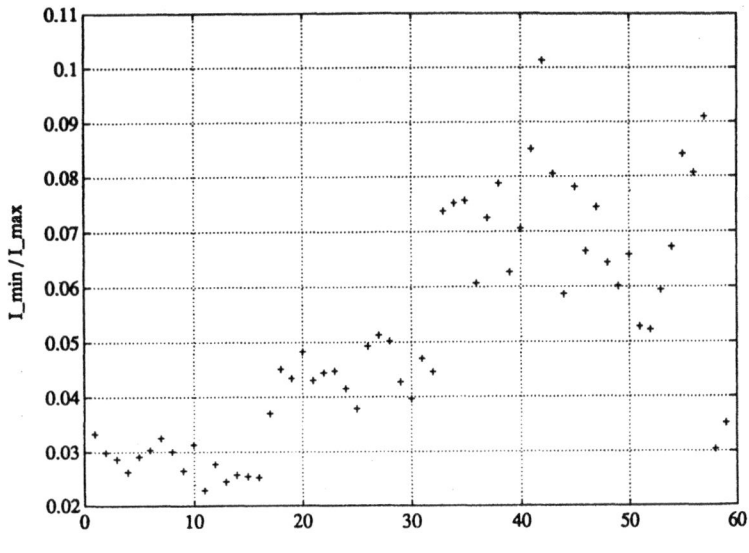

Figure 6: Flame Eccentricity

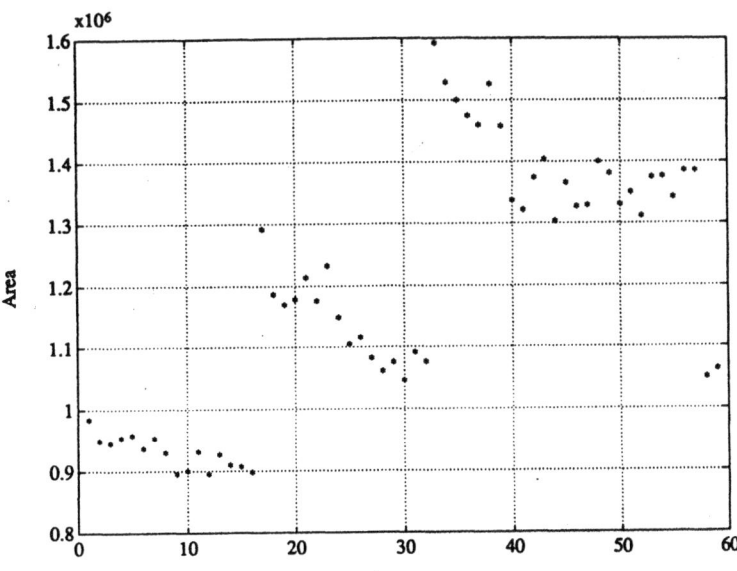

Figure 7: Flame Area

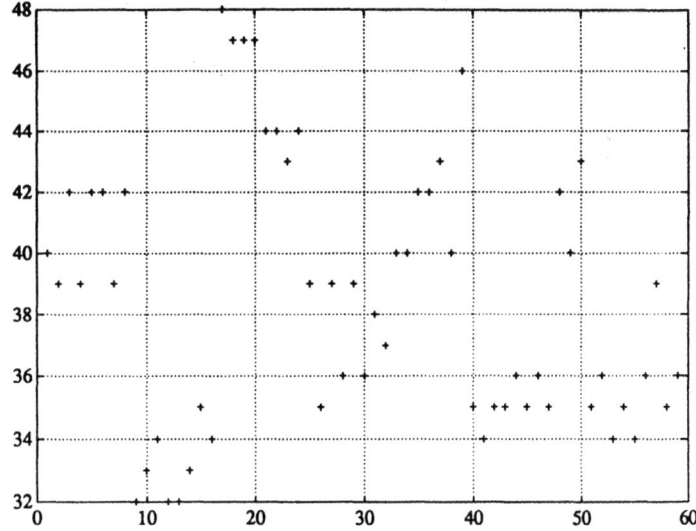

Figure 8: Distance to the Burner (in image coordinates)

455

5 Future Developments

The geometrically based parameters calculated by the Vision System will be validated by comparison with the results of detailed measurements of scalar and velocity characteristics and a performance analysis of their discriminating capabilities will be considered in the future work. The characterization of the flame properties will be complemented by exploitation of the spectral content of the flame radiation.

Once defined a proper set of parameters, either those geometrically motivated or those obtained from spectral radiation analysis, a classification strategy will be defined. For each acquired image, or sequence of images, the classification system will find the correspondent burning operating point. The techniques to be used include the traditional Bayesan methods, heuristic methods, and neural networks.

6 Conclusions

Digitized image analysis has been developed to characterize the size, shape and other important combustion parameters of turbulent flames stabilized in a purpose-built experimental apparatus. The arrangement may include the analysis of one to five flames stabilized at variable distances and at different flow rates and is aimed to allow the development and optimization of a pattern classification methodology for industrial flames.

The results include the definition of geometric features of time averaged flame images, such as symmetry, moments of inertia, length, and area, which are compared with detailed in-flame measurements and shown to be potentially good discriminators of flame characteristics.

7 Acknowledgements

Helpful discussions with colleagues in the Departments of Mechanical and Electrical Engineering during the preparation of the specifications of the work summarized in this report are gratefully acknowledged.

The project is supported from June 1989 by ESPRIT II Programme of the DG XIII of the Commission of the European Communities under the contract EP - 2192. The Project Officer is Dr. A. Adjemian.

References

[1] De, D.S. " Measurements of Flame Temperature with a Multi-Element Thermocouple, " J. Institute Energy, pp.113-116, June, 1981.

[2] Shrokr, M. and Keffer, J.F., " Digital Image Analysis of a Complex Turbulent Wake, " Structure of Complex Turbulent Shear Flow, Ed. R. Dumas and L.Fulachier, pp. 165-174. Springer Verlag, 1983.

[3] Kuo, K.K.,*Principles of Combustion*. John Wiley and Sons, 1986.

[4] Lilja, Gleus, Sutinem, "Image Processing for the Control of Burning Processes," *Processing in Industrial Applications Proceedings*, pp. 81-89, First IFAC Workshop, ESPOO, Finland, 1986.

[5] Dickmans, E.D. and Zapp, A., "Autonomous High Speed Road Vehicle Guidance by Computer Vision," NATO Advanced Research Workshop on Mobile Robot Implementation, Oporto, Portugal, May 1987.

[6] Higham, E.H. "An Assessment of the priorities for Research and Development in Process Instrumentation and Process Control," I. Chem. Industrial Fell. Report. The Institution of Chemical Engineers, 1987.

[7] Iso et al, "Instrumentation and Control for Refining Processes in Steelmaking," *IEEE Control Systems Magazine*, pp. 3-8. October 1987.

[8] Kortele, Majam, "Intelligent Instrumentation of Solid Fuel Power Plants," *IEEE Control Systems Magazine*, pp. 31-35. August 1987.

[9] Lawn, C.J., *Principles of Combustion Engineering for Boilers*. Academic Press, 1987.

[10] Zucherman, L., Kawall, J.G. and Keffer, J.F., "Digital Image Analysis of Turbulent Flame," *Experiments in Fluids*, 6, pp. 16-24, 1988.

[11] Heitor, M.V., Taylor, A.M.K.P., Whitelaw, J.H. "Velocity and Scalar Characteristics of Turbulent Flames Stabilized on Confined Axisymetric Baffles," *Comb. Sci. and Technology*, 62, pp.97-126, 1988.

[12] Costeira, J.P., Victor, J.A., Fernandes, J.P., Sentieiro, J.J.,"VER - A Stereo Vision System for Mobile Robots," IMA - Conference on Robotics, Loughborough, U.K., 1989.

[13] Jadoul, P., "Application d'une Camera CCD a la Thermographie d'un Tour de Verrier, " Séminaire CAPTEURS, Liége, 1990.

ON-LINE QUALITY AND PRODUCTION CONTROL IN DIE CASTING INDUSTRY

Luís Miguel Nazaré   (*)
José Carlos Caldeira   (* *)
Alexandre Diogo Silva   (* *)
João Paulo Baptista   (* * *)
Joaquim Borges Gouveia   (*)
José Manuel Mendonça   (*)

(*)   FEUP-DEEC / INESC-Norte
(* *)   INESC-Norte
(* * *) ISEP / INESC-Norte

1. INTRODUCTION

This paper describes a quality and production control system for the die casting industry. The distributed solution and its integration with a higher level data processing system follows CIM concepts and philosophy.

A set of remote units are connected to an host computer through a work cell level network. Each remote unit acquires from a die casting machine a set of parameters during each complete injection cycle and compares them with a set of so-called typical parameters.

Using mathematical techniques to manipulate the high-frequency sampled curves, showing the pressure in the injection chamber and the piston stroke evolution, a decision on the injected component's quality is made which is then communicated both to the machine operator and to the host computer.

Statistical data concerning quality and production control issues is sent to the host computer to be included in a data base system for further management. This data base system also includes the information about the typical parameters of each die, which are downloaded to a remote unit whenever a new die is installed in the corresponding machine.

The host computer is connected through an Ethernet link to a higher level data processing system, granting users at plant and enterprise-wide control levels access to quality and production control information acquired and processed at shop-floor level.

Using actual data recorded at the plant, corresponding to proper operation and faulty conditions, some tests have already been made to optimize the algorithms enabling the correct decision on the injected component's quality to be made. Fast Fourier Transform (FFT) combined with time analysis algorithms showed very promising results.

A major concern of this project is to develop an integrated and modular product allowing future implementations in similar industries.

2. AN APPROACH TO SHOP-FLOOR QUALITY CONTROL IN DIE CASTING

An efficient quality control process guarantees standard performance of the manufactured products, this being one of the major issues in increasing the product's competitivety in the market.

Quality evaluation of the manufactured components at shop-floor level represents a fundamental step in the manufacturing process, and on-line quality control based on information processing is becoming a must in a great number of enterprises.

In many applications this quality control methodology reduces the risks of visual inspection of the manufactured components, often done as they come out of the production line.

The data processing system used for quality control is to follow an integrated and distributed solution, thus enabling the information acquired to be simultaneously available at area control and plant-wide control levels.

This system can also support on-line monitoring, production control and maintenance management functions thereby paving the way to complete functional integration.

Die casting industry is a sector where on-line quality control using computer based interactive information processing systems is feasible. The work which is now being discussed refers to the very first step of a complete quality control programme implemented in an automotive sub-supplier industry.

Experience shows that the pressure in the injection chamber and the piston displacement are the most significant variables in die casting if we are to make a real time decision on the injected component's quality.

Figure 1 shows the pressure in the injection chamber and the piston displacement during a typical injection cycle, where the relevant short filling time can be seen.

The filling phase (see figure 2), with a duration of a few milliseconds, is the most important phase during an injection cycle concerning the manufactured component's quality evaluation.

The decision on each manufactured component's quality is made by comparing the injection chamber pressure curve and the piston displacement curve during each injection cycle with the corresponding typical curves. This set of typical curves characterizes every die installed in an injection machine.

3. A BRIEF DESCRIPTION OF THE DIE CASTING ON-LINE QUALITY AND PRODUCTION CONTROL SYSTEM

3.1 Hardware and Software Requirements

Figure 3 shows a block diagram representation of an on-line quality and production control system in die casting industry.

A set of remote units are linked to an host computer through a work cell level network. The host computer is connected to the main enterprise data processing system through an Ethernet link. Each remote unit is placed next to an injection machine for acquisition and local processing of relevant data. The requirements for each element of figure 3 are now discussed in detail.

Because of the complex and high rate data manipulation required locally, each remote unit is provided with a set of high performance cards. The "heart" of each unit is a 16-bit microprocessor card with enough RAM and EPROM space, a 12-bit analog data acquisition card, a digital I/O card and a network card implementing the interface between the work cell level network and the remote unit bus. This interface card, specifically developed within the scope of this project, will soon be available at the European market. The whole set is placed inside an industrial cabinet, provided with a monitor and a keyboard for operator interface, installed next to each injection machine.

The host computer is a workstation running several concurrent processes under Unix, namely:
- information interchange between the workstation and the remote units and between the workstation and the main data processing system;
- all the applications needed for quality and production control, such as man-machine graphics interface, statistic and data base management, report generation, etc.

The software drivers interfacing the work cell level network to the host computer bus under Unix are also being developed within the scope of the project.

The work cell level network uses the Bitbus protocol, characterized by a bus structure and a master-slave relationship. This low cost network, once provided with powerful nodes, is an excellent solution for industrial applications where distributed complex data processing is needed together with medium data transfer rates.

3.2 Information Management in Die Casting

When a die is first installed in an injection machine, the corresponding remote unit acquires for each manufactured component a set of pressure and displacement curves and sends them to the host computer through the work cell level network. These curves characterize each corresponding manufactured unit which undergoes afterwards a careful and thorough examination. The results of this inspection are input to the host computer and typical curves and parameters are found for the installed die. These typical curves and parameters are stored in a data base and then downloaded to the corresponding remote unit.

From then on during the manufacturing process the remote unit acquires the pressure in the injection chamber and the piston displacement during each injection cycle and compares them with the typical curves.

A decision on the injected component's quality is made and then communicated to the machine operator. The pressure and displacement curves are displayed on the local monitor superimposed to the corresponding typical curves. This represents a fundamental tool for the injection machine tuning during the manufacturing process.

When a component is rejected the local unit will also communicate to the machine operator the type of fault which has probably occurred. This is obviously relevant to the injection machine maintenance suggesting the utilization of this information within a maintenance management strategy.

The remote unit also acquires from the injection machine a further set of parameters such as temperatures, other pressures and the die support forces, and checks if their values are within allowed tolerances generating alarms whenever necessary.

Statistical results of the injection process are then sent to the host computer and fed to the data base system.

The workstation's operator can monitor in real time each remote unit's status, the number of rejected and accepted manufactured components, the different types of faults and the associated frequency of occurrence and other relevant information. He can also visualize, in real time, the pressure and displacement curves of any remote unit superimposed to their corresponding typical curves. Using the information stored in the data base, quality and production control reports can also be made available to the enterprise management.

4. TESTS AND RESULTS

The success of this project lies on the specific decision criteria enabling each remote unit to make real time decisions on the injected component's quality, and decide if it is to be rejected or not.

A simple algorithm based on time comparison of the acquired curves with the so-called typical curves will be an unreliable solution, considering statistical variance, noise problems, etc.. Instead, a set of algorithms based on FFT techniques and curve comparison time analysis were developed to support the necessary on-line quality control decisions.

4.1 Real Time Constraints

The requirement for substantial processing power at each node is placed by the obvious need of a real time decision making process transparent to the actual production process. For a skilled injection machine operator the system will be granted less than two seconds to manipulate near one thousand data points and produce a decision.

To avoid costly hardware, i.e. the use of a co-processor or even dedicated FFT hardware, some work had to be done to accelerate FFT algorithms.

Extensive use of in-circuit emulation and measuring facilities provided by an universal microprocessor development system helped in achieving a nice trade-off between execution speed and precision.

4.2 Decision Criteria

In order to study and develop quality decision algorithms, a set of actual pressure and displacement curves were acquired in the plant for different dies. The basic methodology is now described.

Figure 4 shows the plotted typical pressure curve for a given die.

Figure 5 shows the modules of the first one hundred components of the corresponding FFT.

Figures 6 and 7 show curves similar to the ones of figures 4 and 5 corresponding to an ordinary manufactured component, and figures 8 and 9 are similar to figures 4 and 5 but now

462

corresponding to a rejected component. This injection cycle was expressly done with an insufficient quantity of raw material. Figures 10 and 11 present the error resulting from the comparison between the corresponding FFT's and that of the typical curve.

In this example, a simple maximum error criteria is enough to decide on the rejection or acceptance of the injected component. Unfortunately there are some cases where a simple criteria as the one just described cannot lead to a reliable decision on the injected component's quality. In such cases FFT combined with time analysis algorithms show very promising results.

This problem cannot however be reduced to this straightforward reasoning. In fact, finding typical curves through preliminary components examination is complex because acceptance ranges need to be defined. Substantial die casting process know-how is required in order to allow cause-effect relationships between certain parameters and the injected component's quality to be established with confidence. The tolerance allowed for the main variables in the quality control process can lead to the definition of a set of n-dimensional acceptance zones which can then allow a decision on the injected components rejection or acceptance to be made on the basis of FFT and time analysis algorithms. The strategy just described is now being undertaken.

5. CONCLUSION

This paper corresponds to a project that INESC is undertaking for a portuguese die casting enterprise. The total effort is about 3 man year for 2 years, and the cost may be estimated in 150 KECUs. From the work done during the first year considerable benefits can be envisaged, as the reduction of the operator's role in the decision making process, the real-time conformance and the contributions to manufacturing optimization, thereby reinforcing the enterprise performance in the quality audits it must undergo. The modular software and hardware architecture allows an open and flexible solution to be attained which can be suited to other die casting industries where on-line quality and production control demand for local real time processing.

The use of powerful remote data acquisition and processing units linked to an host computer by a low cost industrial network, together with the combined utilization of FFT techniques and time analysis algorithms are two major concepts helping to achieve this goal.

Similar strategies could certainly prove useful in other areas, such as plastics, glass and automotive supplier industries.

6. REFERENCES

John Mickowski (1982) "Controlling the Die Casting Process", Die Casting Engineer, July-August, 16-22

Nazaré, L. M., et al (1989) "Sistemas de Aquisição de Parâmetros para Controlo de Qualidade e Produção em Fundição Injectada", ENDIEL'89, Porto

Burrus, S.C. and Parks, T.W. (1975) DFT/FFT AND CONVOLUTION ALGORITHMS, Wiley-Interscience

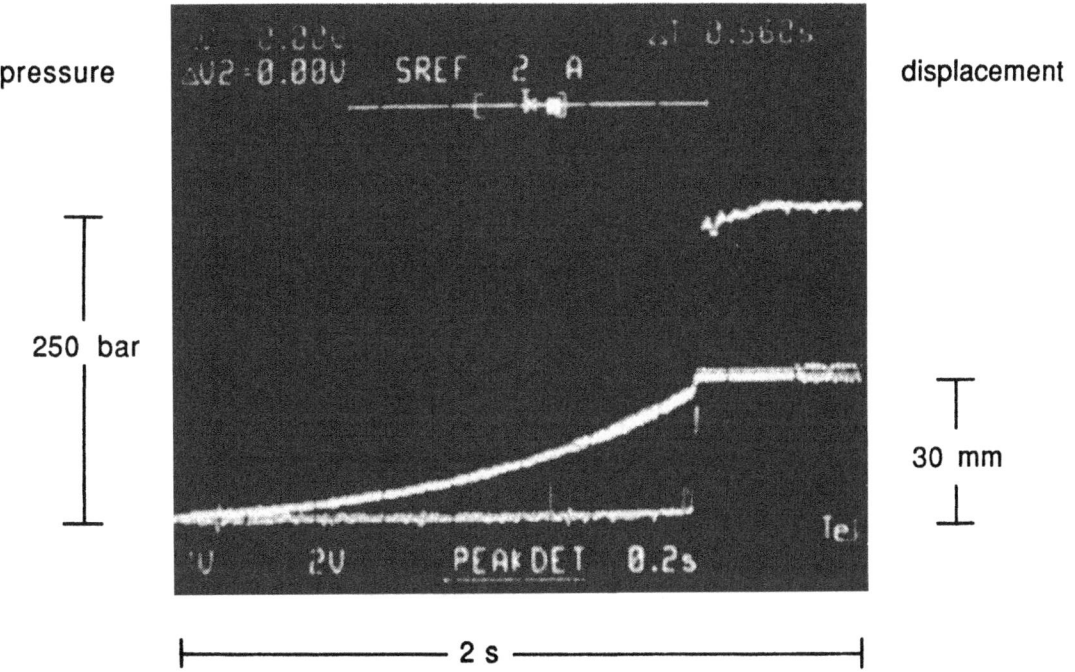

Figure 1 - Injection chamber pressure and piston displacement curves during a typical injection cycle.

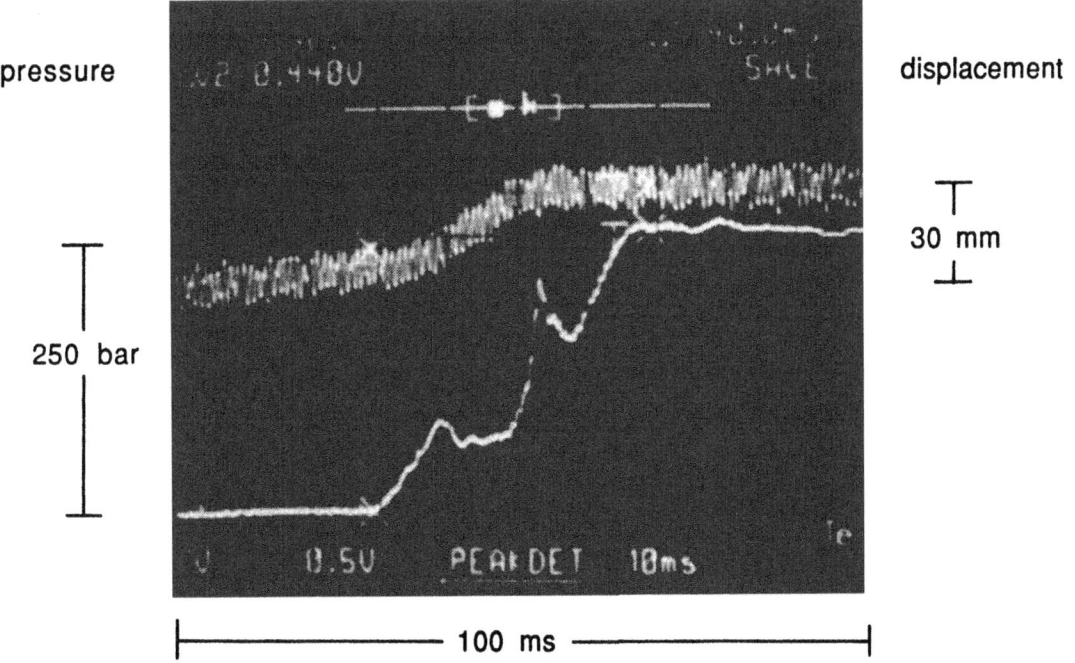

Figure 2 - Injection chamber pressure and piston displacement curves during the filling phase.

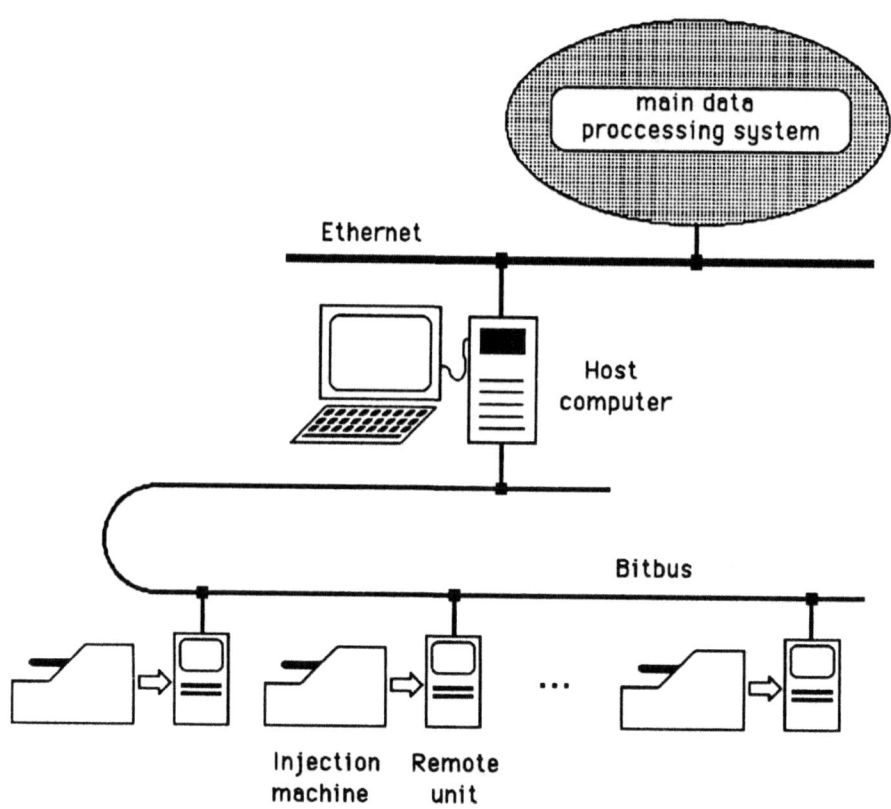

Figure 3 - Block diagram of an on-line quality and production control system in die casting industry.

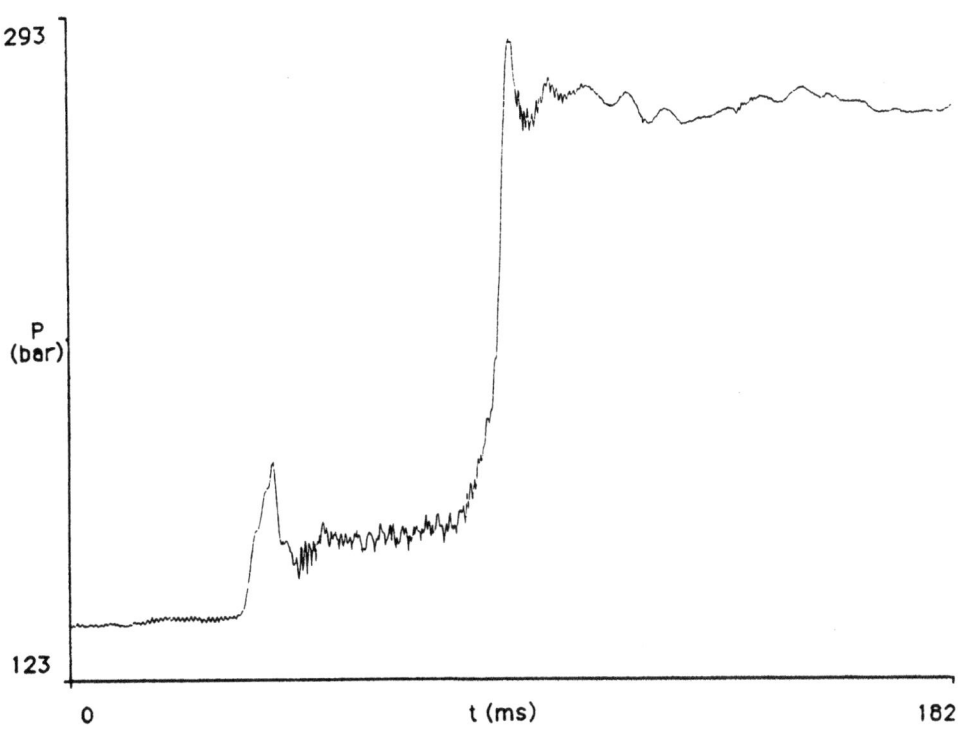

Figure 4 - Typical component pressure curve for a given die

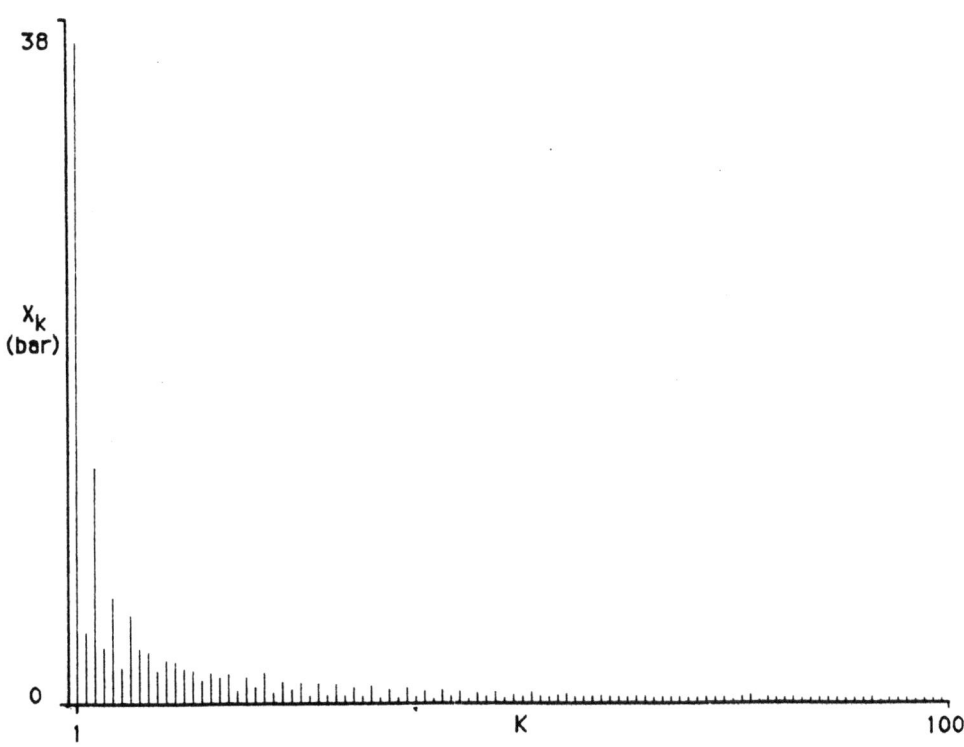

Figure 5 - FFT corresponding to typical component pressure curve
(modules of the first one hundred components)

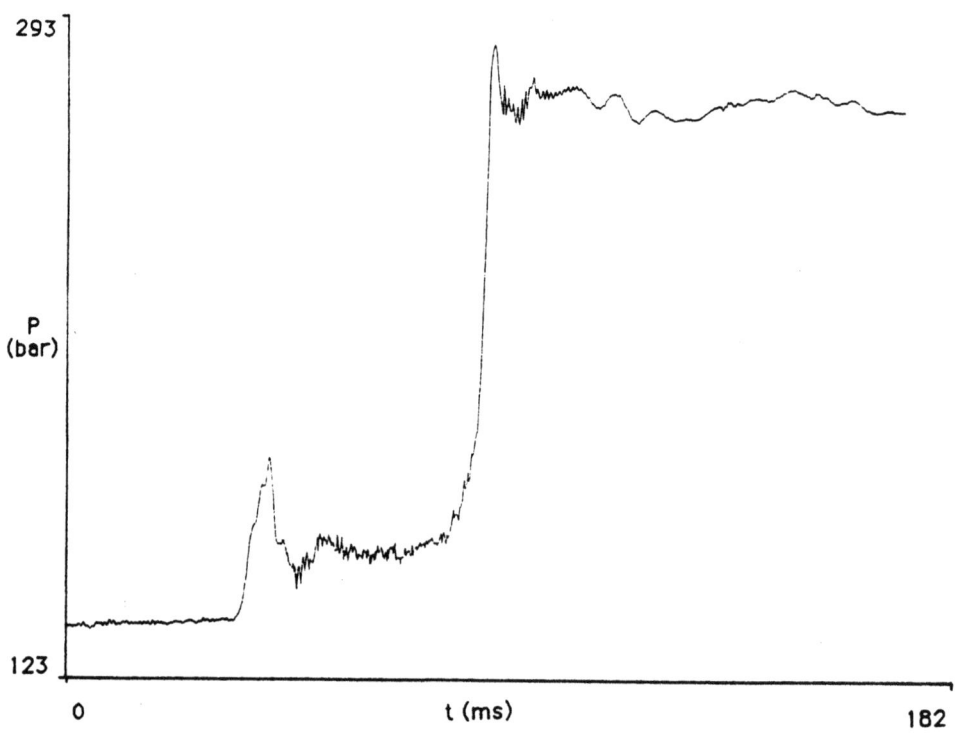

Figure 6 - Ordinary component pressure curve for a given die

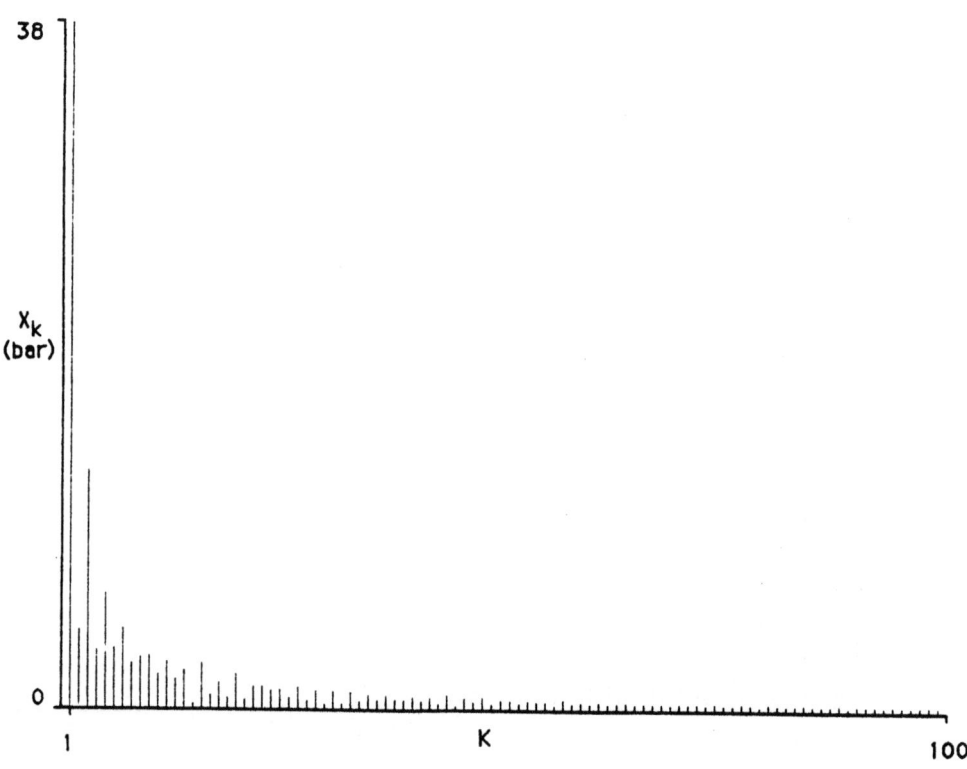

Figure 7 - FFT corresponding to ordinary component pressure curve
(modules of the first one hundred components)

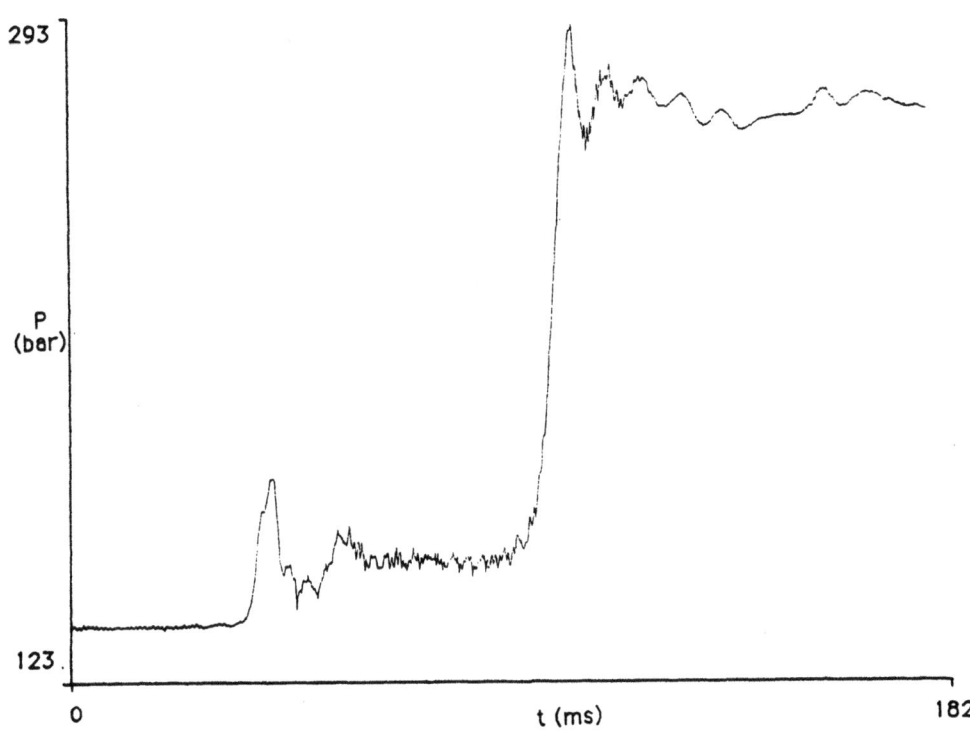

Figure 8 - Rejected component pressure curve for a given die

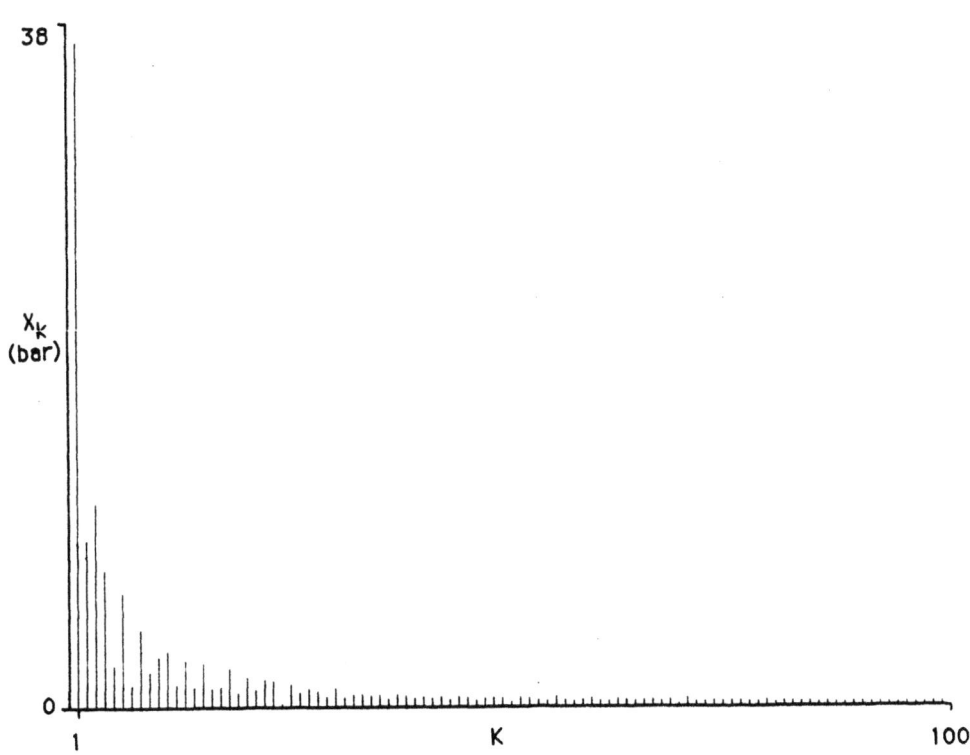

Figure 9 - FFT corresponding to rejected component pressure curve
(modules of the first one hundred components)

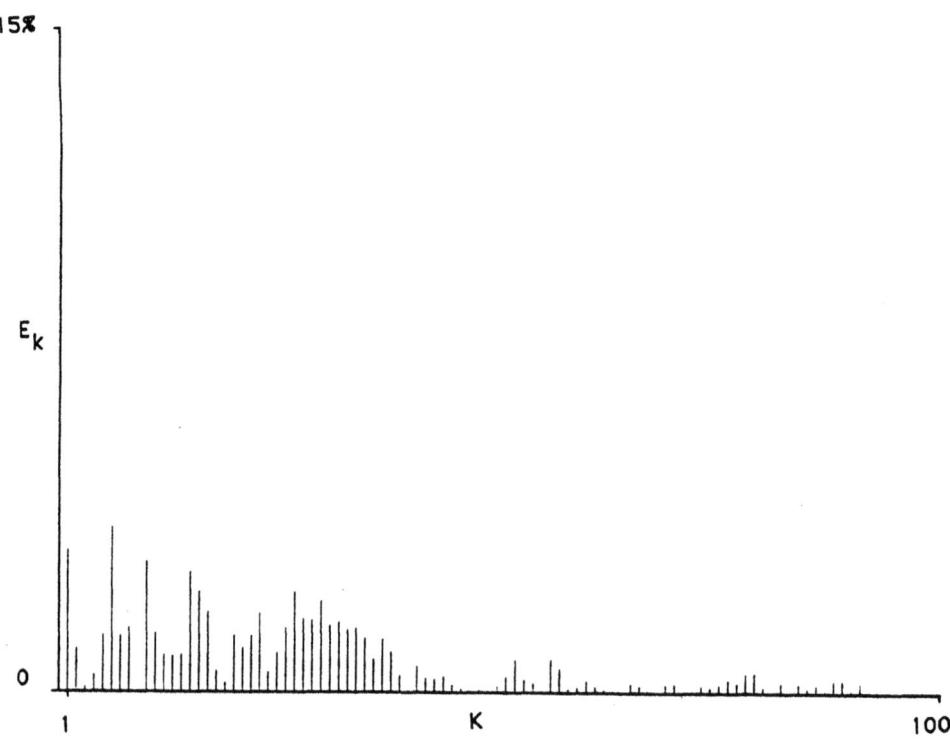

Figure 10 - Errors resulting from the comparison between the ordinary pressure FFT and the typical one

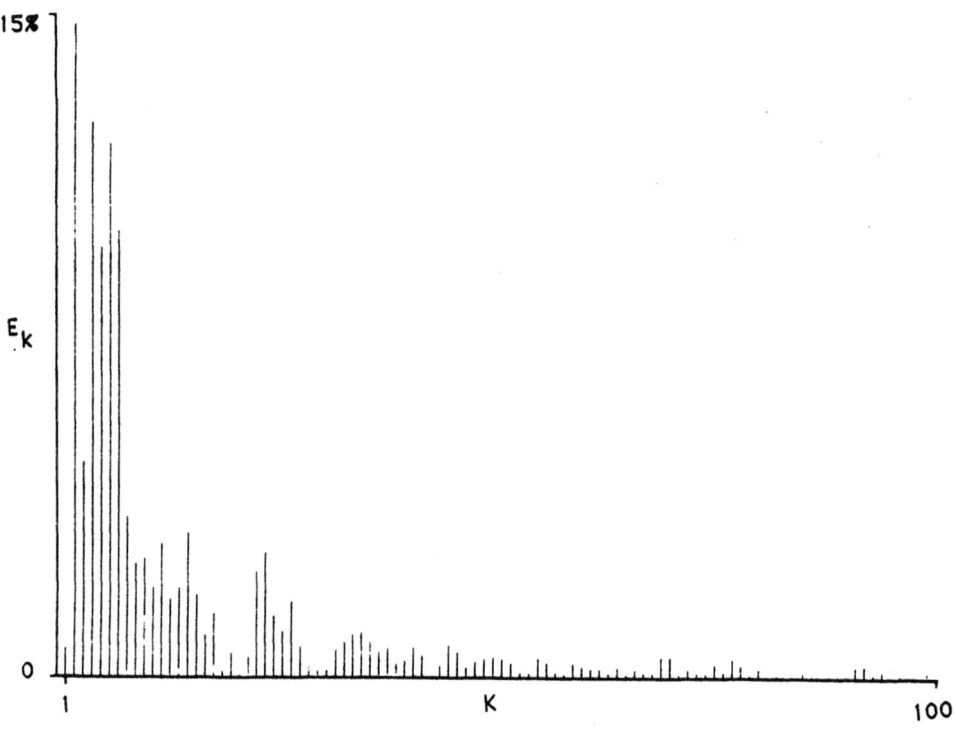

Figure 11 - Errors resulting from the comparison between the rejected pressure FFT and the typical one

Contributor Index